OREGON GEOLOGY

Oregon Geology

6th Edition

ELIZABETH L. and WILLIAM N. ORR

Oregon State University Press
Corvallis

Dedicated to: The Vulgar Boaters

Front cover:
Colorful layered volcanics are exposed on the east bank of Klamath Lake (Photo by William N. Orr)

The paper in this book meets the guidelines for permanence and durability of the Committee on Production Guidelines for Book Longevity of the Council on Library Resources and the minimum requirements of the American National Standard for Permanence of Paper for Printed Library Materials Z39.48-1984.

Library of Congress Cataloging-in-Publication Data

Orr, Elizabeth L.
Oregon geology / Elizabeth L. and William N. Orr.
p. cm.
Includes bibliographical references and index.
ISBN 978-0-87071-681-2 (alk. paper) --
ISBN 978-0-87071-682-9 (e-book)
1. Geology--Oregon. I. Orr, William N., 1939- II. Title.
QE155.O775 2012
557.95--dc23
2012015116

Printed in the United States of America

Oregon State University Press
121 The Valley Library
Corvallis OR 97331-4501
541-737-3166 • fax 541-737-3170
www.osupress.oregonstate.edu

Contents

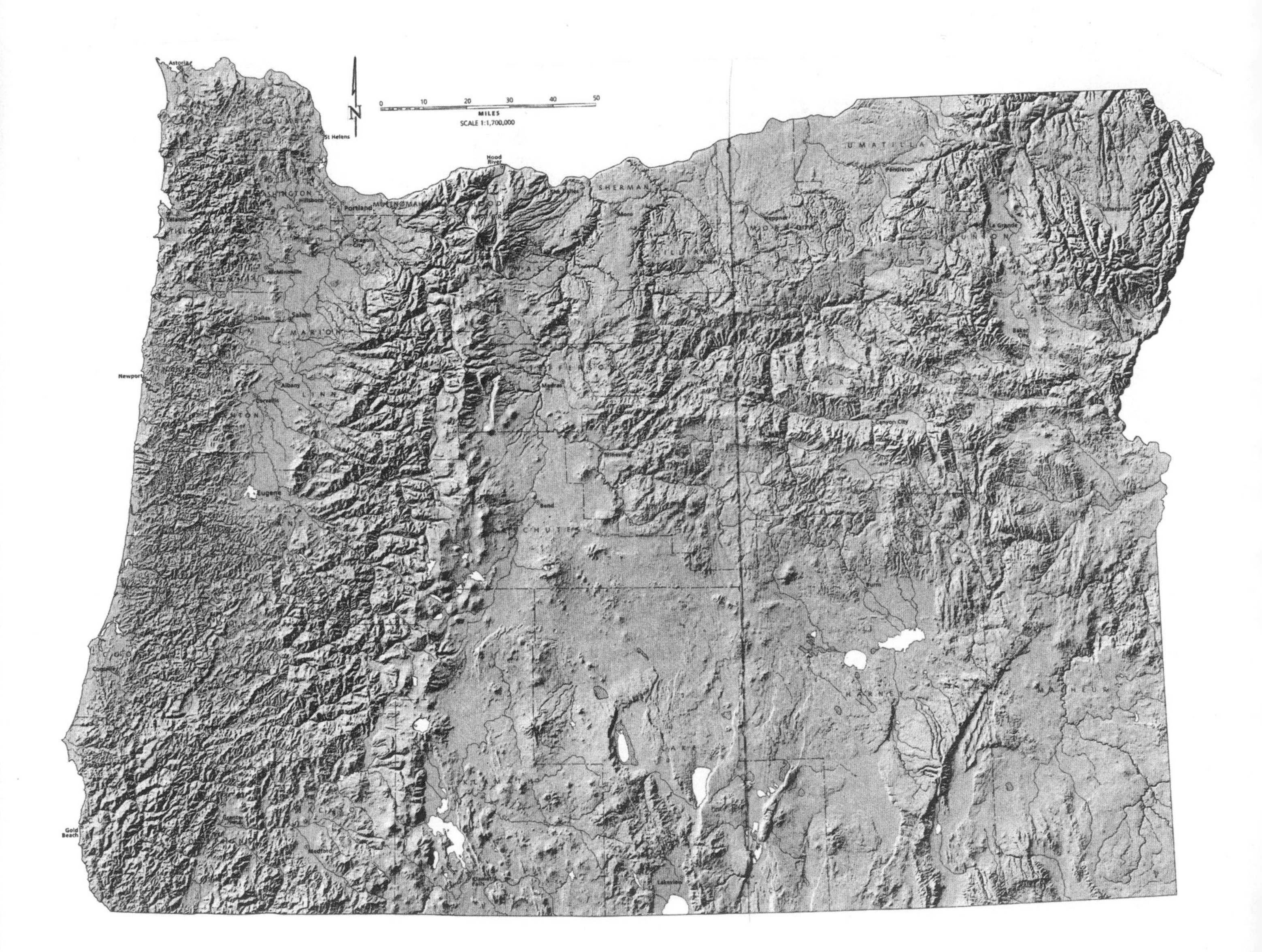

Shaded relief map from Loy, et al., eds., 2001

Preface

In our periodic revisions of *Oregon Geology*, we are invariably challenged by the need to address new ideas and trends in research. Since the last edition in 2000, many revolutionary models and theories have been put forward, leaving us to select those which might have lasting merit and those which should be replaced.

The overall shift in thinking has been away from general geology and economics and toward tectonics. Among the tectonic concepts, hot spots or mantle plumes have emerged as an elegant explanation for much of the volcanic activity east of the Cascades. Our understanding of events surrounding the accretion of terranes has been further refined, and the unexpected discovery of an enormous caldera in central Oregon has provided answers to many of the unsolved Tertiary volcanic issues there.

On a more practical level, we noted an increased emphasis on geologic hazards such as earthquakes and tsunamis as well as an awareness of alterations to the environment that relate to geology. In the area of technology, the use of LIDAR as a geologic tool has been astonishingly effective for recognizing, mapping, and evaluating landslides across Oregon.

Because of our reliance on the labors of others, we felt such efforts should be recognized, so, with some trepidation, we added short biographical sketches and photographs of just a few of the noteworthy individuals.

Acknowledgments

Thanks to the many colleagues who reviewed all or portions of the text and diagrams, thus helping us to avoid errors:

Cal Barnes, Texas Tech University, Lubbock, Texas
John Beaulieu, Oregon Department of Geology and Mineral Industries
Howard Brooks, Oregon Department of Geology and Mineral Industries
William Burns, Oregon Department of Geology and Mineral Industries
Victor Camp, San Diego State University
Julie Donnelly-Nolan, U.S. Geological Survey
Rebecca Dorsey, University of Oregon
Mark Ferns, Eastern Oregon State College
Peter Hooper, Washington State University
Harvey Kelsey, Humboldt State University
Paul Komar, Oregon State University
Vern Kulm, Oregon State University
Todd LaMaskin, University of Oregon
Jason McClaughry, Oregon Department of Geology and Mineral Industries
Alan Niem, Oregon State University
Wendy Niem, Oregon State University
Jim O'Connor, U.S. Geological Survey
Silvio Pezzopane, University of Oregon
Len Ramp, Oregon Department of Geology and Mineral Industries
Bob Reynolds, Central Oregon Community College
Josh Rohring, University of Oregon
Gary Smith, University of New Mexico
Martin Streck, Portland State University
Ed Taylor, Oregon State University
Terry Tolan, GSI Water Solutions, Kennewick, Washington
Tracy Vallier, U.S. Geological Survey
Beverly Vogt, Oregon Department of Geology and Mineral Industries
Bob Yeats, Oregon State University

We utilized the science libraries at the University of Oregon and Oregon State University along with resources from the archives at Oregon State University and the State Archives in Salem. Libraries and local historical societies are always valuable sources for information long forgotten. Indexer Jean Mooney's attention to detail repaired overlooked errors.

Elizabeth Orr – William Orr
Eugene, Oregon, 2012

Life on the Edge

Geology is the story of how the earth has evolved through time. Some processes such as volcanism, landslides, or earthquakes are sudden, highly visible, and catastrophic while erosion, deposition, or changing climates and sea levels are more subtle and lengthy. Because of Oregon's position on the leading edge of a moving crustal plate, a striking diversity of geologic events have gone into molding its topography. Before this complex picture could be deciphered, over a century of field work collecting data on stratigraphic relationships, faulting patterns, volcanic episodes, environments, and dates of rocks had to be compiled. Only then could there be an understanding of the tectonic overprint that drives most of the state's geologic episodes.

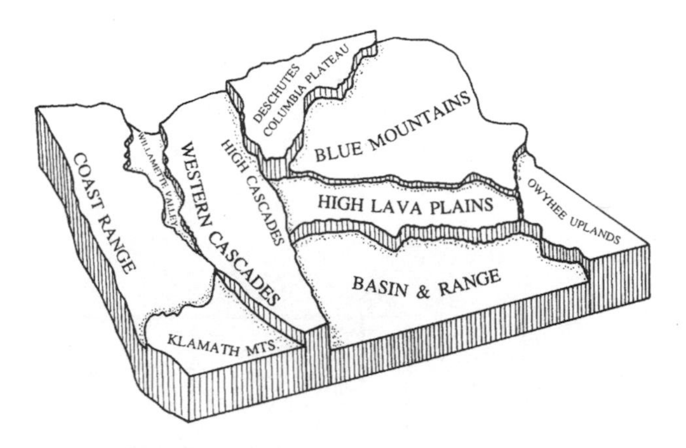

As the details of Oregon's past emerged, it became apparent that the underlying geology was not completely in concert with the physiographic boundaries. Landforms are more obvious and can readily be drawn, in contrast to the outlines for geologic phenomena, which tend to overlap, be buried, or obscured. In spite of this, the geologic content has been adapted to the individual geographic provinces, defined in 1950 by University of Oregon geographer Samuel Dicken. He imposed them on a base map of the state drawn by Erwin Raisz.

Geologists such as Thomas Condon, Joseph Diller, Israel Russell, and Howel Williams took the first steps at interpreting Oregon's geologic terrain. (In this unusual 1898 photo, Diller is sitting beneath a large schist outcrop of Otter Point Formation near Winston, Oregon; courtesy Douglas County Museum, Roseburg)

Drifting tectonic plates

The premise of moving continents or plate tectonics is central to all aspects of Oregon's geology. The notion that continents move was first published in 1915 by its chief advocate Alfred Wegener when he reconstructed the supercontinent Pangaea by matching the shapes of continental margins. Continental drift as a workable theory soon came up short in light of what was known at the time about the structure of the earth's crust. The theory stalled for almost 50 years until marine geologists recognized evidence from ocean floors to support the hypothesis that deep-seated plates, with continents imbedded on their surfaces, are in motion. The new idea of global plate tectonics emerged after studies of rock magnetics and the realization that the plates are not drifting but are spreading along mid-ocean ridges. The adoption of this idea ushered in an entirely new way of looking at the prehistory of the earth as well as at the beginnings of the Pacific Northwest.

Interaction of plates

The advent of plate tectonics was a milestone in geologic thinking. Moving slabs of crust and upper mantle may separate, collide, or grind past one another. Where plates rift and divide, new crust forms. Currently, lengthy continuous chains of undersea volcanic mountains can be found at rifting zones

WESTERN OREGON AND CASCADES

ERA	PERIOD	EPOCH	M.Y.A.*	
CENOZOIC	QUATERNARY	HOLOCENE (RECENT)	.01	COASTAL BOGS-TSUNAMI DEPOSITS ERUPTIONS OF CRATER LAKE AND NEWBERRY VOLCANO ~7,000 Y.A.
		PLEISTOCENE	1.7	SUDDEN EXTINCTION OF LARGE MAMMALS ~11,000 YEARS AGO MISSOULA FLOODS BRING ERRATICS & SILT TO WILLAMETTE VALLEY WILLAMETTE VALLEY WITH WIDESPREAD BOGS, MARSHES ONSET OF GLACIATION AT HIGH ALTITUDES
	TERTIARY	PLIOCENE	5.3	BORING LAVAS BEGIN LIMITED PLIOCENE ROCKS IN OREGON HIGH CASCADE GRABEN OPENS
		MIOCENE	23.8	COASTLINE NEAR PRESENT CONFIGURATION RICH MARINE VERTEBRATE AND INVERTEBRATE FAUNAS
		OLIGOCENE	33.7	LIMIT OF MARINE SEDIMENTATION IN WILLAMETTE VALLEY Temperate MAJOR MARINE CLIMATE SHIFT
		EOCENE	54	Tropical SILETZIA DOCKS AND IS ROTATED WESTERN CASCADE VOLCANICS BEGIN DELTAS & DEEP SEA FANS IN COAST RANGE OVERLAP SEQUENCES IN KLAMATH MTNS. SILETZIA VOLCANICS BEGIN MAJOR PLATFORM
		PALEOCENE	65	LIMITED PALEOCENE EXPOSURES IN OREGON
MESOZOIC	CRETACEOUS		145	MOST TERRANES ACCRETED BY END OF CRETACEOUS MAJOR TRANSGRESSIONS (SEA LEVEL RISE) OVER N.E. & KLAMATHS
	JURASSIC		208	OPHIOLITE DEVELOPMENT IN KLAMATH MTNS. FOSSILS TROPICAL MULTIPLE ACTIVE ARCS (WESTERN KLAMATH & COASTAL TERRANES)
	TRIASSIC		245	OLDEST ROCKS IN KLAMATHS (APPLEGATE LIMESTONES) WESTERN PALEOZOIC AND TRIASSIC BELT & MAY CR. TERRANE
PALEOZOIC	PERMIAN		283	
	PENNSYLVANIAN		320	
	MISSISSIPPIAN		360	
	DEVONIAN		408	
	SILURIAN		438	
	ORDOVICIAN		505	
	CAMBRIAN		590	No rocks known from western Oregon
PROTEROZOIC			2500	
ARCHEAN			4000	
			4500	FORMATION OF EARTH

*Million years ago

along the floors of every major ocean. Subduction takes place when two plates collide, and one descends beneath another. Once ocean crust ages and cools, its density increases, and the older, heavier crust is thrust below or subducted by the overriding slab. Where plates slide past each other, transform faults of epic size and length develop at the boundary between the two, accompanied by destructive earthquakes.

Among the most visible by-products of plate collision and subduction are the build-up of an accretionary wedge, the emplacement of volcanic archipelagos (arcs), and the formation of sedimentary basins. During the subduction process, sediments atop the descending slab are peeled off to accumulate as a jumbled prism or mélange at the outer margin of the upper plate. Associated with this, magma, rising from the lower descending plate, penetrates the upper slab to emerge at the surface as a volcanic chain. Between the accretionary prism and the volcanic archipelago, a forearc basin or depression may develop with a similar backarc basin between the arc and the larger continental mass (craton). Over time, erosion of the volcanic highlands sheds copious amounts of sediment into both basins.

Arrival of terranes

Almost a half billion years in the past, the oldest rocks that would make up Oregon were being

EASTERN OREGON

MYA*

ERUPTION NEWBERRY VOLCANO ~7,000 Y.A.

Glaciers
STEENS, BLUE MTNS. GREENHORN MTN.
ICE SHEETS DEVELOP AT HIGH ALTITUDES
EXTENSIVE PLUVIAL LAKES DEVELOP ACROSS S. CENTRAL OREGON

INCREASINGLY TEMPERATE TO DRY CLIMATE
HIGH CASCADE VOLCANISM BEGINS
LIMITED PLIOCENE EXPOSURES

SILICIC VOLCANIC AGE PROGRESSION ACROSS HIGH LAVA PLAINS

MAJOR CLIMATE SHIFT TO GRASSLANDS FROM FORESTS
DRYING OF E. OREGON BY CASCADE RAINSHADOW
OREGON-IDAHO GRABEN
COLUMBIA RIVER BASALTS BEGIN WITH STEENS SRUPTIONS
RATTLESNAKE
JOHN DAY, MASCAL,
BASIN AND RANGE EXTENSION
fossil rich SUCCESSION OF TUFFS

CROOKED RIVER-WILDCAT-TOWER MTN. CALDERA SEQUENCE BEGINS
CLARNO VOLCANISM BEGINS

LIMITED PALEOCENE EXPOSURES

Major terranes accreted to Oregon by the end of the Cretaceous
OCHOCO FOREARC AND KLAMATH SEAWAY
Marine reptiles

IZEE OVERLAP SEQUENCE IN BLUE MTNS.
WALLOWA, OLDS FERRY TERRANES
Tropical plants
ichthyosaurs, plesiosaurs, and crocodiles

Marine reptiles
Corals
BAKER, WALLOWA, OLDS FERRY TERRANES

OPHIOLITE DEVELOPMENT IN BLUE MTNS.
(BAKER TERRANE)
BAKER TERRANE & WALLOWA TERRANE

SPOTTED RIDGE FM.
Tropical plant fossils

COFFEE CR. LIMESTONE
OCEAN SHORELINE ACROSS WASHINGTON AND IDAHO

OLDEST ROCKS IN OREGON
DEVONIAN LIMESTONES (BAKER TERRANE)

No rocks known in eastern Oregon

OLDEST ROCKS ON EARTH

.01 1.7 5.3 23.8 33.7 54 65 145 208 245 283 320 360 408 438 505 590 2500 4000 4500

*MILLION YEARS AGO

Going back 400 million years, Oregon has had a fascinating history of piecemeal construction, volcanism, and sedimentation even as it was populated by an array of plants and animals.

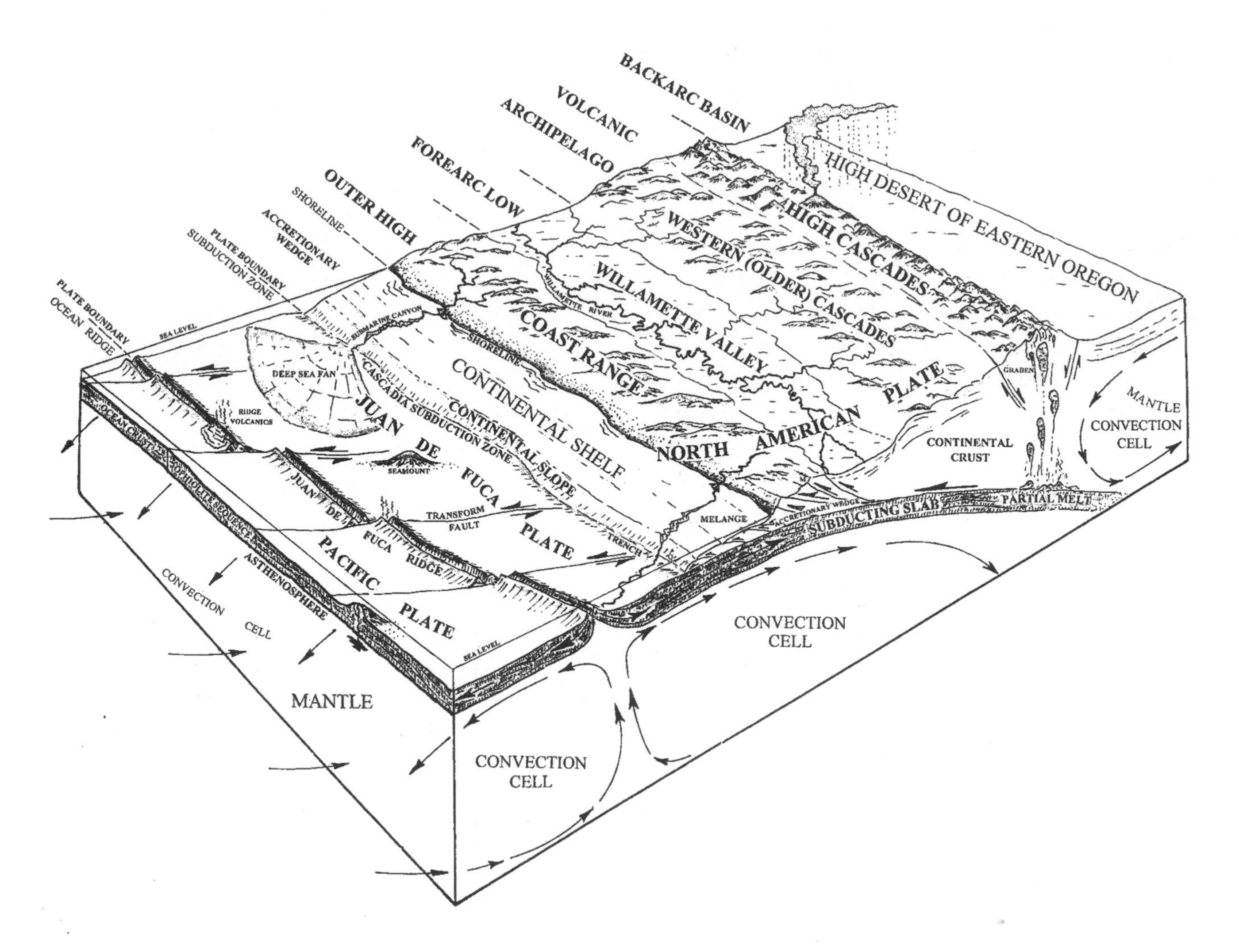

The configuration of the Pacific Northwest adapts easily to global models that address the interaction between tectonic plates. Composed of multiple strips, the continental margin and outer high of the Coast Range, the Willamette Valley forearc low, the Cascade volcanic arc, and the high desert backarc basin fit together to make the parcel that is Oregon.

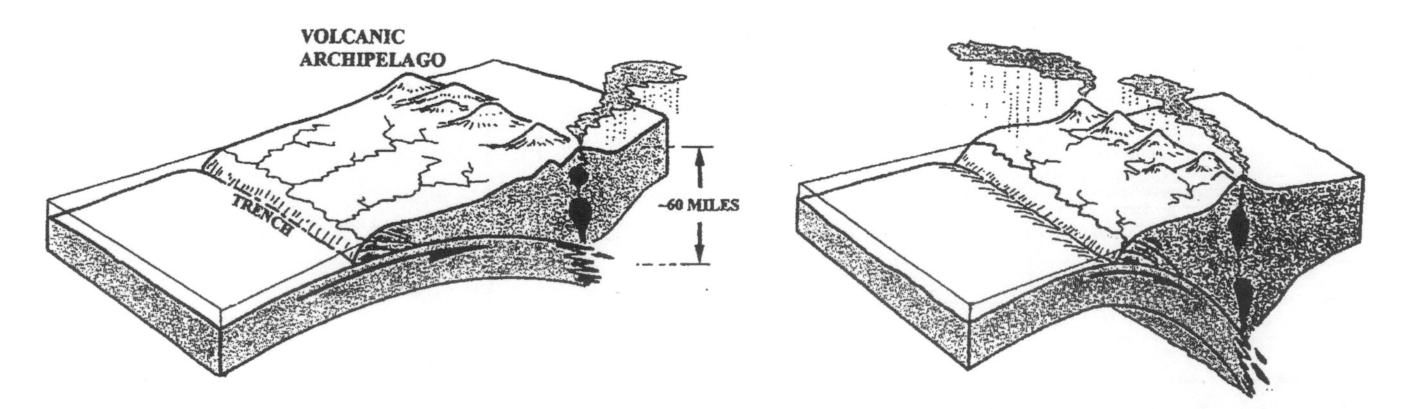

The geographic distance between the subduction trench and volcanic archipelago is a function of the relative age of the crust being subducted and the rate of plate movement. Young, warm crust is buoyant and assumes a low angle of subduction with considerable distance between the arc and trench. Older cooler crust is more dense and droops at a pronounced angle upon subduction, bringing the arc close to the trench.

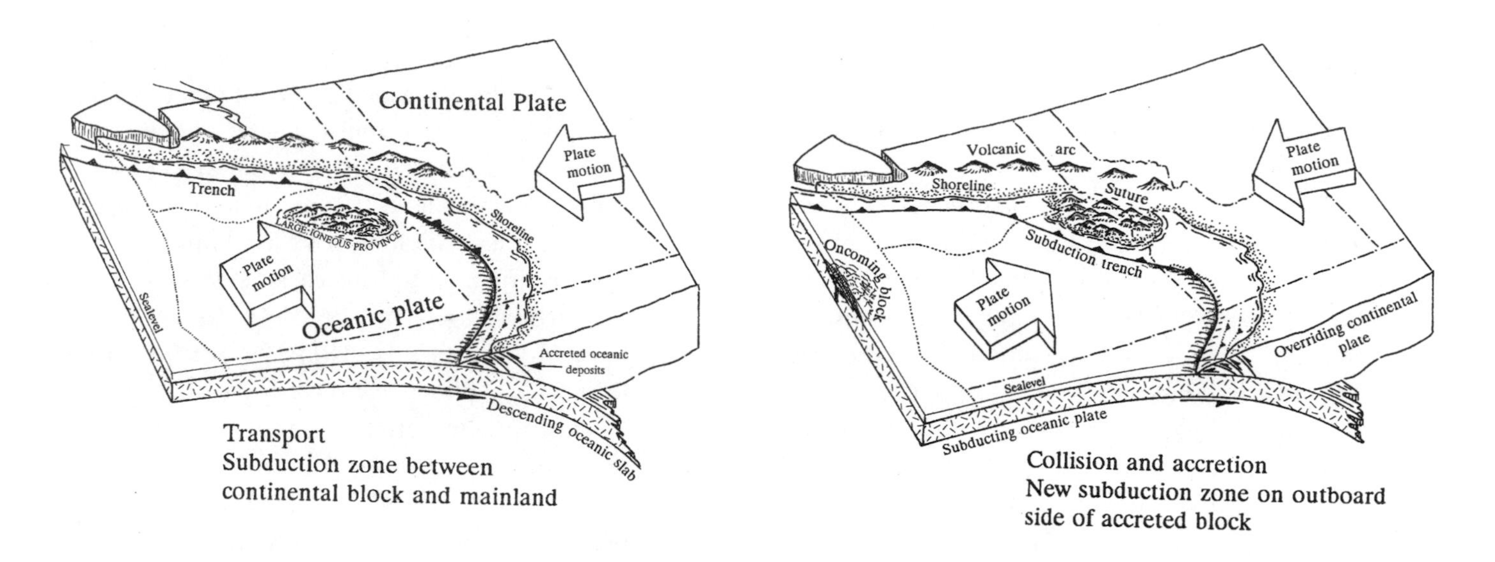

When tectonic plates collide, seamounts, crustal fragments, and island archipelagos are scraped off and accreted to the larger overriding continental block. Following accretion of particularly large tracts, a new subduction zone develops on the outboard side of the oceanic slab.

deposited elsewhere in the Pacific basin or even in distant Asia. Atop subducting plates, assemblages of rocks or terranes, imbedded within the crust, are borne along to be annexed to a larger landmass. The terranes might be volcanic island chains or larger pieces of crust. Defined as recognizable suites of rocks that have a similar geologic origin, terranes are bordered by faults, share a history of sedimentation and displacement, and are distinct from the surrounding strata. After attachment (accretion) to the continent, the terranes are altered and deformed by the heat and pressure of metamorphism as they are being rotated into position. Arriving one after the other, the displaced terranes gradually converged to construct the Pacific Northwest.

Constructing Oregon

Because the foundation bedrock of the state has been assembled piece-by-piece, it is a collage of displaced terranes which originated elsewhere around the Pacific basin. On opposite sides of Oregon, terrane rocks of the Blue Mountains and the Klamath Mountains are the oldest and most accessible, offering the best areas to decipher the state's accretionary history. During the Mesozoic era, successive waves of volcanic island chains traveled eastward and merged with western North America to construct these two provinces. Originating in tropical settings, the crustal pieces brought with them a striking array of fossil fish, large marine reptiles, and fern-like terrestrial plants in association with limestones and reefs.

By the end of Cretaceous time, 65 million years ago, the last of the older terranes had arrived and been accreted. Only the Coast Range block (Siletzia) remained to be emplaced during the early Eocene. With the acquisition of Siletzia, the contours of Oregon were beginning to appear.

Volcanoes, extension, and tectonics

Volcanism and sedimentation, the predominant geologic processes throughout Oregon's Cenozoic interval, are direct by-products of plate tectonic activity. Eocene to Oligocene volcanism coincided with collision of the Farallon and North American plates. While the Farallon slab was being subducted, partial melting at depths up to 75 miles produced low density magma that rose through the overriding North American plate to erupt thick layers of ash and lava over the area that is now the Western Cascades.

Well into the Miocene Epoch, the landscape was reshaped by the construction of vast basalt plateaus in northeast and southeast Oregon and by the spread of incandescent, fast-moving ash flows in the central region. Episodes of volcanism rarely matched worldwide began with immense outpourings of Columbia River flood basalts from a shield volcano in the vicinity of Steens Mountain. The basalt platform would ultimately cover half of

Oregon and large portions of Washington and Idaho. Because they were so voluminous and rapid, tongues of the flows raced across the landscape to the Pacific Ocean where they penetrated the soft coastal sediments.

The basalt eruptions were triggered by movement of the North American plate over a hot spot or mantle plume that is located today beneath Yellowstone, Wyoming. If an area of concentrated heat in the mantle rises, the plume melts the overlying crust and emerges as a volcanic hot spot. As the continent passed over the stationary plume, the balloon-like head flattened and spread out west and northwestward, sending lavas to the Blue Mountains and across the High Lava Plains and Basin and Range.

On the Owyhee plateau, Miocene volcanic centers were concurrent with eruptive phases of the Columbia River basalts. Explosive activity in the McDermitt and Owyhee fields some 17 million years ago was generated by crustal stretching and the proximity of the Yellowstone hot spot. The violent episodes and subsequent collapse of the vents created immense calderas, that are several times larger than those at Crater Lake and Newberry Caldera.

The geologic signature of the Basin and Range and High Lava Plains is the consequence of crustal thinning and volcanism imposed by the interaction between the North American continental and the offshore Pacific plates. As the crust stretched, it eventually broke into a landscape of north-south faulted hills and valleys, characterizing the region today. Volcanism, initiated 10 million years ago, continued into the Recent. The eruptive surges were both age progressive and bimodal. That is, they became younger in age from east to west, beginning at Harney Basin and ending near Newberry Crater. In conjunction with the age progressive phenomenon, the bimodal composition varied from older rhyolitic to younger basaltic lavas.

With waning cycles of the Columbia River basalts, eruptions in the Cascades had shifted as in

Around 50,000 years ago, what is now the Portland-Vancouver metropolitan area saw extensive volcanic activity from the Boring lava field. Looking eastward across Portland toward Mount Hood, just a few of the 80 small volcanoes and cones that compose the Boring field can be seen. (Photo courtesy Oregon State Highway Department)

a wave eastward to construct the High Cascade shield and stratocones during the late Miocene and into the Quaternary Period. Composed of both basalt and andesitic lavas, the High Cascade peaks stand in sharp contrast to those of the eroded older Western range. While most of the High Cascade eruptions in Oregon had diminished or ceased by 30,000 years ago, there have been a number of episodes since that time. Mount Mazama's spectacular explosion, dated at 7,700 years, is the most notable, but Mount Hood and numerous domes in the central chain have been active historically. The 1980 eruption of Mount St. Helens in Washington was the most visible and largest in recent times.

Marine and terrestrial basins

Thick sedimentary layers that cover large areas of Oregon are critical to interpreting the state's geologic past. Changing marine and terrestrial environments are reflected in the stratigraphy and fossils, which have been especially useful for interpreting depositional settings and climate variations.

As part of a worldwide trend during the late Mesozoic, rising seas covered much of Oregon. Some 60 million years ago, only the Blue and Klamath Mountains projected above the surrounding oceans, with a shoreline that ran diagonally from the Klamath Mountains into eastern Washington. Following repositioning of the Farallon, Kula, and North America plates, a lengthy forearc basin connected these two provinces with the Great Valley of California. A thick covering of Cretaceous sediments, which spread across the basins in the Ochoco and Klamath mountains, entombs some of the state's first autochthonous (home-grown) rocks and fossils. This was the high water mark before regional uplift forced the waters to retreat.

The early Tertiary saw the arrival and rotation of the large Coast Range block of Siletzia, the final terrane to be annexed. Situated along the edge of North America and west of the emerging Cascade volcanic arc, Siletzia subsided into a narrow trough even as it was being accreted. Throughout the Eocene Epoch, erosion from both the interior of the continent and the Klamath Mountains shed copious quantities of debris into the basin, which today underlies the continental shelf and Coast Range, Willamette Valley, and Western Cascades. Elevation of the coastal margin and depression of the Willamette Valley brought a shallowing of ocean waters and a reduction of sedimentation during the latest Oligocene.

The subduction of Siletzia beneath the Cascades and eastward toward Idaho generated the Clarno and Challis volcanic eruptions. Lava and ash of the Clarno Formation blanketed large portions of eastern Oregon, overwhelming lakes and streams. Clarno sediments and mudflows, along with those of the successively younger John Day, Mascall, and Rattlesnake formations, preserve remarkable fossil plant and animal remains, which provide a continuous environmental picture from the Eocene through the Miocene.

Climate change

Oregon's geologic record reveals extraordinary shifts in climate, from the tropical humid conditions of the Paleocene and Eocene to glaciation during the Pliocene and Pleistocene. Many of these were global trends and not restricted to the Pacific Northwest.

Often regional climates are a consequence of tectonic activity. Moving from one latitude to another, crustal plates experience profoundly divergent conditions enroute. Ash from explosive volcanism, triggered by plate collision, can obscure the sunlight and cause a rapid drop in temperatures. Alternately, the same episodes may foster a rise in temperatures with the release of greenhouse gasses. Because of its proximity to the Pacific Ocean and the direction of prevailing winds, much of western Oregon's climate is additionally governed by variations in the marine offshore realm.

Worldwide, the Eocene to Oligocene boundary marks a global transition from tropical and semi-tropical climates to those of a more temperate nature. Cooling ocean water and falling air temperatures generated clear floral and faunal modifications, sometimes even leading to extinctions. In western Oregon, changes in the subtropical terrestrial plants and marine invertebrates in the Keasey, Cowlitz, and Eugene formations are particularly notable. In eastern Oregon, the tropical humid environment was transformed with the steady elevation of the Cascade volcanic barrier. By cutting off the moist air

An indirect effect of the cold Pleistocene was cataclysmic flooding as continental ice masses advanced into Washington from Canada. Ice dams, which impounded glacial Lake Missoula, failed periodically, releasing as many as 100 huge floods that scoured eastern Washington and the Columbia River channel on the way to the Pacific Ocean. The volumes of water backed up as temporary Lake Allison in the Willamette Valley, leaving thick layers of silt across the floor. (In the photo, the individual shorelines of Lake Missoula in Montana are visible on the hillsides; courtesy U.S. Geological Survey)

masses from the Pacific, it dramatically altered conditions by the late Oligocene and into the Miocene, resulting in the high desert of today.

Beginning a little less than 2 million years ago, the Ice Ages or Pleistocene Epoch was time of cold temperatures, heavy rainfall, and the rapid build-up of continental ice masses. Worldwide, lower temperatures fostered polar ice caps and lowered sea levels. Throughout this interval, ocean waters rose and fell during cycles of cooler and warmer periods, and glacial erosion deeply etched the land. These trends peaked around 18,000 years ago, after which the earth entered its present-day warm interglacial phase.

While no continental glaciers reached Oregon, thick ice caps covered the mountain ridges, and increased precipitation filled pluvial lakes in the broad fault-bounded depressions of the Basin and Range. These ephemeral lakes, some of which covered hundreds of square miles, provided habitats for herds of mammals, migratory birds, and varieties of fish. Today all of these shallow basins have diminished or dried up completely.

The geologic future of Oregon

The geology perspective means examining the past as well as anticipating the future. To do so, it is necessary to take the long view of modifications to the earth's surface and atmosphere, which are extremely complicated and invariably cyclic. Certainly, this applies to the current public focus on global warming. Glaciologists calculate that, at present, the earth is approaching the end of a 10,000-year interglacial period and should be entering a glacial phase within the next 23,000 years. During these larger episodes, average global temperatures fluctuate frequently, and the last 10,000 years (Holocene) has been anything but stable. If, in the future, glacial conditions prevail and vast ice sheets take up and store ocean water, the Oregon coast will see a substantial drop in sea level. But, on the other hand, with warmer conditions and melting ice, the oceans will rise to invade the land. These are long-term developments that come about over thousands of years.

Caught between converging crustal plates, the Pacific Northwest faces a future of massive earthquakes and tsunamis. Only in the last 20 years has the public become aware of the potential for high magnitude quakes associated with the Cascadia subduction zone. Since then, there has been an ongoing push to compile and analyze data on seismicity. Efforts to explain the current low incidence of Cascadian subduction activity include a number of theories, but to date they fail to explain all possibilities satisfactorily.

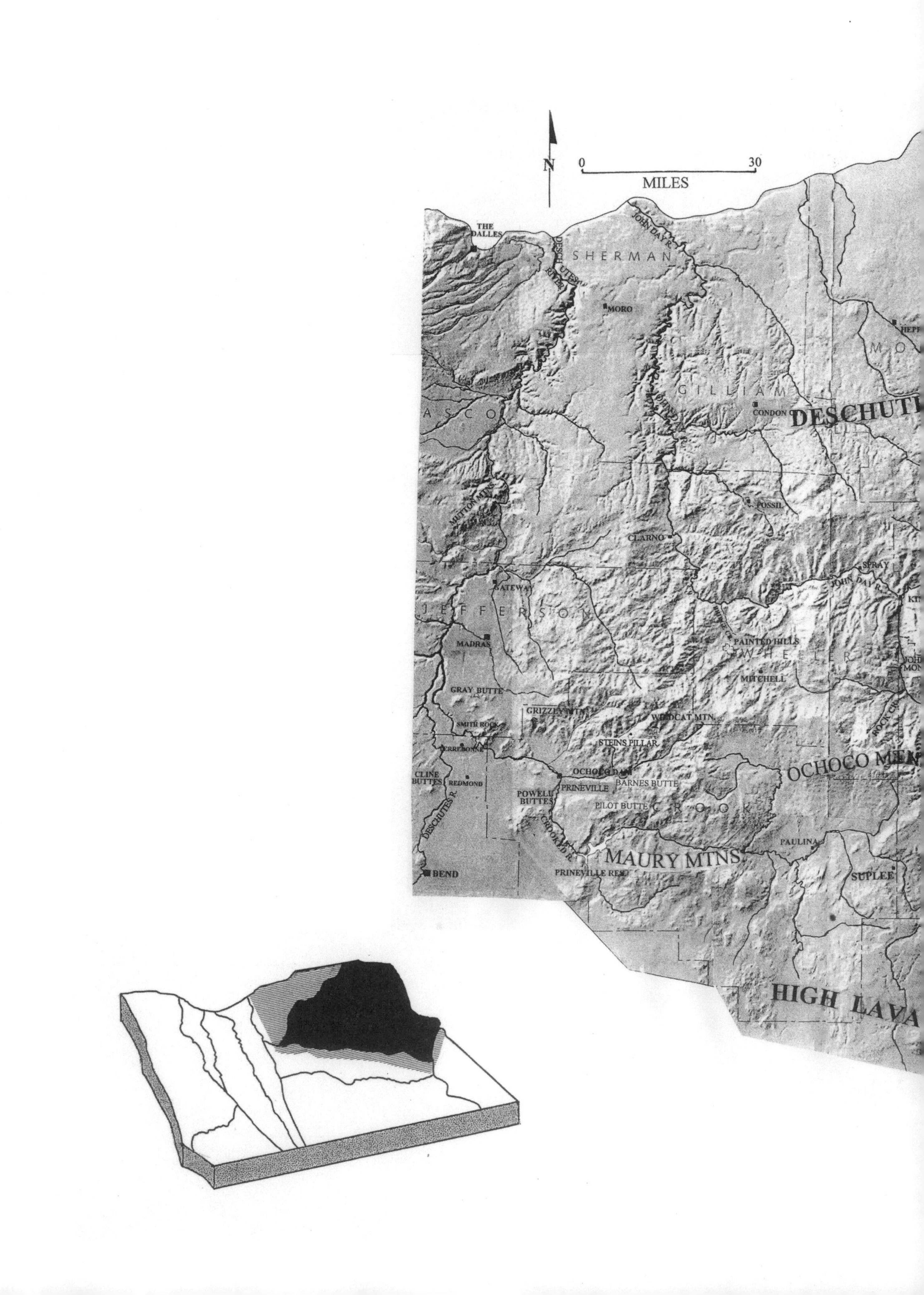
N
0
30
MILES
THE DALLES
SHERMAN
JOHN DAY R.
DESCHUTES RIVER
MORO
GILLIAM
CONDON
DESCHUTE
WASCO
FOSSIL
MUTTON MTNS
CLARNO
SPRAY
JOHN DAY R.
GATEWAY
JEFFERSON
PAINTED HILLS
WHEELER
MADRAS
MITCHELL
GRAY BUTTE
GRIZZLY MTN
WILDCAT MTN.
SMITH ROCK
STEINS PILLAR
TERREBONNE
OCHOCO DAM
OCHOCO MTN
CLINE BUTTES
REDMOND
PRINEVILLE
BARNES BUTTE
POWELL BUTTES
PILOT BUTTE
CROOK
DESCHUTES R.
CROOKED R.
PAULINA
MAURY MTNS
BEND
PRINEVILLE RES.
SUPLEE
HIGH LAVA

Physiographic map (After Loy, et al., eds., 2001)

Blue Mountains

Landscape of the Blue Mountains

The Blue Mountains physiographic province lies almost entirely in Oregon with only small portions projecting into southeast Washington and west-central Idaho. Covering an area of 55,000 square miles, the triangular-shaped region is defined on the east by the Snake River Canyon, on the south by the Owyhee Uplands and High Lava Plains, and to the west and north by the Deschutes-Umatilla plateau.

Geographically, the province is a cluster of ranges of various orientations and relief. Situated in the northeast corner, the Wallowa and Seven Devils mountains are separated from the centrally located Elkhorn, Greenhorn, Strawberry, and Aldrich ranges by the wide flat Baker and Grande Ronde basins. To the west, the Blue Mountains chain divides this province from the Umatilla basin, and the Ochocos lie along the border with the Deschutes River valley.

From the Ochoco Mountains at 6,000 feet in elevation, the landscape rises eastward to glaciated summits of the Wallowa Mountains. An immense oval-shaped range, the Wallowas include a striking array of canyons carved out by ice and water, nine peaks that reach over 9,000 feet, and seven more in excess of 8,000 feet.

The area possesses several extensive watersheds including the 175-mile-long Grande Ronde River, the 144-mile-long Powder River, and the shorter Imnaha, 75 miles in length. Covering a distance of 284 miles from its point of origin in the Blue Mountains, the John Day is Oregon's longest. The northerly-flowing Snake River cuts a deep gash that divides the Wallowa Mountains of Oregon from the Seven Devils peaks in Idaho. A chief tributary of the Columbia River, the Snake is 1,000 miles in length and drains 110,000 square miles in western Wyoming, northwest Utah, northeast Nevada, and southeast Washington.

Past and Present

Early geologic work in the Blue Mountains is tied into the region's mineral wealth and production. Waldemar Lindgren's notable *The Gold Belt of the Blue Mountains* in 1901 was followed by James Gilluly's 1930s reports on gold, copper, and lesser minerals, in which he included a summary of the geology. Clyde Ross's 1938 work on the Wallowa Mountains, conducted under the auspices of the Oregon Bureau of Mines and Geology, treats the geology but excludes the ore deposits. A decade later, Warren D. Smith and John Allen published on the Wallowa Mountains. Smith's University of Oregon students attended the annual field camp there, while Allen was charged with compiling a regional handbook of mining prospects for the Oregon Department of Geology and Mineral Industries (DOGAMI).

William Dickinson and Laurence Vigrass's 1965 DOGAMI bulletin on the Suplee and Izee region remains one of the most definitive to date. Their maps, structure, and stratigraphy for some 35,000 feet of Paleozoic, Mesozoic, and Cenozoic strata outline the geology of 250 square miles in Crook, Grant, and Harney counties. Dickinson taught at Stanford University prior to going to the University of Arizona where he directed a new generation of sedimentologists. Since retirement, he continues to work on plate tectonics and is an authority on the western United States. An emeritus professor from the University of Regina, Vigrass works today with the Saskatchewan Department of Mineral Resources. (Photo courtesy W. Dickinson)

As far back as the 1970s, Tracy Vallier and Howard Brooks began to unravel the evolution of the Blue Mountains in accord with plate tectonics. Their publications during the late 1960s treat the geology of the Snake River, and their U.S.G.S. Professional Papers, edited between 1986 to 1995, summarize current thinking on Paleozoic and Mesozoic island arc accretion, batholith intrusions, and mineral resources.

Howard Brooks' ancestors followed the Oregon Trail to Linn County before they settled in Hazelton, Idaho, near the Snake River canyon, where he grew up. A graduate of the Mackay School of Mines, Brooks began with DOGAMI in 1956, working for 35 years in the Baker City office. Since retirement, he has written a new book on the history of gold mining in the Blue Mountains and volunteers at the National Historic Oregon Trail Interpretive Center in Baker City. (Photo courtesy H. Brooks)

Completing his degree from Oregon State University on the Snake River Canyon in 1967, Tracy Vallier joined the Deep Sea Drilling Project at Scripps Institution of Oceanography before he was assigned to the marine branch of the U.S.G.S. He retired in the late 1990s and currently lectures at several universities. Vallier's life-long interest in Hells Canyon on the Snake River has led to several books that explore the history and geology in combination with his personal experiences there. (Photo courtesy T. Vallier)

George Walker grew up in Palo Alto, California, and graduated from Stanford University in 1948. His involvement in Oregon began as a student when he worked with Aaron Waters on quicksilver deposits in the Horse Heaven Mine. During his career with the U.S.G.S., Walker's projects took him to the most remote areas of eastern Oregon. With his method of what has been called mile-a-minute mapping, he was able to complete 15 minute quadrangles in a week and an entire one-degree by two-degree quadrangle in a field season, much of which was accomplished by driving over unpaved dusty roads. His work culminated in the *Cenozoic Geology of the Blue Mountains*, which he edited in 1990, and his *Geologic Map of Oregon* published with Norm McCloud in 1991. (Photo courtesy C. Walker)

Overview

The geologically unique Blue Mountains are a patchwork of separate blocks (terranes) originating as volcanic island archipelagos and ocean crust in tropical settings well south of their present latitude. Transported during the late Paleozoic and Mesozoic eras, the fragments collided with and were annealed to North America. forming the structural grain of the Blue Mountains.

The Baker, Wallowa, Olds Ferry, and Izee terranes, which became the foundation of the Blue Mountains, include distinctive fossils, sediments, and volcanic rocks that provide an opportunity to study the earliest beginnings of the state. Only in the past 30 years have geologists shed light on the details of major events of the past by mapping, describing, and age-dating components of the terranes.

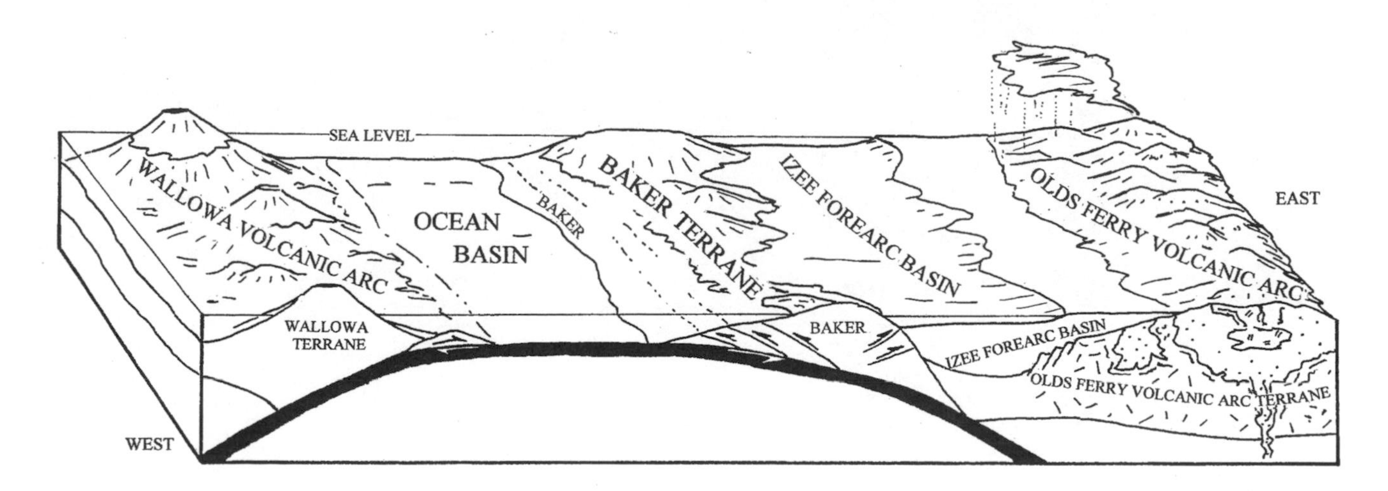

Each of the major Blue Mountains terranes, the Baker, the Wallowa, the Izee, and the Olds Ferry, represents a separate marine setting featuring back-to-back volcanic arcs. (After Brooks, 1979; Brooks and Vallier, 1978; Dickinson, 1995; Dickinson and Thayer, 1978; Dorsey and LaMaskin, 2007; Schwartz, et al., 2010)

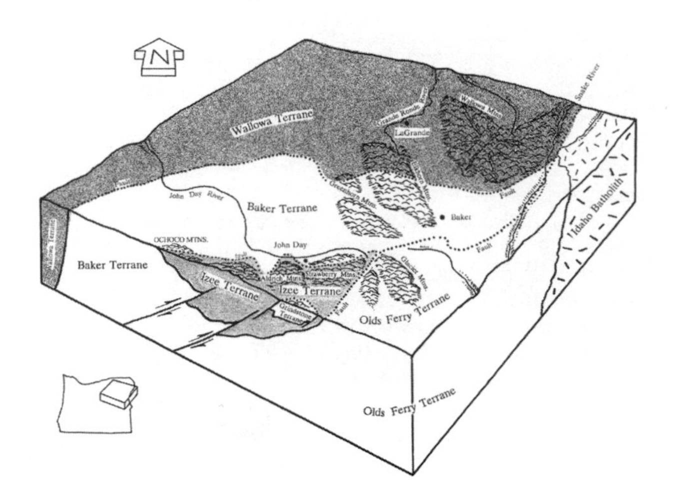

Accreted terranes form the foundation of both the Blue Mountains and the Klamath Mountains. In the Blue Mountains, the rocks span the time interval from the Devonian, 400 million years ago, to the early Cretaceous, 140 million years in the past. Arranged in linear belts that extend southwest to northeast, five suites of rocks identified historically were the Grindstone, the Baker, the Wallowa, the Olds Ferry, and the Izee. (After Ave Lallemant, 1995; Brooks, 1979; Dickinson, 1979, 1995; Dickinson and Thayer, 1978; Follo, 1992; Silberling, et al., 1984; Vallier and Brooks, eds., 1995)

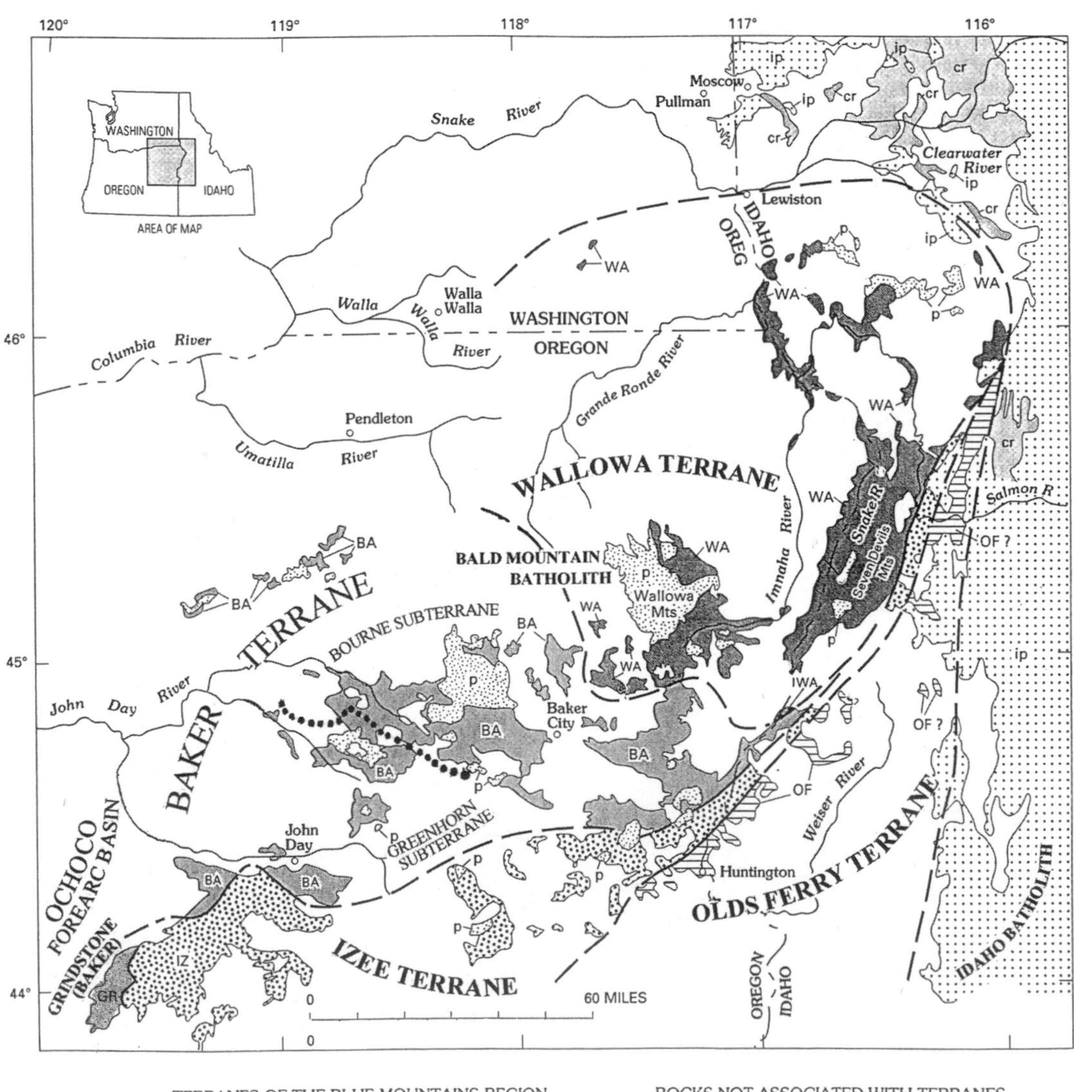

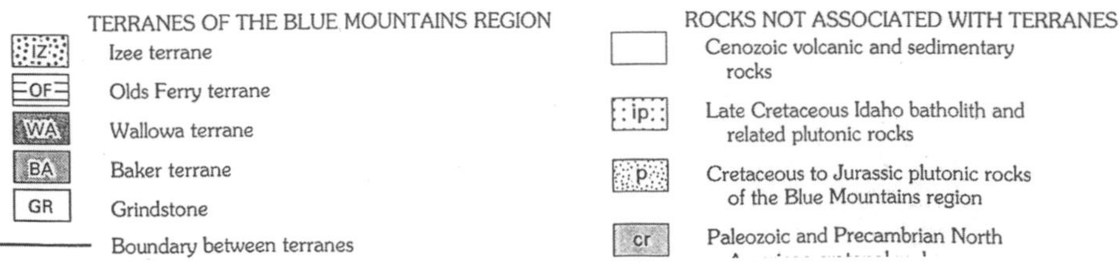

Late in the Mesozoic Era, a vast shallow seaway covered up to three-fourths of Oregon, depositing thousands of feet of mud, silt, and sand. In this environment, marine reptiles and mollusks of every description populated shallow embayments, while a variety of primitive ferns and horsetails grew along the shore.

During the Cenozoic Era the seaway retreated westward following uplift and folding when phases of volcanism alternated with sedimentation. Lakes, streams, and floodplains provided habitats for mammals and plants and fostered conditions that preserved their remains after lavas periodically engulfed the landscape. A temperate Oligocene environment, around 35 million years ago, was followed in the Miocene by eruptions of the Columbia River lavas that built a thick plateau of basalt unmatched in North America.

As the climate grew colder 2 million years ago, widespread glaciation gave the region its final topographic complexion. Glaciers capped mountain tops and descended into stream valleys. At the height of the Ice Ages, nine major glaciers spread throughout, only to melt and retreat with warmer temperatures.

Geology

Paleozoic-Mesozoic

The Blue Mountains province is a collage of displaced Paleozoic and Mesozoic terranes, some from remote oceans, that were covered by Tertiary lavas and sediments. The word terrane applies to suites of similar rocks, separated by faults, and transported from their place of origin. First named and delineated by Howard Brooks, William Dickinson, Tom Thayer, and Tracy Vallier, the individual terranes of the Blue Mountains were steadily modified with increased knowledge of tectonic processes. In 1984, Norm Silberling of the U.S.G.S. revised and refined the boundaries, while the most recent work raises doubts as to whether the Grindstone and Izee are separate terranes.

During the early Paleozoic Era, over 500 million years ago, North America was still part of the emerging super continent Pangaea, which began to break into separate fragments in Triassic time. As the oceanic plates diverged, chains of volcanic islands formed in a tropical (Tethyan) ocean. Carried conveyor-like atop tectonic plates, the island arcs collided with and were affixed or accreted to the West Coast of North America during the late Paleozoic and into the Mesozoic. At that time a vast ocean covered the region that was to become Oregon, and, as each terrane block arrived, it was swept up and attached to the older continental landmass. At the boundary between the plates, layers of sediments, scraped off the lower slab, accumulated as an accretionary wedge.

By the 1970s geologists recognized that the individual terranes making up the Blue Mountains had merged (amalgamated) even before accretion, sharing a history of volcanism, sedimentation, and metamorphism. While details of the interactions among the various terranes are still being hammered out, the consensus is that by the middle to late Triassic Period two large volcanic chains, the Wallowa and the Olds Ferry, were adjacent to the margin of North America. The Baker terrane, separating the two, was an ocean-trench-subduction complex with an accretionary wedge, while voluminous sediments of the Izee terrane were deposited in the seaway between the Baker and Olds Ferry island arcs. In the late Triassic, the Wallowa merged with the Olds Ferry, and the amalgam of all terranes was intruded by granitic plutons and annexed to the continent.

Individual Terranes

Grindstone Terrane

As the oldest and smallest of Oregon's northeastern terranes, the Grindstone is limited to scattered outcrops in Crook, Harney, and Grant counties. Today many geologists no longer regard the tiny Grindstone as an independent terrane but consider it to be part of the much larger Baker assemblage.

The Grindstone suite is composed of early Paleozoic limestone slabs or olistoliths, which became detached from shallow oceanic shelves or volcanic knolls, to slide downslope and mix with late Permian to Triassic rocks in the Izee basin. The Grindstone includes an unnamed 400-million-year-old Devonian sequence overlain by the Mississippian Coffee Creek, the Pennsylvanian Spotted Ridge, and the Permian Coyote Butte formations. During the 1940s, Charles Merriam and Sheridan Berthiaume from Cornell University worked out and named many of the formations, while more

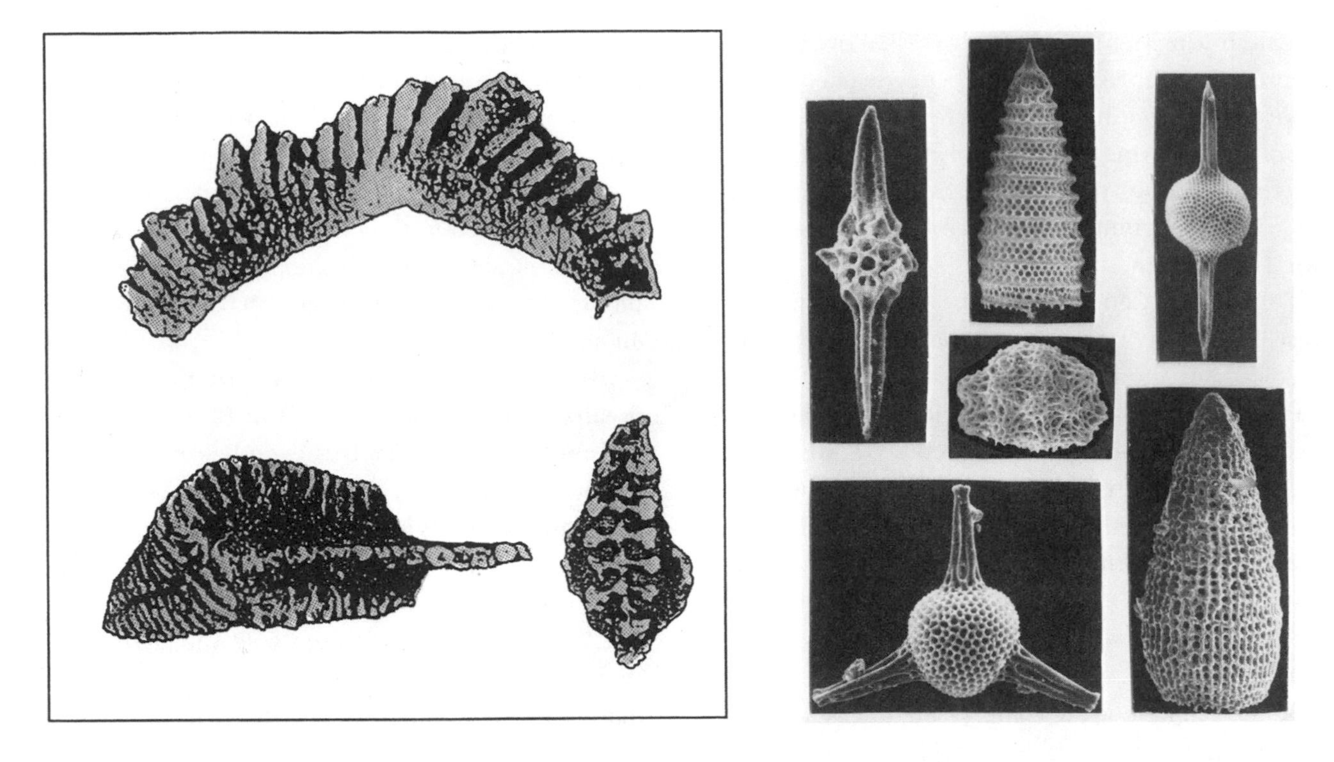

The oldest microfossils known from Oregon are conodonts, the teeth of a primitive Devonian fish found in limestones of the Grindstone terrane (above left). They are less than 1/32nd inch in length. From the Baker terrane, the intricate glassy skeletons of Mesozoic radiolarian microfossils (right) are just 1/64th inch long. (Photos courtesy of N.M. Savage; E. Pessagno)

recently Charles Blome and Bruce Wardlaw of the U.S.G.S. and Merlynd Nestell at the University of Texas refined the stratigraphy and age-dated the rocks using microfossils.

Close to the border between Crook and Harney counties, a 100-foot-thickness of Devonian limestone yields brachiopods, corals and microfossil conodonts. Nearby, over 1,000 feet of limestones, mudstones, sandstones, and cherts of the Coffee Creek Formation includes shallow-water tropical corals and brachiopods, whereas the overlying nonmarine sandstones of the Spotted Ridge Formation record a coastal plain inhabited by ferns and scale trees similar to those in the late Pennsylvanian coal layers of the central and eastern United States. One thousand feet of limestones, cherts, and sandstones of the Coyote Butte make up the uppermost stratum of the Grindstone. This fossiliferous formation yields shallow-water Permian fusulinids, corals, and radiolaria.

Baker Terrane

The Baker terrane curves across Grant, Baker, and Wallowa counties in Oregon to Cuddy Mountain in Idaho. This chaotic mixture of deep ocean sediments and volcanic rocks has been metamorphosed, folded, and intruded by plutons. Separating the Olds Ferry to the southeast from the Wallowa terrane on the north, the Baker incorporates fragments of both and represents a subduction zone and mid-ocean basin. As rocks of the Baker terrane were overridden by the oncoming Olds Ferry, they were tightly compressed into an accretionary wedge. Around the margins of the present-day Pacific Ocean, similar prisms of sediments can be found in severely folded and sheared rocks trapped between colliding plates. Deep sea cherts are especially abundant, and ophiolitic fragments are scattered throughout.

Late Paleozoic and Mesozoic exposures of the Baker contain broken mixed blocks of the Permian Canyon Mountain complex, the Permian to Triassic Burnt River Schist, and the Pennsylvanian through Jurassic Elkhorn Ridge Argillite. The Canyon Mountain complex, which covers an area of about six square miles southeast of John Day, is rich in serpentine and is regarded as a nearly complete ophiolite. An ophiolite suite is a predictable succession of ocean crust with mantle peridotites at the base, followed upward by gabbro, basalt dikes, pillow lavas, and capped by deep sea muds and cherts. Because the rock serpentinite, meaning serpent-like, is easily sheared by faulting, it derives its name from the broken polished surfaces that resemble the texture of snakeskin.

Near Durkee and exposed in the Burnt River Canyon in Baker County, the Burnt River Schist is a 20,000-foot-thick sequence of greenstones, schists, limestones, quartzites, and quartz-rich shales. Limestone and shale pods in the Burnt River Schist yield Triassic microfossils. Where they have undergone the intense effects of heat and pressure, the limestones have recrystallized to marble.

In the vicinity of Baker City and Sumpter, the Elkhorn Ridge Argillite is represented by thick layers of altered mudstones with thinner layers of volcanic ash, chert, and limestones. Fusulinids and corals from the cherts have been dated as Pennsylvanian through late Triassic, but Charles Blome has reported early Jurassic radiolarian faunas in the units near John Day and east of Baker City.

On the basis of its distinctive metamorphic and deformation history, the central part of the Baker terrane was split into two belts in 1995, which were informally named the Bourne and Greenhorn subterranes by Mark Ferns and Howard Brooks. Assemblages of the Elkhorn Ridge Argillite, consistent with the make-up of an accretionary wedge, are

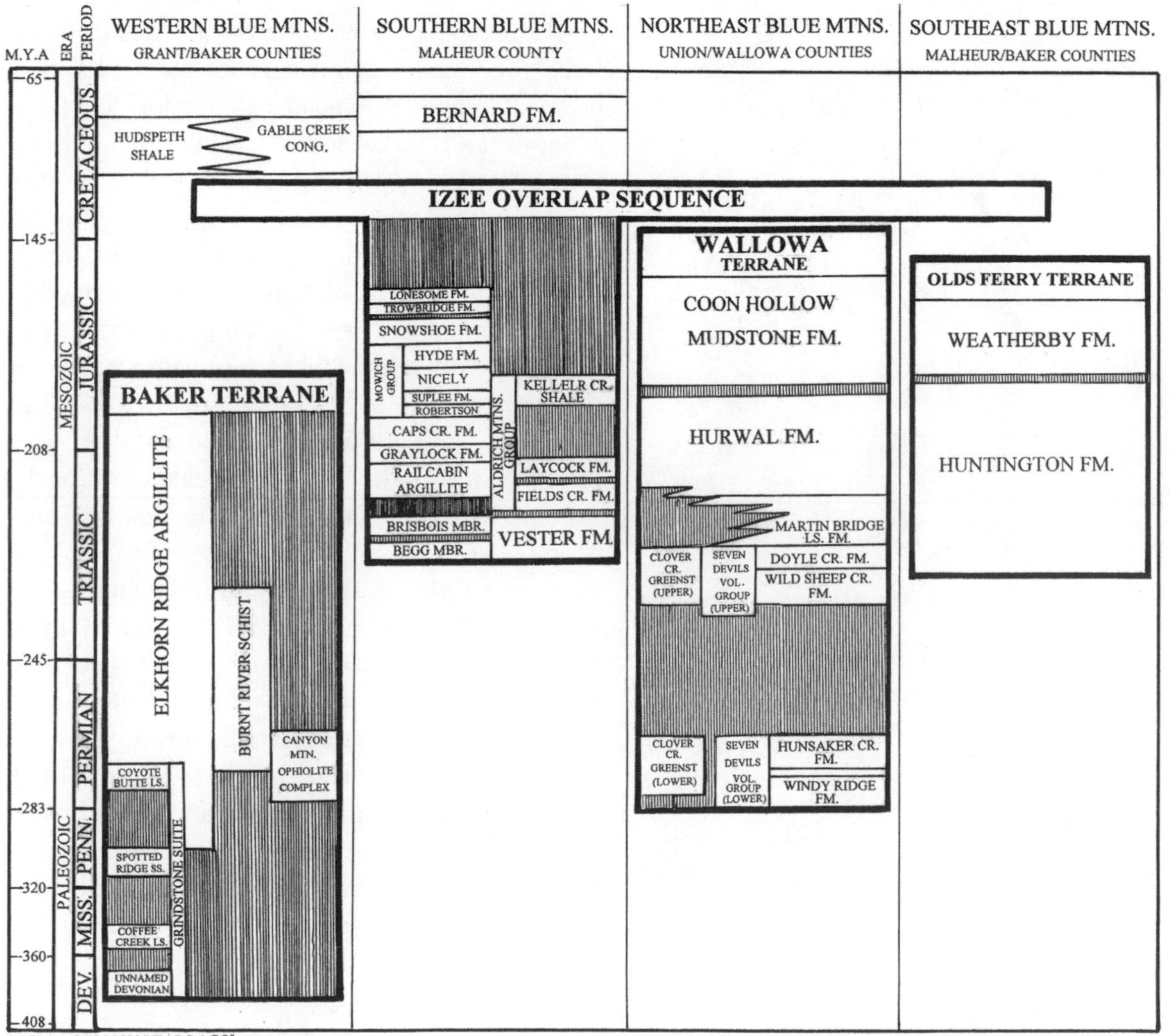

Stratigraphy of the Blue Mountains terranes. (After Armentrout, Cole, and TerBest, 1979; Dickinson and Vigrass, 1965; Dorsey and LaMaskin, 2007; Follo, 1994; Imlay, 1980, 1986; LaMaskin, Dorsey, and Vervoort, 2008; LaMaskin, et al., 2009; Silberling, et al., 1987; Vallier and Brooks, eds., 1986, 1994, 1995; Walker, ed., 1990; White, et al., 1992)

included in the northern Bourne subterrane. The southern Greenhorn is a mélange of serpentine and volcanic sediments. Separated by the east-west Cave Creek fault near Burnt River Canyon, the belts have undergone varying degrees of metamorphism in which the Bourne was buried deep in the collision zone, whereas the position of the Greenhorn was somewhat shallower.

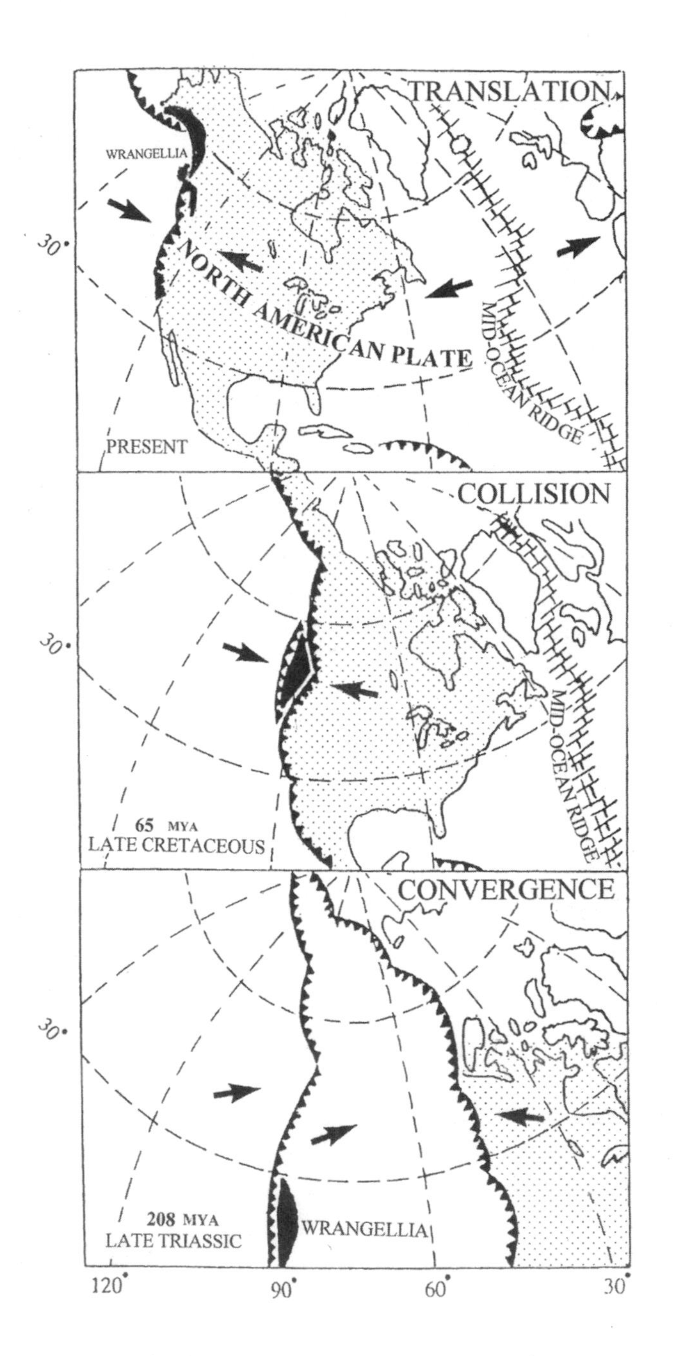

One model for the convergence, collision, and translation of Wrangellia shows that the North American plate, moving in a northwesterly direction, captured and swept up the Wrangellia superterrane before shifting the individual slices along faults toward Alaska. (After Brooks, 1979; Jones, Silberling, and Hillhouse; 1977; Kays, Stimac, and Goebel, 2006; Moore, 1991; Stanley, 1987)

Wallowa Terrane

The Wallowa is the northern-most terrane of the Blue Mountains province, underlying the Wallowa Mountains in Oregon and the Seven Devils range in Idaho. Isolated exposures appear north of Walla Walla, Washington, and south of Lewiston, Idaho, while continuous sections can be traced for over a hundred miles in the Snake River Canyon.

The puzzle of the Wallowa volcanic island arc has undergone modifications, and several models have been presented to explain its complexities. David Jones of the U.S.G.S. found striking similarities between Mesozoic rocks and fossils in scattered slivers from the Blue Mountains in Oregon and those of the Wrangell Mountains in Alaska, leading him to propose the concept of a superterrane Wrangellia. During the Triassic Period, Wrangellia was a single crustal block or a large igneous province situated in the tropical western Pacific. Transported eastward atop a tectonic plate, it collided with and was accreted to North America. After attachment, Wrangellia was sheared up by large-scale faults that translated or conveyed the individual pieces northward to Canada and Alaska, a distance of 1,500 miles.

George Stanley at the University of Montana believes the Wallowa terrane originated as an island archipelago within a larger volcanic center but not as part of a superterrane. By mid-Jurassic time, the edifice was close to North America and associated with other Blue Mountains terranes. Stanley's studies of Triassic fossils in the Wallowa package suggest that it may have an endemic linkage with the Alaskan Wrangellia but that it shows little affinity with other North American fragments.

The Wallowa terrane is composed of a diversity of Permian to Jurassic volcanic and sedimentary rocks of the Seven Devils Volcanic Group, the Clover Creek Greenstone, the Martin Bridge, and the Hurwal formations. Named by Tracy Vallier in 1977,

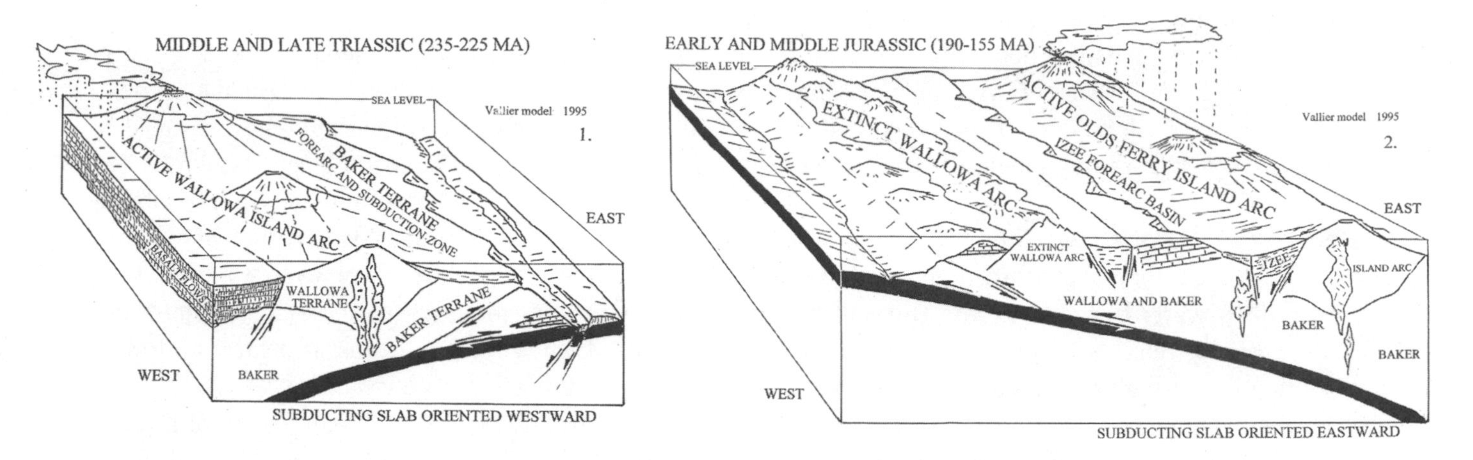

One scenario illustrating the evolution of the Blue Mountains features a major reorientation of the subduction slab, which shifted from west-dipping during the Triassic to east-dipping in the Jurassic. (After Vallier, 1995)

the oldest Seven Devils Group was divided into the Windy Ridge, the Hunsaker Creek, the Wild Sheep Creek, and the Doyle Creek formations. These rocks partially correlate with the Clover Creek Greenstone, a 4,000-foot-thick sequence in northern Baker County that contains cool-water boreal Permian brachiopods in its lower intervals but only scarce Triassic fossils in the upper portion. Volcanic flows of the Windy Ridge Formation, erupted in a submarine environment, are limited to exposures along the Snake River at the Oxbow, whereas the shallow-water Hunsaker Creek Formation was deposited in a basin adjacent to a volcanically active but eroding island arc. Exposed along the Snake and Salmon river canyons as well as in the Seven Devils and Wallowa mountains, the Hunsaker Creek has well-preserved crinoids, Permian bryozoa, and mollusks in limestone lenses. Volcanic rocks of the Wild Sheep Creek and volcaniclastics of the Doyle Creek accumulated in a forearc or interarc basin. Limestones between lava flows contain the Triassic deep-water flat clam *Daonella*.

During the late Triassic to early Jurassic, the Martin Bridge and Hurwal episodes record a variety of

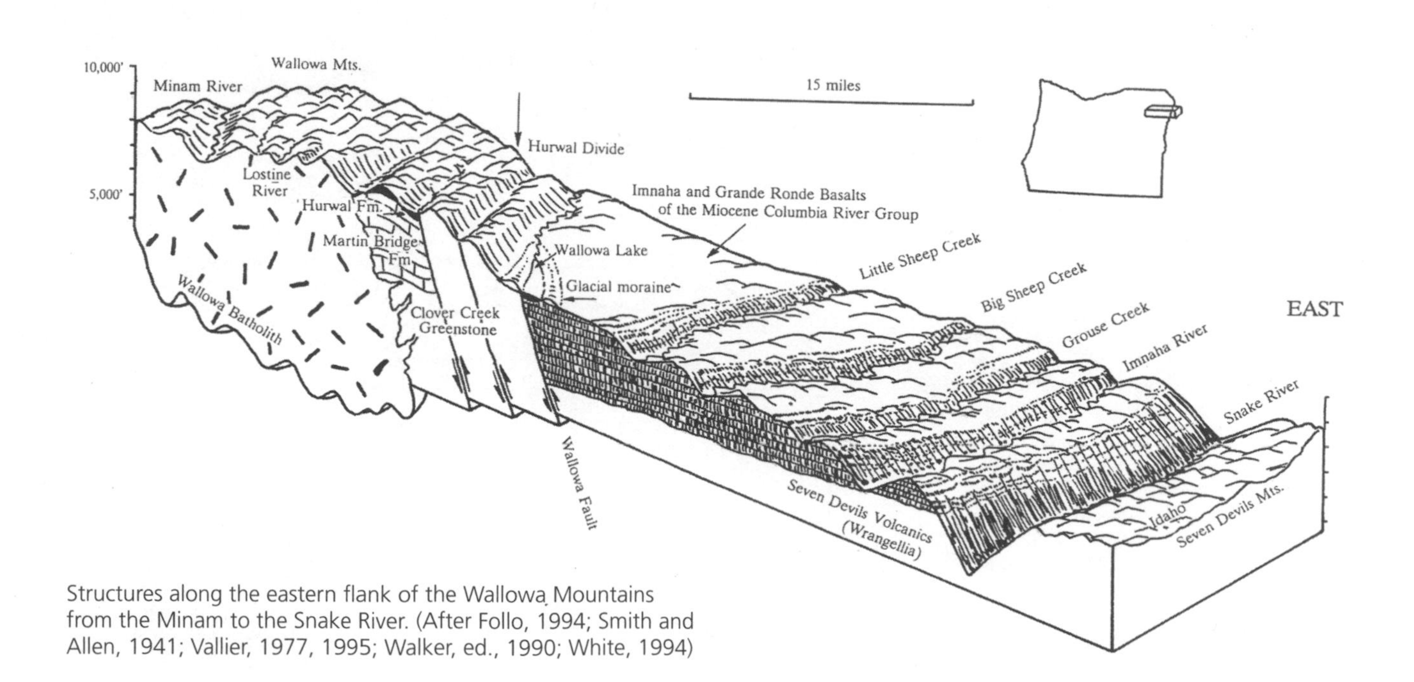

Structures along the eastern flank of the Wallowa Mountains from the Minam to the Snake River. (After Follo, 1994; Smith and Allen, 1941; Vallier, 1977, 1995; Walker, ed., 1990; White, 1994)

marine environments. Following a lull in volcanic activity, Martin Bridge limestones accumulated atop a shallow shelf dotted with reefs. An abrupt change to a deeper oceanic slope and basin is recorded by coarse-grained breccias and conglomerates present in the strata.

Renewed volcanism and submergence of the Martin Bridge carbonate platform coincided with a shift to deep-water turbidites of the Hurwal Formation. Erosional channels cut into Hurwal mudstones filled with conglomerates and breccias, some of which may have been derived from the Baker terrane. Fossil mollusks, corals, and marine reptiles within the Martin Bridge and Hurwal are remarkably similar to faunas of the western Pacific, central Europe, and the Himalayan Mountains, reflecting the exotic nature of the displaced Wallowa terrane.

Exposed along the Snake River, Jurassic Coon Hollow mudstones, sandstones, and conglomerates, deposited atop the uneven eroded surfaces of the Wild Sheep Creek and Doyle Creek formations, are the uppermost layers of the Wallowa terrane. Fragmentary plants and occasional invertebrates, carried by streams onto deltas and floodplains, record nonmarine and nearshore conditions and a temperate rainy climate.

Compressional forces that accompanied the accretion process in late Mesozoic time intensely deformed the Wallowa terrane, locally overturning entire folds.

Izee Terrane

Originally designated as a separate terrane, Izee sediments have been recently interpreted as deposits in a forearc basin between the Wallowa and Olds Ferry volcanic arcs. The assemblage is regarded as an overlap sequence, covering the older rocks of the Wallowa, Baker, and Olds Ferry terranes. This suggests that at least part of Izee sedimentation took place after amalgamation of the older crustal fragments, which leaves its status as a fault-bounded terrane in doubt.

A continuous 12-mile-thick succession of Triassic to Jurassic rocks depicts varying depths and distinct environments for the Izee. On the basis of structure, petrology, and fossils, the strata were divided into numerous formations by Dickinson and Vigrass. Ammonites and radiolaria, inhabiting a Tethyan, tropical marine environment during the late Triassic, were supplanted by invertebrates of a boreal nature by the middle Jurassic, evidence that the terrane had moved from warmer to cooler latitudes. Exposures are limited in areal extent from Suplee and Izee in Crook and Grant counties eastward to the John Day inlier in Wheeler County. The older rocks are revealed in the inlier or erosional window cut through a covering of younger rocks.

Within Izee strata, the mid-Triassic Vester Formation is separated from the Triassic to Jurassic Aldrich Group by the northeast-southwest trending Poison Creek Fault. Both have a selection of cephalopods,

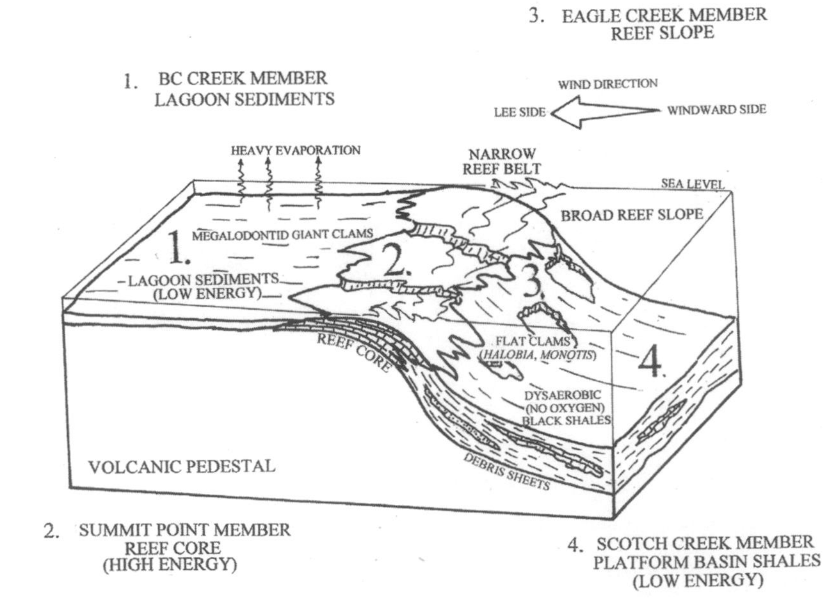

Some of the best exposures of early Mesozoic shallow-water limestones in North America are found in the Martin Bridge intervals. The lagoon and reef settings are rich in mollusks, corals, crinoids, and microfossils, whereas the deep-water fauna consists of radiolaria, flat clams, ammonites, and marine reptiles. Utilizing changes in deposition, George Stanley divided the Martin Bridge into four members, the BC Creek, the Summit Point, the Eagle Creek, and the Scotch Creek, representing reef lagoon, reef core, reef slope, and a deep water basin respectively. Gigantic clams (bivalves) in the BC Creek lagoon lived in gregarious colonial assemblages. (After Stanley, McRoberts, and Whalen, 2008)

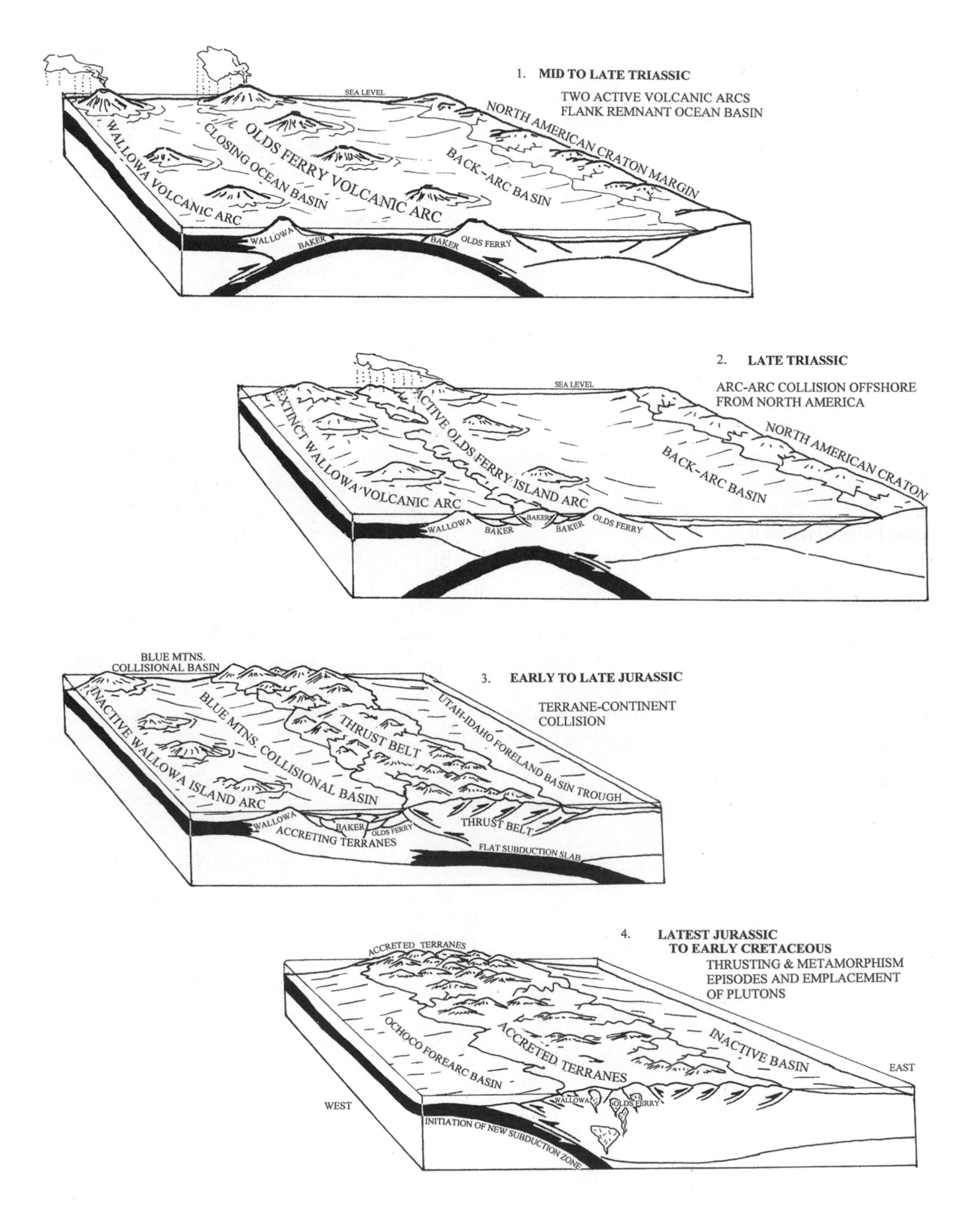

Sequential models reconstruct intervals of sedimention, volcanism, plate collision, and terrane accretion in the Blue Mountains between the Triassic and Cretaceous. (After Dorsey and LaMaskin, 2007; Lamaskin, et al., 2009)

brachiopods, and pelecypods. The presence of thick-shelled oysters in the earliest Begg Formation suggests the shallow waters of a marine shelf and shoreline. The Jurassic Mowich Group, which is less than 500 feet thick in most places, includes the sandy nearshore Suplee Formation and limestone reefs of the Robertson Formation that are largely made up of tightly packed colonies of plicatostylid clams (*Lithiotis*). Comprising the upper layers of the Izee basin, the Snowshoe, Trowbridge, and Lonesome formations reach thicknesses of 4,000 feet. Ammonites are particularly numerous in the Snowshoe, but fossils are scarce in the other two.

Thrusting of the Izee terrane that began during the late Jurassic gradually intensified into the Cretaceous. The intrusion of granites along with episodes of folding greatly complicates the interpretation of Izee tectonic history.

Olds Ferry Terrane

The Olds Ferry terrane extends toward the northeast in a curved alignment of isolated exposures from Huntington in Malheur County to Cuddy Mountain, Idaho. Triassic to lower Jurassic Huntington and the middle Jurassic Weatherby formations of this terrane were described by Howard Brooks and more recently refined by Todd LaMaskin. Along the flanks of the Olds Ferry island arc, volcanic flows, tuff, breccia, and conglomerates of the Huntington Formation interfinger with sandstone and siltstone beds bearing ammonites and clams. A distinctive limestone layer in the lower Huntington reflects a brief quiet period between volcanic eruptions. In contrast to other Blue Mountains terranes, Olds Ferry rocks show only minimal folding toward the end of the Jurassic.

Above the Huntington, the Weatherby Formation is exposed in the Ironside Mountain inlier. The Weatherby consists of volcaniclastic sedimentary rocks, shallow marine limestone pods with ammonites, and nonmarine conglomerates covered by black marine shales and turbidites eroded from the Olds Ferry volcanic arc. The Connor Creek fault, which separates the Weatherby from the Baker terrane, can be traced in a southwesterly direction for almost 30 miles across the Snake River from Idaho.

In their overview, Rebecca Dorsey and Todd LaMaskin divided the volcanics and sediments of the Blue Mountains into two distinct phases or megasequences, thick covering layers that represent separate well-defined lengthy stages during the evolution of the province. The transition between the two megasequences reflects a significant change in the regional tectonic setting from multiple small, arc-related basins to a single large-scale marine trough.

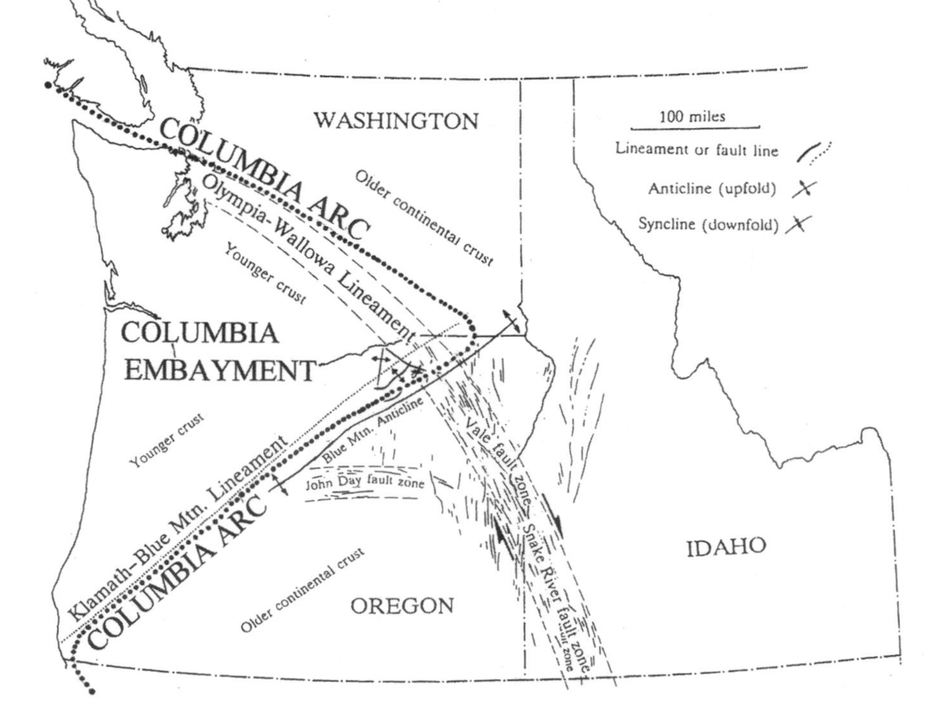

The Olympic-Wallowa and Klamath-Blue Mountains lineaments are enormous features that run for hundreds of miles. The OWL strikes southeast-northwest from Puget Sound to the Snake River and parallels the Wallula and Eagle Creek faults. It crosses the southwest to northeast-trending Blue Mountains anticline, a major Tertiary structure from Powell Buttes in Crook County almost to Lewiston, Idaho. The anticline folded upward in the Oligocene, around 35 million years ago, recording strong northwest-southeast compression. The Klamath-Blue Mountains lineament strikes northeast-southwest for a distance of 400 miles and is slightly north of the Blue Mountains anticline. (After Lawrence, 1976; Mann and Meyer, 1993)

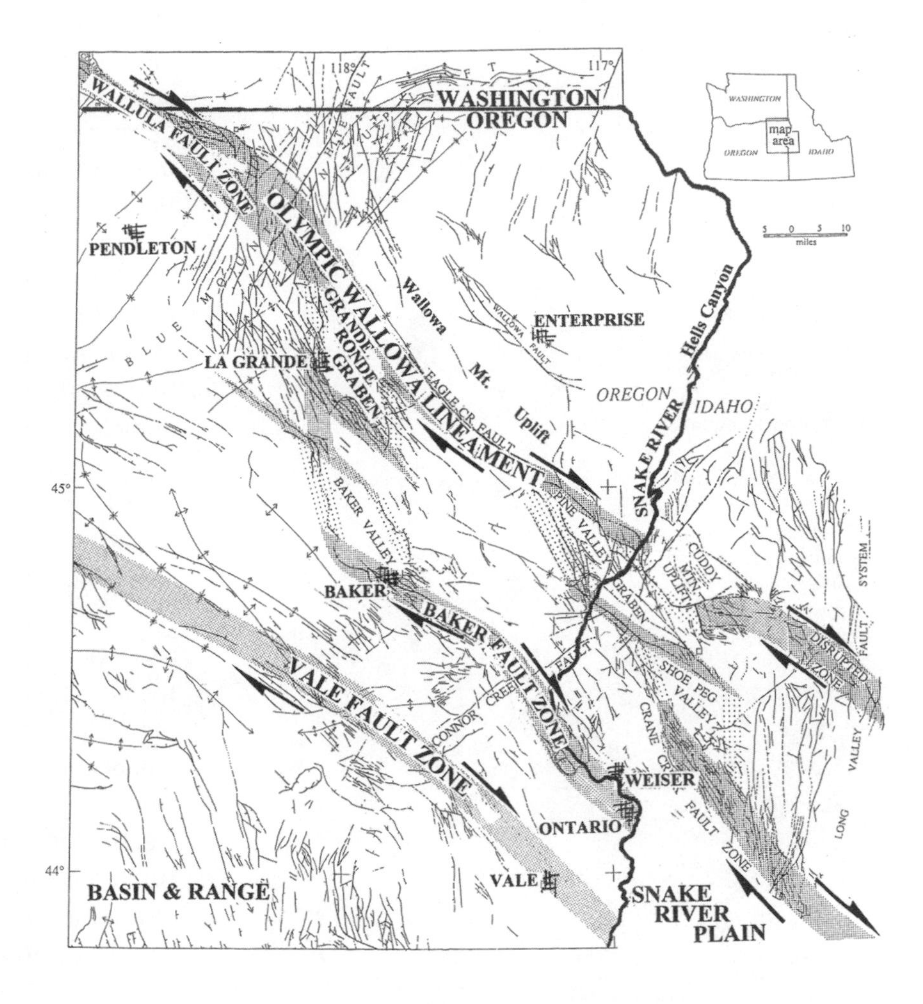

Elongate pull-apart basins or grabens such as the Baker, Grande Ronde, and Pine Creek valleys develop along megashears. They are the product of east-west crustal extension south of the Olympic-Wallowa lineament where thinning (extension) is greatest. (After Hooper and Conrey, 1989; Mann and Meyer, 1993; Riddihough, Finn, and Couch, 1986; Walker and Robinson, 1990)

Batholiths and Granitic Rocks

From the Paleozoic to Mesozoic eras, the Blue Mountains terranes were intruded by batholiths and smaller plutons and altered by heat and pressure. A mass of coarse-grained crystalline rock, batholiths invade the crust and cool slowly. Tracy Vallier divided the bodies by age into the late Paleozoic and Triassic plutons of the Wallowa and Baker terranes, the middle Triassic to early Jurassic of the Olds Ferry, and the later Jurassic to early Cretaceous batholiths, which intruded all of the terranes. In the Elkhorn Mountains, the 144-square-mile Bald Mountain batholith is smaller than the Wallowa batholith at the core of the Wallowa Mountains, which covers 324 square miles and is the largest in Oregon.

Faulting and Structure

Lineaments are visible on the surface but are of such magnitude that they can be seen best on high-altitude aerial photographs or satellite images. Although poorly understood, they are thought to be megashears, lengthy strike-slip faults above some type of boundary or structure buried deep in the crust. Lineaments are expressed along interconnected faults, folds, and grabens or by more subtle features such as the straight alignment of stream valleys or volcanoes.

Cretaceous—Marine Seaway

With the end of accretion in the early Cretaceous, a shallow seaway spread over wide expanses of the Northwest. The shoreline meandered diagonally from the Klamath Mountains to the Blue Mountains and into Washington and Idaho. Among a diversity of inhabitants in the Oregon ocean, the many species of coiled, shelled cephalopods or ammonites are useful in age-dating the rocks. Reptiles such as *Ichthyosaurs*, *Plesiosaurs*, and primitive crocodiles occupied the marine niche, along with flying *Pterosaurs* that fed on fish.

Late Cretaceous sediments atop the older rocks of the Baker and Izee terranes were deposited in a lengthy forearc trough stretching from the Ochoco Mountains in central Oregon to the Klamath

A lengthy Cretaceous basin stretched from the Ochoco Mountains in central Oregon, to the Klamath Mountains in the southwest, and on to the Great Valley of California. (After Dorsey and Lenegan, 2007; Housen and Dorsey, 2005; Kleinhans, Barcells-Baldwin, and Jones, 1984; Wyld and Wright, 2001)

Tertiary stratigraphy of the Blue Mountains. (After Armentrout, et al., 1988; McClaughry, Gordon, and Ferns, 2009; McClaughry, et al., 2009; Retallack, Bestland, and Fremd, 2000; Walker, ed., 1990)

Mountains and the Great Valley of California. Similarities between the ammonites and clams in the Bernard, Hudspeth, and Gable Creek formations from the eastern part of the state to those of the Hornbrook at Grave Creek in Jackson County support this picture. In Wheeler County, coarse-grained conglomerates, mudstones, and turbidites of the Gable Creek and Hudspeth formations are revealed in the Mitchell inlier, a 70-square-mile window through the younger volcanic covering. These were immense deep-sea fan deposits, whereas fossils in exposures at Bernard Ranch imply shallower water.

Paleomagnetic studies of Cretaceous rocks in the Ochoco basin indicate that they were originally situated at 40° latitude. This, in turn, indicated to Bernie Housen of Western Washington University and Rebecca Dorsey that the basin may have been adjacent to the Sierra Nevada Mountains of California during the Cretaceous and that the Blue Mountains are a link between the Sierras and British Columbia.

Cenozoic

By the end of the Mesozoic Era and the beginning of the Cenozoic, 66 million years ago, the sea had withdrawn to the west, while the Cascades had yet to develop. During the accretion process, the Blue Mountains had been elevated well above sea level with intensive erosion as well as ongoing, but intermittent, volcanic activity. Paleocene strata are known only at two locations, a flora at Denning Spring in Umatilla County and freshwater diatoms from Imbler in Union County, neither with a formational designation. In contrast, the Eocene through Miocene epochs are characterized by extensive lava, ash, and sediments rich with fossils.

Eocene

The eastern Oregon climate during the Eocene Clarno interval from 54 to 40 million years ago was sub-tropical to tropical. Andesitic and basaltic lavas from volcanic cones scattered throughout the western Blue Mountains, along with lahars, multicolored soils, and tuffaceous sediments, make up the blend of the Clarno Formation. The tectonic setting of Clarno volcanism is not well understood, but the timing of the eruptions matches that of the Challis activity in Idaho, leading to speculation that the two are related to a magmatic arc extending diagonally northeast by southwest across Oregon, Idaho, and Montana. Peter Hooper's 1995 model for Clarno volcanism appeals to thinning of the crust where extensional processes and regional stretching brought the heated mantle close to the surface.

When volcanic ash mixes with water it forms lahars, which mobilize and flow like syrup on even the slightest slope. Lahars and individual uninterrupted deposits of the Clarno reach thicknesses of 1,000 feet in Wheeler County. A variety of plant and animal remains, entombed in the mudflows, typify ancient stream and lake habitats, where avocado, cinnamon, palm, and even bananas grew in a frost-free climate. A diversity of large land mammals such as rhinoceros and brontothere, along with the sheep-like oreodons, shared the habitat with bear-like predators. A peculiar aquatic rhinoceros, not unlike a modern hippopotamus, wallowed in the shallows, where it was preyed upon by crocodiles. The Clarno is perhaps most famous for its many species of tropical nuts, fruits, and seeds, found in exposures just east of the old townsite of Clarno. Many are so well-preserved that they appear modern.

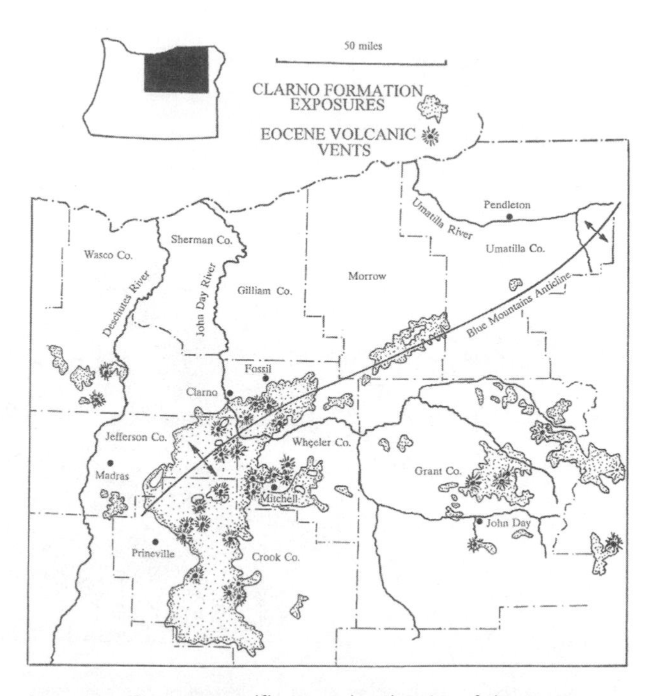

In eastern Oregon, mudflows and volcanics of the Eocene Clarno Formation are distributed across vast areas from Grant to Jefferson and Wasco counties. (After McClaughry, et al., 2009; Robinson, Walker, and McKee, 1990; Smith, et al., 1998; Walker and Robinson, 1990)

Oligocene-Miocene

Clarno volcanic episodes diminished with the onset of the John Day period, spanning the late Eocene to early Miocene. John Day rhyolitic lavas covered an area of some 35,000 square miles in Jefferson, Crook, Wheeler, and Grant counties with incandescent gas-charged clouds of ash, which cooled as the distinctive ignimbrites.

The superb quality and quantity of both plant and animal fossils from the John Day Formation are almost unmatched. Sluggish streams and quiet lakes, which encouraged plant growth and attracted

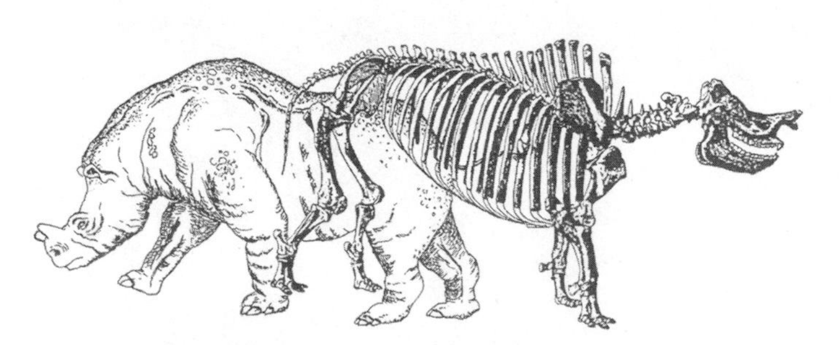

Living in eastern Oregon during the Eocene, brontotheres were similar to a rhinoceros in bulk and structure. (After Orr and Orr, 1999)

An Eocene palm leaf *Sabalites* from the Clarno Formation measures over two feet in length. (Photo courtesy Oregon Department of Geology and Mineral Industries)

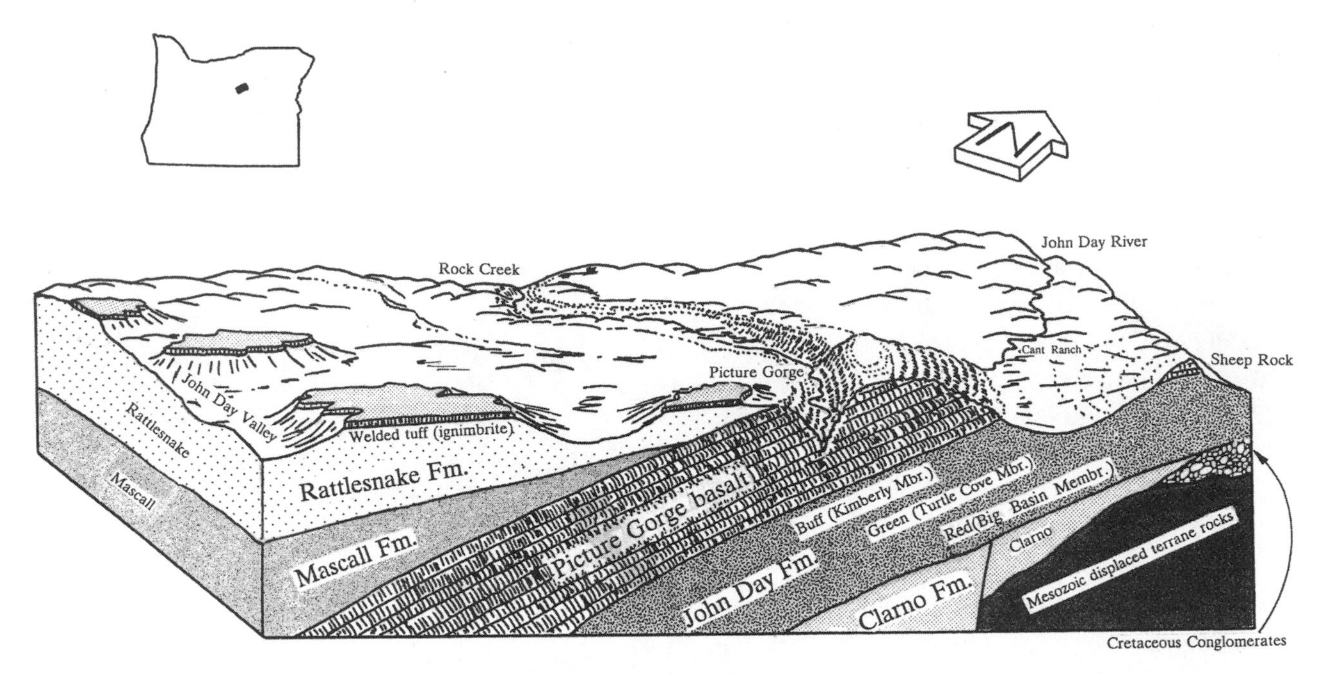

Changes in the John Day climate are reflected by the nature and color of the sediments. In 1901, John Merriam, at the University of California, divided John Day strata into three members based on color. His units were refined in 1972 by Richard Fisher and John Rensberger of the University of Washington into the highly oxidized, deep red claystones of the oldest Big Basin member, the pea green Turtle Cove tuffaceous claystones, and the cream-to-buff-colored tuffs and conglomerates of the youngest Kimberly and Haystack Valley sequence. Fossil plants, animals, and soils chronicle paleo-environmental variations as lakes and rivers gave way to open grasslands. On the basis of fossil soils (paleosols), Greg Retallack of the University of Oregon further separated the Big Basin Member into three units. (After Baldwin, 1976)

animals, provided optimum conditions for preservation. Because they only date back between 18 to 39 million years, most of the plants and animals look somewhat familiar, although none of the species precisely corresponds to those living today.

Crooked River Caldera

The question of the volcanic sources for the Clarno and John Day formations has long been the topic of speculation and controversy. Cascade vents or smaller buttes toward the west were proposed, while the notion of a single volcano or caldera was discounted. Recent mapping on the extreme western edge of the Blue Mountains by Mark Ferns and Jason McClaughry of DOGAMI led to the discovery of three large-scale caldera complexes. They identified the oldest Wildcat Mountain in Crook County at the crest of the Ochoco Mountains, the Crooked

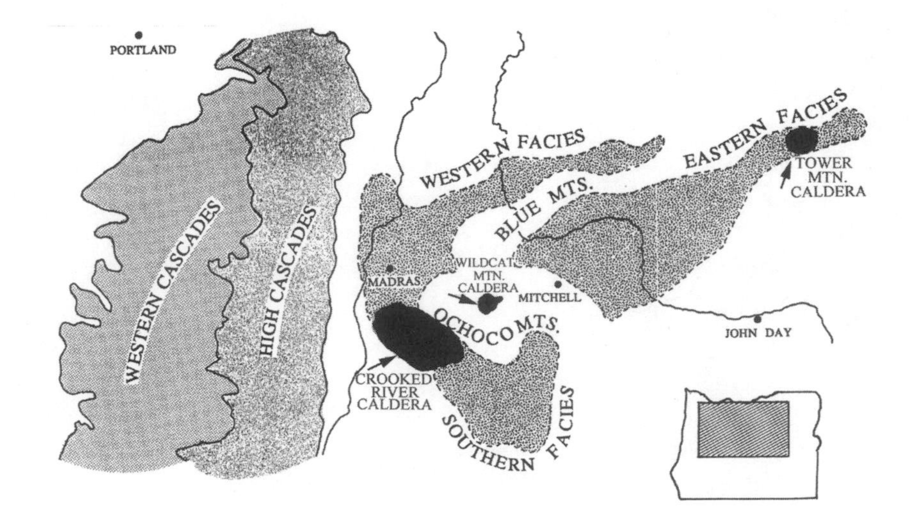

The John Day Formation has been further divided into three geographic areas: the eastern facies along the John Day River, the southern on the Crooked River, and the western on the Deschutes River. (After Merriam, 1901; Fisher and Rensberger, 1972; Retallack, Bestland, and Fremd, 2000; Woodburne and Robinson, 1977)

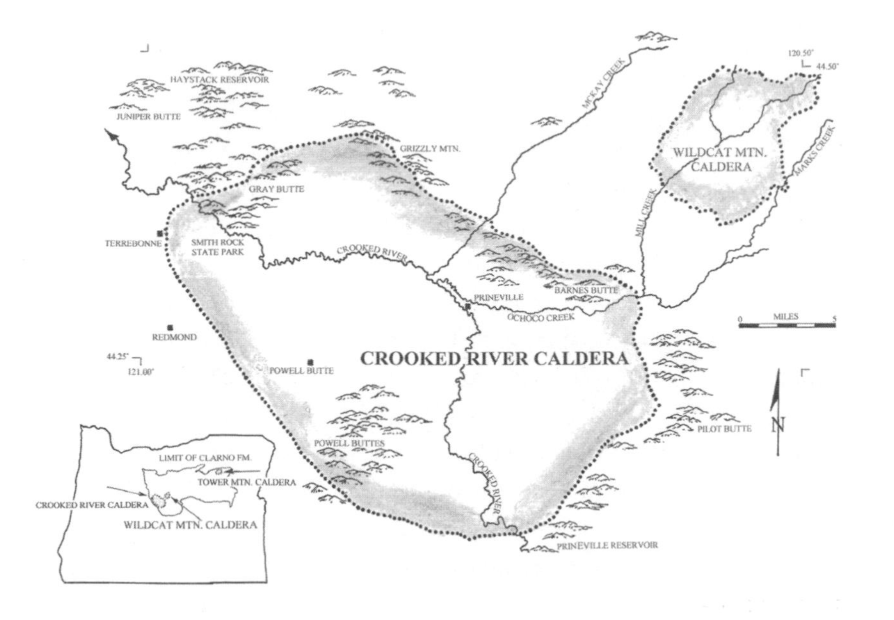

Volcanics of the Eocene Clarno and Oligocene John Day formations have been traced to three large calderas in central Oregon, the Wildcat, Tower Mountain, and Crooked River. (After Ferns and McClaughry, 2006; Fisher and Rensberger, 1972; McClaughry, et al., 2009; Walker and Robinson, 1990)

River caldera near Prineville in Jefferson and Crook counties, and the youngest Tower Mountain near Ukiah in Umatilla County.

A thick pile of rhyolitic tuff, the Wildcat Mountain caldera, dated at 43.8 to 36 million years old, is the foremost structure in the Ochoco volcanic field. Formed with the eruption of Steins Pillar Tuff, the deeply eroded caldera lies 12 miles northeast of Prineville. Rocks of the Ochoco field are equivalent to those of the Clarno Formation.

John Day eruptions from numerous vents in the Tower Mountain and Crooked River volcanic complex are dated around 28 to 29 million years ago. Roughly 10 miles wide, the Tower Mountain caldera has a ponded core of tuff and is surrounded by a ring of rhyolite domes. These rocks are equivalent to the eastern facies of the John Day.

Lavas from the Crooked River caldera were bimodal, producing both basalts and rhyolites of the John Day Formation. Between the Prineville Reservoir to the southeast, Gray Butte to the northwest, and the Ochoco Mountains to the east, the Crooked River caldera is 25-by-18-miles in diameter and projects into the Deschutes-Umatilla plateau. This

Following the eruption and collapse of the Crooked River caldera, rhyolitic flows, domes, and dikes from the Powell Buttes, Gray Butte, Barnes Butte, and Ochoco Reservoir centers were emplaced along the rim of the structure. (In the photograph looking north-northwest along the Crooked River canyon, Powell Buttes is at the top left; photo courtesy Condon Collection).

enormous structure is bounded by the Klamath-Blue Mountains lineament and is positioned at the junction of the Brothers Fault zone and Sisters fault systems.

Its construction began 29.5 million years ago with explosive tuff from Smith Rock, an Oligocene facies of the John Day Formation. Collapse of the caldera was followed by rhyolite flows at Ochoco Reservoir and emplacement of Gray Butte, Powell Buttes, Grizzly Mountain, Juniper, and Pilot buttes. Barnes Butte and other domes may be smaller calderas within the larger, older Crooked River center. Debris from the Crooked River eruption and collapse can be traced over 10,000 square miles, primarily to the northeast of the main caldera.

Miocene

Over a period from 17 to 6 million years ago, multiple flows of Miocene Columbia River basalts poured westward from cracks and fissures in the Blue Mountains and adjacent Washington and Idaho. Obscuring the older formations and creating a broad platform over eastern Washington and northeast Oregon, the lavas overlapped each other until only the highest of the older outcrops projected above the layers.

The individual flows have been identified by their chemical and mineral composition and traced back to the northwest trending Monument and Chief Joseph dike swarms. In the John Day Valley, Monument fissures fed the dark-gray, coarse-grained Picture Gorge basalt, which filled deep canyons and smoothed out the topography. Geologists speculate that the ancestral Blue Mountains uplift confined the spread of Picture Gorge flows to the southern Columbia plateau. An estimated 21,000 dikes from the Chief Joseph field in the Wallowa Mountains erupted the Imnaha basalts, followed by the Grande Ronde and Wanapum.

The sheer breadth and volume of the Columbia River lavas tend to overshadow middle to late Miocene volcanic events elsewhere in the area. The andesitic Sawtooth Crater, Strawberry Mountain, and Dry Mountain volcanoes each evolved as separate edifices, which, in turn, line up with a fourth volcano of the same composition at Gearhart Mountain in the Basin and Range. Along this northeastward-trending zone, the centers are spaced at regular 100-mile intervals, but their distribution is still to be explained. The vents do not seem to be related to subduction, but they may be situated above at a major crustal boundary or

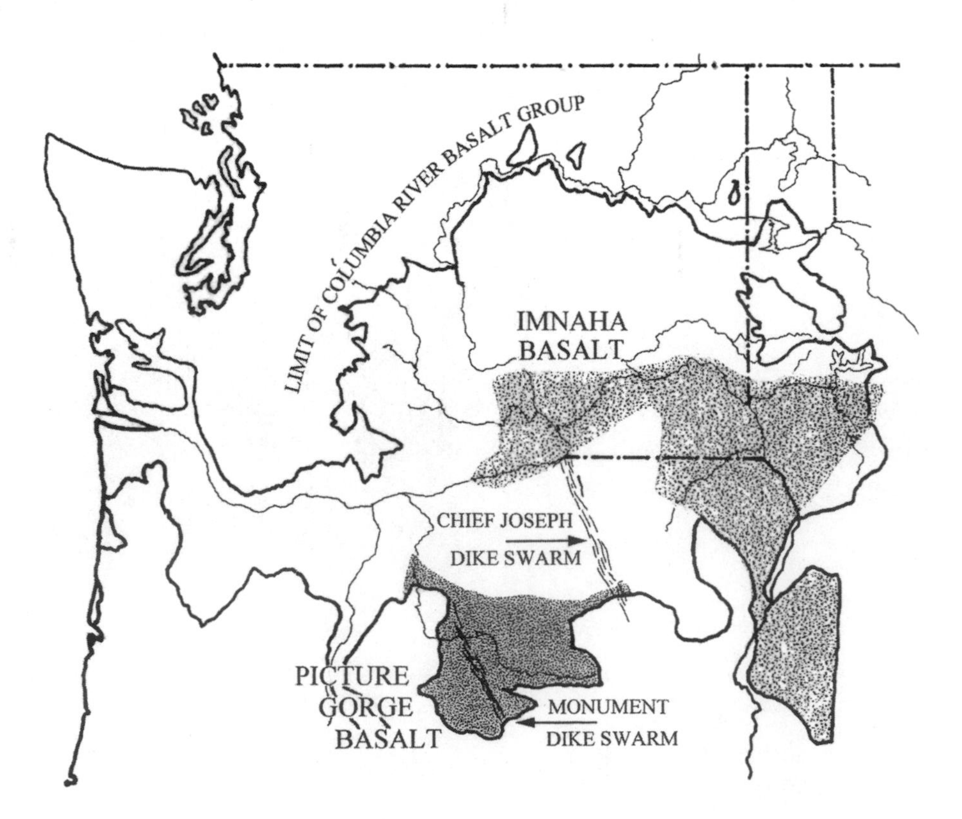

Sources of the Miocene Columbia River basalts in the Blue Mountains have been traced to the Chief Joseph and Monument dikes. (After Anderson, et al., 1987; Hooper, 1997; Tolan, et al., 2009; Walker, ed., 1990)

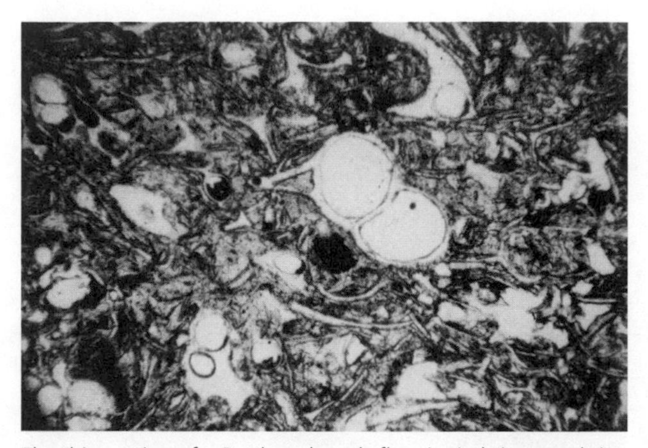

The thin section of a Rattlesnake ash-flow ignimbrite reveals its composition as a mass of broken glassy shards. (Photo courtesy Oregon Department of Geology and Mineral Industries)

anomaly. The characteristic stiff, slow-flowing andesites, produced at these sites, are of a type that typically accompanies very explosive eruptions.

Middle Miocene lavas at the southern extent of the province originated from a north-south line of dikes and vents beginning at Dooley Mountain and Unity Reservoir and continuing through the Lake Owyhee and McDermitt volcanic fields in the Oregon-Idaho graben.

By middle to late Miocene time, the shoreline of the Pacific Ocean had retreated westward, and the rising Cascade Range brought increasingly drier and temperate climates to eastern Oregon. Volcanism was confined to local eruptions of lava and showers of ash. With headwaters in the Blue Mountains, streams collected and redeposited volcanic debris into isolated depressions atop the Columbia River basalts, providing ideal sites for the burial and preservation of all manner of fossils. The Mascall and Rattlesnake formations in Crook and Grant counties and the Juntura and Drewsey formations in Harney and Malheur counties are known for their vertebrate assemblages.

Hoofed mammals, rodents, and predators of the Mascall Formation, in association with broadleaf plants, suggest an open woodland and savannah. Above the Mascall, sands, gravels, and silts of the 5-to-8-million-year-old Rattlesnake Formation have fewer fossils, but many are quite modern in aspect, although camels, rhinoceros, and mastodon give the formation an exotic complexion.

At the southern edge of the province, abundant middle Miocene vertebrates have been recovered from mudstones, volcanic tuffs, and sandstones of ancient lakes and streams near Juntura. Ephemeral lakes filled and then drained periodically when rivers were dammed by lava and ash. Many of the bones from the Juntura and Drewsey formations are broken and show wear, reflecting postmortem water transport and redeposition.

Pliocene sediments in Oregon are nearly as rare as those of the Paleocene, but Jay Van Tassel from Eastern Oregon State University has located and described a remarkable assemblage of diatoms, fish, invertebrates, and mammals from Baker City which date to that epoch.

Pleistocene—Carving the Mountains

Glaciation impacted eastern Oregon during the Pleistocene or Ice Ages beginning a little less than 2 million years ago and winding down about 11,000 years in the past. This period was marked by cycles of glacial advance and retreat, each episode lasting about 150,000 to 200,000 years, as moving ice straightened and deepened stream valleys. Continental glaciation, where huge ice sheets expanded southward, was more typical of the Great Lakes and upper Mississippi Valley. In western North America, continental ice sheets pushed into Washington, Idaho, and Montana, but in Oregon only ice caps and valley glaciers were present.

With the exception of the Ochoco Mountains, all of the ranges in the Blue Mountains display the unmistakable signs of alpine glacial striations on bedrock, layers of till, moraines, and strings of lakes gouged in valley floors. The higher basins are distinctively U-shaped in cross-section, the result of incision by slow-moving ice. Great symmetrical piles of till or moraines, left by melting glaciers, are easily identified by the poorly sorted mixture of angular pebbles, boulders, and clay. Perhaps the most dramatic are the lateral and end moraines, which dammed Wallowa Lake, now adjacent to the small community of Joseph. Joseph itself is built upon an ancient outwash plain, characterized by irregular hummocky ground that spread out in front of melting ice.

At the height of the Ice Ages, the Wallowa Mountains had nine large glaciers, each more than 10 miles long, which radiated outward from the center. The longest were the Lostine glacier at 22

Serrated ridges (arêtes) and amphitheatre-like cirques of the Wallowa wilderness are signs of glacial topography. (Photo courtesy U.S. Forest Service)

miles, the Minam at 21 miles, and the Imnaha at 20 miles. Averaging 1,000 feet in thickness, these ice masses extended down the valleys but failed to cover the ridges above 8,500 feet, thus a true ice cap did not develop. As late as 1929, the last glacier on the ridge above Glacier Lake in the Eagle Cap Wilderness was 800 feet long, 60 feet wide, and 24 feet thick. Today only small stagnant sheets of ice occupy a few of the high cirque basins.

Geologic Hazards

Earthquakes

The Snake River canyon experiences periodic earthquakes measuring up to 6.0 on the Mercalli intensity scale. A Mercalli estimate is based on eyewitness reports and damage and not on precise seismograph measurements that are recorded on the Richter scale.

Between the Brownlee Dam and Homestead on the Oregon side and Cuddy Mountain in Idaho a wide zone of northwest-striking faults, associated with pull-apart grabens and basins, is one of Oregon's most active seismic regions. The earliest records of quakes here in 1913 and in 1916 caused minimal damage. Geologists have concluded that seismicity along the Snake River is associated with the Olympic-Wallowa lineament, which parallels the Pine Creek Valley and Brownlee faults. Seismologists with the U.S.G.S., Gary Mann and Charles Meyer have reported that a number of moderate-sized earthquakes took place between 1981 and 1984 near Brownlee Dam, at the approximate intersection of the Pine Creek Valley and Cuddy Mountain faults.

Landslides

Landslides are a product of factors that affect the stability of a slope, and in the geologically complicated Blue Mountains differences in rock types, fracturing, long-term undercutting by creeks and rivers, and periodic seismic activity are all part of the equation. Although the moderate rate of precipitation here reduces the incidence of slides, it is not unusual for a combination of conditions to result in rock avalanches.

Along the Snake River and Hells Canyon, Tracy Vallier notes that some of the most spectacular visible features are underlain by earthflows. At Wild Sheep Rapids, Granite Rapids, Johnson Bar, and Big Bar, Pleistocene slide debris surged into the riverbed from steep walls. The Rush Creek landslide north of Hat Point moved 20 million cubic meters of material into Hells Canyon from the Idaho side, building a dam nearly 400 feet high. Even though of uncertain date, rubble from the slide overlies accumulations from the Bonneville flood but is covered by Mt. Mazama ash, placing the date from 15,000 to 7,700 years ago.

Unusually heavy precipitation in 1984 generated the notorious slide at Hole-in-the-Wall south of Sparta in Baker County. The slow-moving event

Debris from an ancient landslide at Pittsburg Landing on the Snake River was smoothed over and reworked by the subsequent surging waters of the Bonneville flood. (Photo courtesy T. Vallier)

began in August, when large fissures and cracks, opening high on a slope above the Powder River, initiated slumping which continued to accelerate. Highway 86 was completely covered with rocks and debris, even as the slide made its way across the river, where blockage created a 20-foot-deep reservoir.

Decomposing lavas and tuffs of the John Day and Deschutes formations are particularly prone to mass movement along steep canyon walls of the Deschutes and Crooked rivers. The older landslides are obscured by vegetation, whereas the more recent ones are bare of plants and soil cover. Near Prineville, some of the slide-prone areas have been confused with stable lava flows and have become fashionable for the placement of housing.

Since completion by the Bureau of Reclamation in 1920 for irrigation and flood control, the Ochoco Reservoir on the Crooked River has been plagued by leakage and structural problems related to its positioning on landslide deposits. The right abutment is secured in Clarno and John Day rocks, while the left is anchored in younger layers of the Deschutes Formation. Because of continuous seepage related to sinkholes in the slide debris, the dam was emptied in 1993, modified, repaired, then refilled.

A combination of melting snow and heavy rainfall throughout Oregon during February, 1996, caused a landslide near Summerville at the north end of the Grande Ronde Valley. Part of an ancient slide, the debris moved from Pumpkin Ridge eastward toward Highway 82 and the river. (Photo courtesy R. Carson).

Natural Resources

With the exception of the copper in volcanic layers of the Wallowa terrane, economic minerals in the Blue Mountains such as gold, silver, and zinc are primarily associated with veins in rocks of the Baker terrane. Hot fluids, impregnated with metals, were emplaced into fractures during the Mesozoic intrusion of granitic plutons.

Gold

This province has produced about three-fourths of Oregon's gold, most of which occurs in a strip 40 miles wide and 120 miles long from the John Day to the Snake River and into Idaho. As noted by Howard Brooks, there are countless placer workings within this belt and more than 14 lode mines. Lode gold in veins is mined by tunneling along the tabular bodies, whereas placers are worked by sluicing with water and panning in streams. Because gold is seven times heavier than quartz and almost twenty times heavier than water, it settles on the deepest levels of a streambed.

Mining substantially changed the face of eastern Oregon, which had no permanent settlements in the years before Henry Griffin and his party from Portland made the original discovery of placer gold west of Baker City in 1861. Diggings in the Canyon Creek drainage were so rich that within a year a tent city of 1,000 men appeared, and shortly thereafter prospectors were working in most of the gold-bearing areas of the Blue Mountains and Idaho.

Water was needed for flushing out gold, and when that commodity became scarce extensive ditches were dug to bring it to the placers. The Auburn ditch, parts of which are still used by the municipality of Baker City, was completed in 1863. The Rye Valley ditch was finished in 1864, and the 100-mile-long El Dorado in 1873 brought water from the head of Burnt River to the Malheur district.

Lode gold was discovered in 1862 at the Virtue mine on the Powder River, where a 10-stamp mill was constructed to process the ore. Stamp mills, employing heavy steel rods, pulverized the rock to extract the gold. The Virtue and Sanger mines in the Wallowa Mountains and the Connor Creek mine near Lookout Mountain dominated lode gold production until the early 1900s when operations at the Cornucopia mines began. Ranking as the top lode producer, the Cornucopia district was most active from the late 1930s to closure in 1941. The values of the gold extracted would be about $350 million at today's prices.

Early gold totals for the Blue Mountains can only be estimated because there are no exact figures, and much of it was sent to the mint at San Francisco where it was credited to California. Placer mines had an approximate yield of 630,500 ounces in comparison to lode operations, which totaled just over 825,300 ounces between 1900 and 1965, after which amounts dropped off dramatically.

Data on private companies is no longer available to the public. Since the 1990s, recreational operators

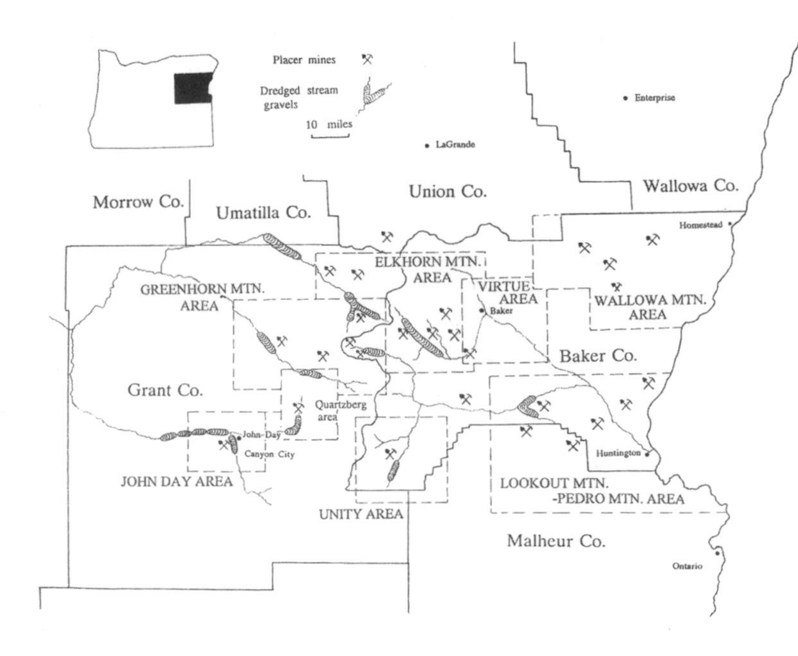

Major gold mining districts in the Blue Mountains (After Brooks, 2007; Brooks and Ramp, 1968)

By processing the gravels, bucket line dredges such as the one used in Sumpter (above right) in 1913 substantially aided gold production. Almost 100 years later, the still-exposed 2,600 acres of dredge tailings in Sumpter Valley (above left) are slowly being covered by vegetation. (Photos courtesy Oregon Department of Geology and Mineral Industries and Oregon State Highway Department).

using small portable dredges have accounted for most of Oregon's gold recovery.

Silver and Copper

Silver and copper ores are typically removed with gold as part of the extraction process, and in the Blue Mountains silver production peaked between 1900 and 1965, the most lucrative period. Beginning in 1897, copper was mined from the Permian Hunsaker Creek Formation at the Iron Dyke Mine on the Snake River. In the 1930s, output from the Iron Dyke was the highest in the state, after which the mine went through several changes in ownership, reduced production, and then closure. It reopened for a few years in 1979, when 20,000 ounces of gold, 40,000 ounces of silver, and 1,900 tons of copper were extracted.

In June, 1913, one of the state's largest nuggets, weighing 80.4 ounces, was recovered in placer gravels near Susanville in Grant County. Today the five-inch-long specimen, on display at the U.S. Bank of Oregon in Baker City, would be worth almost $160,000. (Photo courtesy Oregon Department of Geology and Mineral Industries)

Mercury

In the Clarno Formation, mercury sulfide or cinnabar is associated with intrusive volcanic plugs and along fault planes. The Horse Heaven district in Jefferson County, responsible for the highest yields in the state, operated between 1934 and 1958, when the reserves were exhausted. Overall, the mine produced 17,214 flasks of quicksilver. Mercury processing has the potential to release high levels of toxic contaminants, and since 2002 state and federal agencies have begun a program to inventory and evaluate historic mine sites.

Thundereggs is a popular name for geodes, agate-filled gas pockets. Thundereggs were designated as the official state rock by the 1965 Oregon legislature. (Photo courtesy Oregon Department of Geology and Mineral Industries)

Thundereggs

The gem-like thunderegg, a spherical nodule or geode, is not a rock but a body of opal, agate, or chalcedony with a knobby drab exterior. Thundereggs form in a variety of ways, but the process typically begins when a hollow cavity, produced by a gas bubble in highly viscous lavas, is lined or filled with crystals. Ranging from less than one inch to over four feet, most geodes average close to a baseball in size. Geodes can be found in many of the volcanic terrains, but they are especially common in rhyolitic lavas of eastern Oregon, where some commercial collecting is permitted.

Industrial rock. Diatomite is mined from the Miocene Juntura and Drewsey formations in Malheur and Harney counties. Diatoms are siliceous single-celled aquatic plants, which accumulated on the floor of ancient freshwater lakes. White to buff in color, diatomite beds vary from a few inches to 20 feet in thickness. First mined in 1917, diatomite is used as an abrasive, as an absorbent, as a filler in paints, for cat litter, and for filtering chemicals, drinking water, and liquid food products. Eagle-Pitcher Minerals, Inc., near Vale, is the major Oregon producer.

Bentonite claystones, along with zeolites, are extracted from open-pits in the Miocene Sucker Creek Formation of Malheur County and from the John Day Formation in Crook County. Derived from decomposed volcanic ash, bentonite is highly expansive when wet. Known commercially as fuller's earth, it is marketed for cat litter, as sealants, and for numerous industrial purposes. Teague Mineral Products, supplier of bentonites, also mines zeolites. Zeolites have a limited market, serving to control odors, to stabilize toxic waste, as an animal food supplement, and for a variety of agricultural applications.

Geothermal Energy

The potential for geothermal energy in the Blue Mountains is only moderate in contrast to that of the High Lava Plains, the Basin and Range, or the Cascades. In the Grande Ronde Valley, commercial spas and public facilities have been opened at thermal springs and wells near Hot Lake, Cove, and Union. Waters at Hot Lake are the warmest at 185° Fahrenheit, while those at the other two resorts average 100° or less.

Thermal waters follow narrow conduits along fault zones to reach the surface. This has the effect of localizing the hot waters so that near Hot Lake, for example, a cold water well is less than 150 feet from the heat source for the spa.

Surface and groundwater

The Blue Mountains is a region of modest ground and surface water with rainfall averaging 25 inches a year. Much of the supply is depleted by consumptive use for irrigation, and occasional summer downpours add little. The groundwater picture is tied into the structure of large sedimentary basins, underlain by Columbia River basalts, and the surrounding high mountains of basalts and granite intrusions. Both the sediments and basalts are the primary aquifers, while the granites characteristically have very low permeability and yield only small quantities of water. Interlayers and faults within the Tertiary lavas may supply a steady flow but can be slow to recharge and are prone to decline regionally. Sedimentary layers in the valleys can supply copious amounts of good quality water.

Beneath the floor of the Grande Ronde and Baker valleys, the Columbia River basalts are widespread and the most heavily developed aquifer. Many of the agricultural wells, which penetrate the basalt, have been deepened repeatedly as the upper layers are depleted. In addition to agriculture, municipal wells for Baker, LaGrande, and Union rely on these units. Sedimentary fans, deposited where the Grande Ronde River, Ladd Creek, Powder River, and other perennial streams enter the flat lowlands, serve as extensive aquifers. Often several hundred feet thick, they can produce as much as 2,500 gallons a minute of good quality water.

Primary aquifers in the John Day basin occur within flows of the Miocene basalts or in Quaternary alluvium along the John Day and tributary stream valleys. Alluvium in the upper reaches of the river provides water for shallow wells.

The availability of surface water in all portions of the province varies. High winter and spring runoff, accompanied by melting snow, is followed by low summer flows and negligible precipitation, when usage is greatest. Small headwater creeks draining the uplands tend to be intermittent, and during late summer they are reduced to isolated pools or dry up entirely. Even the larger Grande Ronde and Powder rivers are greatly diminished. Storage reservoirs help to supply current demands, but they fail to meet all the needs for agriculture, wildlife, and water standards.

Geologic Highlights

The Wallowa Mountains

Variously called the Granite Mountains, the Powder Mountains, the Eagle Mountains, or the Switzerland of Oregon, the rugged peaks of northeast Oregon were officially designated as the Wallowa Mountains by the U.S. Board of Geographic Names. Wallowa comes from the Nez Perce language and means a series of stakes arranged in a triangular pattern for catching fish, as were used in the Wallowa River. The Wallowas extend into Idaho, where they are known as the Seven Devils Mountains. Between these two ranges, the Snake River has carved the deep Hells Canyon gorge.

Shaped by separate episodes of Ice Age glaciation, a diversity of peaks in the Wallowa Mountains reach heights of nearly two miles above sea level and produce a sharply serrated skyline. Southwest of Wallowa Lake, Matterhorn Mountain, composed of blue-white marble, rises to 10,004 feet opposite Sacajawea Peak at 9,880 feet.

At three and one-half miles long and three-fourths of a mile wide, Wallowa Lake is the largest of the 100 lakes in the Blue Mountains province. Nearly 4,440 feet in elevation, it is hemmed in by lateral moraines of glacial till that date back only a few thousand years.

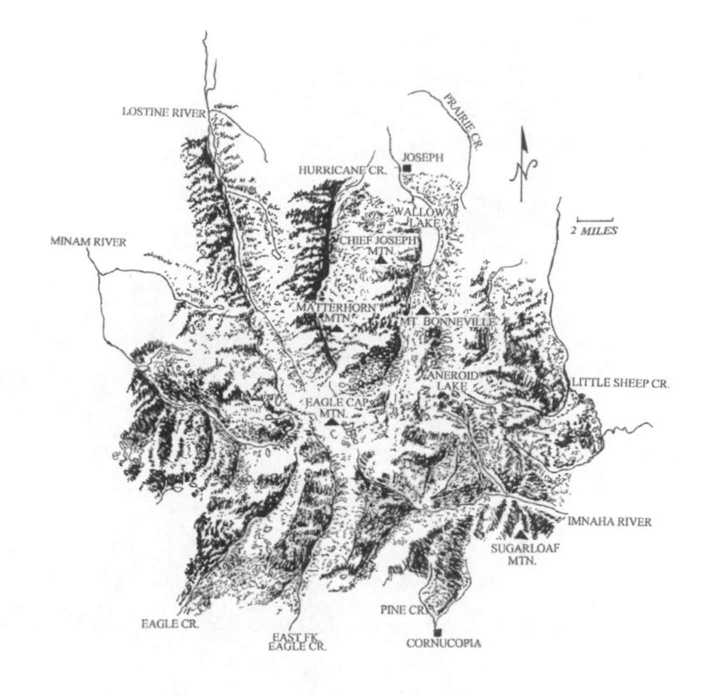

Heavily scalloped by alpine glaciers, the Wallowa Mountains are among the most rugged in Oregon.

Steins Pillar

About 20 miles east of Prineville in Crook County, Steins Pillar is one of the most imposing of the columns eroded from layers of welded tuff. The original spelling was Steens Pillar, named after Major Enoch Steen of the U.S. Army who explored this region for road development in the 1860s. As often happens with place names, it was misspelled so frequently that the incorrect version became official.

In this southward view, the community of Joseph lies at the bottom right on the irregular glacial outwash plain. Penned in by glacial moraines, Wallowa Lake is in the center, and Mt. Bonneville is toward the back. (Photo by D.A. Rahm; courtesy Condon Collection)

An erosional remnant of volcanic ash-flow tuffs of the Clarno and John Day formations, Steins Pillar stands 350 feet high. Oxidizing iron from the volcanics has stained the surface yellow-brown. (Photo courtesy Oregon State Highway Department)

The construction of Steins Pillar began about 40 million years ago with incandescent ash and pumice of the Clarno Formation. These were, in turn, covered with John Day volcanic and sedimentary debris. Upon cooling, the layers split into long vertical joints, and, over millions of years, erosion enlarged the cracks to isolate the steep columns seen today. The height of each pillar imparts some idea of the incredible volume of ash that has been stripped away.

Officer's Cave

Although the Blue Mountains are not known for caves, one of Oregon's most noteworthy geologic features is Officer's Cave south of Kimberly in Grant County. Named after the homesteading Floyd Officer family, it is among the largest developed in non-karst terrain in North America. Unusual processes of erosion and landslides in soft clays and silts carved out the cave during heavy rain storms, when surface and groundwater water scoured steep gullies and removed entire underground sections. Huge broken blocks of the John Day Formation, which have fallen from the roof, lie tipped in all directions.

John Day Fossil Beds National Monument

The 14,400 acres of the John Day Fossil Beds National Monument, which reveal the Cenozoic history of the southwestern edge of the Blue Mountains, was set aside as a federal preserve in 1974. A new display center opened at the Sheep Rock station in 2004, providing a concise overview of the regional geology and paleontology.

At three different areas in the park—Clarno, the Painted Hills, and Picture Gorge in Grant and Wheeler counties—visitors are treated to a fascinating array of 44-to-6-million-year-old animals and plants, entombed in colorful volcanic and sedimentary layers.

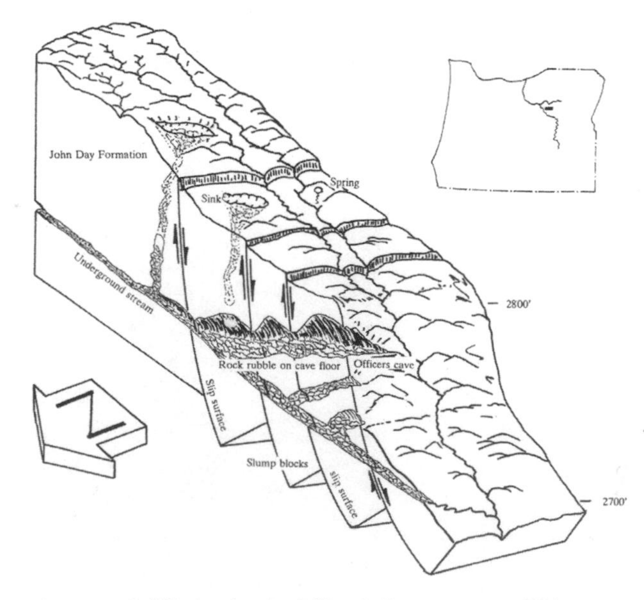

A maze of differing levels, Officer's Cave was over 700 feet long when first explored in 1914, but since then it has increased substantially because the soft John Day Formation is particularly susceptible to erosion. Examined in 1975, it measured over 1,500 feet in length. Underground channels follow the east-west direction of nearby Officer's Cave Ridge. (after Parker, Shown, and Ratzlaff, 1964)

Near Dayville, the headquarters of the park at Picture Gorge and Sheep Rock tells the story of life here 18 to 30 million years ago. Picture Gorge received its name because of ancient Indian pictographs on the canyon walls, while Sheep Rock is so-called because at certain angles the projecting cap of Columbia River Basalt resembles an animal head. The green and buff-colored volcanic tuffs of the middle and upper John Day beds make Sheep Rock one of the most familiar sights in eastern Oregon. (Photo courtesy Oregon State Highway Department)

North of the main park headquarters, Cathedral Rock provides a good look at the middle John Day green Turtle Cove Member. Near the skyline, the 28-million-year-old Picture Gorge ignimbrite appears as a dark layer tens of feet thick. (Photo courtesy Oregon State Highway Department)

Approximately 15 miles southwest of the town of Fossil, the Eocene Clarno unit is well-known for its tropical suite of fruits, nuts, and vertebrate bones, the oldest in the park. The mudflows or lahars are particularly distinctive, eroding into steep cliffs, spires, and columns that follow vertical fracture lines in the softer sediments.

The badlands physiography at the Painted Hills location, six miles northwest of Mitchell, reveals variegated exposures of rust-red soils alternating with yellow and light brown tuffs of the John Day Formation. The John Day is famous for both vertebrate and plant fossils. Among these, remains of the sheep-like *Oreodon*, herds of which inhabited grassy savannas, are commonly encountered, as are leaves of the *Metasequoia*, for which the temperate Bridge Creek flora is famous.

Grand Canyon of the Snake River

Even though the exact geographic length of the Grand Canyon of the Snake River (Hells Canyon) varies, Tracy Vallier's excellent guide to the geology and topography of the area places it as that stretch from the Oxbow to the mouth of the Grande Ronde River. From the summit of He Devil Peak in Idaho, Hells Canyon has an average depth of 8,000 feet, substantially greater than the Grand Canyon of the Colorado. Hat Point, on the Oregon side, offers a panorama of vertical cliffs, a sharply cut chasm, and narrow ledges high above the river.

Dated at 250 million years, the oldest rocks exposed in the canyon walls are Permian, deposited long before the Snake River existed. Broken and shattered by folding, the layers were later buried by Miocene lava flows. Faulting produced the deep

The scenery today in the Snake River canyon resulted from erosion of the successive basalt flows. As the area was slowly elevated, tributary rivers from the mountains incised down through the deposits. For almost two-thirds of its route through the gorge, the river dissects the dark-colored Imnaha basalts, which reach thicknesses of 6,000 feet. (Photo courtesy Oregon State Highway Department)

Carved by the Crooked River into vertical pillars and precipitous cliffs, Smith Rock lies near the western rim of the Crooked River caldera. (Photo courtesy of Oregon State Highway Department)

intermontane basins and raised mountains, which received a final molding by Pleistocene ice and the Bonneville flood some 15,000 years ago.

Smith Rock

The pinnacles, walls, and cliffs at Smith Rock State Park, seven miles northeast of Redmond, are nationally known by climbers, hikers, and geologists. Named for its discoverer John Smith, the sheriff of Linn County in the mid 1880s, the Smith Rock edifice rises over 400 feet above the Deschutes River floor, has an elevation of 3,200 feet, and is visible for miles.

At the junction of three physiographic provinces, Smith Rock displays successive layers of colorful tuffs, eroded into fantastic shapes. The monolith is one of several rhyolitic domes along the northwestern margin of the Crooked River caldera that sent lavas, ash, and mudflows of the John Day across the landscape 30 million years ago. Much later, the ancestral Crooked River cut its channel down through the tan, red, and green John Day strata to shape the vertical walls of Smith Rock.

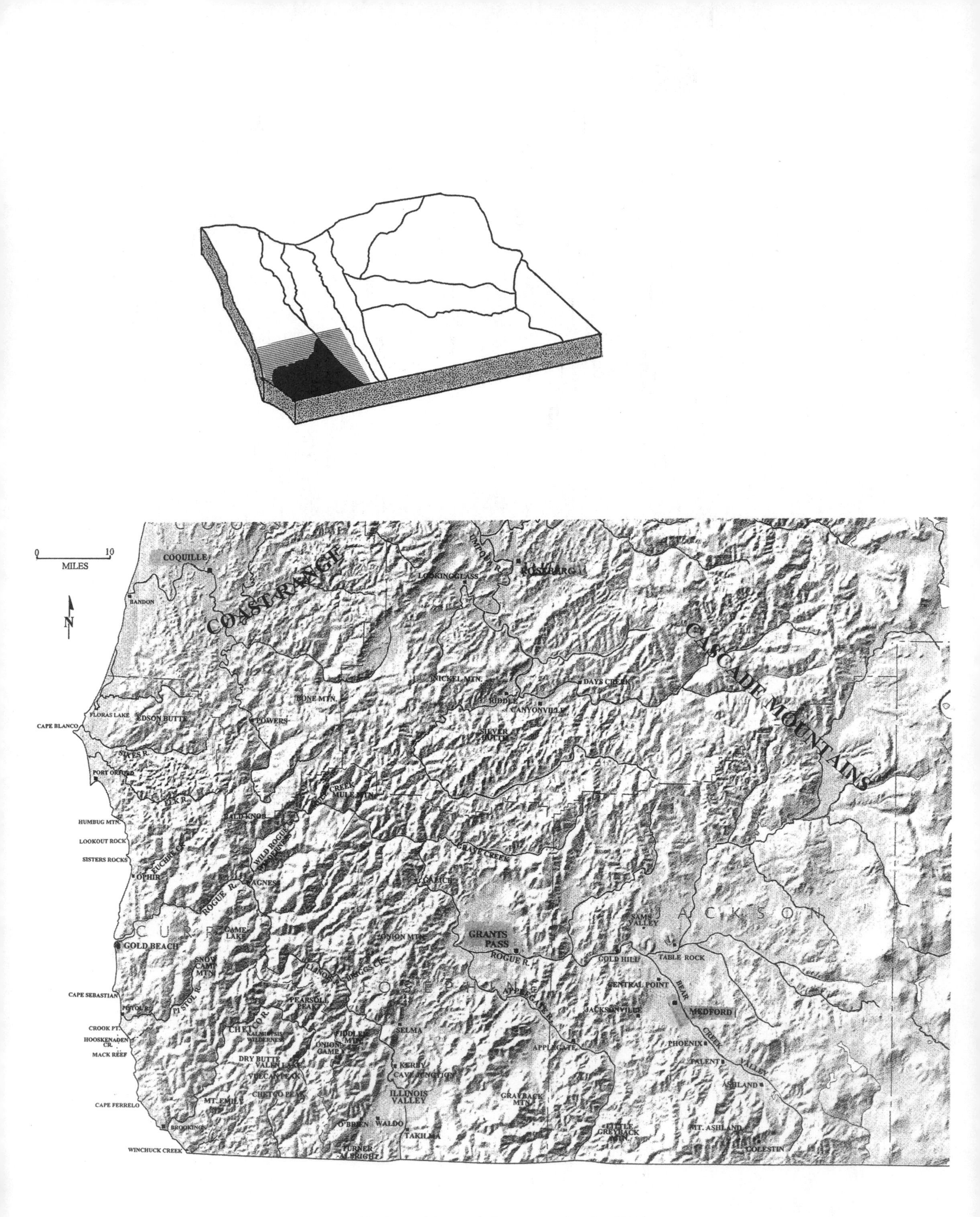

Physiographic map (After Loy, et al., eds., 2001)

Klamath Mountains

Landscape of the Klamath Mountains

The Klamath Mountains is an elongate north-south trending physiographic province extending a distance of 250 miles from a narrow point near the southern tip of the Willamette Valley to the northern end of the Sacramento Valley in California. The Oregon Klamaths are bordered on the north by the Coast Range, on the west by the Pacific Ocean, and to the east by the Cascades. A component of the Klamath Mountains, the Siskiyou peaks span the Oregon-California border.

Although the region boasts deep canyons, steep slopes, and high ranges, overall the summits exhibit a uniformity of relief. Mt. Ashland at 7,530 feet is the highest point in the Oregon Klamaths, while in California slightly higher elevations are attained. Inland south of Medford, the Siskiyou Mountains rise to 4,000 feet.

From Port Orford southward, rocky headlands project directly into the sea all the way to Ophir, but from there a narrow sandy beach some 60 to 80 feet wide reaches to the mouth of the Rogue River at Gold Beach. Except for thin strips of coastal terraces, precipitous cliffs are continuous into California.

In Oregon, the province is drained by the Klamath, Rogue, and Chetco rivers, all three of which empty into the Pacific. Following a winding path from its headwaters at Upper Klamath Lake, the Klamath River cuts through the mountains into California, then westward. The Rogue originates on the slopes of the Cascade Mountains at Boundary Springs, then flows southwesterly to Grants Pass. After dropping 5,000 feet, the river collects water from its major tributaries the Applegate and Illinois before entering the ocean at Gold Beach, a distance of 215 miles. The smaller Chetco heads in the Kalmiopsis Wilderness then continues southwesterly to meet the ocean at Brookings.

Past and Present

Detailed work on the Klamath Mountains, one of Oregon's most fascinating geologic areas, was begun by Joseph Diller in the late 1800s and early 1900s, but it was almost 50 years later that Francis Wells led U.S.G.S. teams to examine and produce maps of many of the quadrangles. In 1961 a colored geologic map of the western part of the state was authored by Wells and Dallas Peck. Reports by David Jones during the 1960s resolved some of the problems, as did Robert Dott's regional surveys, which expanded and revised the stratigraphy in 1971. John Koch's study of coastal Mesozoic stratigraphy, and the Cretaceous correlation by UCLA professor William Popenoe were pivotal contributions.

Supplementing previous compilations with his own mapping, Porter Irwin's tectonic synthesis of the northern California Coast Range and Klamath Mountains in 1960 was an enormous step forward. Since then, knowledge about the tectonics and stratigraphy has been expanded and refined by Mary Donato, who examined the Applegate Group, by Calvin Barnes on plutons, and by David Howell, M.C. Blake, Norm Silberling, and others on tectonostratigraphic terranes. The 1990 special paper, edited by David Harwood and M. Meghan Miller, summarizes the Paleozoic and Mesozoic paleogeography. Additionally, a 2006 geologic overview, edited by Arthur Snoke and Calvin Barnes, along with the 2008 paper on ophiolites, volcanic arcs, and batholiths edited by James Wright and John Shervais, continues to unravel the Klamath knot.

In 1971, Robert Dott (foreground) was one of the first to apply the revolutionary new idea of plate tectonics to the southwest Oregon coast, a task still ongoing 40 years later. Dott made the decision to investigate the complex relationship between sedimentation and tectonic activity while working for Humble Oil and Refining Company in 1955. Receiving his PhD from Columbia University, Dott took a professorship at the University of Wisconsin, where, with his graduate students, he devoted much of his time to the marine sedimentology and stratigraphy in southwestern Oregon, culminating in his 1971 *Geology of the Southwest Coast*. While retired from teaching, Dott completed two new books, the latest being the *Roadside Geology of Wisconsin.* (Photo courtesy M. Chan)

Born near rural Pomeroy, Washington, Ewart Baldwin worked in a sawmill and as a copper miner to pay his way through college. Completing a degree at Cornell, he returned to the west to spend four years with DOGAMI surveying the Coos Bay coal resources. Baldwin joined the staff at the University of Oregon in 1947, where he and his many students mapped over 4,000 square miles of southwest Oregon during the next 40 years. In addition to publishing on the regional geology, he made major revisions to the Eocene stratigraphy. Editions of his *Geology of Oregon*, first issued in 1959, continue to be used. Baldwin died in 2009. (In the 1960 photograph, Baldwin is dwarfed by turbidites of the Roseburg Formation; photo courtesy Condon Collection).

William Porter Irwin's work on the Klamath Mountains province began in the 1950s with a compilation that covered 19,000 square miles in California. In his 1960 *Geological reconnaissance of the northern Coast Ranges and Klamath Mountains, California* he was the first to delineate the lithic belts, laying the groundwork for the concept of tectonically emplaced terranes. Born in 1919 in Springfield, Illinois, Irwin completed his graduate work at the California Institute of Technology. During his career with the U.S.G.S. his studies, which refined the tectonic evolution of western North America, recognized and described distinct terranes in the western Paleozoic and Triassic belt. He retired in 1988 but is still involved in research. (Photo taken in 1974; courtesy California Division of Mines)

Overview

The Klamath Mountains are composed of accreted terranes that were originally crustal slabs or volcanic archipelagos during the early Paleozoic to Mesozoic. Formed in an ocean setting, the island arcs were carried on plates to collide with the North American landmass. As successive terranes arrived, they were thrust beneath one another and imbricated eastward like tilting shingles on a roof. Clockwise rotation of the terranes during the middle and late Mesozoic accompanied the intrusion of crystalline rocks. Refinements to the relationships between terranes is ongoing, but separate belts with accompanying subterranes have been distinguished by age, petrography, geochemistry, metamorphism, and fossils.

About 140 million years ago, Cretaceous seas covered much of what is now western Oregon, depositing sediments in a broad forearc basin that extended for hundreds of miles from California to British Columbia. Regional uplift in the Klamath Mountains shifted the shoreline north and westward and brought widespread erosion to the province throughout the Tertiary.

With cooling during the Ice Ages, just a few small glaciers developed on Chetco Peak, but most were restricted to the higher California Klamath range. Elevation of the coast during the Pleistocene produced stair-step terraces that are prominent north of Coos Bay. Rapid uplift and subsequent erosion of the land are reflected by the stacks, arches, and reefs along the shoreline.

The long history of ocean basin rifting, faulting, and the intrusion of terrane rocks has enriched the province with a diverse assortment of economic minerals. From the 1850s onward, mining played a major role in the settlement of southwest Oregon, and the extraction of gold, silver, and other ores has contributed millions of dollars to the economy.

Geology

Paleozoic-Mesozoic

The Klamath region grew progressively when rocks of oceanic origin, assembled a distance offshore, were accreted to the western margin of North America. As the volcanic island slabs were rafted eastward in the late Mesozoic, they were folded and faulted upon collision with the larger continent, then thrust beneath each other, with the oldest to the east and the youngest to the west. The province provides an ideal venue to study and apply the concept of displaced tectonic terranes, which are fundamental to understanding West Coast geology.

Much of the work in clarifying the terrane concept in the Klamath Mountains can be attributed to Porter Irwin, who introduced the word *terrane* in 1972 when he recognized four distinct units within the Western Paleozoic and Triassic belt. Basing his divisions on the ages of radiolarian microfossils in the rocks, he was aided greatly by the work of Emile Pessagno at the University of Texas, Dallas, and Charles Blome of the U.S.G.S. Terranes are defined as separate groups of strata reflecting ocean floor environments, volcanic archipelagos, and microcontinents, each bounded by faults and distinguished by sequences of distinctive rocks and fossils. An important paper in 1980 by Peter Coney, David Jones, and James Monger of the U.S.G.S. pulled together much of the earlier work to show that the region is a collage of multiple terranes.

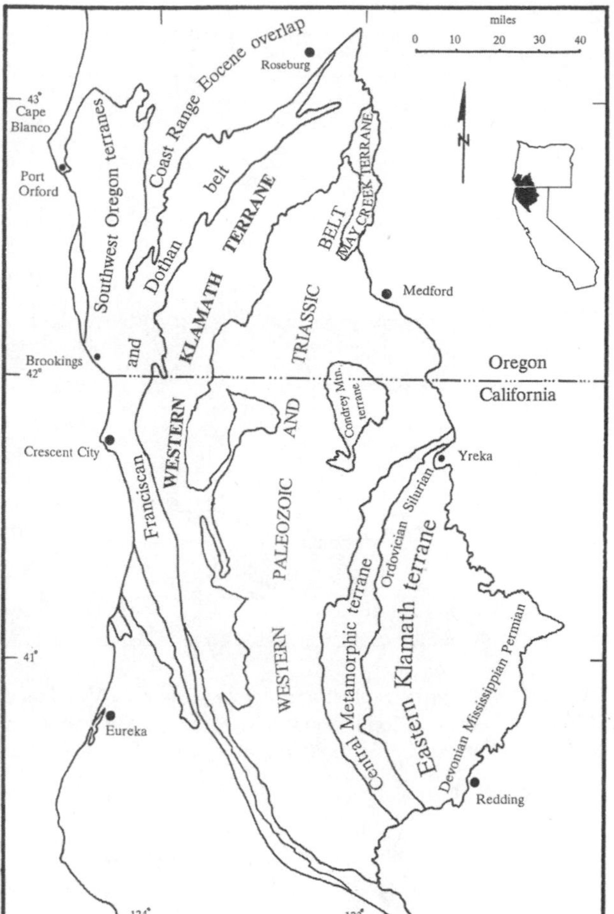

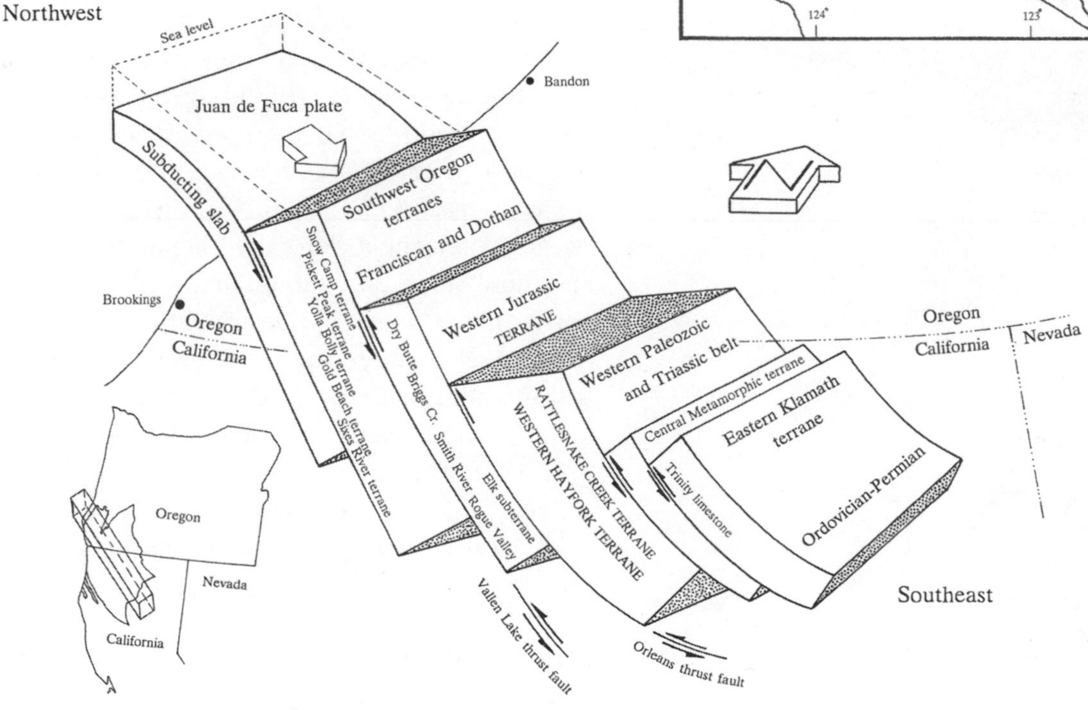

The Klamath Mountains province in southwest Oregon and northwest California is the product of the accretion of ocean crust and volcanic island arcs. Curving in a northeast to southwest direction, the oldest terranes are inland and the youngest are toward the Pacific Ocean. From east to west, they are the eastern Klamath, the central metamorphic, the western Paleozoic and Triassic belt, the Condrey Mountain, the May Creek, and the western Klamath terranes. The Snow Camp, Pickett Peak, Yolla Bolly, Gold Beach, and Sixes River are the most westerly. The eastern Klamath and the central metamorphic terranes are confined to California. (After Blake, et al., 1985; Irwin, 1994; Silberling, et al., 1987; Snoke and Barnes, 2006)

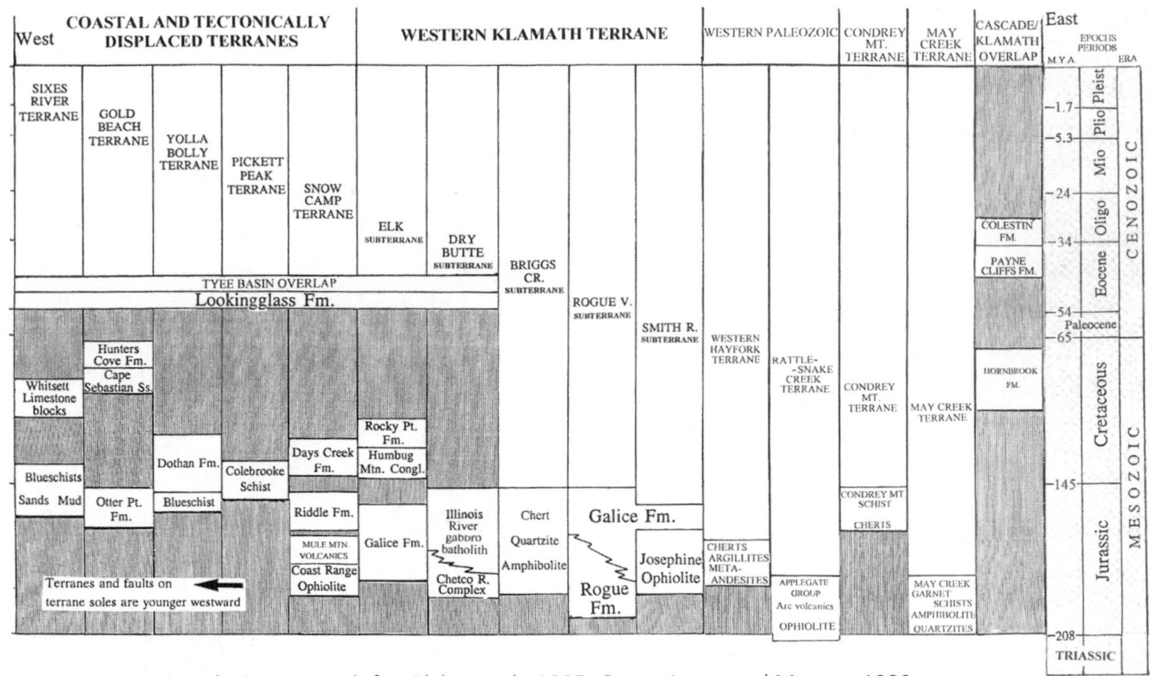

Klamath Mountain terrane stratigraphy in Oregon. (After Blake, et al., 1985; Coney, Jones, and Monger, 1980; Donato, Barnes, and Tomlinson, 1996; Harper, 2006; Harper, et al., 1994; Harwood and Miller, 1990; Irwin, 1972; Irwin, Jones, and Kaplan, 1978; Silberling, et al, 1987; Snoke and Barnes, 2006; Wright and Wyld, 1994; Yule, Saleeby, and Barnes, 2006)

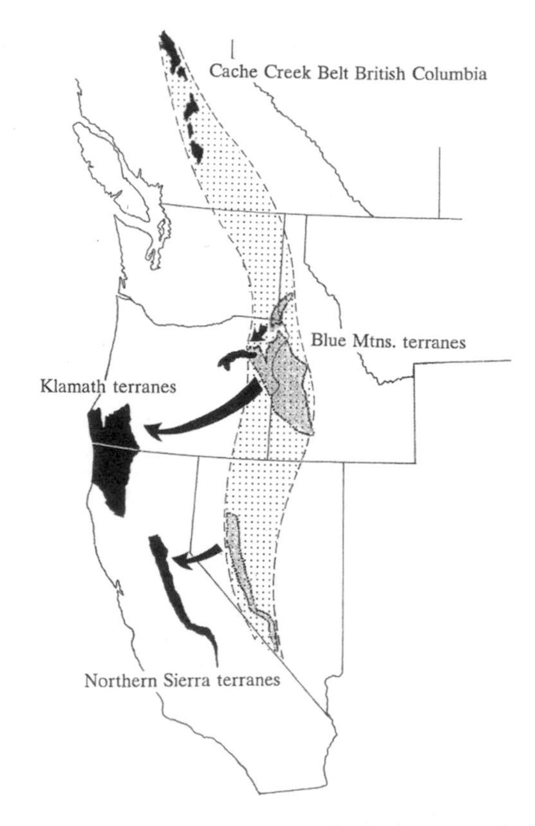

An early model illustrates how the individual terranes along the West Coast from Canada to California may have been rotated into position between the Mesozoic and Cenozoic. The arrows show the movement of the terranes westward from their original positioning. (After Jones, Silberling, and Hillhouse, 1977; Kays, Stimec, and Gobel, 2006; Mankinen, and Irwin, 1990; Miller, 1987)

A comparison of the fossils and rock layers and the degree of metamorphism in the Klamaths, the Blue Mountains, and the Sierra Nevada Mountains with fragments in Mexico led geologists in the 1960s to theorize that these areas were initially part of a megabelt, Wrangellia. Today, however, the consensus is that there is no connection between the terrane rocks of the Klamath Mountains and those of the Blue Mountains.

The oldest strata in the Klamath province are in the Yreka subterrane of the eastern Klamath belt in California, where Ediacaran, jelly-fish-like fossils from the Neoproterozoic, 640 million years ago, were used to date the rocks. In Oregon, the Applegate Group has Triassic conodonts (microfossils of fish teeth) that place its age around 210 million years ago.

Individual Terranes

After delineating and mapping terrane rocks, the process of grouping the many isolated pieces was the next step in reconstructing their geologic relationships. Curving northeast by southwest, the individual fault-bounded terranes have been divided into separate subterranes and formations.

Western Paleozoic and Triassic Belt

Stretching a distance of 200 miles from near Canyonville, Oregon, to the South Fork of the Trinity River in California, the extensive western Paleozoic and Triassic belt reaches widths of 50 miles. It was accreted in middle late Jurassic then rotated to its current configuration during the late Mesozoic to early Cenozoic. Since being described by Irwin in 1960, it has been subdivided into four terranes: the Rattlesnake Creek and the western Hayfork in Oregon, and the Fort Jones and North Fork in California.

Comprising a volcanic island archipelago environment, the western Paleozoic and Triassic belt is a collage of cherts, shales, volcanics, and limestones, perforated by intrusions. Despite its name, the belt has yielded tropical fossil radiolaria from Triassic to Jurassic age in California, although in Oregon only Jurassic radiolaria have been extracted from these sediments in Josephine and Jackson counties.

Rattlesnake Creek and Western Hayfork Terranes

In Oregon, rocks of the Western Paleozoic and Triassic Belt were placed in the enigmatic and widespread Applegate Group by Irwin in 1972. Later revisions by Mary Donato and coworkers have shown that this unit is lithologically indistinguishable from the Rattlesnake Creek and Western Hayfork terranes.

Heterogeneous rocks of the Rattlesnake Creek terrane represent an ophiolitic mélange overlain by ocean arc rocks. Serving as the basement or foundation for later volcanic regimes, the Rattlesnake Creek was covered by the middle Jurassic Western Hayfork volcanic arc, then intruded by plutons. The Eastern Hayfork was thrust over the Western Hayfork in the vicinity of the Wilson Point fault in northern California.

May Creek and Condrey Mountain Terranes

Exposed near Medford, the Jurassic May Creek and Condrey Mountain schists were formerly interpreted as part of the Applegate, but, after reviewing the geochemical and radiometric data, Donato excluded both from that group. Remnants of a backarc basin that was thrust beneath the Applegate Group, the May Creek and Condrey Mountain schist are considered to be individual terranes. The May Creek is ophiolitic and has been heavily distorted and altered by heat and pressure to the high grade metamorphic amphibolite suite.

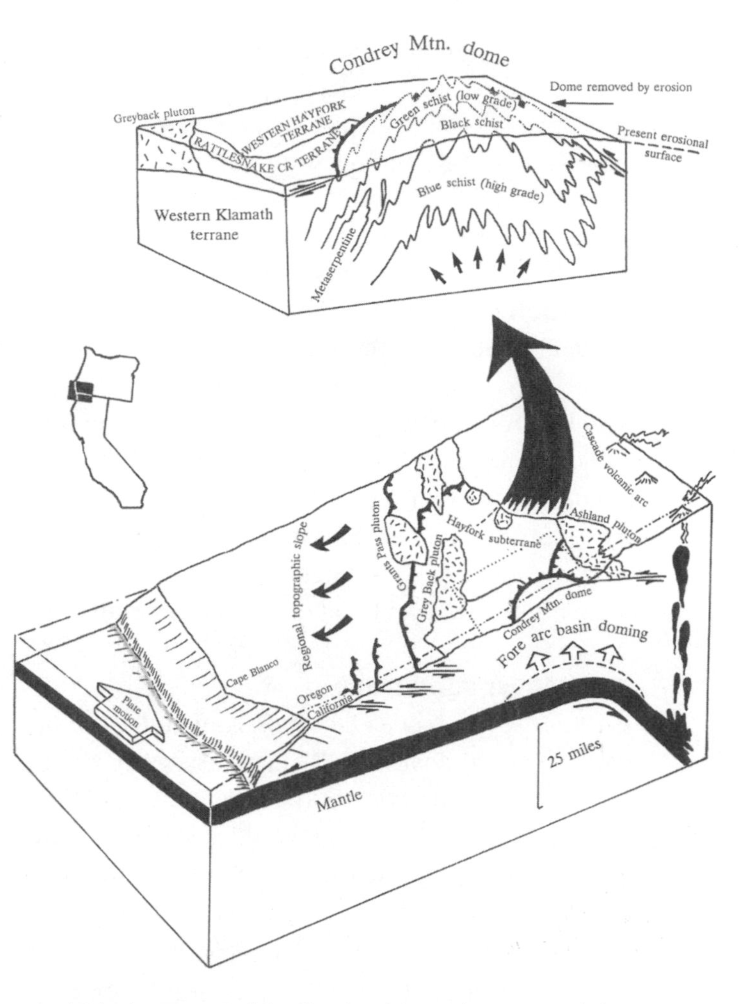

An intriguing aspect of the Condrey Mountain exposure is its distinct doming, believed to have developed during the Miocene. As the Juan de Fuca plate was being subducted beneath North America, lateral pressure between the two caused the bowing or doming, compressing the rocks by as much as five percent and raising them over four miles in elevation. Pressure for this doming may have come from a change in the direction of the subducting plate and the position of the Condrey Mountain dome in front of the Cascade volcanic arc. (After Donato, Barnes, and Tomlinson, 1996; Donato, Coleman, and Kays, 1980; Kays and Ferns, 1980)

The Condrey Mountain terrane is a circular exposure, which projects from Ashland into California. It formed during the late Jurassic between 146 to 148 million years ago when a heated oceanic slab and cooler ocean sediments, brought together by thrusting plates, were altered to schist. The body of the Condrey Mountain is composed of an outer low-grade greenschist layer, a middle graphite-rich blackschist, and a blueschist core, all of which have been compressed into tight folds. Because of its advanced metamorphic state, the Condrey Mountain is difficult to correlate with rocks regionally. Erosion of overlying Rattlesnake Creek created a window into the Condrey Mountain schists.

Western Klamath Terrane

For 200 miles along the western edge of the Klamath Mountains, the western Klamath terrane, originally known as the western Jurassic belt, was constructed along the North American margin and accreted during the Nevadan orogeny, 155 to 145 million years ago. It includes the Smith River, the Rogue Valley, the Briggs Creek, the Dry Butte, and the Elk subterranes. The Smith River subterrane has been the focus of considerable attention because of its paleo-environment and economic minerals.

Smith River subterrane

The Smith River is one of the most convoluted subterranes within the western Klamath assemblage with a suite of rocks known as the Josephine ophiolite, covered, in turn, by three-mile-thick sandstone and shale deposits in the Galice basin. Studies of the radiolaria suggested to Emile Pessagno at the University of Texas, Dallas, that this terrane developed far to the south at a tropical latitude before being transported to its present position in southwest Oregon.

Formed within the Galice interarc basin, the Josephine is one of the largest and most complete ophiolite sequences in the world. Rich in magnesium and iron and altered to serpentine in most exposures, the Josephine is famous for its massive sulfides, enriching the Turner-Albright mining district on the southern Oregon border.

Greg Harper, a consultant in San Francisco, has compared the Josephine and Smartville ophiolitic belt with that in the Coast Range of California and Oregon. Dated about 165 million years ago, the Coast Range ophiolite occurs as fragments from southern California into the Rogue Valley, but the Oregon Josephine and the Smartville in the Sierra Nevada Mountains run discontinuously along the western edge of the Klamaths for 120 miles. The relationship between the various ophiolites is unclear, but all may have originated in suprasubduction settings.

Interpreted as a component of an interarc basin, sediments of the Galice Formation lie above the Josephine ophiolite. Geochemical evidence shows that the Galice detritus was derived from two sources. Thick turbidites were carried by river systems from previously accreted terranes of the North American craton, while the younger volcaniclastic material was shed from the Rogue-Chetco volcanic arc, demonstrating that new covering sediments might have originated from several sources.

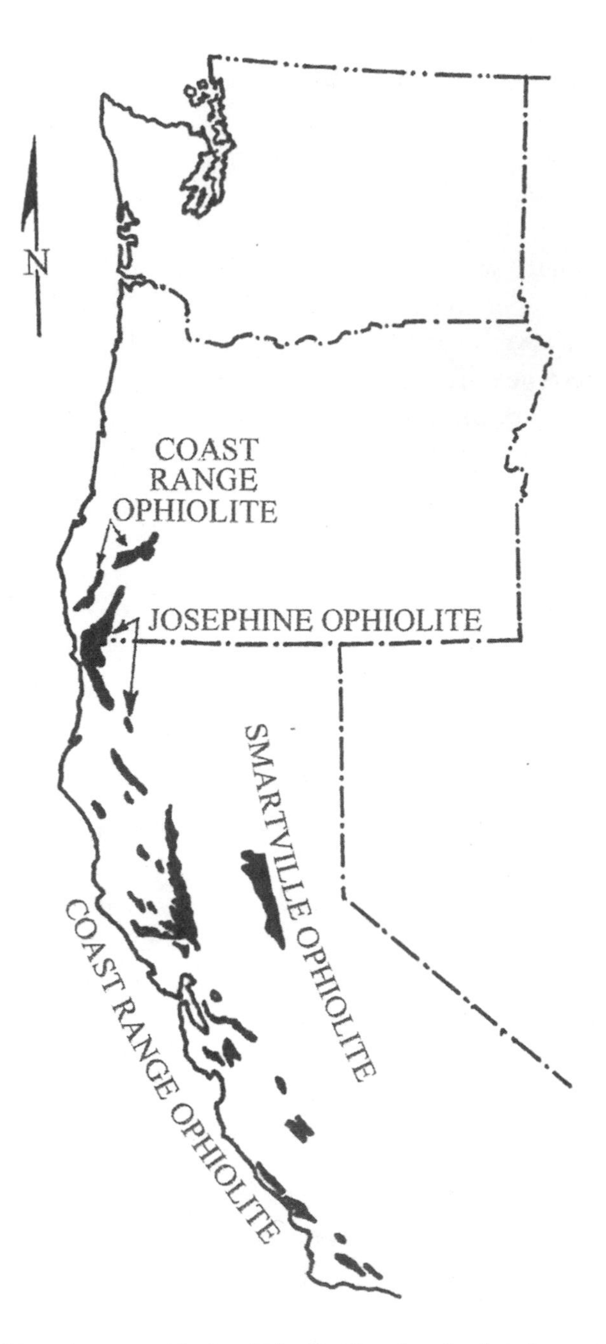

Although geographically within 12 miles of each other, two Oregon ophiolitic belts may have different sources. The Josephine developed during rifting of a basin close to the western margin of North America, whereas the Coast Range ophiolite may have exotic origins. (After Dickinson, Hopson, and Saleeby, 1996; Harper, Giaramita, and Kosanke, 2002)

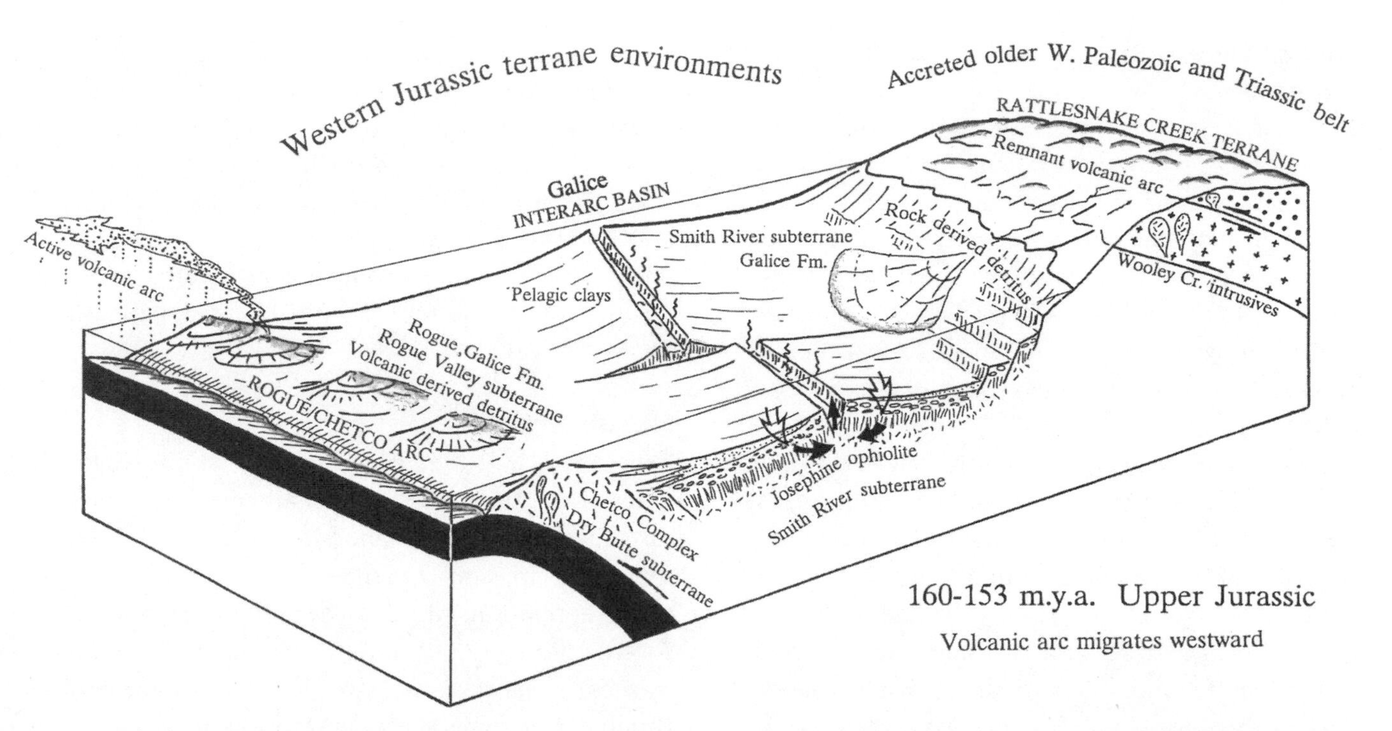

From the late Triassic through Jurassic intervals, a marginal ocean basin along the West Coast featured rifting of the older Rattlesnake Creek accreted terrane, the active Chetco-Rogue volcanic arc, the Josephine ophiolite, and the Galice interarc basin. Sediments eroded from both the Rogue-Chetco arc and the North American continent were deposited in the basin. (After Frost, Barnes, and Snoke, 2006; Yule, Saleeby, and Barnes, 2006)

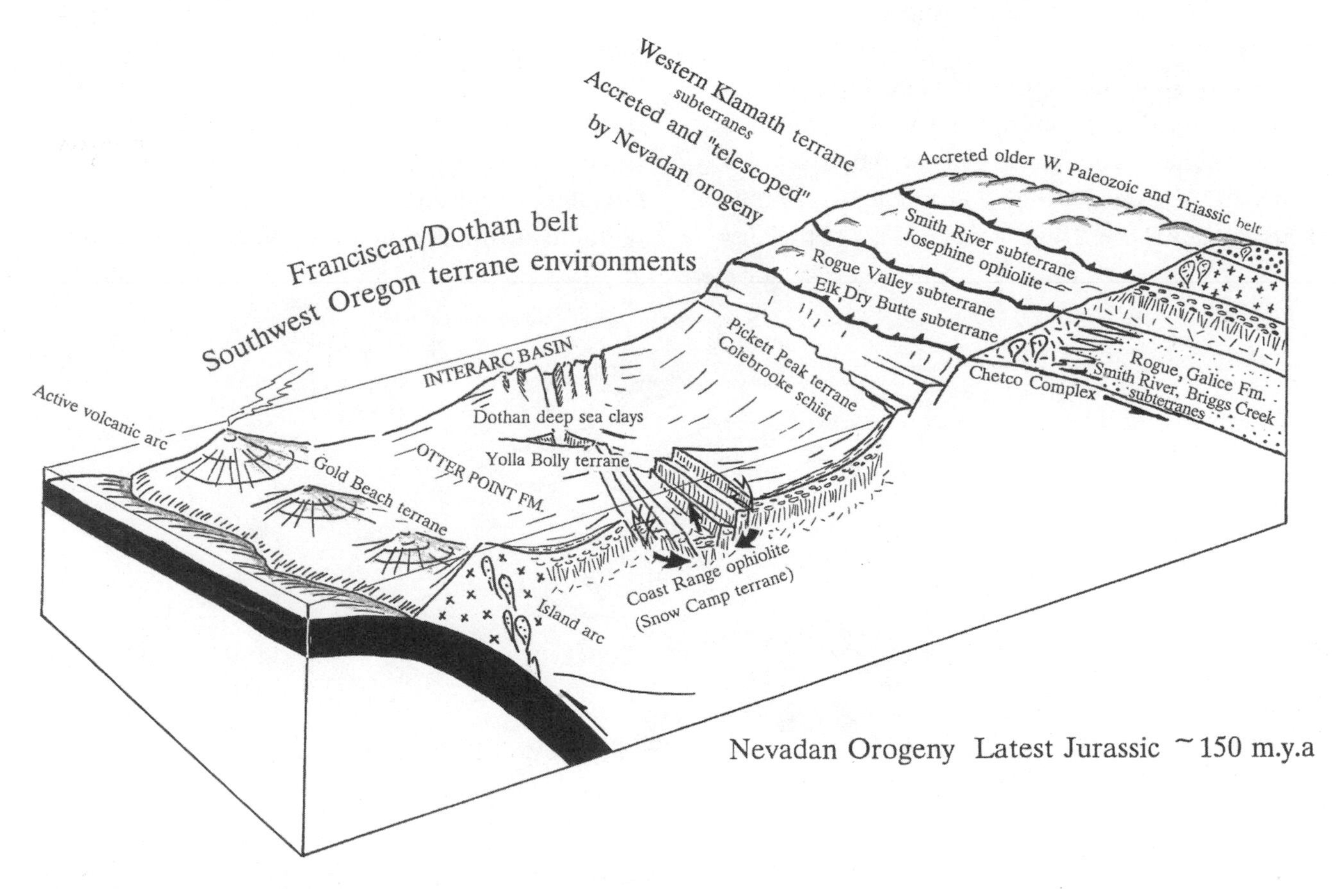

The Nevadan mountain-building episode in the latest Jurassic saw faulting and folding accompanied by accretion of the Rogue Valley and Dry Butte subterranes and volcanic eruptions of the Gold Beach terrane. (After Blake, et al., 1985; Silberling, et al., 1987; Snoke and Barnes, 2006)

Rogue Valley — Briggs Creek — Dry Butte subterranes

The east-west imbrication of the Rogue Valley, Briggs Creek, and Dry Butte subterranes preceded deformation of the rocks during the Nevadan mountain-building phase (orogeny). A five-mile-wide strip just northwest of the Smith River subterrane, Rogue Valley rocks consist of undersea volcanic flows and ash of the upper Jurassic Rogue Formation covered by shales and sands of the Galice Formation.

Within the Rogue Formation, two elongate lithospheric blocks were recognized and informally named by J. Douglas Yule at California State University as the Onion Camp complex and the Fiddler Mountain olistostrome. Tentatively dated using radiolaria, Triassic to Jurassic rocks are exposed from Onion Camp to the Rogue River, while chaotic slabs of the olistostrome, which slid downslope in a marine setting, are exposed along ridges at Fiddler Mountain. Yule concluded that these ocean crustal blocks denote a transition zone between the Rogue-Chetco volcanic arc and the Josephine ophiolite.

The elongate Briggs Creek subterrane, which includes folded and altered garnet-bearing amphibolites, has been construed as the basement on which the Rogue Valley volcanic arc was built. Thrust beneath the Briggs Creek, rocks of the Dry Butte subterrane form an arcuate belt along the western margin of the Oregon Klamaths. Predominantly igneous, this deeply eroded late Jurassic subterrane consists of the Chetco River (Illinois River pluton) complex. Originally the Briggs Creek and Dry Butte were defined by Blake in 1985 as separate subterranes, but currently they are viewed together as middle to late Jurassic components of the Rogue Valley arc.

Elk subterrane

Exposed over 20 miles northward of the main outcrops of the western Klamath terrane, sandy turbidites, shales, and andesitic lavas of the Elk subterrane are similar to those of the upper Jurassic Galice Formation. Also known as the Elk outlier, it consists primarily of the Galice Formation covered by coarse marine gravels of the early Cretaceous Humbug Mountain Conglomerate and Rocky Point Formation. Broken fossil mollusks and plant fragments in these strata record a high-energy water environment along a rocky coast.

In mapping the Elk outlier along the lower Elk River, Mario Giaramita at California State University found an area composed almost entirely of parallel dikes and a hydrothermally altered serpentinite melange, both consistent with the lower sections of an ophiolite. He surmised that this package might be a remnant of the Josephine ophiolite, even though

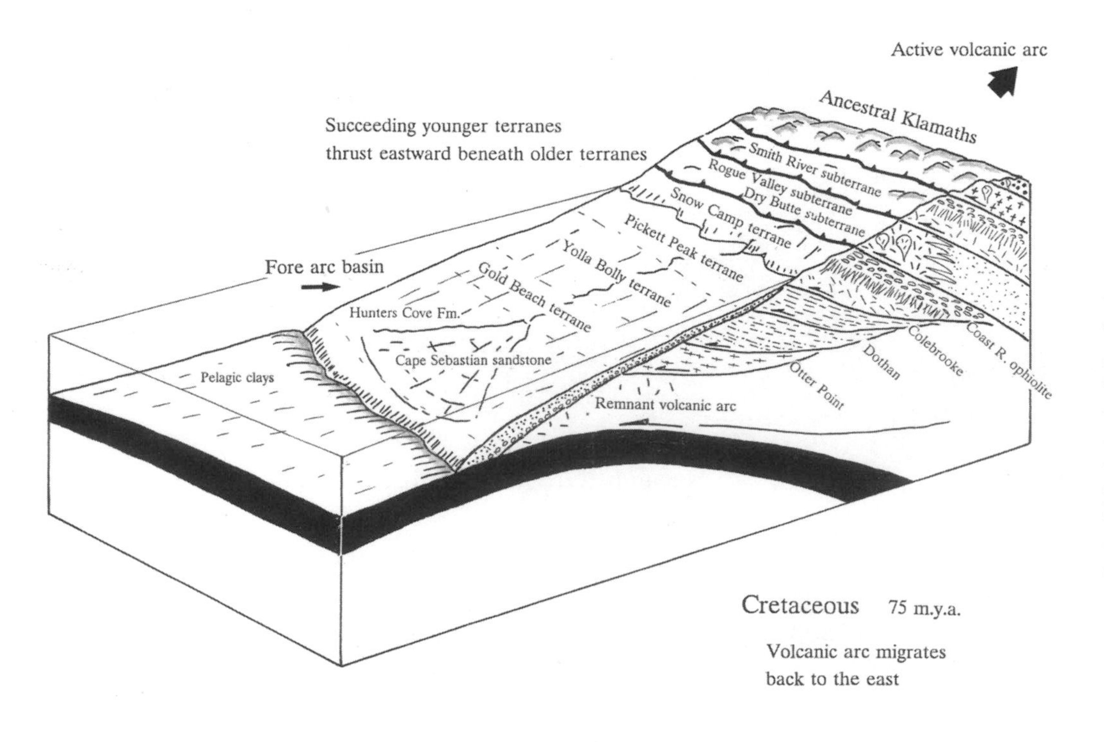

The Snow Camp, Pickett Peak, Yolla Bolly, and Gold Beach are small coastal terranes that were thrust eastward beneath the western Klamath terrane during the latest Jurassic to Cretaceous, 150 to 75 million years ago. In the Cretaceous forearc basin, they were covered by marine sediments of the Hunters Cove and Cape Sebastian formations. (After Blake, et al., 1985; Dott, 1971; Garcia, 1982; Harper, 2006; Silberling, et al., 1984)

Mule Creek canyon on the Rogue River follows the Rogue thrust sheet for two miles. In this stretch, the river provides a swift, if somewhat treacherous, ride to boaters, rafters, and kayakers. (Photo courtesy Condon Collection).

it is some distance away. This conclusion would be consistent with the northward displacement of the terranes by faulting during the Cretaceous Period.

Smaller Terranes

Five slightly younger terranes, the Snow Camp, the Pickett Peak, the Yolla Bolly, the Gold Beach, and the Sixes River, are separated from the overlying western Klamath terrane by thrust faults of early Cretaceous age. Located along coastal southwest Oregon, these slices have been transported northward from sites in California to their present locations.

Snow Camp Terrane

Named by M.C. Blake in 1985, the Snow Camp is a disjunct group of rocks that includes the Coast Range ophiolite, the late Jurassic Mule Mountain volcanics, and a covering of Jurassic Riddle Formation and Cretaceous Days Creek conglomerates, silts, and sands. Recently the Riddle and Days Creek have been given the resurrected Myrtle Creek formational name and compared to rocks in the Franciscan unit of California. Inspection of ophiolites and rocks in California reveals a close correlation with those of the Snow Camp, indicating that this terrane may have been displaced northward by faulting as much as 180 miles from the Sacramento Valley.

Coast Range ophiolites at both Game Lake and in the Wild Rogue Wilderness area were examined in 2002 by Harper, who found that the Game Lake is similar enough to the Josephine ophiolite to suggest that it might be a remnant of the Josephine and not of Coast Range affinity. If that is the case, both may have originated in the same interarc basin. First mapped in 1979 by Len Ramp, the sheeted dikes cutting the Wild Rogue ophiolite were also reviewed by Harper, who placed it in the Coast Range cluster.

Above Paradise Lodge on the Rogue River, the Blossom Bar shear zone, which cuts through the Coast Range ophiolite, creates some of the most difficult rapids for rafters. This mylonite interval, up to one-half mile wide, consists of crushed rocks that have been heavily altered by intense pressure. West of the shear, pillow lavas in the upper part of the ophiolite are fractured, while to the east the ophiolite is sliced up by three-foot-wide sheeted dikes.

Exposed along the northern flank of the Klamath Mountains, the late Mesozoic Riddle and Days Creek formations cover the Coast Range ophiolite. The oyster-like clam *Buchia* and ammonites, common in alternating beds of sandstones and siltstones of the Cretaceous Days Creek, are similar to faunas on the west margin of the Sacramento Valley in California.

Pickett Peak Terrane

In California, the Pickett Peak terrane includes the South Fork Mountain schist and the Valentine Spring Formation, whereas in Oregon it is represented by the Colebrooke schist, first described in detail by Robert Coleman of the U.S.G.S. Found in just two Oregon locations—between Agness and Lobster Creek and at the top of Edson Butte in Curry County—Pickett Peak rocks are similar to those of the Galice Formation. Metamorphosed in the early Cretaceous, the Colebrooke is a blueschist that was originally a mixture of tuffs, cherts, and pillow lavas from a deep sea environment. Lying close to the continental margin, it was thrust into its present position beneath the western Klamath and Snow Camp terranes.

Yolla Bolly Terrane

Bounded by the Canyonville fault zone to the north and the Valen Lake thrust on the southeast, the Yolla Bolly terrane stretches from the Sixes River of Oregon to the South Fork Mountains in California. Of late Jurassic to early Cretaceous age, it has been divided into east and west sections, both of which include the distinctive Dothan Formation. In the Dothan forearc basin, sands, muds, and deep-water cherts, carried downslope by turbidity currents, were derived from continental as well as volcanic arc sources. Mineralization in faulted metamorphosed rocks of the Dothan has produced gold-bearing quartz veins.

Gold Beach Terrane

The westernmost Gold Beach terrane, near the town of the same name, is an unusual mélange of both late Jurassic Otter Point and late Cretaceous Cape Sebastian and Hunters Cove formations. The deep-water Otter Point was originally interpreted as peripheral to a volcanic arc, but Michael Garcia from the Hawaii Institute of Geophysics concluded that the extensively sheared and broken volcanic rocks and sediments of this formation were deposited as part of an accretionary wedge along a subduction zone. The cohesive block of the Gold Beach terrane originated hundreds of miles further south along the continental margin before being transported by faulting from California in the early Tertiary.

Jurassic and Cretaceous fossils are rare in coastal outcrops, but two localities south of Port Orford have produced remarkable finds. The dark gray mudstones of the Otter Point at Sisters Rocks are notable for marine fossils of Jurassic age, yielding the skull and mandible of an ichthyosaurian reptile, squid-like belemnites, and mollusks.

Plutonic peridotite (olivine-rich) rocks are common to the Kalmiopsis Wilderness, where they underlie Pearsoll Peak, the highest point in the upper Chetco River basin at 5,098 feet. Not only is the unique chemical nature of the terrane strata responsible for the minerals, the rocks also host a number of rare plant species not found elsewhere in the state. Because of this, the Big Craggies Botanical Area was proposed for withdrawal from mining in the late 1960s as were 80,000 acres of the Kalmiopsis in 1983. (Photo courtesy Condon Collection; taken by W.A. Long)

Overlying these sediments, broken shells and hummocky stratification of the Cape Sebastian sandstone and Hunters Cove siltstone reflect storm wave conditions with deeper sands and shales carried by turbidites. Oregon's only known dinosaur remains came from Cape Sebastian sandstones when the pelvis and sacrum (fused backbone) of a duckbill (Hadrosaur) were extracted in the 1970s by David Taylor at Portland State University.

Sixes River Terrane

The Sixes River terrane, exposed just north of Cape Blanco and in a small area south of Roseburg, was previously mapped by Robert Dott as the Otter Point Formation of the Gold Beach terrane. However, it was distinguished from the Gold Beach by Blake in 1985 on the basis of Jurassic and Cretaceous mudstones, sandstones, and conglomerates, which are studded with huge blocks of blueschist and the distinctive high-grade metamorphic rock, eclogite. Near Roseburg, slabs of the middle Cretaceous Whitsett limestones with corals, deep-water shales, and pillow lavas are also part of this terrane. The Sixes has been compared to the central metamorphic terrane of northern California and, like the Gold Beach, it may have been displaced northward in the Cretaceous or early Tertiary.

Rotation of Terranes

As they were accreted or annexed to North America, terranes of the Klamaths were subjected to episodes of clockwise rotation, intrusion by plutons, and faulting. Rifting of the slabs led to the formation of depositional basins and the development of ophiolites. The intricate geology of the Klamath Mountains is the sum of these multiple processes.

In the Pacific Northwest, individual terranes experienced clockwise rotation dating back to the latest Triassic or early Jurassic and continuing into the early Tertiary. An examination of the alignment of magnetic mineral crystals in plutonic intrusions provides a record of their initial orientation and can be used to determine the timing, direction, and degree of rotation.

Comparing paleomagnetic data from the Klamaths, the Blue Mountains, and the Sierra Nevadas, Edward Mankinen and William Irwin demonstrated that most rotation took place when the various terranes were offshore and before they were annexed to the North American continent. They found the paleomagnetic data from the eastern Klamath terrane of California was the most complete, showing that rotation was ongoing even as Jurassic strata were being deposited and that most of the clockwise motion was completed by the early Cretaceous. A smaller degree of Tertiary rotation was attributed to subduction of the Juan de Fuca plate and Basin and Range extension.

Since both the Klamath and Blue mountains have experienced clockwise rotation, early models proposed that there might have been a basement continuity between the two provinces, but that notion no longer has any validity.

Plutons and Granitic Rocks

Igneous stitching plutons, which cooled and crystallized at depths, intruded all belts of the Klamath Mountains binding the terranes together. Plutons vary in size from exposures covering less than one square mile to the large batholiths extending over 40 square miles or more. Larger plutons form at or near the boundaries of converging or diverging tectonic plates, where the magma invades fractures in the rock, whereas smaller dikes and sills appear along rifts or where the brittle crust has been stretched or thinned. The distinctive age, composition, and tectonic significance of each intrusive body allows conclusions to be drawn about the regional geologic history. For example, the timing of an intrusion can be established by determining whether it cuts across or has been cut by faulting.

The oldest plutons of the Klamath Mountains are the Forks of Salmon in California and Squaw Mountain in Oregon, which intruded the western Hayfork terrane. The Oregon Wooley Creek suite and coeval magmatism of the Chetco-Illinois complex were next in age, followed by the youngest western Klamath and post-western Klamath granodiorite intrusions.

Dated at 154 and 156 million years ago, the Jacksonville and White Rock plutons are the youngest in the Wooley Creek suite, while the Wimer and Thompson Ridge are the oldest. The Ashland pluton of biotite granite is about 150 square miles in size. Approximately two-thirds lies in Oregon, where Mt. Ashland is its major exposure. The large Grayback, which underlies the area from Jacksonville across the border into California, intruded the western Hayfork and Rattlesnake Creek terranes of the western Paleozoic and Triassic belt.

The Chetco suite, dated at 160 to 155 million years ago, includes the Illinois River gabbro and the Chetco volcanics and is roughly synchronous with the Ashland, Grayback, and Wimer plutons. Intruding the Dry Butte subterrane before amalgamation, the Chetco complex forms the plutonic roots of the Chetco-Rogue arc.

Western Klamath plutons, emplaced during and after the Nevadan orogeny between 144 to 150 million years ago, are exposed at Gold Hill and Buckskin Peak in Oregon. Because of the high temperature of the magmas, substantial quantities of the Galice Formation sandstones and shales may have been assimilated by the plutons during emplacement.

The early Cretaceous Grants Pass pluton, at 139 million years, is one of the youngest intrusives in

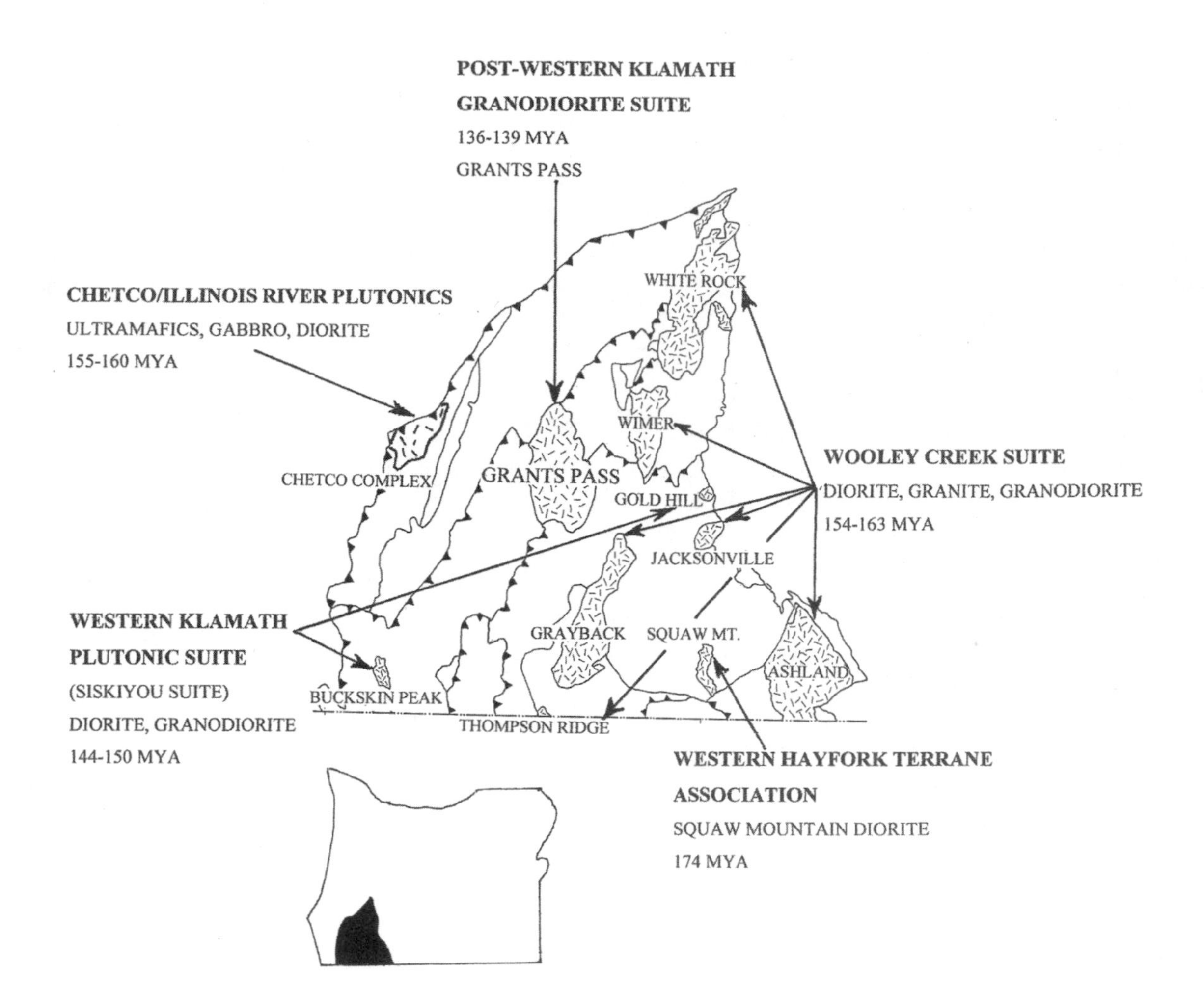

Groups of plutons, emplaced in the Klamath Mountains province, have been loosely arranged by rock composition and age. Examining isotopes and trace elements in the intrusives, Charlotte Allen at the Australia National University and Calvin Barnes at Texas Tech University organized them into six major northeast-southwest curving belts. The five in Oregon are the western Hayfork, the Wooley Creek, the Chetco-Illinois River, the western Klamath, and the post-western Klamath suites. The western Hayfork includes the Squaw Mountain; the Wooley Creek incorporates the Jacksonville, the White Rock, the Ashland, the Grayback, the Wimer, and the Thompson Ridge plutons; the Chetco-Illinois River includes the Chetco complex; the Gold Hill and Buckskin Peak are in the western Klamath plutonic belt, and the Grants Pass pluton is within the post-western Klamath group. (After Allen and Barnes, 2006; Hotz, 1971; Irwin, 1985; Johnson and Barnes, 2006)

this province, and only the Shasta Bolly and the Yellow Butte in California were emplaced later. Displaying 85° of clockwise rotation, the granodiorite Grants Pass pluton intruded the Smith River, Rattlesnake Creek, and Rogue Valley subterranes after amalgamation had taken place.

Faulting and Structure

At the end of the Jurassic and into the early Cretaceous periods, thrust faulting and folding accompanied terrane collision. Historically, this mountain building episode has been called the Nevadan orogeny because it was thought to be synchronous with similar events in the Sierra Nevada region further to the south.

Immense north-northeast-trending thrust faults separate the tectonic terranes in the Klamath Mountains. Of these, the 250-mile-long Valen Lake fault was active during the late Cretaceous. First recognized by Ramp, this tectonic feature is connected to a major fault zone from Douglas County southward into California. Irwin interprets this thrust as marking the western boundary of the Klamath Mountains province.

Ophiolites—Ocean Crust

Commonly occurring in the Klamaths of Oregon and California, ophiolites are layers of ocean crust developed in backarc or interarc basins. The term *ophiolite* was first published in 1969

Roughly contemporaneous, the Orleans and Madstone Cabin faults were active between 155 and 145 million years ago, even though deformation continued for another 15 million years along the Orleans thrust fault. Exposed on the Klamath River in California and into southwestern Oregon, the Orleans fault system led to burial of the Smith River subterrane beneath the Rattlesnake Creek terrane when the rocks suffered an estimated 60 miles of telescoping. Identified and mapped near Vulcan Peak in Curry County by Len Ramp, the Madstone Cabin thrust pushed the Chetco volcanic complex beneath the Smith River subterrane. (After Blake, et al., 1985; Harper, 2006; Harper and Wright, 1984; Harwood and Miller, 1990; Pessagno and Blome, 1990; Ramp, 1975; Snoke and Barnes, 2006)

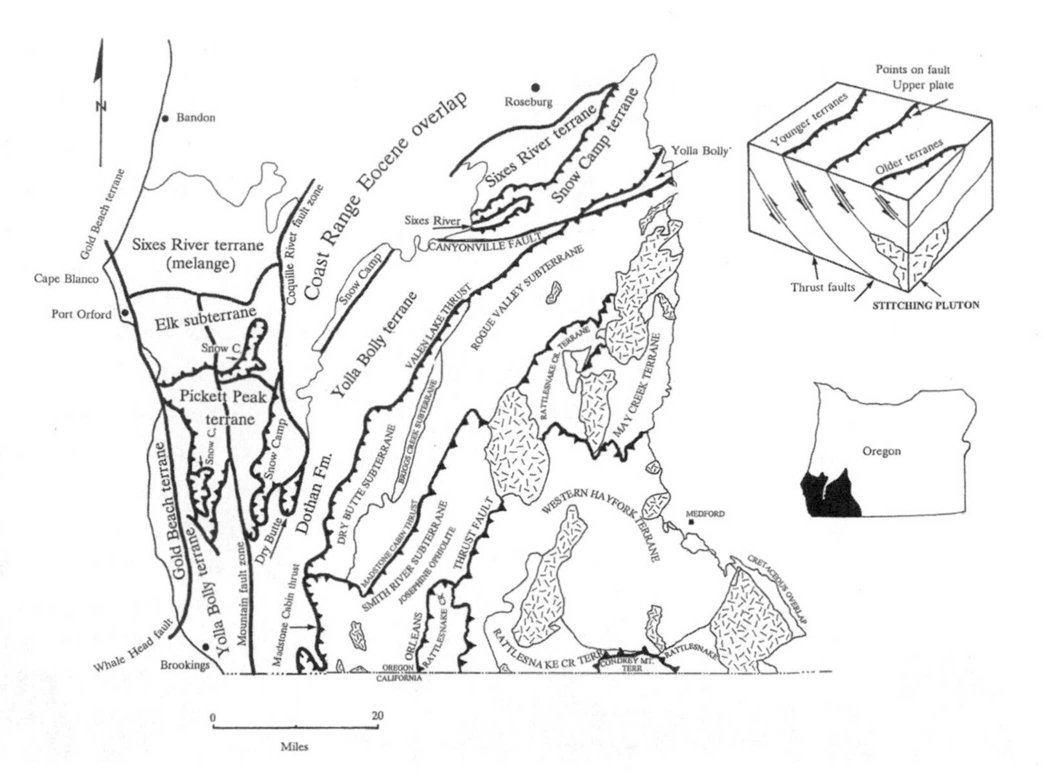

by Stephen Bezore, then at the University of Oregon, in describing rocks along the northern California coast. Reaching thicknesses of three miles, ophiolites follow a remarkably set pattern in structure and composition. At the base, dark-colored ultramafic rocks of mantle peridotite are overlain by gabbros. These in turn grade upward into sheeted dikes without an apparent host rock. Pillow basalts, extruded underwater onto the sea floor, cover the dikes. The pillow-shaped blobs of lava are capped by cherts and pelagic clays with fossils of radiolaria and foraminifera that are typically found today in ocean sediments at depths of several thousand feet.

The upper pillow basalts of an ophiolite are highly porous, and the convection of mineral-laden super-heated seawater through the fractured rocks promotes the precipitation of massive base-metal sulfide deposits, enriching them with silver, gold, platinum, iron, copper, and zinc. Nickel and chromium ores are primarily associated with the deep peridotite layer near the base.

Basalts, gabbros, and related ultramafics in ophiolites are typically altered to form the low-grade metamorphic rocks greenstone and serpentinite. As the main component of the rock serpentinite,

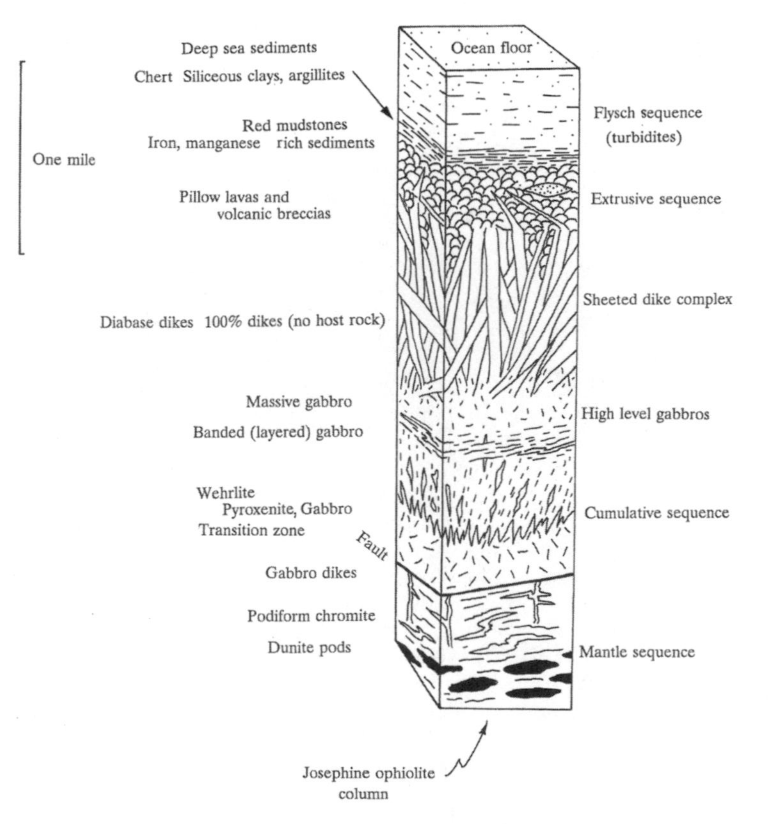

Ophiolites are complex layers of ocean crust organized in predictable sequences. The arrangement is significant as it offers a chance to examine first-hand crust and upper mantle rocks. The famous Josephine ophiolite of the Klamath Mountains is responsible for much of the mineral wealth in that region. (After Harper, 1984; Harper, et al., 1985)

the mineral serpentine readily fails under pressure, hence its designation as ophiolitic, a term referring to the snakeskin-like faulted texture.

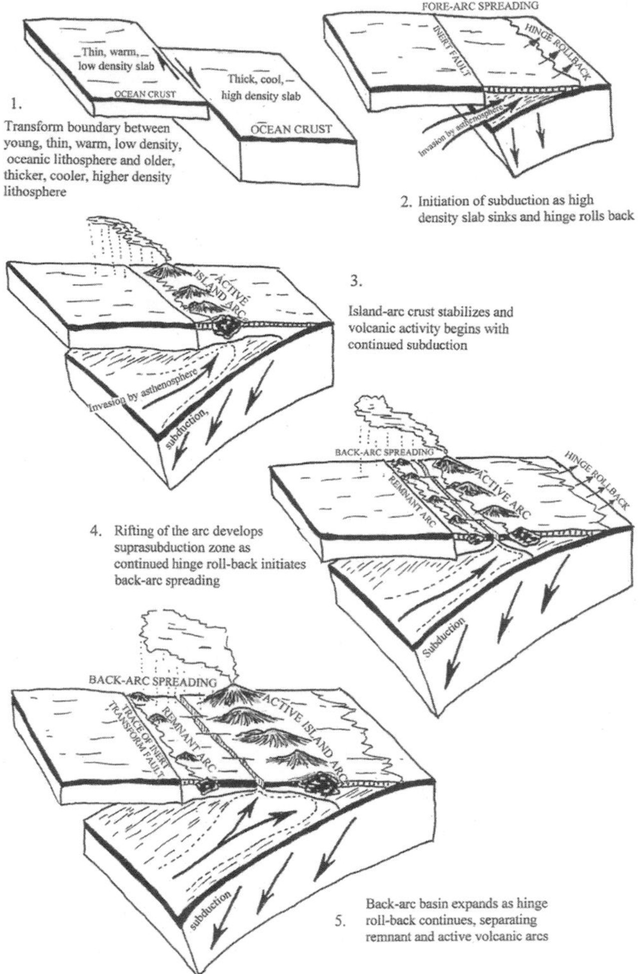

In the 1960s, intact ophiolites were thought to be generated exclusively at mid-ocean ridge spreading centers. This notion has been challenged by Rodney Metcalf at the University of Nevada and John Shervais of Utah State University, who propose that whole ophiolites usually form in backarc spreading centers adjacent to active volcanic island arcs and in proximity to continental landmasses. In their suprasubduction model, the onset of subduction is initiated with hinge rollback and invasion by the upper mantle (asthenosphere). As subduction continues, the back arc basin spreading center produces ocean crust (ophiolite) behind an active volcanic arc. (After Bloomer, 1995; Metcalf and Shervais, 2008; Shervais, 2001; Stern and Bloomer, 1992; Weins and Smith, 2003)

Cretaceous—Marine Basins

In the Klamath Mountains, profound changes in tectonic style, volcanism, and sedimentation between the late Cretaceous and early Eocene were generated by repositioning between the offshore Kula, Farallon, and North American plates. A lengthy forearc basin connected the Ochoco Mountains in southcentral Oregon with the Hornbrook in the Klamaths and the Great Valley of California.

In the Hornbrook basin, 4,000-foot-thick sands, shales, and conglomerates grade upward from freshwater streams, calm lagoons, and beach sands, to open ocean strata, reflecting a deepening or transgressing Cretaceous seaway, a trend that was taking place worldwide. Advancing eastward, the ocean inundated much of Oregon. Named in 1956 by Dallas Peck, the Hornbrook Formation was subsequently divided into five members by Tor Nilsen, of the U.S.G.S. In his definitive monograph, Nilsen surmised that the basin was open to the proto-Pacific toward the northwest and that erosion of the uplifted Klamaths was the major source for Hornbrook sediments.

Cenozoic

With accretion and rotation of the Coast Range block (Siletzia) to the continent, a lengthy shelf subsided west of the emerging Cascade volcanic arc and north from the Klamath Mountains to British Columbia. The basin filled with muddy

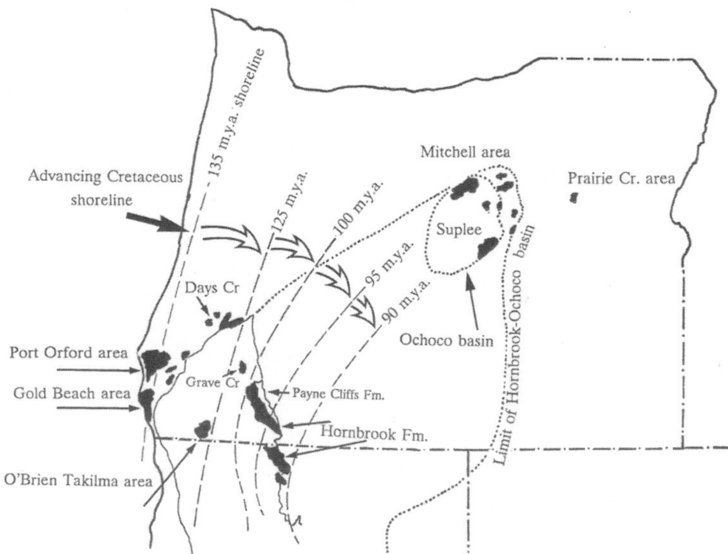

Ocean waters advanced eastward during the Cretaceous Period bringing open marine environments across western and south central Oregon and providing habitats for varieties of invertebrates. (After McKnight, 1984; Miller, Nilsen, and Bilodeau, 1992; Nilsen, 1984; Peck, et al., 1956)

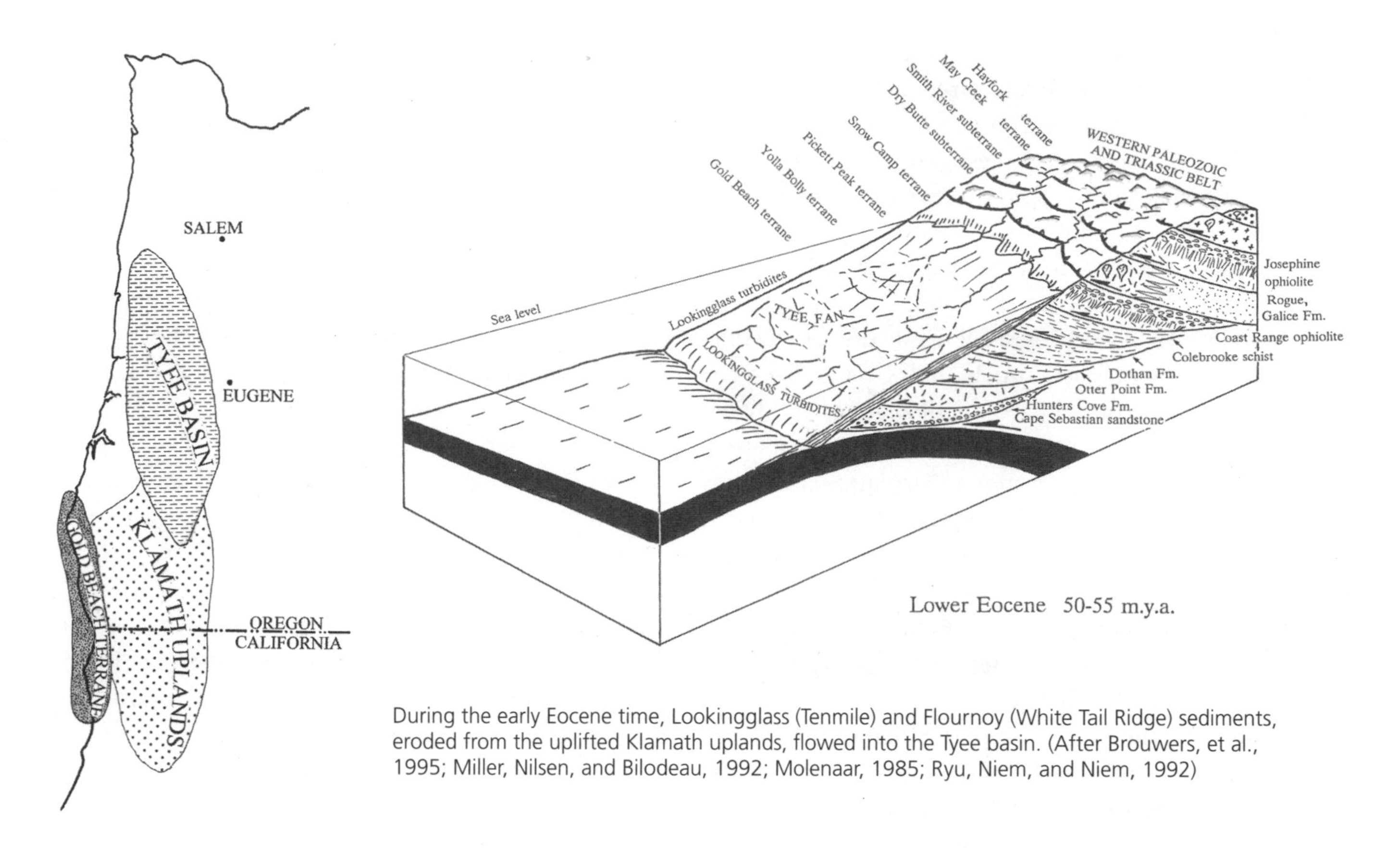

During the early Eocene time, Lookingglass (Tenmile) and Flournoy (White Tail Ridge) sediments, eroded from the uplifted Klamath uplands, flowed into the Tyee basin. (After Brouwers, et al., 1995; Miller, Nilsen, and Bilodeau, 1992; Molenaar, 1985; Ryu, Niem, and Niem, 1992)

deep-sea fans of the Eocene Tyee Formation and a thin veneer of mica-rich sands and silts of the Lookingglass (Tenmile) and Flournoy (White Tail Ridge) formations. Turbidity currents repeatedly surged across the Tyee depression, where shoreline plants accumulated with tropical marine invertebrates. Beginning in the Cretaceous and continuing into the Cenozoic, uplift of the Klamath Mountains pushed the shoreline northward, restricting the seaway in this province.

An extensive erosional surface across the Klamath Mountains was noted by Diller, who proposed that during the late Miocene there was a level peneplain from northwestern California to southwestern Oregon with a second Bellspring flat expanse to the west. He based his concept on the even topography of the summits and ridges, which he saw as evidence of a uniformly high dissected platform. For many years this notion was dismissed as overlooking the complex landscape of the Klamaths. In 1997, however, careful mapping of the uplands by Porter Irwin validated Diller's interpretation of a Klamath-Bellspring plateau that had been uplifted and eroded late in the Cenozoic. (Photo shows the view across Roseburg toward the level horizon; from Diller, 1898)

The mapping, description, and naming of Eocene strata in southwest Oregon by pioneering geologist Joseph Diller in 1896 formed a solid basis for subsequent investigations. Covering a phenomenal 1,000 square miles in just three field seasons, Diller and his assistants with the U.S.G.S. provided the first insights into regional stratigraphy and physiography. Some 70 years later, Ewart Baldwin and his students at the University of Oregon remapped much of the same area in greater detail. In his 1974 publication, Baldwin revised the Cenozoic stratigraphy and structure, subdividing Diller's Umpqua Formation into the Roseburg, Lookingglass, and Flournoy. In 1992, the previously discarded term *Umpqua* was resurrected as the Umpqua Group and assigned to the combined Lookingglass and Flournoy formations by In-Chang Ryu, Alan Niem, and Wendy Niem at Oregon State University.

South of the Tyee basin in the Bear Creek Valley, late Eocene non-marine conglomerates and sands of the Payne Cliffs Formation overlie Cretaceous marine Hornbrook sandstones and mudstones. Recording some of the earliest volcanic pulses of the Western Cascades, the Payne Cliffs is covered by volcanic debris of the Oligocene Colestin Formation, much of which was transported by streams originating high in the range.

Pleistocene

Small glaciers developed in the Oregon Klamath Mountains during the Pleistocene, although more extensive ice sheets were present in the higher California portion of the range. Evidence of glaciation is most visible in the upper Chetco River watershed, where Len Ramp mapped U-shaped valleys, amphitheater-like cirques, tarn lakes, and moraines at altitudes close to 3,000 feet.

Uplift of the coastal margin and fluctuating global sea levels during glacial and interglacial periods produced prominent raised marine terraces. Where they are 200 to 300 feet in elevation from Coos Bay to Port Orford, a number of these surfaces can still be distinguished despite being narrow and dissected. Southward from Ophir to Gold Beach, however, the terraces have been fragmented and deeply eroded by streams. In the vicinity of Pistol River, sand obscures the surfaces, but at Crook Point the ocean waters have swept them clear. Near Brookings the platforms reach close to 125 feet, but they diminish toward the California border.

Geologic Hazards

Earthquakes

Early catalogs of historic earthquakes on the Pacific Coast were compiled by E.S. Holden, Sidney Townley and Maxwell Allen, Perry Byerly, and J.W. Berg and C.D. Baker. Taken together, these cover the period from 1769 to the late 1950s. More recent data can be found in the 1995 earthquake history by Ivan Wong and Jacqueline Bott, and with Pacific Northwest networks at the University of Washington, Oregon State University, and the U.S. Geological Survey.

The 1873 Crescent City quake damaged buildings in Port Orford, and the Cape Blanco lighthouse vibrated briefly in 1896. Ashland felt its first notable shock in 1891 and light tremors a few years later. Similar sporadic quakes at Medford, Phoenix, and Grants Pass involved only slight motion, but those centered in northern California are more frequent and stronger. The 1954 event in Eureka, California, event registered 6.6 on the Richter scale and caused minor damage in Brookings.

Major earthquakes on the southern coast are associated with the offshore Cascadia subduction trench or with the Mendocino and Cape Blanco fracture zones. An examination of the evidence from historic subduction quakes shows they occur in 300-to-600-year-cycles. The 1996 *Earthquake Hazard Map for Oregon,* edited by DOGAMI geologists Ian Madin and Matthew Mabey, rates the Klamath coast as an area high concern, meaning that considerable widespread damage is possible if a subduction earthquake were to take place.

Several variables are involved in predicting the next big quake. The pre-instrument Crescent City earthquake, estimated at 7.3 magnitude, was thought to have been triggered along the Cascadia subduction zone, but on closer inspection Ivan Wong, a private consultant, concluded that it was related to motion along the Gorda plate. Some geologists now consider that the plates are temporarily locked up but that a very strong quake may happen within the next 100 years.

Landslides

Landslides are not infrequent in the Klamath Mountains, a province dominated by faulting, precipitous gradients, and failure-prone incompetent rocks. It has been estimated that 10 to 30 percent of the region has been subjected to slides, slumping, or more massive earth flows in excess of one square mile. Most involve clay-rich soils along slip surfaces and slopes in fractured serpentinites or deformed mélange rocks, where clear-cutting, road building, vegetation removal, and rainfall may initiate the movement.

In most instances, slides fail to generate a great deal of attention or concern unless human traffic is impeded, but the job of keeping roadways open, especially those paralleling the coast, is never-ending. Of varying severity, debris flows along Highway 101 pose particular problems, since this is the main connecting north-south route.

While a knowledge of the different causes of erosion is instrumental to interpreting geologic hazards, that information has been available but largely ignored in the Northwest where timber harvesting has long occupied a central economic role. An early study of landslides in the upper Elk and Sixes river watersheds by Margaret McHugh then at Oregon State University utilized aerial photos taken over a 37-year period to show the relationship between landslides, rock types, clear-cutting, and road building. When compared to occurrences on undisturbed forested terrain, her results determined that timber harvests increased slides and debris flows by a factor of seven; but road construction had a substantially greater impact. Generated by road work, the 1970 Six-Soldier slide near Selma in Josephine County, as well as the ongoing Arizona Inn landslide, substantiates McHugh's conclusions.

The notorious Arizona Inn landslide near Lookout Rock south of Port Orford has recurred frequently since work began on Highway 101 in 1938. In 1954, some 500,000 cubic yards of material cascaded downslope, and in 1993 a mass of four million cubic yards traveled from a high rock cliff into the Pacific Ocean, closing the road for two weeks. The slide took place in a sheared fine-grained sandstone and mudstone melange, where slow ongoing movement was accelerated by heavy rains.

Of the 30 major slides across Highway 101 between Gold Beach and Brookings, the largest occurred along Hooskenaden Creek moving a square mile of debris over 165 feet during a single event in 1958. In the early 1970s, out-of-state developers proposed to place a hotel on the low sloping slide material, but Curry County planners coordinated efforts with state geologists, voiding the project.

Coastal rock-falls are more frequent than those inland, but Oregon's most deadly slide near Canyonville in Douglas County killed nine men

South of Port Orford, major landslides involving hundreds of cubic yards of sheared mudstones and sandstones move downslope from high ridge tops to the ocean with ongoing regularity. Where the toe of the slide reaches the ocean, the underlying support is removed, and the activity may continue for years. (Photo courtesy Condon Collection)

Basing his conclusions on the size of the growing trees, Len Ramp estimated that the damming of scenic Valen Lake by an ancient slide in the Dothan Formation occurred about 300 years ago. That date is significant in that it marks the most recent subduction earthquake in the Pacific Northwest. During winter storms, movement in Dothan rocks can strip off soil and vegetation, sending the mass downhill. (Photo courtesy Condon Collection)

in January, 1974. Exceptional rainfall onto the snowpack, slopes, and thick fractured Tertiary rocks caused the disaster. A Pacific Northwest Bell crew was repairing a cable when 15,000 cubic yards of material cascaded down from the canyon wall to engulf them.

Flooding

When heavy rainfall or snow combine with warm temperatures, flooding and landslides result. The excess of water can damage property or cause loss of life, and the Klamath Mountains are not exempt, in part because houses and other structures are frequently placed in high-risk areas. The coast is especially vulnerable to ocean flooding or tsunamis that could reach far inland. Paul Komar examines beach processes in detail, reporting that tsunami waves averaged 10 feet in height after the March, 1964, Alaska earthquake, which brought flooding to Gold Beach and the lowland at Brookings. A similar wave after the 2011 quake in Japan inundated the Brookings bay and port.

In spite of flood-control dams, rivers overflow their banks with some regularity. Damaging floods are almost annual events, but historically high waters have been more severe in the Rogue Valley between Gold Hill and Grants Pass and along Bear Creek. Following two major storms, the largest recorded floods in 1861 and 1890 destroyed Gold Hill, and Ashland was inundated by Bear Creek in the winters of 1996 and 2009. The Rogue River destroyed Dodge Bridge (above) on the Sams Valley Highway during the statewide devastating Christmas flood of 1964. (Photo courtesy Oregon State Archives)

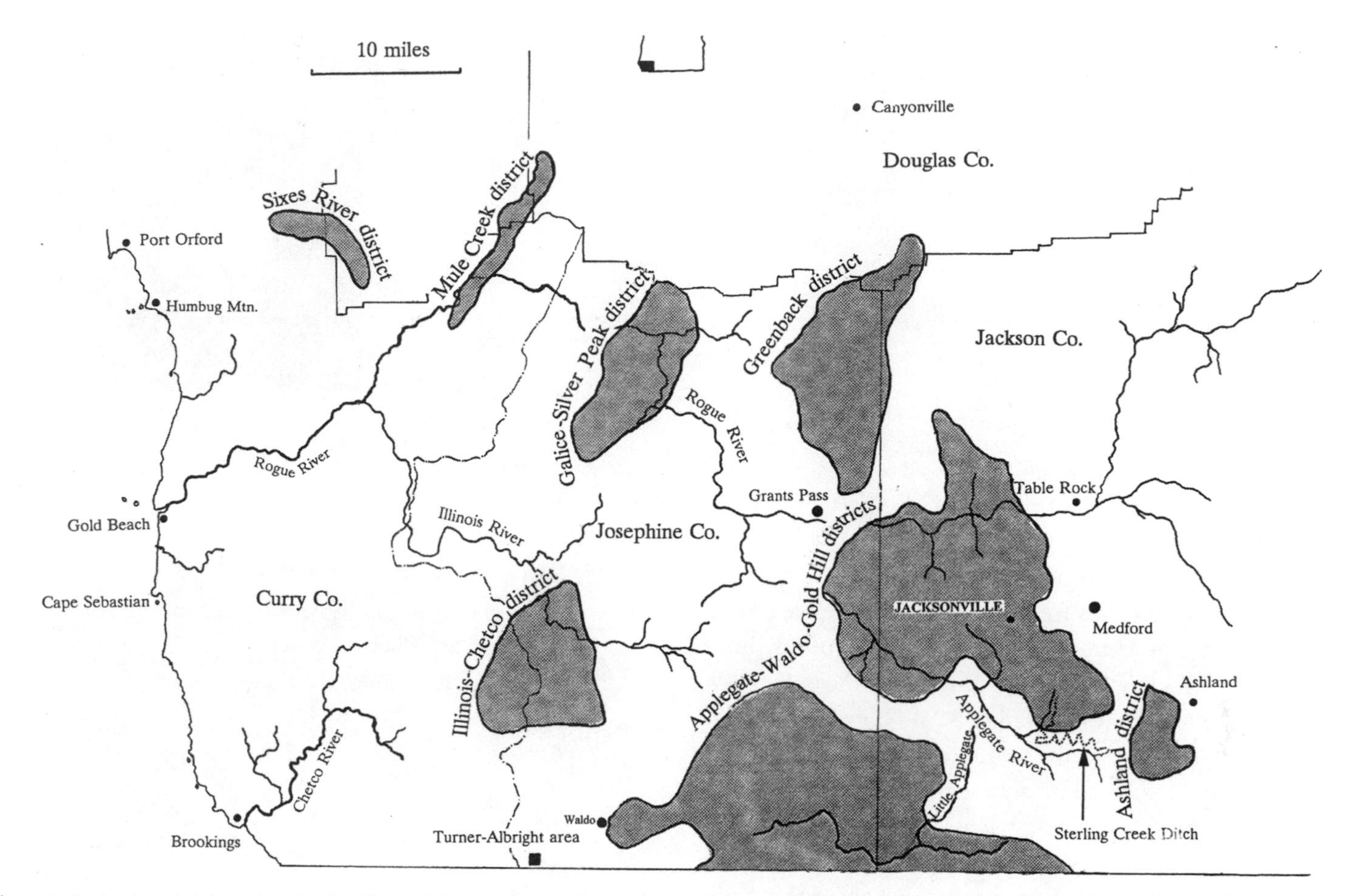

The principal mineral-rich regions in the Klamath Mountains are located near Galice and Silver Peak, in the Greenback district, near Ashland, in the Gold Hill-Applegate-Waldo area, and in the upper Illinois and Chetco river watersheds. Except for the smaller Sixes River coastal mines, all of these utilized both lode and placer workings. (After Brooks and Ramp, 1968; Ferns and Huber, 1984; Purdom, 1977)

Natural Resources

Long recognized for its mineral wealth, the Klamath province has productive deposits of gold, silver, copper, nickel, and chromite, as well as smaller amounts of other ores. Their presence results primarily from mineralization during tectonic plate evolution and secondarily to the intrusion of plutons. Most of the minerals are found in sedimentary and volcanic rocks of the Rattlesnake Creek terrane and Josephine ophiolite in association with the Ashland, Gold Hill, and Grants Pass plutons.

In the once remote and undeveloped sections of southwest and northeast Oregon, gold and silver provided the impetus for building the first communities and roads. But the story of mining is also the history of Oregon through the 1930s, and Howard Brooks and Len Ramp chronicle the events with their *Gold and Silver in Oregon*. After researching details of the mining story, they compiled a book that pulls together the production, localities, and geology, while outlining the beginnings of the industry.

Mineral resources played a major role in the formation of the Oregon State Department of Geology and Mineral Industries (DOGAMI). Created by the legislature in 1937 during the Great Depression, the department was charged with encouraging the unemployed to take up mining, and a percentage of its budget was distributed to participants in that program. Oregon saw large numbers of people willing to take a chance.

Gold and silver

Over 75 percent of the gold in the Klamath Mountains came from placers, where stream gold was washed out of gravels by hydraulic methods necessitating the construction of ditches and flumes to bring water to the diggings. The ruins of many of

Earl K. Nixon, DOGAMI's first director, was a mining engineer, born in New England, who had lived in many areas of the world. When appointed to the position, Nixon was working at the Esterly mine near Waldo in southwest Oregon. The popular Nixon initiated a wide range of projects to promote the mining economy. He began the *Ore Bin* magazine, later *Oregon Geology*, publishing ongoing research in the state. His first bulletin reviewed the state's mining laws and set the department's course through the mid-1970s, when its aim was to establish a geologic basis for the state's future economic potential. Nixon resigned in 1944. (Photo by Kenneth Phillips; courtesy Geologic Society of the Oregon Country).

Len Ramp, who was hired as a field geologist for DOGAMI in 1951, worked out of the Grants Pass office for 36 years. Ramp's ancestors had come along the Oregon Trail in 1853 and settled near Salem. Receiving his degree from the University of Oregon, Ramp concentrated on the geology and mineral resources in southwest Oregon. His bulletins on the Chetco River, Douglas, Josephine, and Curry counties, and on a variety of industrial minerals added significantly to the limited information in those regions. Ramp lives near Grants Pass and contributes his retirement time to the Crater Lake ski patrol. (Photo courtesy L. Ramp with his great-grandson)

the old workings can still be seen. As the placers were depleted, gold was traced back to lode sources. Lode mining peaked from 1879 to 1908, but, of the more than 150 operations, only six supplied the bulk of the gold recovered, estimated at $7 million. Active mining continued into the 1940s, and today recreational panning is enjoyed by gold seekers.

Mining began when a party of men, most of whom were from Illinois, discovered placer gold along the Illinois River in 1850. This set the stage for a rush the following year. By December thousands had reached Jackson and Josephine counties, which became the most populous area in the state.

Placers were located on Grave and Galice creeks in the Galice-Silver Peak district, a narrow belt about five miles wide and 15 miles long through the Rogue River basin. But significant lode processing was not initiated for another thirty years when a 100-ton furnace and more than 3,000 feet of tunnels at the Almeda Mine made it the longest underground operation. Mineralization in altered volcanics and intrusions into the Rogue, Galice, and Dothan formations was the source of the ore. Both the Galice and Silver Peak produced copper in addition to the other ores, and the Almeda operated a small copper smelter intermittently from 1911 to 1917.

When the Japanese owners of the Silver Peak mine went bankrupt in 1995, the facility was abandoned. Ten years passed before attention was focused on the millions of gallons of acidic waters draining annually from the waste piles, transforming the creeks and rivers to a shiny bronze color. After concerted efforts by environmentalists, the

A process called heap-leaching, in which a solution of sodium cyanide or other chemicals is employed to extract gold from low-grade ores, is not permitted in Oregon today. This 1914 photograph shows cyanide vats (center, back) at the Benton mine in the Galice district. (Photo courtesy Geologic Society of the Oregon Country)

Near Waldo large placer operations used hydraulic elevators to access pits dug below the water table. Covering more than 30 acres, the pits at the Llano de Oro or Esterly Mine were active until 1945. They are now filled with groundwater and known as Esterly Lakes. (Photo taken in the 1920s; courtesy Oregon Department of Geology and Mineral Industries)

At Jacksonville, gold occurs in discrete pockets close to the surface. The richest of these, Gold Hill was discovered in 1857, but the ore was so concentrated that the miners had difficulty breaking the outcropping rock. The vein went back only 15 feet but produced $700,000. In this 1904 photo, David Briggs holds a chunk of nearly pure gold from the Briggs pocket near the Oregon–California state line. (Photo courtesy *Grants Pass Courier*)

federal government placed the site on the Superfund list in 2007.

Northeast from Grants Pass, the Greenback district contributed significant gold ores from ophiolitic layers of the Galice Formation and from mineralization along faults around the Grants Pass batholith. On Jumpoff Joe and Grave creeks, placers were active in 1883, where the largest dredge in the history of Josephine County still operated 50 years later. Processing 115 acres of gravels upstream from Leland, the behemoth constructed such an enormous mass of loose gravel that access to fresh bedrock was blocked, and operations were halted. Equipped with 65 buckets, this electrically powered machine could handle 5,000 cubic yards in 24 hours. Except for the North Pole Mine in Baker County, output from the Greenback was the highest of any in Oregon at $4 million.

The broad Applegate-Waldo-Gold Hill district, which stretches along the southern Oregon border in Jackson and Josephine counties, yielded copper, gold, silver, and pyrite from ophiolites of the Rattlesnake Creek terrane. The Sterling Creek placer, south of Jacksonville, had the highest output after construction of a 23-mile-long ditch supplied Little Applegate River water. By 1914, the mine had produced more than $3 million in gold.

Extracted from rocks of the Applegate Group (the Rattlesnake Creek terrane) and intrusions of the Ashland batholith, gold veins southwest of Ashland were accessed by over two miles of tunnels. The operation produced close to $1.5 million between 1886 and 1939, but activity declined after that.

In the Smith River subterrane, ores in the upper Illinois and Chetco rivers are associated with the

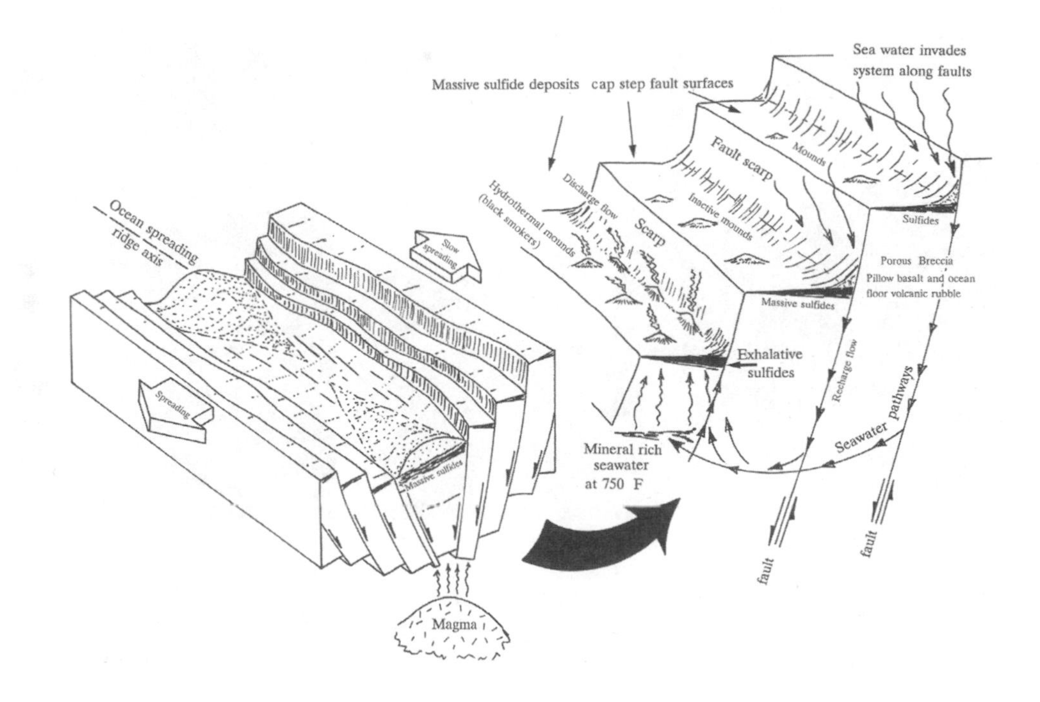

A reconstruction of the paleo-environmental setting of the Turner-Albright massive sulfides near Takilma (Waldo) shows hydrothermal deposits precipitated when super-heated mineral-laden sea water solutions circulated through fractures and faults to emplace ores. (After Harper, et al., 1985)

Galice Formation and the Josephine ophiolite. Output was highest at placers on Josephine and Briggs creeks, where the gold-bearing gravels could easily be washed, but incomplete records from both placer and vein deposits make it impossible to estimate how much gold was recovered. In 2010, the Chetco River Mining and Explorations company filed to dredge around 20 miles of the unspoiled Chetco River within the Kalmiopsis Wilderness area, although environmental groups are appealing.

The Mule Creek and Sixes River mines are located in the steep, rugged landscape of northeast Curry and southeast Coos counties where repeated attempts at mining have only been marginally profitable. The Coast Range ophiolite in the Snow Camp terrane was the source of small quartz veins in the Sixes River drainage, but along Mule Creek sulfides in shear zones of Dothan Formation volcanics yielded limited quantities of low grade gold ore.

Copper

Near Waldo in Josephine County, copper at the Queen of Bronze mine was discovered as early as 1860, but significant quantities were not marketed until 1904, when a small smelter was constructed. The Queen of Bronze exploits the Turner-Albright sulfides within the Josephine ophiolite. After the Iron Dyke in Baker County, it was the second largest copper producer in Oregon until mining ceased in 1933. Exploration continues intermittently, and subsequent drilling programs in the 1970s and 1980s indicate the possibility of additional mineral reserves.

In the Siskiyou Mountains immediately south of the Oregon border, deposits in the Blue Ledge mine on Copper Butte were discovered in 1898 and actively worked through World War I. Contaminated with arsenic, cadmium, copper, lead, and other chemicals, the tailings are on 700 acres of private land, but in 2010 the U.S. Environmental Protection Agency took over jurisdiction for cleanup of the site.

Nickel

Mined near Riddle from weathered peridotites, nickel was one of Oregon's most important economic minerals. From deep within ophiolitic layers, peridotites decompose quickly upon surface exposure, leaving a nickel-rich laterite residue, which is relatively easy to excavate.

Chromite

Like nickel, chrome ore as chromite occurs in ophiolites throughout the Klamaths. West of Grants Pass in the Illinois Valley, the Oregon Chrome Company removed a total of 49,000 tons of ore between 1917 and 1958, while in the upper Chetco River at Sourdough Flat the greatest production came from the Pearsoll Peak ultramafic plutonic body. In 2003, the Oregon Resources Corporation, an Australian company, filed to develop a chromite strip mine

The only nickel mined in the United States came from Nickel Mountain, a few miles west of Riddle in Douglas County. Begun in 1881 and expanded to include a power plant (above left, 1902), sawmill, and housing for workers, the facility was steadily developed, but only marginally productive, until permanently closed in 2003. Subsequent habitat restoration under DOGAMI's reclamation program has improved roughly half of the acreage. (Photos courtesy Oregon Department of Geology and Mineral Industries and Douglas County Museum)

At the terminus of the California-Oregon railroad, 15 miles southwest of Grants Pass, Waters Creek was the shipping point for copper and chromite. (Photo taken in 1917 by Elizabeth Miller; courtesy Oregon Department of Geology and Mineral Industries)

on 2,000 acres between Charleston and Bandon. Government geologists have determined that the project would not generate toxic waste, but despite local opposition, the company began work in 2011.

Surface and Groundwater

The rugged mountainous section of the Klamaths is heavily forested but thinly settled, and it is only in the central and southcentral valleys of Ashland, Medford, Grants Pass, and Kerby where almost all of the agriculture, most of the population, and the majority of water users are centered. The basin from Ashland through Medford is the driest, receiving less than 20 inches of rainfall a year, even though nearby Mount Ashland sees over 50 inches. Some of the highest peaks exceed that.

Overall the region is well-watered by its rivers and streams, however, flows are variable and seasonally controlled and greatly reduced by diversions for agriculture. Unlike the Applegate and Illinois, the Rogue is replenished by rainfall and snowmelt and sustained throughout the year by large springs, which issue from joints in fractured Tertiary lavas. Emerging from the western slopes of the Cascade Mountains near Crater Lake, the Rogue and tributaries are the most far-reaching, discharging 8 million acre-feet of water annually. One acre foot of water is equal to one-third of one million gallons or 325,851 gallons of water.

Aquifers, which are hydraulically connected to surface flows, also diminish during droughts, and groundwater is generally limited in the valleys. Yields from metamorphic and sedimentary rocks are only sufficient to satisfy domestic needs. Aquifers are recharged through precipitation, stream infiltration, and irrigation, although on steep slopes runoff is high and infiltration low.

Situated above harder, impermeable bedrock, the groundwater table in coastal terraces is limited and shallow. Water sources are less than 50 feet in depth, but the productive zones are thin (less than 10 feet) and can be quickly depleted. Pollution comes from both drainfield sewage and the possible intrusion of seawater. By contrast, river gravels along the lower stretches of the Rogue and Chetco have a high storage potential, and individual households enjoy ample supplies from wells.

Inland, the largest municipalities rely on multiple water facilities, including intake mains from the Rogue and smaller creeks, elaborate pumping

Water usage began in 1850 by miners, who diverted streams to their diggings through lengthy wooden flumes and ditches. It wasn't unusual for the piping systems to be taken over to provide water for communities and farms, which expanded even while mining diminished. (Photo courtesy The Irwin-Hodson Company, Portland, 1902)

systems, miles of canals, numerous storage reservoirs, and purchases from water districts. Coastal cities also tap surface flows backed up by reservoirs.

In recent years four dams were removed on the Rogue River, and several on the Klamath River are slated to come down by 2020. The Gold Ray dam near Medford and the Savage Rapids, utilized by Grants Pass and built by the Bureau of Reclamation, were taken out.

Geologic Highlights

Oregon Caves National Monument

Known for its marble colonnades, Oregon Caves National Monument was established in 1909 and is managed currently by the National Park Service. In 1874 the entrance was discovered by Elijah Davidson, who followed his dog through thick underbrush. Approximately three and one-half miles long from entrance to exit, the cavern consists of a variety of shapes, crystals, and graceful forms built of calcium carbonate dripstone. While most caverns in North America are developed in limestone, the Oregon solution caves in marble make this feature unique.

Table Rocks

A few miles north of Medford, two prominent flat-topped mesas, known as upper and lower Table Rocks, stand close to 800 feet above the valley floor. Overseen by the Bureau of Land Management, Table Rocks has been designated an Area of Critical Environmental Concern.

Seven million years ago, lavas from Olson Mountain, a volcanic vent near Lost Creek Lake, filled the narrow upper canyon of the ancestral Rogue River, then spread out through the wide Medford valley, where they cooled and hardened. In Pliocene and Pleistocene time, intensive erosion of the surrounding softer sediments exposed the paleo-stream valley cast in lava, which now stands out in relief in what is called reverse or inverted topography. These two mesas are distinctive because their curving shape preserves the ancient meandering pathway

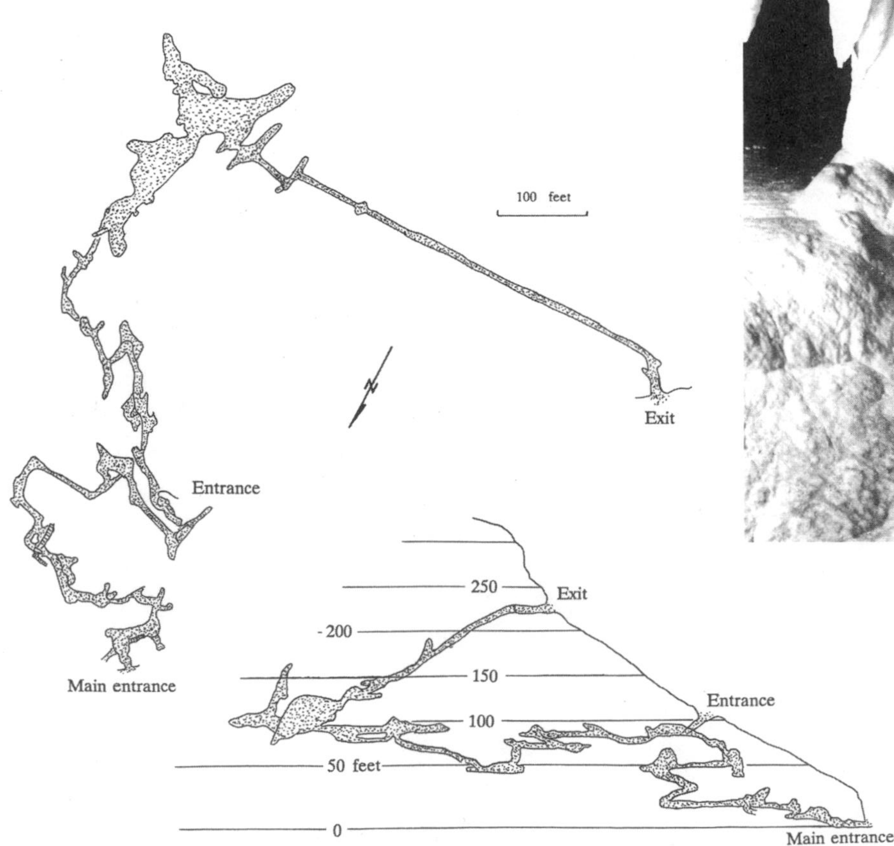

Dripstone stalactites, stalagmites, and columns of calcium carbonate in the Oregon Caves National Monument have been fashioned from metamorphosed Triassic limestones of the Applegate Group. Cracks and joints were widened into tunnels by percolating groundwater, which precipitated icicle-like stalactites suspended from the ceiling and stalagmites rising from the floor. In cross-section, the tunnel rises over 200 feet from entrance to exit. (Photo courtesy Oregon Department of Geology and Mineral Industries)

With a distinctive horseshoe-shape when viewed from the air, the surface area of upper Table Rock is one square mile, while lower Table Rock is slightly smaller. Both are capped by an 125-foot-thick layer of resistant dark-grey Miocene basalt, which overlies sandstones, conglomerates, and mudstones of the Eocene Payne Cliffs Formation. The above view of Table Rocks from the west shows lower Table Rock in the foreground. (Photo by Tim Townsend; courtesy Condon Collection)

of the river. With the realization that the surrounding land surface was once at the same elevation as Table Rocks, it is possible to calculate the erosive rate of the Rogue River that stripped away the surrounding sediments.

Sea Stacks, Arches, and Reefs

Landforms along Oregon's Pacific border from Cape Blanco south to Brookings bear little resemblance to those on the north coast. In the Klamaths, resistant Jurassic and Cretaceous rocks have been dissected to carve an extremely rough landscape of headlands and offshore stacks, shoals, and reefs, with few of the sandy pocket beaches and bays typical in Lincoln and Tillamook counties. Erosive processes, which create the topography, depend on the hardness of the rock, on wave patterns, and on sea levels.

From Cape Blanco southward, the most notable of the projecting promontories are Port Orford, Humbug Mountain, Sisters Rocks, Cape Sebastian, Crook Point, and Cape Ferrelo. Cape Blanco, Port Orford, Sisters Rocks, and Crook Point are braced by Jurassic Otter Point sandstones, mudstones, and volcanics. Cape Ferrelo is composed of sandstones, basalts, and cherts of Dothan Formation, whereas Humbug Mountain is coarse Cretaceous conglomerates and sandstones.

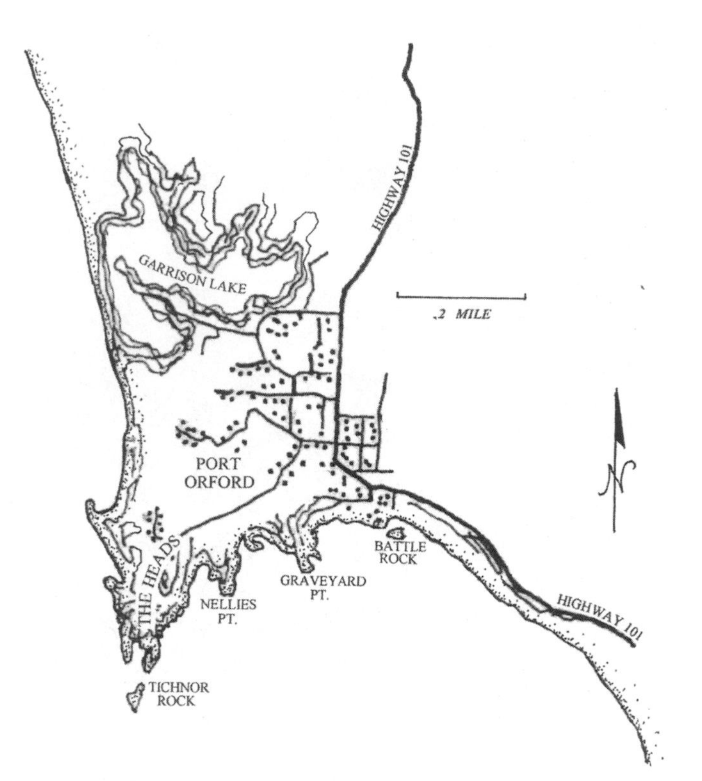

Along Highway 101 at Port Orford, the elongate Battle Rock, projecting 30 to 40 feet above the water, is a striking natural landmark. During a conflict with Indians in 1851, settlers took refuge atop the projection until they could safely retreat to the community at Reedsport. Westward around the cove, the cliffs at Graveyard Point, Nellies Point, and The Heads are all armored by rocks of the Otter Point Formation. The Heads mark the southern end of a Pleistocene terrace.

A short distance from Port Orford, Humbug Mountain is the highest point on the southern coast. The rounded promontory overlooks the ocean and offers a sweeping view to anyone willing to climb the three-mile-long trail. A coarse bed of boulders at the base of the mountain grades upward into sandstones of the early Cretaceous Humbug Mountain Conglomerate. (Photo courtesy Oregon State Highway Department)

g
l
where
r
sion, the isolated rock masses remain as islands, arches, and knobs. Many are clustered in groups above the water level, others are exposed at low tide, while some are lined up in reef belts. The sea stacks are scattered in varieties of extraordinary shapes and sizes, having imaginative names such as Yellow Rock, Saddle Rock, and Devils Backbone.

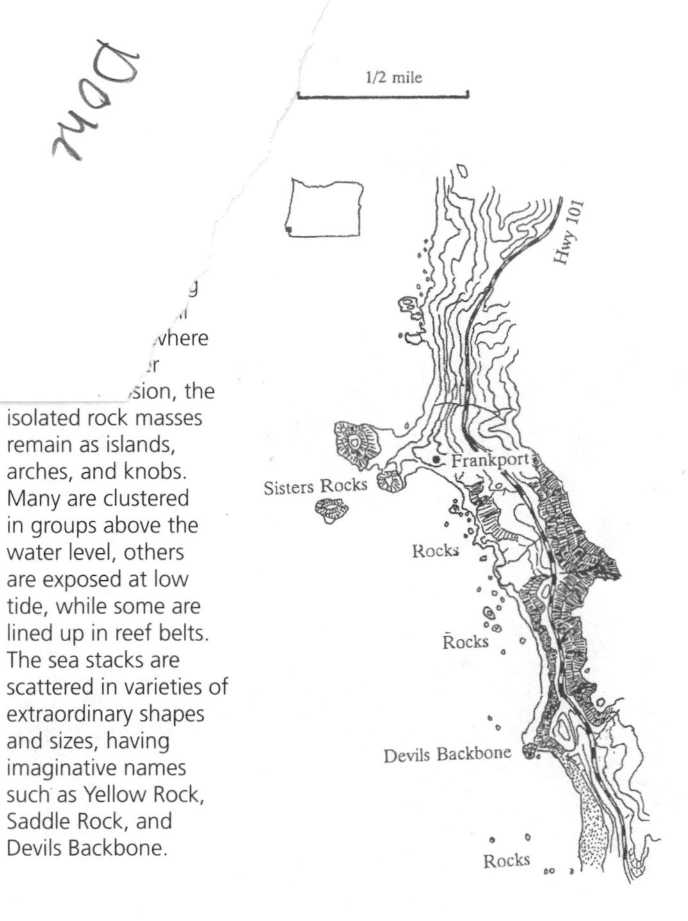

The massive 900-foot-thick sandstones and siltstones of the Cape Sebastian and Hunters Cove formations, forming precipitous cliffs at Cape Sebastian, are also Cretaceous. Oregon's only dinosaur, the fragment of a backbone, came from Cape Sebastian sandstones near Gold Beach.

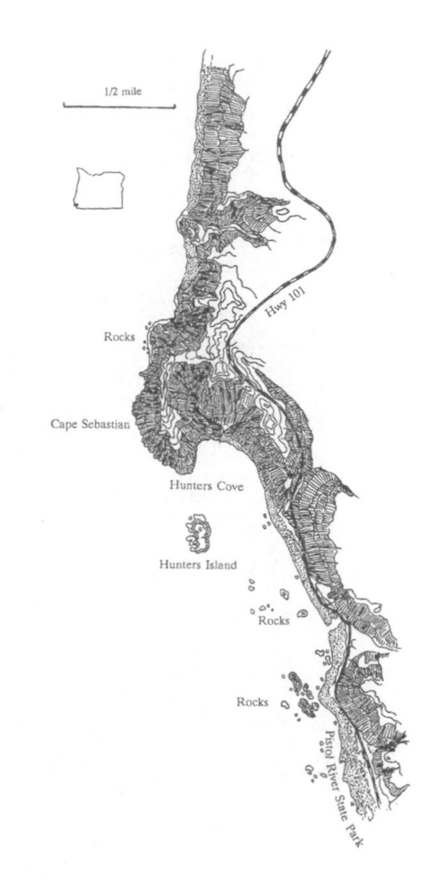

South toward Brookings, Arch Rock Point in Boardman State Park and Natural Bridge were cut through by waves. Goat Island is the largest islet of the Oregon Islands National Wildlife Refuge and the state's largest offshore land mass. Rising 184 feet above sea level, the total refuge consists of 21 acres of dark gray sandstones, siltstones, and volcanics of the Dothan Formation. The community of Brookings is similarly armored with Dothan rocks.

A field of sea stacks, which make up Mack Reef south of Crook Point, are eroded features of the Otter Point Formation. At Mack Reef, a complex of parallel stacks line up with Crook Point. Mack Arch, a 325-foot-high monolith, is the remnant of a reef through which a tunnel was eroded by persistent wave action. (Photo courtesy E.M. Baldwin)

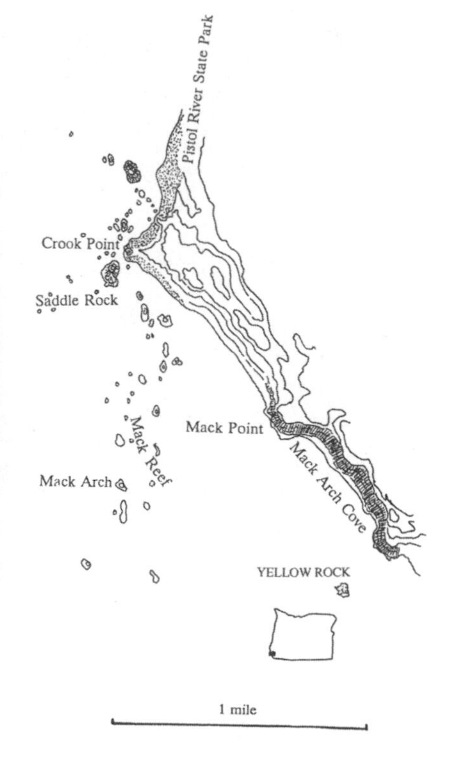

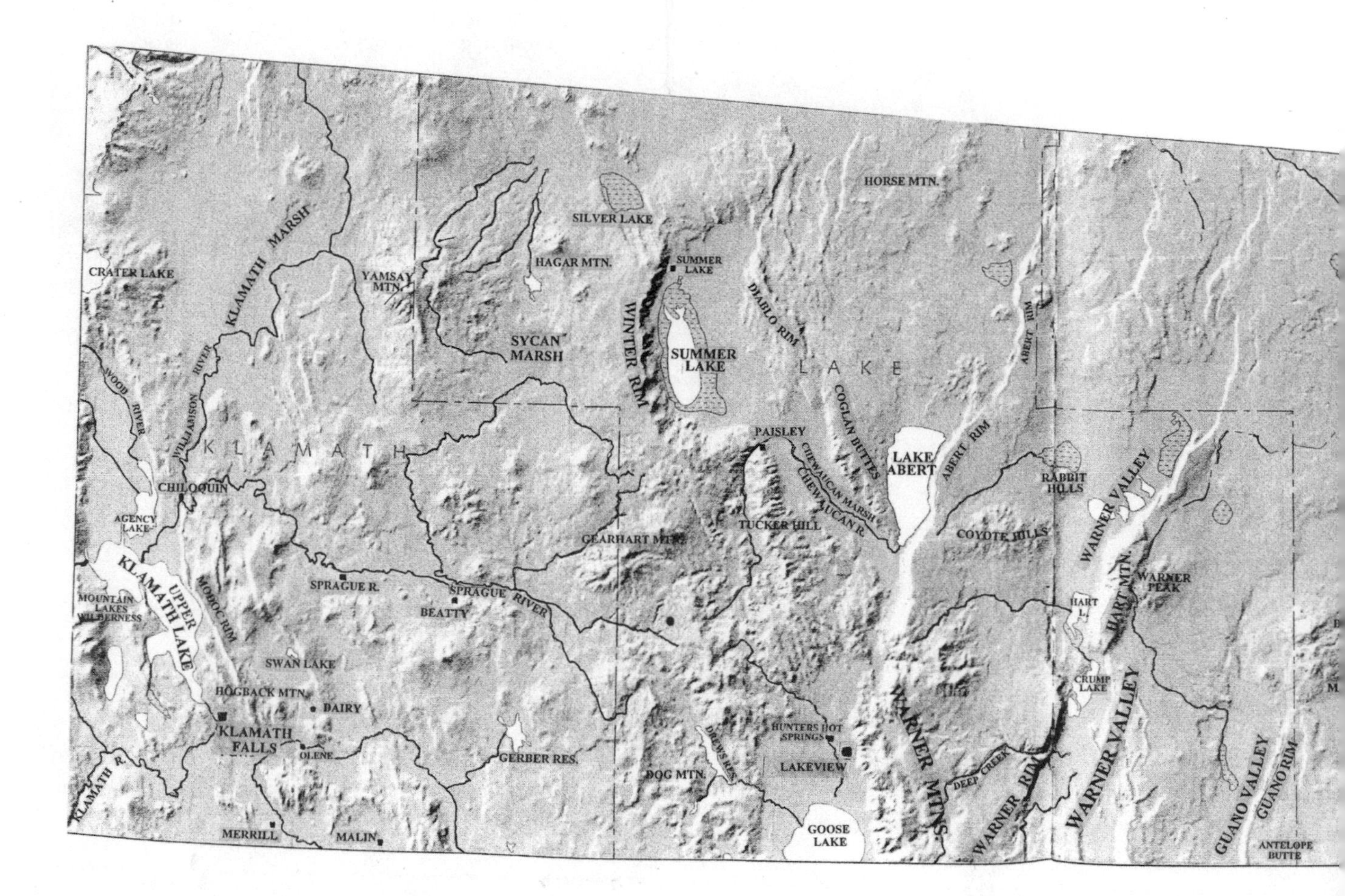
CRATER LAKE
KLAMATH MARSH
YAMSAY MTN.
HAGAR MTN.
SILVER LAKE
SYCAN MARSH
SUMMER LAKE
WINTER RIM
SUMMER LAKE
DIABLO RIM
HORSE MTN.
LAKE
COGLAN BUTTES
ABERT RIM
PAISLEY
CHEWAUCAN MARSH
CHEWAUCAN R.
LAKE ABERT
ABERT RIM
RABBIT HILLS
WARNER VALLEY
COYOTE HILLS
TUCKER HILL
GEARHART MTN.
WOOD RIVER
WILLIAMSON RIVER
KLAMATH
CHILOQUIN
AGENCY LAKE
UPPER KLAMATH LAKE
MODOC RIM
MOUNTAIN LAKES WILDERNESS
SPRAGUE R.
SPRAGUE RIVER
BEATTY
SWAN LAKE
HOGBACK MTN.
DAIRY
KLAMATH FALLS
OLENE
KLAMATH R.
MERRILL
MALIN
GERBER RES.
DOG MTN.
DREWS RES.
HUNTERS HOT SPRINGS
LAKEVIEW
GOOSE LAKE
WARNER MTNS.
DEEP CREEK
WARNER RIM
HART MTN.
WARNER PEAK
HART L.
CRUMP LAKE
WARNER VALLEY
GUANO VALLEY
GUANO RIM
ANTELOPE BUTTE

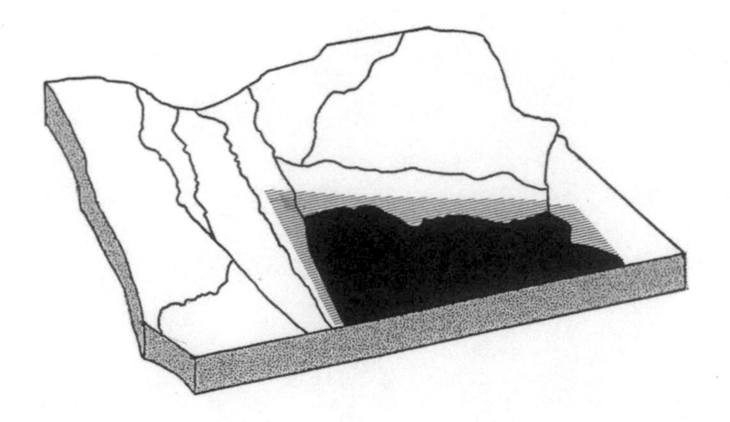

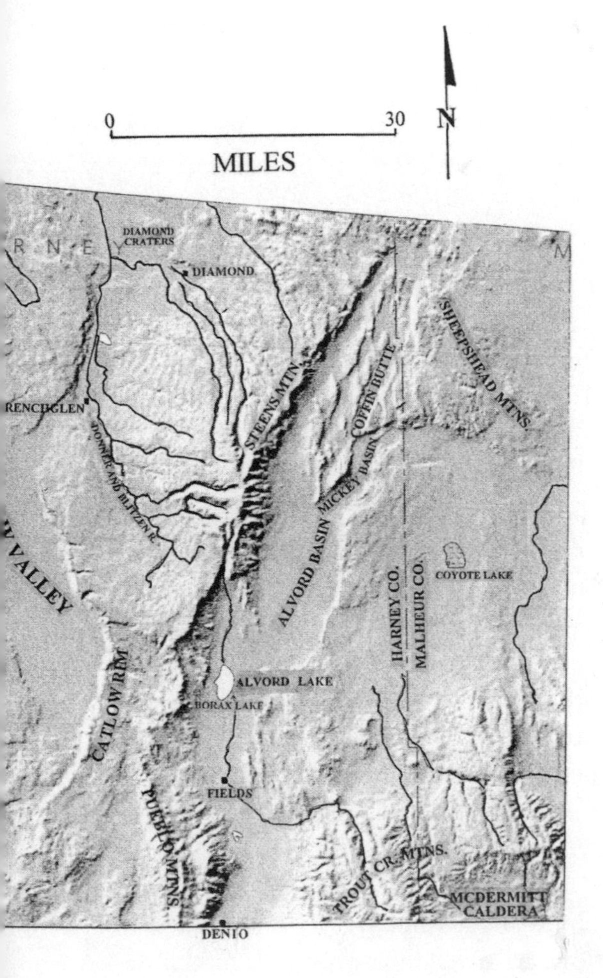

Physiographic map (After Loy, et al., eds., 2001)

Basin and Range

Landscape of the Basin and Range

The Basin and Range across southcentral Oregon is only a small northern portion of a vast region that covers parts of Utah, Nevada, Idaho, Oregon, Arizona, California, New Mexico, and Mexico, encompassing an area of 300,000 square miles. The Oregon Basin and Range is bordered on the west by the Cascade Mountains, on the north by the High Lava Plains, and to the east by the Owyhee Uplands.

As the name implies, the Basin and Range is made up of long, narrow, north-south mountains alternating with broad valleys. From west to east, the most prominent geographic features are Klamath Lake, bordered by Modoc Rim and Hogback Mountain, the Summer Lake and Lake Abert basins, enclosed by Winter Rim and Abert Rim, Warner Valley and Hart Mountain, and the easternmost Catlow Valley, Steens Mountain, and Alvord basin. Steens Mountain is the highest in the province at 9,670 feet, followed by Warner Peak at 8,065 feet, while Hart Mountain is slightly lower at 7,724 feet. Overall, the elevation averages 4,000 feet.

The basins themselves vary in size as do the lakes filling the lowest portions. Klamath Lake is the longest at 50 miles as compared with Summer Lake at 20 miles. Both are about 10 miles wide. Some of the older dry lakebeds are grassy meadows that become shallow lakes during wet years.

A distinguishing feature of the province is that almost all of the streams have reduced watersheds which empty toward the interior. The Chewaucan River drains 620 square miles, Thomas Creek covers 325 square miles, and the Ana River, just six miles long, is fed by springs. Only the Klamath River indirectly reaches the Pacific Ocean. From a beginning in Upper Klamath Lake, it follows a torturous route through northern California to enter the ocean south of Crescent City. Its watershed encompasses 12,000 square miles, and it is augmented to a considerable extent by its tributaries the Wood, the Williamson, and the Sprague.

Past and Present

The Great Basin remained largely unknown until the 1840s when engineer and surveyor John C. Fremont explored and named the province, describing the topography and recognizing that its drainage was internal. Some thirty years later, G.K. Gilbert, a geologist with the U.S.G.S., whose work was primarily in Utah and Nevada, applied the names *Basin and Range* and *Basin Ranges*.

Overview

The Basin and Range is a tectonically youthful, regionally uplifted province with a thin crust and high heat flow. This area in Oregon is distinguished by the multiple processes of crustal extension, faulting, and accompanying phases of volcanism. East-west stretching, that began in the Miocene, is in concert with the movement of tectonic plates along the western margin of North America.

Extension of the crust led to faulting and volcanism. As the crust stretched and thinned, it fractured, allowing lavas to break through. Within the past 10 million years, complex zones of both northwest- and north-northeast-trending faults developed in response to extensional processes. A broad belt of Miocene strike-slip faults cut northwest, while Quaternary north-northeast normal faults give the province its distinctive physiography of high escarpments and intervening basins.

Volcanism began with eruptions of the 17-million-year-old flood basalts at Steens Mountain, triggered by the movement of the North American

Israel Russell, who accompanied G.K. Gilbert, went on to a career with the Survey. Russell's interests were glacial geology, geomorphology, and hydrology, and in central Oregon he published on pluvial and present-day lakes and water resources. Born in 1852, Russell received a civil engineering degree in 1872 from the University of the City of New York and taught for a time at the School of Mines, Columbia College, even later becoming a professor of geology at the University of Michigan. His *Lakes of North America* is one of the first works to describe the geology of lake basins. Russell died in 1906. (In the 1890 photo, Russell is standing third from the left with a U.S.G.S. group on the Alaskan Malaspina glacier moraine; courtesy U.S. Geological Survey)

Working at Fossil and Summer lakes in 1936, Ira Allison's pursuits in the Basin came some years after those of Russell. One of the first staff members in the newly established Geology Department at Oregon State University, Allison's focus on geology and education kept him at Corvallis for 62 years. Born near Chicago in 1895, he obtained his PhD in 1924 from the University of Minnesota. His main interest was the Pleistocene Epoch, and he completed classic works on Fossil Lake, Lake Chewaucan, and Fort Rock Lake. Publications on the significance of glacial erratic dropstones in the Willamette Valley gave additional evidence to the theory of remarkable Ice Age floods. In 1988 a geomorphology professorship was established in Allison's name at Oregon State University shortly before he died. (Photo courtesy Oregon State University, Archives)

plate across a mantle plume or hot spot. The plume sent eruptions northward as far as the Columbia plateau and activated explosive rhyolitic eruptions that prograded across the Oregon Basin and Range and High Lava Plains.

During the Pleistocene Epoch, a cooling climate and increased rainfall filled the lowlands with vast pluvial lakes and brought glaciers to Steens Mountain. The ephemeral lakes attracted migratory birds and herds of mammals, whose abundant fossil remains litter the sandy flat valleys.

While mineral resources in the province are sparse, mercury and uranium have been extracted in Lake County, as have sunstones and opals.

Geology

Paleozoic-Mesozoic

In southcentral Oregon, the foundation or basement rocks lie buried beneath layers of younger volcanics and sediments. Late Paleozoic and Mesozoic terranes, enormous slabs of rock that were moved on continental plates and accreted to the West Coast, are obscured by a mantle of late Tertiary sediments and volcanic debris.

Although evident in the Blue Mountains, exposures of underlying terrane rocks are limited to the fringes of the Basin and Range in Oregon, and it is necessary to step over into Nevada to make comparisons and conclusions about the pre-Cenozoic history. In the Pueblo and Trout Creek mountains that straddle the Oregon-Nevada border, metamorphic rocks and granitic intrusions have been

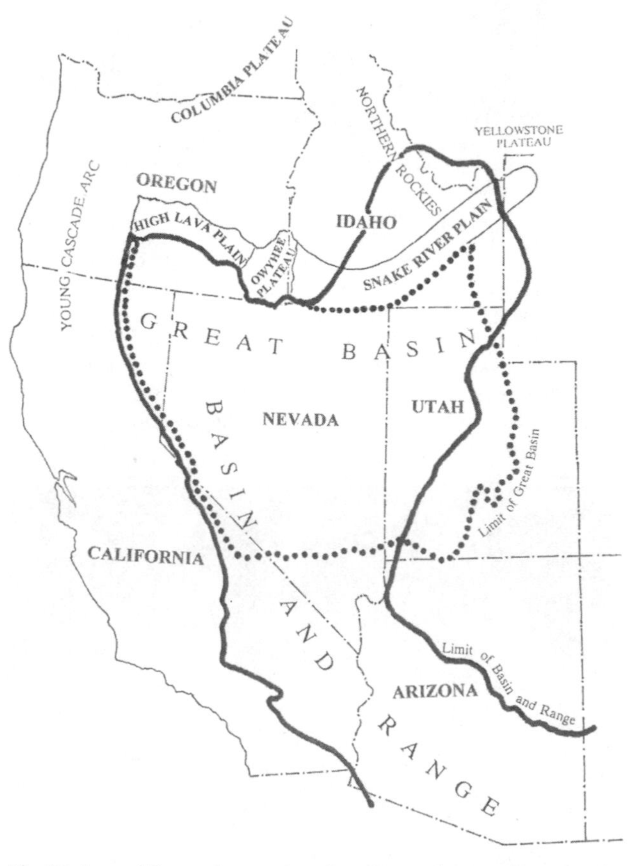

The Basin and Range is a vast region across the southwestern United States covering portions of Nevada, California, Idaho, Oregon, Arizona, Utah, and New Mexico, extending into Mexico. The Great Basin, which takes in only the northwestern portion of that area, is characterized by internal drainage and north-south mountains separated by broad structural valleys.

assigned to the Jurassic or older periods. Nearly identical rocks in Nevada have been dated with some confidence as Paleozoic to late Triassic, suggesting that terrane rocks may form the basement

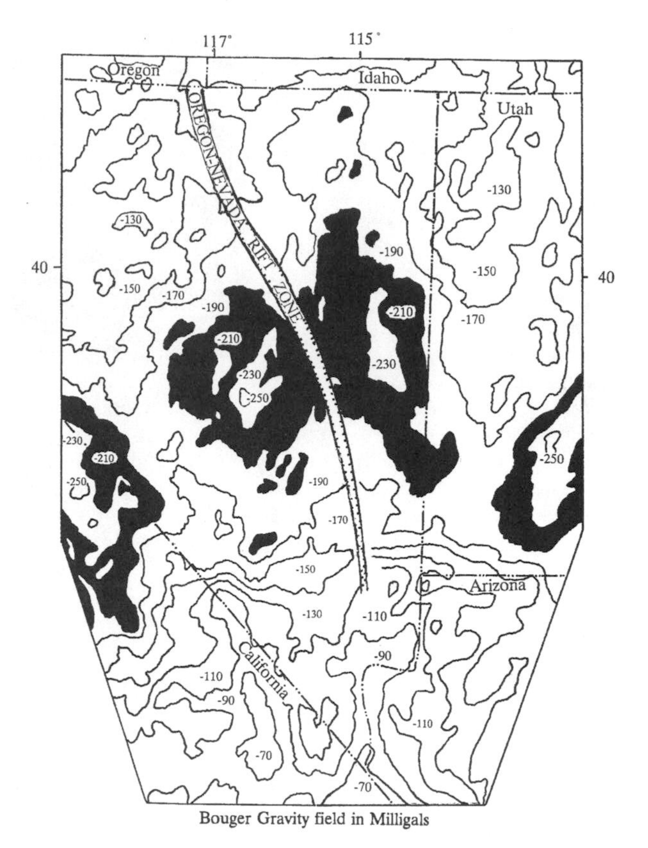

Stretching in this province is reflected by a distinctive pattern of gravity anomalies in Nevada and Oregon. Over a low (dark areas), there is less gravitational attraction as compared to that over a gravity high (light areas). The gravity map for the basin shows remarkably similar images east and west of the of the 117° meridian, which denotes the northern end of the Nevada rift zone where it enters Oregon. (after Eaton, et al., 1978)

North of Klamath Falls the shear flat face of a rock quarry displays slickenslides and striae characteristic of movement along a fault. (Photo courtesy Oregon State Highway Dept.)

here. On the other hand, geophysical work suggests that the crust beneath the Cascades thins toward central Oregon but thickens again beneath the Owyhee Plateau, indicating that the Miocene volcanics underlying the Oregon Basin and Range may, in fact, constitute the oldest strata.

Cenozoic Extension, Faulting, and Volcanism

The structure of the Basin and Range is characterized by extension, faulting, and magmatism that can be traced to the interaction of crustal plates. As the Farallon, Kula, and Pacific tectonic plates converged along the North American margin, the Farallon and Kula slabs slowly sank beneath the continental landmass and largely disappeared, while the Pacific plate expanded. Today the north-northwest motion of the Pacific plate at more than two inches a year with respect to North America can be linked to extension and right lateral strike-slip faulting (dextral shear) of the mainland.

Accommodating the motion between the two plates, a broad belt of concentrated strike-slip faults known as the eastern California shear zone was named in 1990 by Roy Dokka and Christopher Travis of Louisiana State University. They show that the belt may project east of the Sierra Nevada Mountains and into central Nevada and Oregon, where it is known as the Walker Lane zone. In a 2010 paper, Kaleb Scarberry and coauthors at Oregon State University speculate that the High Cascade graben may be the northern limit of the Walker Lane shear.

Extensional Tectonics

The most profound geologic event in the Basin and Range is the east-west stretching or extensional phenomenon, which thinned the crust and expanded the northernmost portion of this province by as much as 17 percent during the past 15 million years, 39 percent since 37 million years ago, and 72 percent since 50 million years ago. With extension, the crust was pulled apart and thinned until it began to fail. The resultant faults provided routes for mineral-laden hydrothermal fluids and lavas to reach the surface as hot springs and volcanic eruptions.

Faulting and Structure

Extension and crustal thinning generated complex networks of faults and fractures in the Basin and Range, which were linked to a major rift system that included the Columbia River Basalt dike swarms, the Oregon-Nevada rift, and the opening of the Oregon-Idaho and western Snake River grabens. Beginning 12 million years ago, northwest strike slip faults and north-northeast normal faults modified the topography of the province.

Broad northwest-trending zones of strike-slip faults that run for hundreds of miles across the Great Basin and High Lava Plains were active as late as 7 to 8 million years ago. First named and described by George Walker and Robert Lawrence, then at Oregon State University, the McLoughlin, the Eugene-Denio, and the Brothers fault zones are enormous structural features that parallel the Olympic-Wallowa lineament. Each is composed of a network of smaller overlapping faults which show 150 feet of maximum displacement with spacing of about a mile between them. The poorly-understood

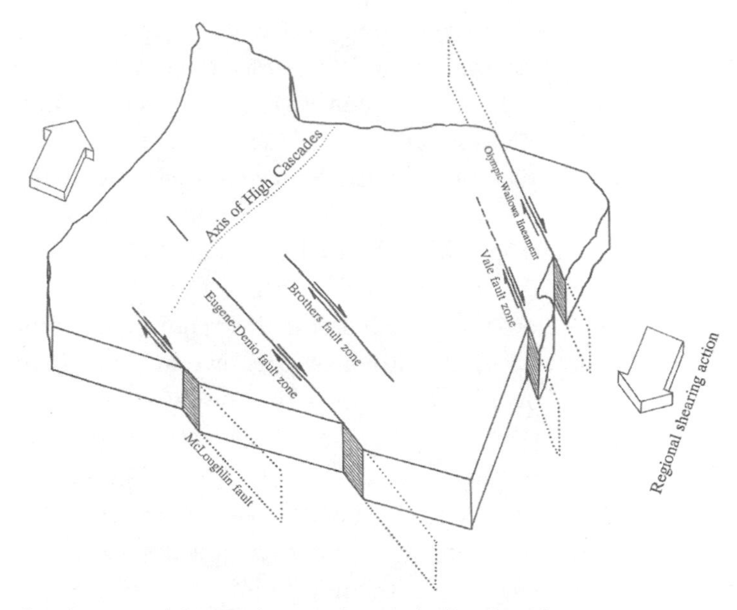

Crossing central Oregon in a northwesterly direction, the McLoughlin, the Eugene-Denio, and the Brothers faults are wide networks of smaller overlapping faults. (After Lawrence, 1976; Pezzopane, 1993; Walker, 1969, 1973)

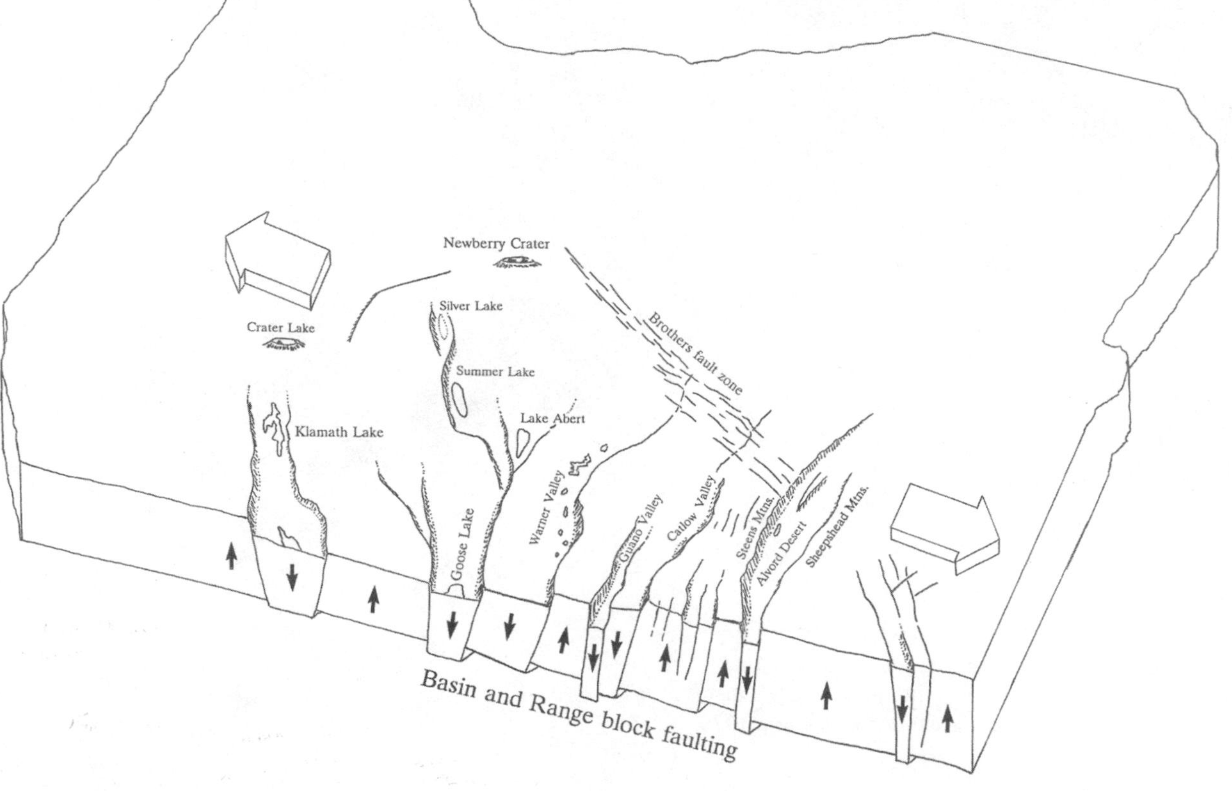

Extensional tectonics produced north-northeast faults that give the province its character of elongate tilted blocks (horsts) and valleys (grabens) that widen northward. With extreme extension and thinning of the crust, central Oregon may have expanded to twice its original width. (After Christiansen and Yeats, 1992; Eaton, 1979; Pezzopane, 1993; Pezzopane and Weldon, 1993; Scarberry, 2007; Scarberry, Meigs, and Grunder, 2011; Wells and Heller, 1988)

McLoughlin and Eugene-Denio belts post-date and offset the axis of High Cascade volcanoes. The Brothers fault zone, defining the northern boundary of the Basin and Range, has slowed to the point of deactivation since the Miocene in response to the north-northeast faults, which terminate against it.

North-northeast block-faults, dated at 6 million years ago, cut the older northwest fault belts. The younger faults, which define the major topographic character of the province, are several miles apart and show displacements of thousands of feet. Lying at right angles to the stretching direction, most are normal faults, where one side has uplifted to tilt the mountain block adjacent to a deeper valley. The ranges are elongate, and the result is that the basins tend to be open at the north end.

A study by Scarberry and coauthors showed that extensional faulting migrated in a wave from central Nevada toward the Nevada-California border and into southcentral Oregon. Summarizing the timing and progression of faulting, they proposed that the locus shifted westward from the Steens Mountain fault around 10 million years ago, to the Abert Rim fault at 8.7 million years ago, and to the present-day Cascade arc around 5 million years ago.

In Nevada, the initial block-faulting stage was succeeded by listric or detachment faulting. In this case, normal faults curve at depth to a shallow angle or even horizontally as the crust is extended. Listric faulting typically yields a topographic pattern where the tops of the fault-block surfaces slope away from a central axis, in contrast to randomly tilted blocks.

To determine ages and movement along the north-northeast faults in central Oregon, Silvio Pezzopane and Ray Weldon at the University of Oregon looked at the offset of late Quaternary deposits. Fresh fault scarps and vertical displacement of Pleistocene lake sediments were recorded at Goose Lake, Warner Valley, Abert Rim, Slide Mountain-Winter Ridge (Ana River), and Steens Mountain. With the exception of the 1968 Adel earthquake, which was generated by motion along the Warner Valley fault, activity at these localities dates back around 2,000 years.

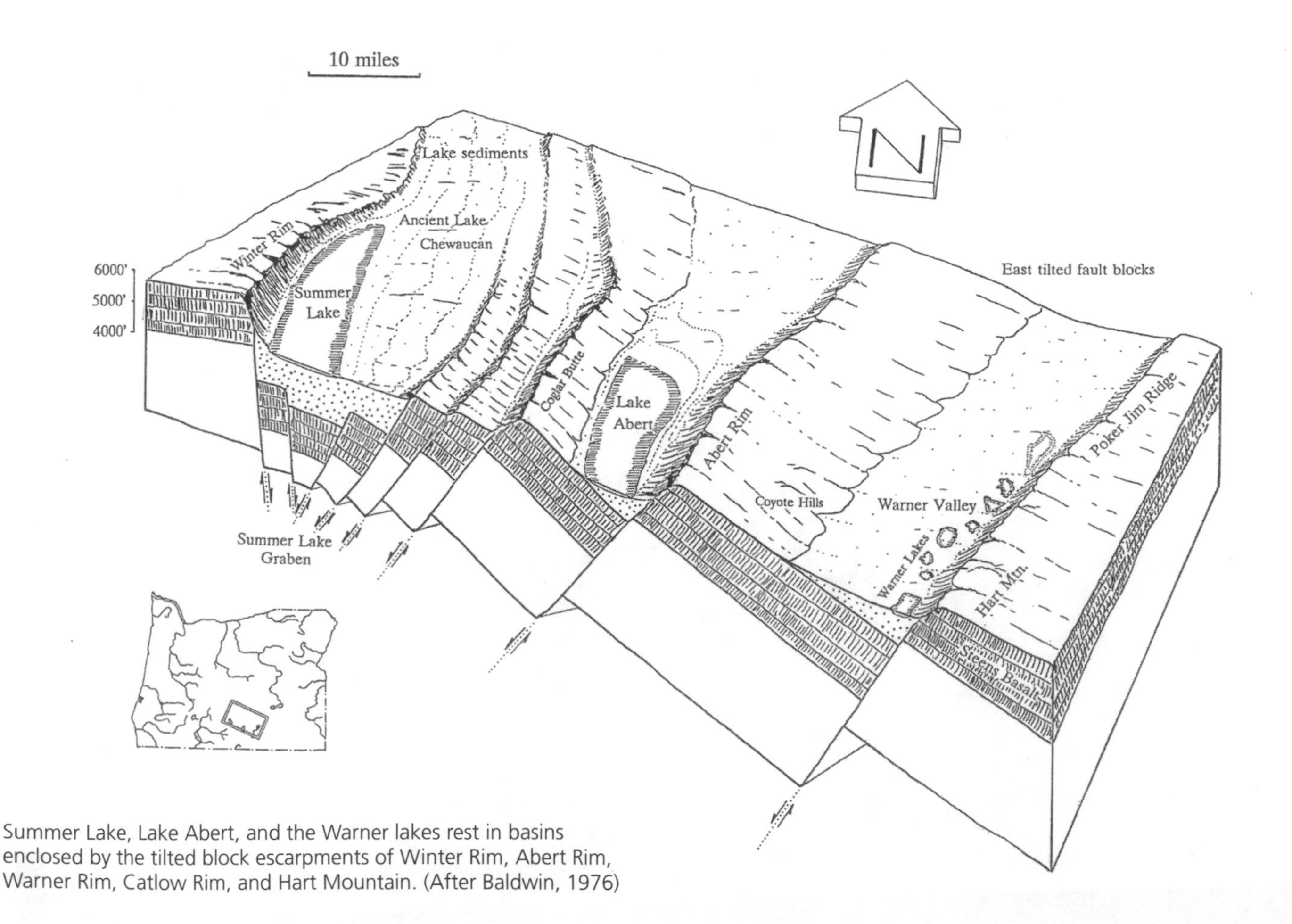

Summer Lake, Lake Abert, and the Warner lakes rest in basins enclosed by the tilted block escarpments of Winter Rim, Abert Rim, Warner Rim, Catlow Rim, and Hart Mountain. (After Baldwin, 1976)

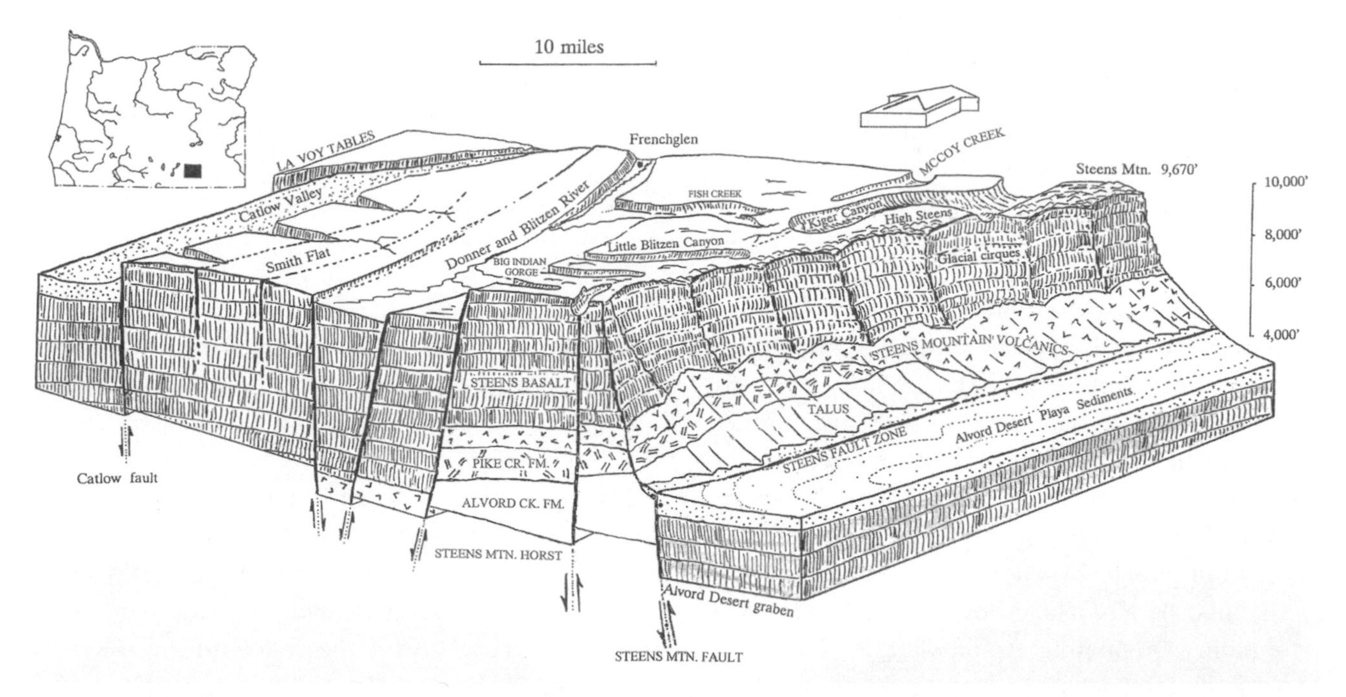

The long-lived Steens Mountain fault has been active over the past 10 million years. Severely carved by Pleistocene ice masses, the up-thrown block of the Steens Mountain escarpment displays the classic features of glacial erosion. (After Baldwin, 1976; Evans and Geisler, 2001)

At the base of Abert Rim, fault scarps displace late Pleistocene and Holocene lake and alluvial deposits. In this view toward the north, the faulted blocks of Abert Rim are clearly delineated. (Photo courtesy E. Baldwin)

Flood Basalts and Age Progression

Extension and faulting, that traversed the province, were accompanied by surges of volcanic activity. Late Tertiary lavas, which covered large regions in eastern Oregon, have been attributed to a mantle plume, while Quaternary volcanism has been associated with both crustal extension and a residual hot spot. The oldest eruptions from the Coleman Hills and Rabbit Hills volcanic complex along Abert Rim were dated at 29.3 to 21.7 million years ago. These bimodal rhyolitic and basaltic lavas formed much of the early Miocene landscape well before the extrusion of Steens flows 17 million years ago.

Volcanism on the eastern margin of the Oregon Basin and Range began in the middle Miocene with flood basalts from source vents near Steens Mountain. Originating from what is interpreted to be a huge shield volcano, the voluminous flows form a subprovince of the Columbia plateau. While subduction may have played a role in Steens volcanism, evidence is lacking for the older idea of back-arc spreading. Current thinking instead emphasizes the importance of a mantle plume.

A plume is a section of the mantle where heat becomes concentrated so that it moves upward as a hot spot. When the westward-moving North American crustal plate was passing over a hot spot, the plume rose and spread along the base of the crust to delaminate the overlying layer and emerge as Steens basalts. After being deflected northward to the Columbia plateau, the plume tail erupted across the Snake River Plain of Idaho and into the Yellowstone Plateau in Wyoming, where it is situated today. Currently, the heat source and origin of the Yellowstone hot spot, precisely how it works, its size, and depth are poorly understood. It may be much larger and deeper than previously thought, extending in excess of 500 miles into the mantle.

The concept of hot spots and mantle plumes was first proposed in the 1960s when several volcanic provinces were linked to their presence. Volcanic hot spots are known from dozens of sites around the globe, and recognition of these centers is a matter of tracing a chain of volcanoes back to its source or oldest vent. This is the case with the Hawaiian Islands, where intermittent volcanic activity from a mantle plume has left a string of eruptions.

The 17-million-year-old Steens lavas spread southward to the Pueblo Mountains and west to Abert Rim. The thickness varies widely because of the uneven subsurface enveloped by the basalts. In places it is well over one-half mile deep, covering as much as 20,000 square miles, for a total volume of almost 8,000 cubic miles. In the vicinity of Alvord Creek, the "great flow" of Steens Mountain is over 900 feet and is characterized by individual columns,

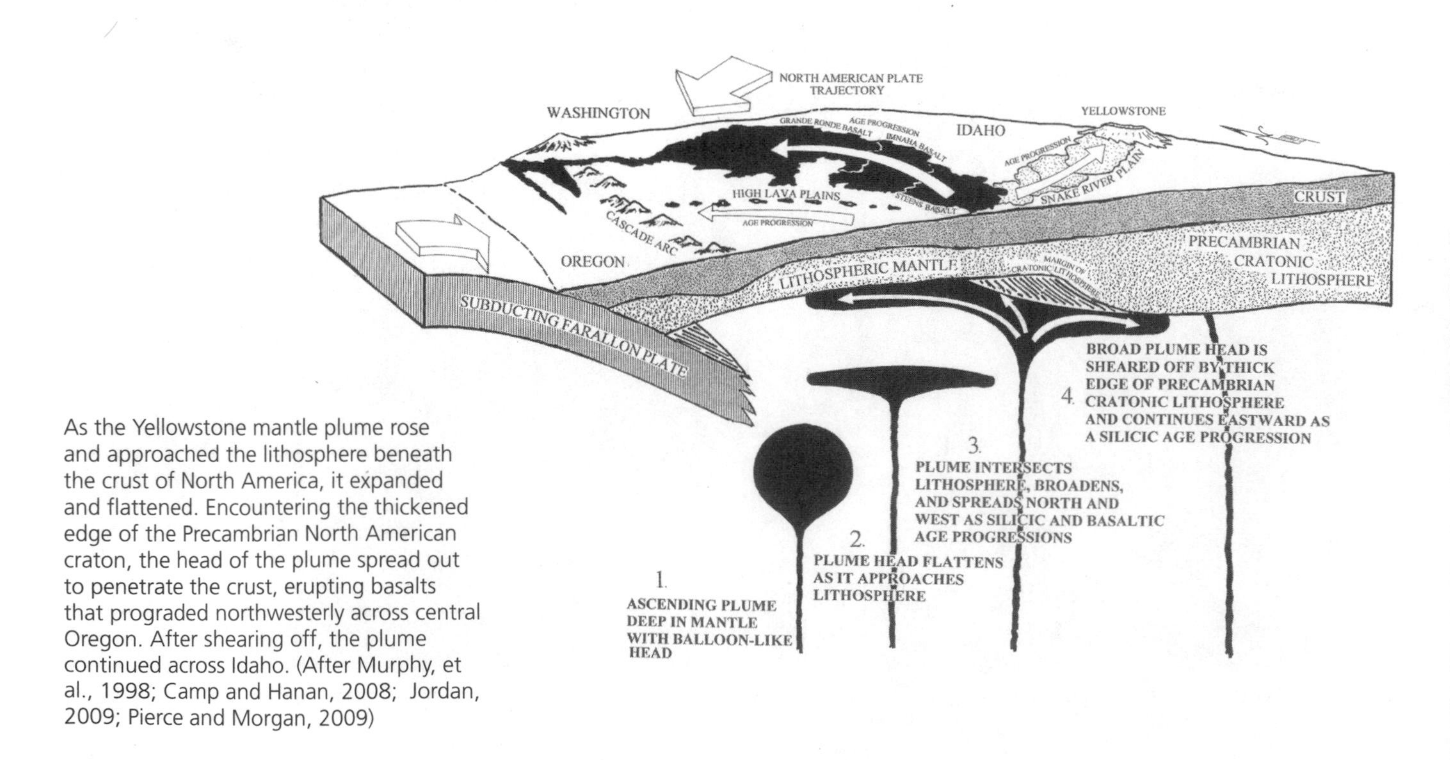

As the Yellowstone mantle plume rose and approached the lithosphere beneath the crust of North America, it expanded and flattened. Encountering the thickened edge of the Precambrian North American craton, the head of the plume spread out to penetrate the crust, erupting basalts that prograded northwesterly across central Oregon. After shearing off, the plume continued across Idaho. (After Murphy, et al., 1998; Camp and Hanan, 2008; Jordan, 2009; Pierce and Morgan, 2009)

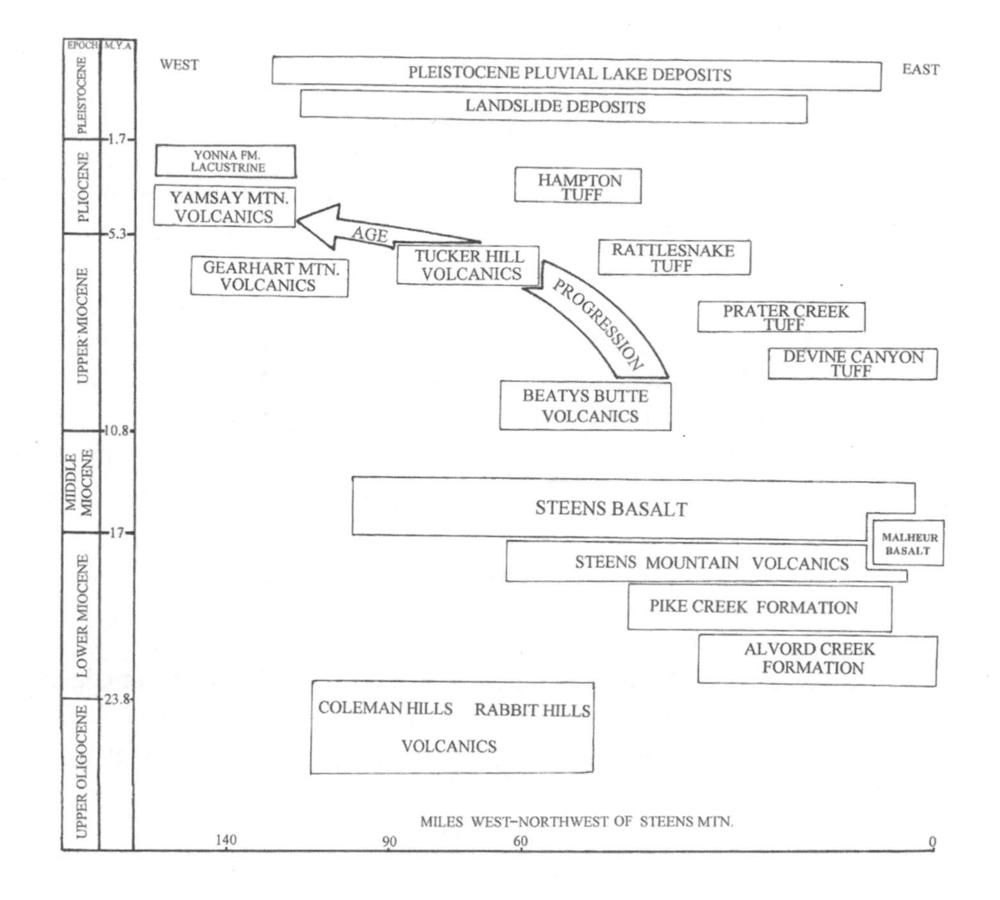

Tertiary volcanic stratigraphy of the Basin and Range (After Camp, Ross, and Hanson, 2003; Scarberry, Meigs, and Grunder, 2010; Walker, 1979)

Glaciation along the front of Steen Mountain (looking south) is illustrated by the well-developed cirque valleys. (Photo courtesy Oregon State Highway Department)

which measure as much as five feet across and rise to 300 feet. Victor Camp estimates that the greatest volume of the Steens erupted in less than 50,000 years, even though sporadic activity might have continued over a 200,000-year-interval.

Noting the similarity in age and chemistry between the Steens and the Columbia River basalts, Camp assigned the Steens as the oldest formation within the Columbia River Basalt Group. Moreover, he considers the Malheur Gorge basalt to be a sequence of the lower (oldest) Steens layers. The Malheur Gorge flows, which erupted 16.5 to 15.7 million years ago, correspond to the Imnaha Basalt and the Grande Ronde Basalt and produced around 15 cubic miles of lava from vents along the Middle and South forks of the Malheur River.

Two roughly parallel volcanic series, one in the Oregon Great Basin and the other in the High Lava Plains, erupted from southeast to northwest. Between 10.5 to 5 million years ago, the volcanic wave propagated at the rate of 20 miles every million years, slowing to 9 to 18 miles every million years after that. The northern belt in the High Lava Plains has twice the number of vents as in the southern area. Bimodal volcanism, which accompanied the progression, was characterized by a change in the composition of the lavas from rhyolitic to basaltic.

In the late Miocene, eruptions at Beatys Butte in the Basin and Range were the most easterly and oldest at 10.4 years ago. Those at Tucker Hill occurred at 7.4 million years ago, and the westerly Pliocene Yamsay Mountain eruptions were 4.7 years in the past. The eruptions from the Pliocene Yamsay shield cone were also bimodal, beginning with basaltic then rhyolitic lava. But near the end of its cycle, small amounts of basalt again were extruded from its flanks. At Gearhart Mountain, basalts are overlain by a veneer of andesitic lavas.

Pleistocene Glaciers, Lakes, and Precipitation

The Pleistocene Epoch opened almost 2 million years ago with increased precipitation and low temperatures, alpine glaciers, and pluvial lakes. This environment provided a rich habitat for hoofed

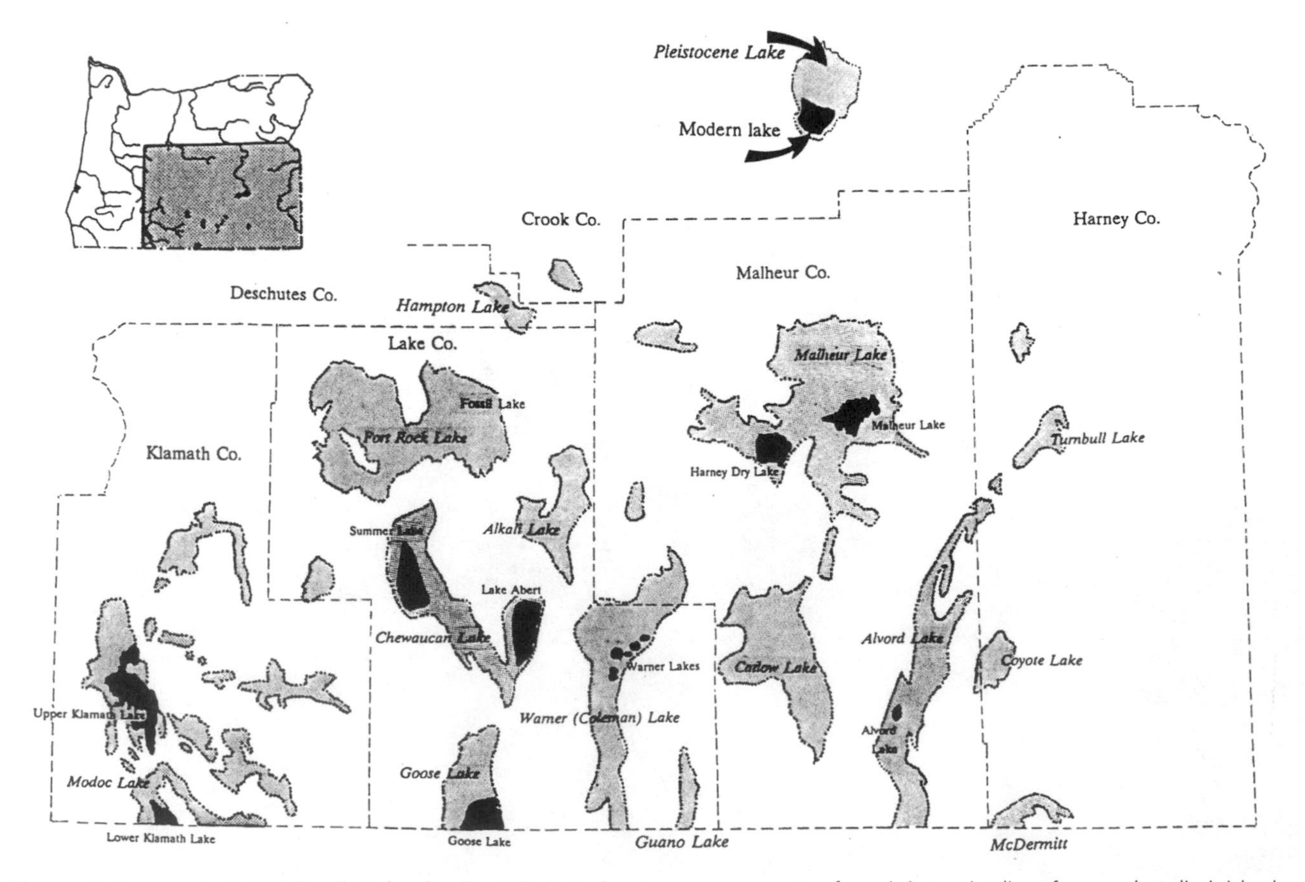

The small playas and pluvial lakes that dot the Great Basin today are mere remnants of much larger bodies of water that diminished rapidly at the end of the Ice Ages. (After Allison, 1982; Dicken and Dicken, 1985; Dugas, 1998; Negrini, 2002)

mammals, giant sloths, predators, fish, and birds, whose skeletal remains are exposed in the old lakebeds.

Dependent on precipitation, ephemeral lakes change dramatically with variations in climate, and today during an interglacial warm phase, the water bodies are shrinking. In the Ice Ages, nine pluvial lakes occupied central Oregon. The largest of these was Lake Modoc covering over 1,000 square miles. Smaller in size, Lake Chewaucan, Lake Coleman, now Warner Lakes, and Alvord Lake each spread over 500 square miles.

Pleistocene lakes, some of which were long-lived and extensive, persisted over considerable areas of the Great Basin. Old terraces and wave-cut benches clearly mark former water levels and shorelines of the lakes, but exploration with boreholes or ground penetrating radar yields an even clearer picture of beaches, deltas, spits, channels, and other geomorphic features. The identification and mapping of paleo-shorelines at Summer Lake, Lake Chewaucan, and Lake Alvord confirmed the stages of fluctuating water levels and climates.

In 2000, boreholes drilled into the Summer Lake basin revealed environmental changes throughout the past 250,000 years. Andrew Cohen at the University of Arizona analyzed the ostracods (flea-sized aquatic arthropods), pollen, water chemistry, and lake sediments to demonstrate that alterations in salinity, inflow, and biologic productivity correlate with global cooling and the waning and waxing of the North American continental ice sheet. Temperatures dropped substantially between 250,000 and 165,000 years ago, then briefly warmed from 100,000 to 89,000, followed by a cooling trend from that time onward. At California State University, Robert Negrini's research on paleo-climate data from Quaternary sediments in lakes Modoc and Chewaucan in Oregon and Lake Lahontan in Nevada showed similar results.

As recently as 7,700 years in the past, Summer Lake and the Chewaucan Marshes were impacted by the catastrophic explosion of Mount Mazama (Crater Lake). Even though 60 miles from the volcanic source, the basin was covered with tremendous volumes of pumice and ash, which created a dune field in the northeast corner of Summer Lake and along lower Chewaucan Marsh.

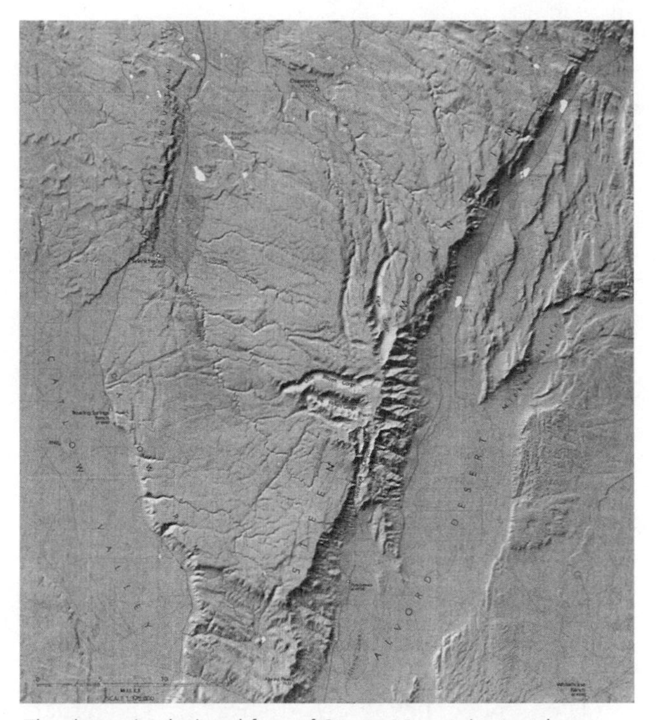

The dramatic glaciated face of Steens Mountain stretches northward along the margin of the province. Cooling during the Pleistocene fostered the onset of extensive ice masses on the Steens, the only range in the Oregon Great Basin that saw glaciation. Because of the steep gradients and greater precipitation at higher altitudes, the ridgeline has been thoroughly dissected by erosion. In cases where glaciers cut into the east rim of the mountain, the crest migrated westward. Working on both sides of a rock face, ice left only thin-walled canyons, but, when moving slowly through a basin, it carved symmetrical U-shaped valleys. (Courtesy University of Oregon)

By mapping pluvial Lake Alvord in 2007, David Wilkins and William Clement from Boise State University concluded that the ongoing ebb and flow of waters eastward through Big Sand Gap into Lake Coyote probably maintained Lake Alvord at a fairly constant level. Deron Carter at Central Washington University further suggested that large floods from Lake Alvord, some 13,000 years ago, burst into Coyote basin and from there through the Crooked Creek and Owyhee River drainages to the Snake River.

Lake Modoc in the Klamath basin expanded and contracted periodically during the Pliocene and Pleistocene, and at one time the water may have discharged into the Klamath River and from there emptied into the Pacific Ocean. Layers of volcanic ash and debris record some of Oregon's only Pliocene sediments, easily recognized in road cuts by the even laminations of the Yonna Formation.

Although scarce, a variety of fossils are present in the strata.

Geologic Hazards

A considerable expenditure of research, time, and money is currently underway on geologic hazards throughout the Pacific Northwest. In central Oregon, however, earthquakes and landslides are infrequent, and available recorded data are often lacking, thus directing the focus toward the western portion of the state.

Earthquakes

Holocene earthquakes in this province tend to be moderate in magnitude and average once every ten years. Seismic activity is associated with a clusters of young north-northeast trending faults projecting from the eastern California shear zone.

There have been few studies on Quaternary faulting and related earthquakes here. In 1993, Silvio Pezzopane, then at the University of Oregon, used aerial photographs, field mapping, and published studies to locate and identify active faults. By characterizing potential quake sizes and frequencies of recurrence, Pezzopane calculated a 10 percent chance of a seismic event of magnitude 7.0 taking place in central Oregon every 1,000 years.

In the early 1900s, Paisley in Lake County experienced several small shocks, but in 1923 and 1925 strong quakes at Lakeview along the Abert Rim fault registered a 6.0 intensity. A 1968 earthquake swarm near Adel, related to the Warner Valley fault, measured 5.1 on the Richter scale. Over a ten-day period, the shaking damaged chimneys, produced cracks in the ground, and increased flows at hot springs. Within the Chewaucan basin a total of 12 earthquakes have been recorded historically. The largest in 1999 had a 4.3 magnitude.

Earthquakes near Klamath Falls originate from faults along the Klamath graben. Mapped in 2008 by George Priest, the east and west faults, which delineate the graben, extend almost to Crater Lake and are part of a regional network responsible for at least 12 significant events over the past 50 years. There were moderate quakes near Klamath Falls in 1947 and again in 1949, but one on September 20, 1993, with a magnitude of 3.9, was followed by a second shaking measuring 5.9 on the Richter scale.

Landslides and slumping generated by the 1993 Klamath Falls earthquakes were visible over an area of 150 square miles surrounding the epicenter. Boulders such as this one, which fell on Highway 190 at Upper Klamath Lake, rolled downslope from steep cliff faces. One fatality resulted when falling rocks struck a car on Highway 97 near Modoc Point. (Photo courtesy Oregon Department of Geology and Mineral Industries)

Diminishing aftershocks lasting several months were centered in the Mountain Lakes Wilderness on the west side of the graben. Felt as far away as Coos Bay, Eugene, and Chico, California, the quakes resulted in two deaths and extensive structural damage.

In connection with the quakes, flows from springs increased by 20 percent, and well levels rose about six to seven feet. Water in some private wells became grayish and smelled of rotten eggs, but geothermal water temperatures were unchanged.

Landslides

Holocene slides are not common in this province, and most are debris flows caused by summer downpours. However, there is evidence of gigantic landslides in thick volcanic and tuffaceous sedimentary rocks along the Winter Ridge-Slide Mountain escarpment bordering the Summer Lake basin. By characterizing the deposits, Thomas Badger, Washington State Department of Transportation, and Robert Watters at the University of Nevada concluded that the slides took place during the late Pleistocene and that the large volumes of debris, the sudden nature of the activity, and the lengthy time frame of occurrences indicated a strong shaking.

S N

1 km

Neopluvial shoreline

Along the west side of Summer Lake, the Bennett Flat and the Foster Creek landslides occurred from 900,000 to 17,000 years ago. These were single events, while the Punchbowl slide (in the photo) occurred in multiple episodes, most recently 10,000 years ago. (Oblique view from U.S.G.S. digital elevation model; Courtesy T. Badger)

Natural Resources

Uranium and Mercury

Although a score of metallic ores have been known for years in the Oregon Basin and Range, only uranium and mercury have been mined. About 200 tons of uranium ore were extracted from volcanic rocks at sites northwest of Lakeview. Shipments in 1955 were the first uranium ores marketed in the state, but the processing mill closed ten years later. Located in the Fremont National Forest and overseen by the U.S. Department of Agriculture, the abandoned mines are subject to clean-up under federal and state regulations. Contaminated ponds and radioactive stockpiles pose hazards to residents, recreational users, wildlife, and to surface streams and groundwater. In 2005 the U.S. Environmental Protection Agency began a remedial action program, capping the most polluted soils, neutralizing the acidity of the ponds, and recontouring. However, a clean-up plan for the plume of contaminated groundwater, which is spreading through a shallow regional aquifer, remains to be implemented.

Borax

Along the eastern Steens Mountain escarpment, hot springs in the Alvord basin contain sodium bicarbonate and sodium chloride along with concentrations of fluoride, arsenic, and boron compounds. Of these, borax salts were mined from Borax Lake, an 800-foot-wide pool that discharges water at 97° Fahrenheit. Between 1898 and 1907, the 20 Mule Team Borax Company produced 1,700 tons annually.

Removing borax in wagons drawn by teams of mules, the 20 Mule Team Company became the Rose Valley Borax Company after a legal struggle over the name. Borax house, on the shore of the lake, was built with blocks of borax sod. Lake water directly behind the building is fed by hot artesian springs impregnated with the borax salts. Some ruins of the old operation still remain. (Photo courtesy Condon Collection).

Gemstones

The state gem, the heliote or sunstone, is a calcium-rich, clear oligoclase feldspar that varies in color from orange, yellow, or red, to green and blue. The color depends on the amount of hematite in the stone, ranging from a low of 20-parts-per-million for yellow to 200-parts-per-million for red. The most valuable are the deep orange to red sunstones, some of which are priced over $1,000 per carat. Three of Oregon's sunstone localities are privately owned, and only the one on Bureau of Land Management property in Lake County is open to the public. Here these gems are referred to as "Plush diamonds," after the nearby town of Plush. Commercial producers in Harney County send the uncut stones to Asia for faceting before they are sold worldwide.

Precious fire and blue opals are also part of the state's gemstone resources. With mines in Morrow and Lake counties, Oregon ranks as the seventh largest producer in the United States. Opals are a hydrated form of quartz derived primarily from volcanic ash, which crystallizes at low temperatures. The annual income from both opals and sunstones is close to $1.5 million.

The Crump Geyser, which was activated in 1959, began with considerable violence when the subsurface waters were penetrated by a well drilling operation. (Photo courtesy Oregon Department of Geology and Mineral Industries)

Geothermal energy

Because it is a province of high temperature gradients, a thin crust, and numerous faults, the Great Basin is distinguished by its thermal resources. Extension and thinning brings heated rocks into contact with near-surface waters, which emerge as hot springs through faults and fractures. Similarly, hydrothermal fluids may come into contact with magma at depths before the circulating waters reach the surface. Heat flow varies considerably from place to place, but higher temperatures are found in the mountains, whereas lower temperatures characterize the valleys.

In 2007 the Oregon legislature passed a renewable energy bill that identified geothermal resources as marketable. Explorations by the U.S. Department of Energy and DOGAMI are focused on seven sites with high potential. Most are in the Great Basin and Owyhee regions.

In the Alvord basin, thermal waters at Alvord Hot Springs, Mickey Springs, and Borax Lake can be identified by high mounds of precipitated calcium carbonate, built up where they emerge from faults along the base of scarps. With temperatures averaging 160° Fahrenheit, waters at Alvord Springs flow eastward from the Steens Mountain fault into a small bath house, whereas Mickey Springs releases water through vents into pools and a mud pot. Just north of Borax Lake, springs emerge from fractures at 220°, before entering the lake.

Possible geothermal exploration in the Alvord Basin by the Anadarko Petroleum Company led to the purchase of Borax Lake by The Nature Conservancy in 1982 to protect the endangered Borax Lake chub (fish). In 2000 the Bureau of Land Management further restricted geothermal development with the Steens Mountain Cooperative Management and Protection Act.

North of Lakeview and Warner Valley, Crump Geyser and Hunters Hot Springs lie in a 50-foot-wide zone of faults along Warner Rim. Hunters Hot Springs has been drilled and cased. About every 30 seconds, the geyser-like pulsating water spurts 40 or 50 feet in the air, issuing from the ground at 180° Fahrenheit. The waters and surroundings have been commercially developed. Arsenic from the mine tailings near Lakeview have been detected in the spring water.

At Klamath Falls, the Geo-Heat Center at Oregon Institute of Technology utilizes local geothermal resources for practical applications. Initially the Institute harnessed the heat to warm a swimming pool and classrooms, but in 2009 OIT drilled into a fracture zone a mile below ground, tapping into 300° water for powering a plant to heat the entire campus. A recent paper by John Lund and coauthors from the Institute provides an overview of Oregon's geothermal development.

Surface and Groundwater

The Oregon Basin and Range is semiarid with precipitation ranging from less than eight inches annually in the lowlands to 30 inches at the higher reaches. Water is supplied by snowmelt, rainfall, or springs, but much of it either percolates into the shallow playas or is lost to evaporation.

Almost all streams in the province have no outlet to the sea. The exception is the Klamath River that exits to the Pacific Ocean. Many of the smaller creeks, especially those in the dry interior, are intermittent. The larger rivers, with headwaters on the ridge tops, are perennial, and flows are augmented by groundwater stored in porous Tertiary volcanic and sedimentary rocks. Aquifers within interbeds between fractured basalt flows, which are recharged from uplands, are the most reliable. Over the volcanic rubble, thick sedimentary layers of mud, sand, gravel, and diatomite that accumulated in Pleistocene lakebeds are covered by thin alluvial deposits. With low permeability, these sources are marginal, and saline water is often encountered in wells drilled near the playas.

An overview and field guide in 2009 by Michael Cummings from Portland State University examines the impact of the 50-mile-long Steens Mountain scarp on ground and surface water storage, evaporation, and geothermal conditions. The

During infrequent wet years, precipitation ponds in the Alvord basin, which was occupied by a 12-mile-wide and 70-mile-long lake during the Ice Ages. Today a glistening white alkali crust, which rims the water, is residue left after summer evaporation. (The photo is looking east from Steens Mountain across Alvord basin; Courtesy Condon Collection).

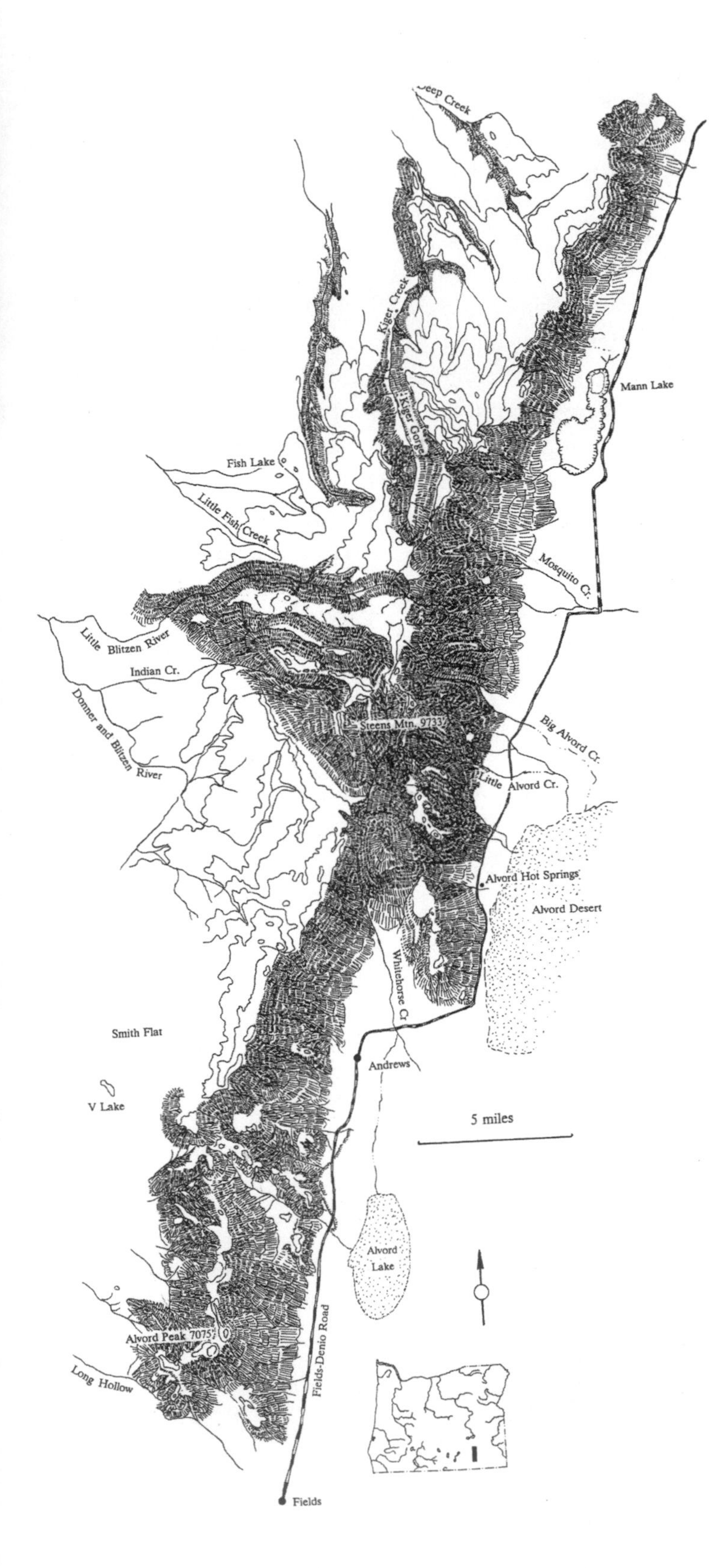

A spectacular view of Steens Mountain glacial and fault topography can be seen when driving north along Highway 395 from Fields near the Oregon border. Pleistocene valley glaciers scalloped the layers into U-shaped canyons and sharp ridges at Kiger Gorge, Little Indian Canyon, and Little Blitzen Canyon. Basins, contoured by glaciation, are occupied by shallow lakes.

numerous snow-fed streams, which emerge from the east and west slopes, supply hot springs and cold artesian wells along with domestic and irrigation demands.

A 2007 study by Marshall Gannett of the U.S.G.S. provides a picture of the groundwater hydrology of the Upper Klamath basin in both Oregon and California. Of the three formerly large water bodies, Upper and Lower Klamath and Tule lakes, only Upper Klamath remains as an important source of surface water for irrigation. The other two have been drained and used for agriculture. Historically, surface water supplied domestic and irrigation needs, but groundwater pumping has increased by 50 percent in the past 10 years, bringing a 15-foot decline in water levels.

Currently experiencing severe deficiencies in both ground and surface water, the Klamath basin lacks an overall hydrogeologic plan by the Oregon Water Resources Department. Farmers, who had welcomed government reclamation projects for irrigation, discovered in the 1990s that they didn't actually own "federal" water. This has led to the present tangle of water rights, political disputes, and additional "solutions." Today, the buying, selling, borrowing, and juggling of water rights from place to place or from person to person are being tried in an effort to apportion the dwindling resource. Gannett surmises that the water table may recover with a reduction in pumping.

Geologic Highlights

Basin and Range topography is a mixture of precipitous escarpments, gently sloping meadows, and flat lakebeds. Sagebrush, saltbush, and other xeric shrubs of the high desert country contrast with the lush watered grasses growing in the broad depressions, all of which give the landscape an arresting variety.

Mountain Ranges

Five north-south-trending ridges, Modoc Rim, Winter Rim, Abert Rim, and Hart and Steens mountains break the province into basins. Of these the Hart and Steens mountains are the most prominent. Rising over a mile above the desert floor, Steens Mountain is a large fault block (horst), where layers of thick flood basalts, sediments, and volcanic ash from eruptions in Harney basin have been fractured into several pieces. The northernmost portion tilted toward the west as the block rose, giving that side a gentle slope into Catlow Valley, whereas the abrupt eastern face drops precipitously into the Alvord desert.

Hart Mountain, which merges northward with Poker Jim Ridge, is part of a fault block complex whose sheer western wall projects more than 3,000 feet above Warner Valley. The Hart Mountain Antelope Refuge, dedicated in 1936, encompasses 215,000 acres and is overseen by the federal government. The refuge is home to pronghorn antelope and mule deer, varieties of birds, rodents, and predators.

On Steens Mountain, erratics were carried by glacial ice from snow-covered heights to be dropped some distance from the original outcrop. Many are large in size, as emphasized by the hat on the rock. (Photo from Evans and Geisler, 2001)

Looking north into Kiger Gorge, the flat-bottom and U-shape of a glaciated valley are unmistakable, as are the distinct flows of Steens Basalt in the walls. Along the rim on the top right, back-to-back glaciers cut through at The Big Nick. (Photo courtesy Oregon State Highway Department)

The Warner Valley is occupied today by Bluejoint, Stone Corral, Campbell, Flagstaff, Hart, and Crump lakes, as well as by other small remnants of what was originally a much a much broader water body. At Bluejoint, runoff from the low hill to the right sends enough water into the playa to grow a sparse stand of bluejoint grass. (Photo taken in 1936; courtesy Oregon State Archives)

Lakes and Valleys

Between the high mountains, lakes of varying sizes occupy the depressions. With no outlets, the lakes are maintained by rainfall and small amounts of water issuing from springs and snow-fed streams. When the water vanishes during periods of low precipitation, the lakes may break into a string of ponds, or the valley may even become a dry meadow. Since the late Pleistocene, water levels have fluctuated considerably.

Ancient Lake Modoc, which once reached nearly 75 miles in length, has receded to the present-day disconnected Upper and Lower Klamath lakes and Tule Lake in California. The broad expanses of Yonna and Langell valleys are dry portions of Lake Modoc.

Pluvial Lake Chewaucan, whose remnants are Summer Lake, Lake Abert, and the Chewaucan Marshes, was originally almost 375 feet deep. The marsh at the north end of Summer Lake is fed by Ana River.

Alvord and Guano lakes

On either side of Steens Mountain, the mud floors of Alvord and Guano basins have been broken into gigantic fissures with polygonal patterns. Ranging in size from 50 to 1,000 feet long and up to 15 feet deep, huge cracks developed in the clay of the dry lakebeds after water levels dropped enough to cause shrinkage and contraction. The fissures form irregular shapes so large that they may not be obvious when standing nearby. The mud polygons on Guano Lake are up to 100 feet across making them the largest structures of this type known in North America.

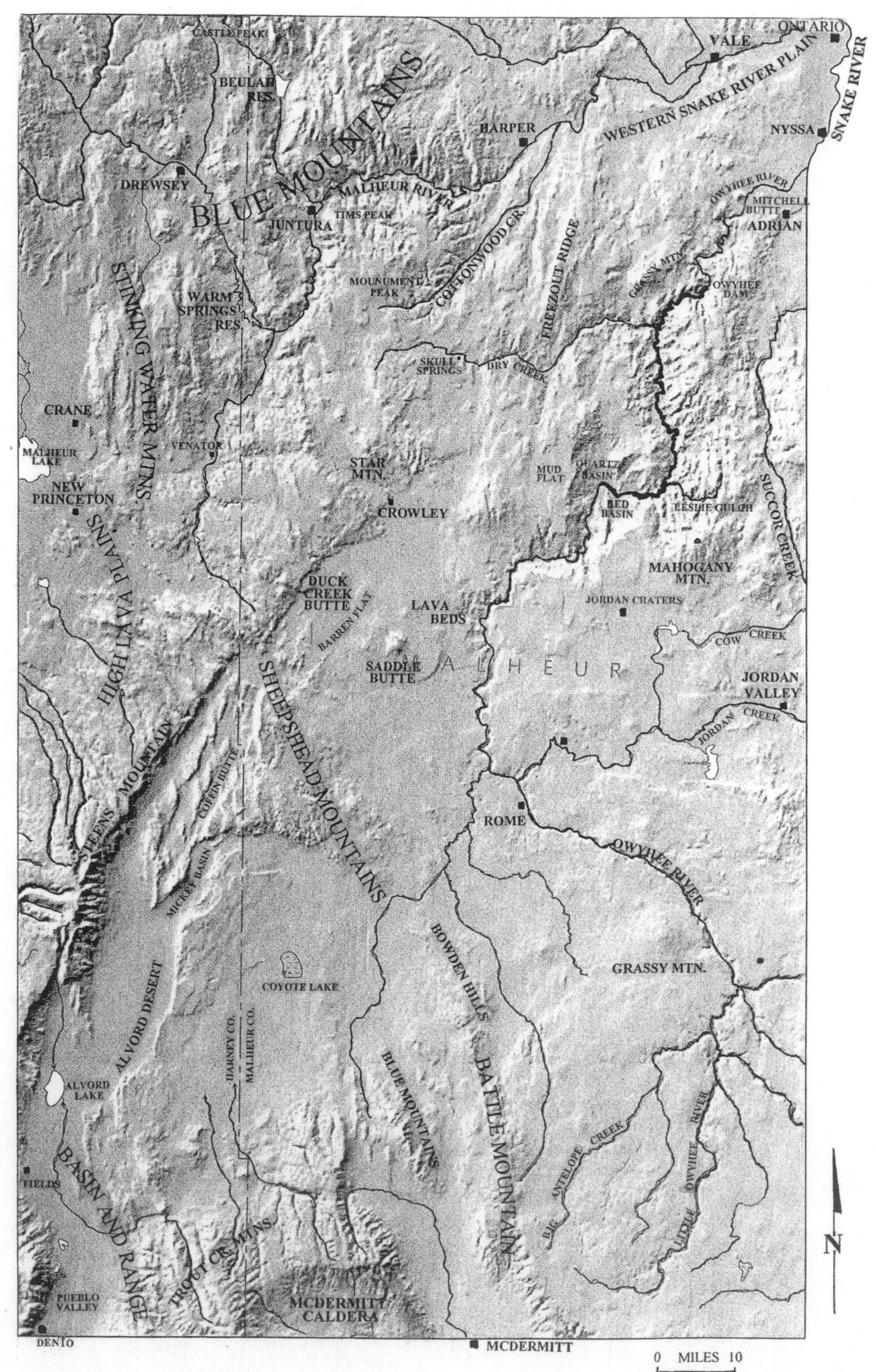

Physiographic map (After Loy, et al., eds., 2001)

Owyhee Plateau

Landscape of the Owyhee Plateau

Covering the extreme southeastern corner of Oregon, the Owyhee physiographic province reaches into southwestern Idaho and northern Nevada. In Oregon, the plateau lies almost completely within Malheur County. It is bounded on the north by the Blue Mountains, to the east by the Snake River Plain, and on the south and west by the Lava Plains and the Basin and Range.

While frequently placed within the northern Great Basin, the Owyhee exhibits a different topography. It is actually is a high plateau with a thick covering of nearly flat-lying lavas and sediments which are deeply incised by canyons and punctuated by ranges. The most notable are the Owyhee Mountains which reach around 8,000 feet at War Eagle and South Mountain in Idaho and continue to the 6,522-foot-high Mahogany Mountain, the highest point in the Oregon province.

Stream drainage is external, and most waterways exit the province through the Snake River system. The Owyhee River traverses the area from south to north, whereas the Malheur River occupies just a small region along the northern margin. With a watershed just over 6,000 square miles in Oregon, Idaho, and Nevada, the Owyhee River experiences modest rainfall and intermittent conditions in the upper stretches. The Owyhee and its tributaries the Little Owyhee and Antelope Creek cut downward from elevations of 5,000 feet in the uplands to approximately 2,000 feet at the point where the mainstem merges with the Snake River near Nyssa. Succor Creek is not a branch of the Owhyee, but it originates in the Owyhee Mountains near the Oregon-Idaho boundary. Flowing northward through a deep canyon, it enters the Snake near Homedale, a distance of 50 miles.

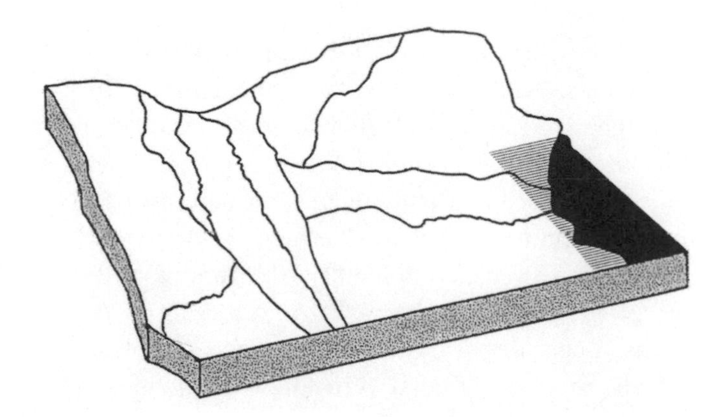

The Owyhee River furnishes irrigation to support a diversified agriculture that includes sheep, cattle, grasses, and an assortment of vegetables.

Past and Present

Geologic exploration of the Owyhee uplands lagged far behind that in other areas of Oregon. This was partly due to its harsh climate, the rough black basalt landscape, its lack of gold or silver discoveries, and its physical isolation from wagon routes crossing the state. In 1884, Israel Russell of the U.S.G.S. described the broken volcanic tableland and noted stream drainage patterns. Ultimately, an examination of the geology revolved around the projected water needs for farms and ranches. Following a Bureau of Reclamation proposal to construct the Owyhee reservoir for irrigation, Kirk Bryan of the U.S.G.S. outlined the stratigraphy and named many of the formations in 1928, while Warren D. Smith examined the regional faulting.

Larry Kittleman's work came after a thirty-year gap, and twenty years later Mark Ferns added to much of the geology. Since then, James Rytuba and James Evans of the U.S.G.S., Victor Camp at San Diego State University, and Michael Cummings at

Treating a part of Oregon long neglected geologically, Laurence Kittleman's PhD dissertation in 1962 deciphered the Cenozoic stratigraphy and structure of the Owyhee plateau. Born in Colorado Springs and growing up in Denver, his interest in geology began at age 10 when he collected rocks and taught himself the names of extinct mammals and dinosaurs. During his career, Kittleman worked for the U.S. Atomic Energy Commission before joining the Museum of Natural History at the University of Oregon. He retired in 1977 but continues to pursue interests in sedimentology and petrology while living in Eugene. (Photo courtesy L. Kittleman)

A native of Medford and raised in nearby Phoenix, Mark Ferns grew up in the shadow of the Eocene Payne Cliffs Formation. His appointment in the Baker City office of DOGAMI in 1979 came after graduation from the University of Oregon. During his years in eastern Oregon, Ferns mapped and published on many quadrangles of the Owyhee province. A culmination of field studies in central Oregon led to identifying the remarkable Crooked River caldera in 2006. His other interests include mining and volcanogenic mineralization, rhyolitic volcanism, and the relationship between rocks and plant species. Ferns retired from DOGAMI in 2010 to take a faculty position at Eastern Oregon University. (Photo courtesy M. Ferns; taken in 2010)

Portland State University published on the volcanic fields, glaciation, and aspects of Steens Mountain, while David Lawrence at Boise State University produced an overview of the Succor Creek area.

Overview

The Owyhee plateau chronicles Cenozoic volcanism, faulting, graben formation, and sedimentation. These episodes began with the eruption of early Miocene flood basalts near Steens Mountain in the Basin and Range province. By the middle to late Miocene, the volcanic style had altered to explosive rhyolitic eruptions, that created immense calderas at the McDermitt and Lake Owyhee fields. A final volcanic stage, which began in the Pleistocene and continued as late as 4,000 years ago, spread lava across the southeast corner of the province.

Volcanic eruptions are only part of the geologic picture. Throughout the Miocene, north-south extensional faulting in response to Basin and Range stretching created the Oregon-Nevada rift. This belt of dikes, faults, and lavas merges to the north with the Oregon-Idaho graben. As the graben or depression subsided, it was broken into smaller structural basins where stream and lake sediments accumulated. Because of favorable conditions for habitation, animals and plants flourished, and their fossil remains are well-represented in the sedimentary layers.

Evidence of glaciation during the Pleistocene is sparse, but cold weather and torrential rains created a string of pluvial lakes atop the horizontal lava surfaces. As the climate warmed, arid basins are now all that remain of the temporary bodies of water.

The hydrothermal emplacement of minerals into fractured volcanic and sedimentary layers has given the province its suite of unusual ores, such as mercury and uranium. Even though gold is present, it has not received the mining interest seen elsewhere in Oregon. Geothermal resources are substantial enough to be exploited commercially, while limited ground and surface water must be carefully utilized.

Geology

Paleozoic-Mesozoic

Whereas accreted Paleozoic and Mesozoic terrane rocks crop out along the borders of the Owyhee uplands, the most profound geologic events took place in the middle to late Cenozoic, when vast quantities of volcanic and sedimentary deposits buried the older basement rocks.

Cenozoic

Flood basalts, ash-flow tuffs, and volcaniclastic sediments (volcanically shattered) modified the Owyhee landscape as layers reaching 6,000 feet in thickness were deposited from the early Miocene into the Holocene. Changing tectonic patterns initiated Miocene volcanism, in which the individual phases can be differentiated by the timing of the eruptions and by the chemical composition of the lavas. Beginning 17 million years ago, eruptions of Steens flood basalts were followed by violent

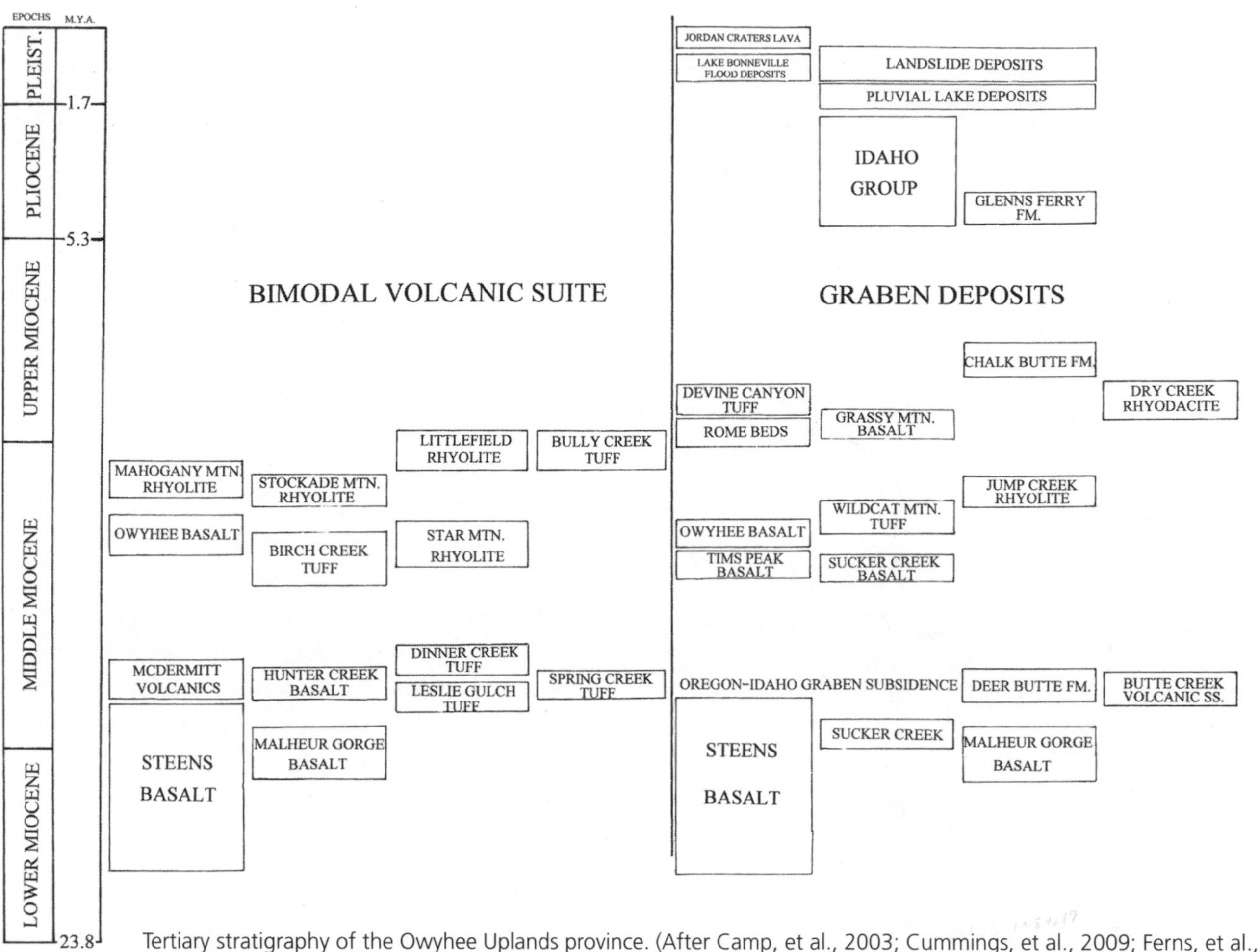

Tertiary stratigraphy of the Owyhee Uplands province. (After Camp, et al., 2003; Cummings, et al., 2009; Ferns, et al., 1993; Ferns, Evans, and Cummings, 1993; Kittleman, et al., 1973)

caldera-forming explosions in the McDermitt and Owyhee fields. Subsequent calc-alkaline Owyhee basalts 14.5 million years ago and a 9,000-year-old eruptive phase added an additional covering across the plateau.

Flood Basalts and Calderas

Episodes of Miocene Steens basalts are so extensive in the Owyhee uplands that it can be considered a geologic subprovince of the Columbia plateau. Even though the source vents were centered over a large shield volcano at Steens Mountain in the adjoining Basin and Range, the lavas spread across this region and into southwest Idaho. In the Owyhee Canyon, Steens basalts attain thickness of 1,500 feet but thin elsewhere in the province. In 2010 Victor Camp concluded that the Steens is the oldest formation of the Columbia River Basalt Group.

The Steens eruptions were generated as the westward-moving North American plate was passing over a rising plume in the mantle. As the Yellowstone plume made contact with the lithosphere of the North American craton, it spread out, melting the overlying crust to erupt the basalts. After encountering the thickened edge of the continent, the plume head was sheared off and deflected north-northwesterly toward the Columbia plateau. The main track of the plume continued across the Snake River Plain in Idaho and into Wyoming.

McDermitt and Lake Owyhee Volcanic Fields

Around 16 million years ago, volcanism shifted toward the McDermitt and Lake Owyhee fields in southeast Oregon, southwestern Idaho, and northwest Nevada. Close to the Steens basalt in age, the McDermitt and Lake Owyhee vents produced immense deep calderas. Explosive rhyolite ash flows triggered caldera collapse and accompanying graben development. Orientation of multiple north-south dikes associated with the episodes indicate that the

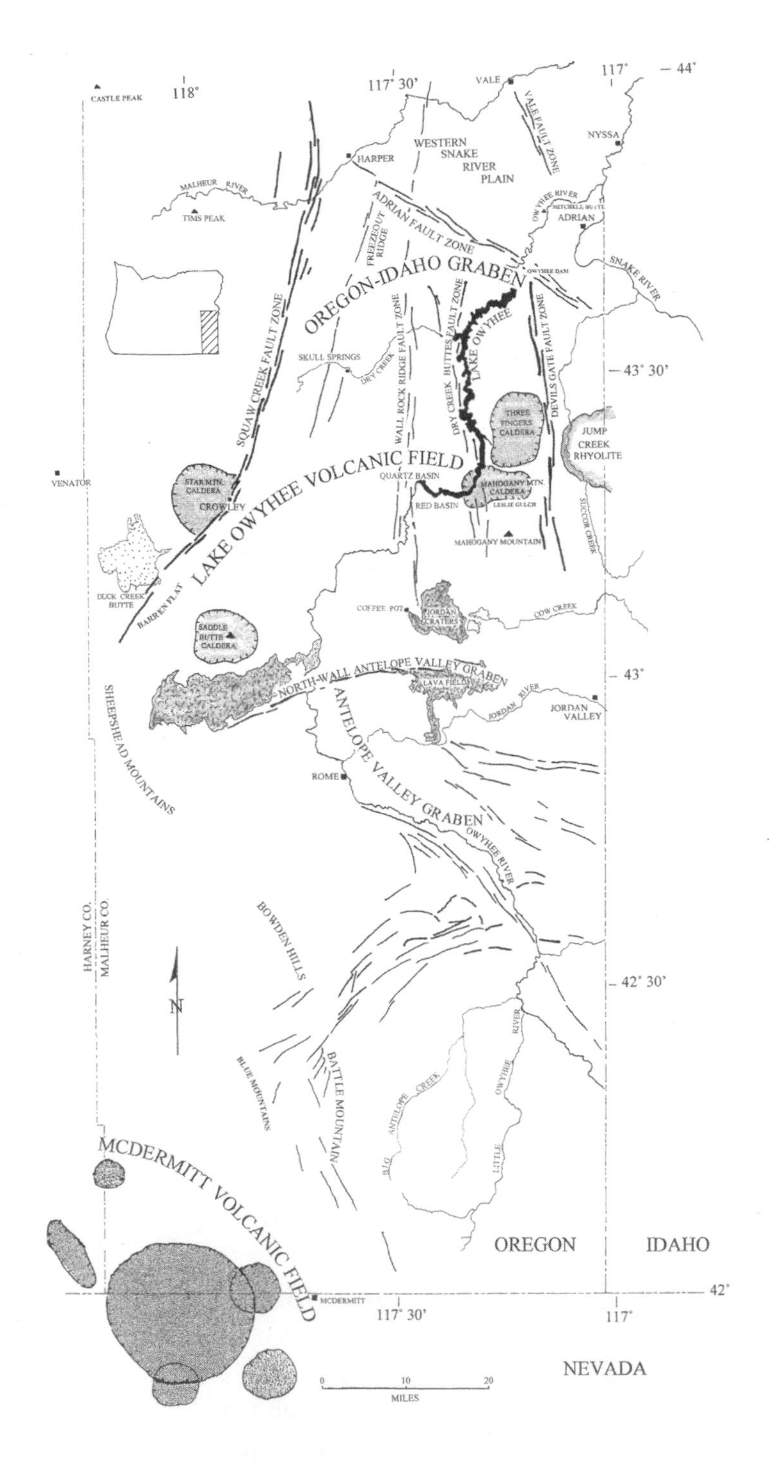

Lying almost wholly within Malheur County, the Owyhee province includes the Antelope Valley, Oregon-Idaho, and Snake River grabens as well as the McDermitt and Lake Owyhee volcanic fields. (After Carlson and Hart, 1987; Ferns, 1988; Ferns and Williams, 1993; Ferns, Evans, and Cummings, 1993; Rytuba, 1994; Rutyba and McKee, 1984; Rytuba, et al., 1990)

latest Steens basalts as well as the McDermitt and Lake Owyhee rhyolites may be related to Basin and Range extension.

Erupting in a relatively short one-million-year time span, the McDermitt volcanic field covers over 8,000 square miles. A summary paper by James Rytuba and Edwin McKee at the U.S.G.S. describes seven ash flow sheets of high-silica rhyolite. The eruption of each flow produced a large subsiding caldera 20x30 miles in diameter, which consists of four overlapping structures.

Paralleling the western boundary of the North American craton, the McDermitt volcanic complex is near the northern end of the Oregon-Nevada rift (northern Nevada rift). The rift is a zone of grabens, aligned dikes, and lava flows that extends over 500 miles from southern Nevada into Oregon and records extensional stress in the Basin and Range. Generated by the Yellowstone plume that produced the flood basalts on the Columbia River plateau, the dikes and volcanics in the rift show an age progression with the younger structures found in the south.

North of the McDermitt complex, the Lake Owyhee volcanic field erupted about 15.5 to 15 million years ago with extensive ash-flows from Mahogany Mountain, Three Fingers, and Saddle Butte calderas. Five flow sheets from Mahogany Mountain make it the most prolific. With a north rim extending 1,000 feet above Leslie Gulch, the edifice of Mahogany Mountain is the remnant of a cone, 10 miles in diameter, which erupted the Leslie Gulch tuffs, whereas the circular Three Fingers caldera is eight miles in diameter and produced the Spring Creek tuffs. Within the depression, Three Fingers rock, the highest feature, is the vestige of a rhyolite neck.

Close to 15 miles across, the most southerly Saddle Butte caldera, responsible for the Hunter Creek basalt, is almost completely buried beneath younger flows. Source dikes for the Hunter Creek lie along the western border of the Oregon-Idaho graben, and Camp considers these lavas to be chemically indistinguishable from flows of the Grand Ronde Basalt. Castle Peak at the northern end of the Lake Owyhee field collapsed with the eruption of the Dinner Creek tuff. Filling the entire Castle Peak caldera and spreading over 1,500 square miles, the tuff is an important marker bed that is easily traced by

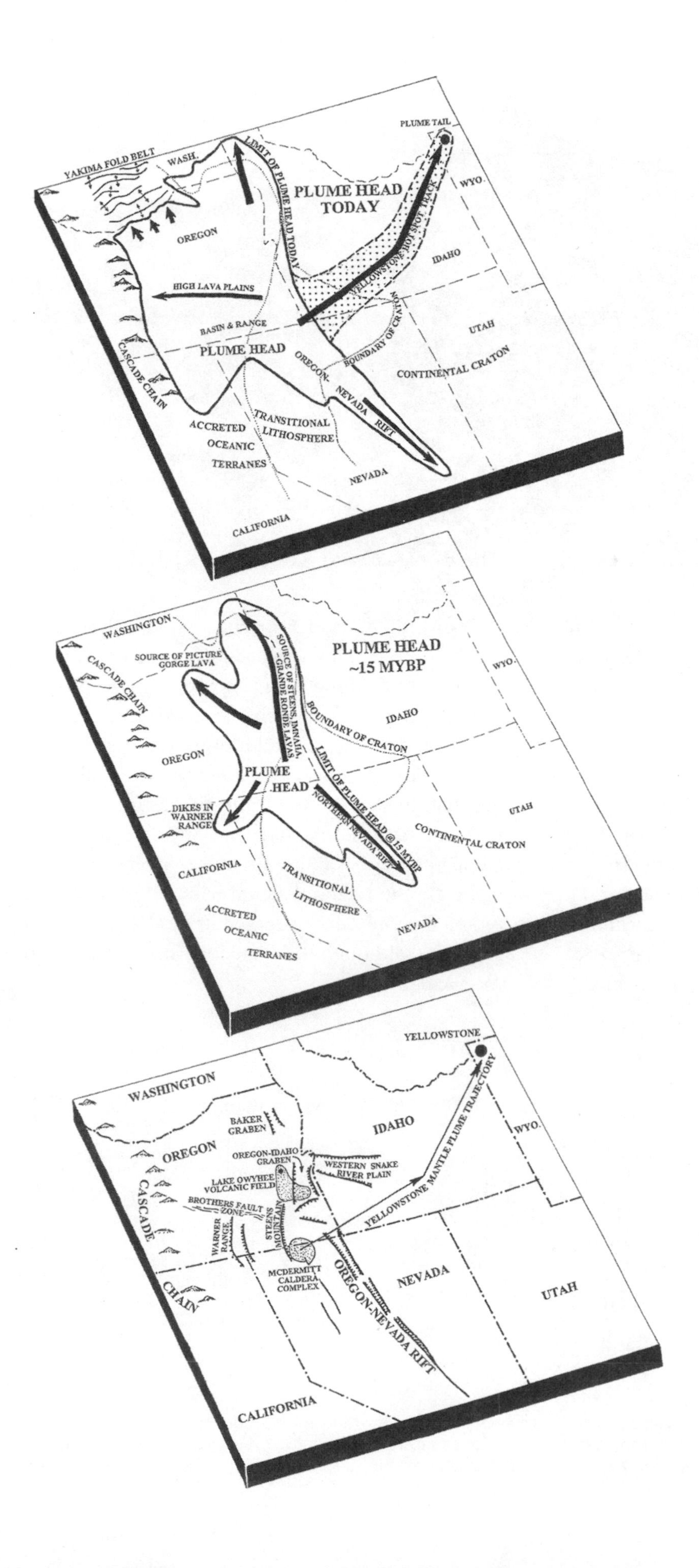

The Owyhee province is the focal point for impact of the Yellowstone hot spot. Most of the Miocene volcanic activity in eastern Oregon is the consequence of the mantle plume encountering the cratonal edge and flattening beneath the lithosphere (crust and mantle). (After Camp, 1995; Camp and Ross, 2004; Christiansen, Fougler, and Evans, 2002; Cummings, 2000; Glen and Ponce, 2002; Zoback, et al., 2004)

Small volumes of Tims Peak olivine-rich basalt erupted during a short interval around 13.5 million years ago after waning of the caldera-forming rhyolitic lavas. Southeast of Juntura, Tims Peak overlooks a plateau underlain by basalt flows. (Photo from Kittleman, et al., 1965)

the distinctive ledges on both sides of the Malheur River. As a comparison, the 15-mile-diameter Lake Owyhee calderas are roughly twice as large as those at Crater Lake and Newberry Volcano.

Faulting, Grabens, and Basalt

Structurally, the Owyhee plateau is defined by the Oregon-Nevada rift on the southeastern margin, by the Basin and Range scarps and valleys to the east and west, and by the Paleozoic to Mesozoic terrane rocks of the Blue Mountains and the Vale lineament to the north. In this small corner of Oregon, many of these features originate in the adjoining provinces.

With initiation of Miocene rhyolitic flows, the north-south striking Oregon-Idaho graben, aligned with extensional faults and dikes of the Oregon-Nevada rift, subsided adjacent to the North American Precambrian margin. Evolution of the basin began 15.7 million years ago, preceding collapse and faulting that divided the floor into subbasins. Ash flow tuffs and basalts, released from fractures and vents along the edges of the structure, spread into the depression to be interlayered with lake and stream sediments. Subsidence ceased around 10.5 million years ago, although the smaller basins continued to receive debris.

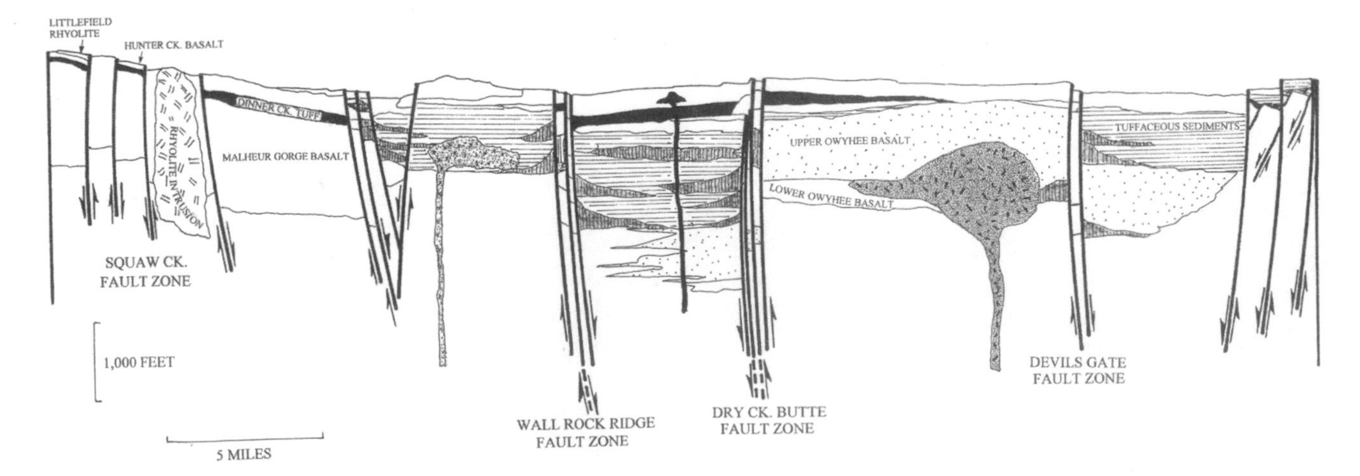

A cross-section drawn along 42°30′ latitude shows faulting and sedimentation within the 40-mile-wide, 60-mile-long Oregon-Idaho graben. Mapping by Mark Ferns and Michael Cummings identified the huge depression, which is truncated by the Snake River Plain and the east-west trending Antelope graben. (After Camp, et al., 2003; Cummings, et al., 2000; Evans, Cummings, and Ferns, 1993; Ferns, et al., 1993)

Owyhee Basalt

Around 14.5 million years ago, eruptions of calc-alkaline Owyhee Basalts from craters, cinder cones, and small shield volcanoes along the Oregon-Idaho graben marked a change in the volcanic regime. Over 17 separate flows are exposed near the Owyhee reservoir. Camp has interpreted the Owyhee Basalts to have resulted from crustal extension and thinning rather than from the presence of a mantle plume. He also notes that they are chemically distinct from the Columbia River Basalt Group.

Lake and Stream Environments

Volcanism was accompanied by regional subsidence and punctuated by quiet periods during which freshwater basins and wetlands hosted populations of plants and animals. Periodically overwhelmed by pyroclastic flows and lava, the flora and fauna were preserved in tuffaceous and fluvial sediments of the Miocene Sucker Creek and Deer Butte formations and the Butte Creek Volcanic Sandstone.

The Owyhee region is renowned for its fossilized fish, vertebrates, and plants that reflect the ancient environments. Beautifully preserved leaves of hardwood trees signify a rainy more temperate climate than the arid conditions of today. Hoofed mammals, elephants, and a multitude of rodents occupied plains and wooded slopes, while aquatic rhinoceroses and giant beaver resided in the wetlands.

Of the various Miocene floras in Oregon, those from the Sucker Creek Formation are among the most abundant with over 60 genera of pollen, leaves, fruit, and flowers. Vertebrate fossils are equally plentiful, and the remains of small mammals show ready adaptation to a diversity of habitats, even to the changes brought on by volcanic disturbances. Dated at 16.7 million years ago, the Sucker Creek Formation was eventually covered by younger flows of the Owyhee Basalts.

At Skull Springs and Quartz and Red basins, fossilized skeletal fragments of beaver, rhinoceroses, and the small, three-toed horse *Merychippus,* along

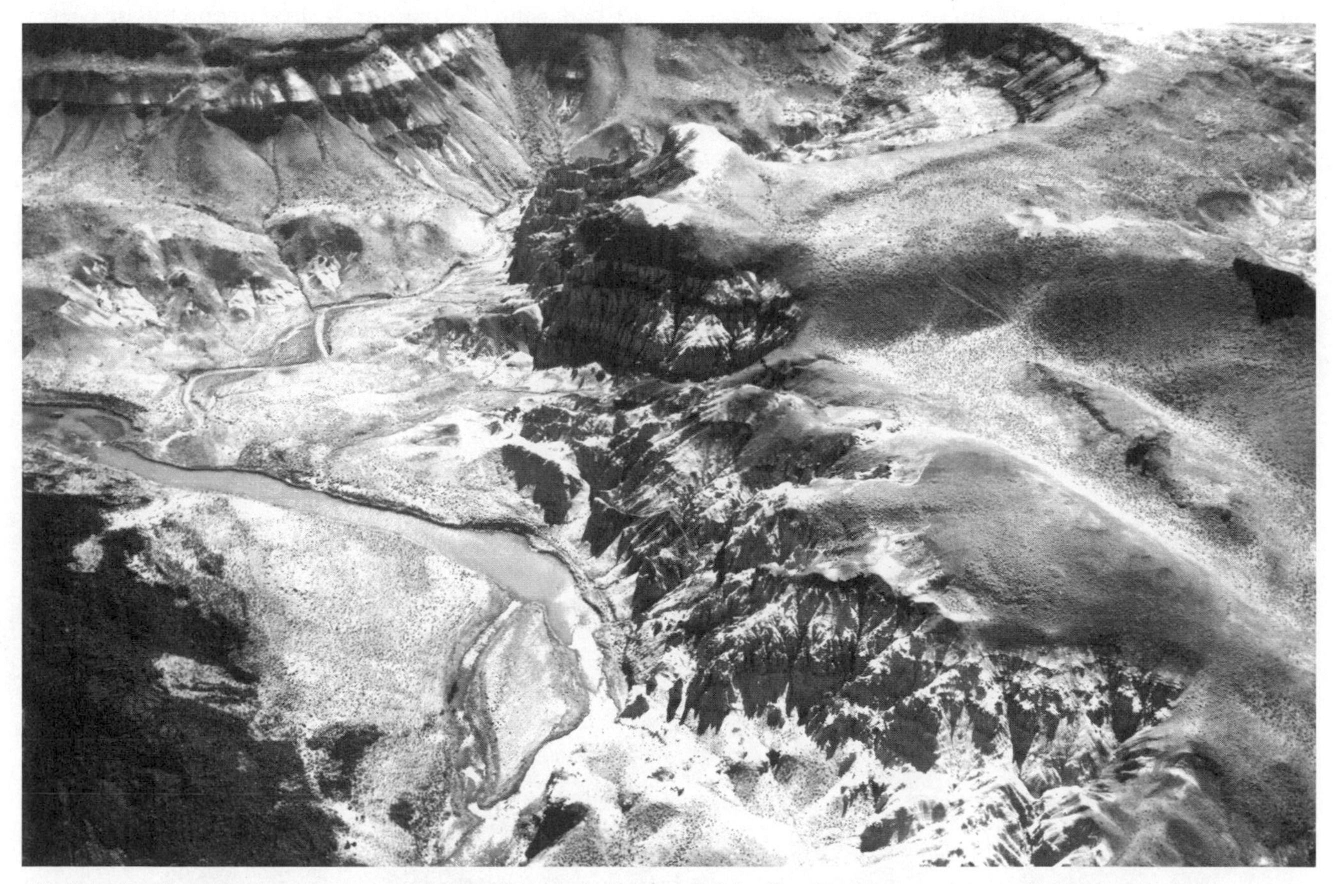

Chalk basin, along the upper Owyhee River near Rome, displays a colorful yellow to white assortment of late Miocene Devine Canyon ash-flow tuffs and lakebed deposits. Representing the southern-most reaches of Miocene Lake Idaho, the sediments are comparable to the Rome Beds. (Photo courtesy Oregon State Highway Department).

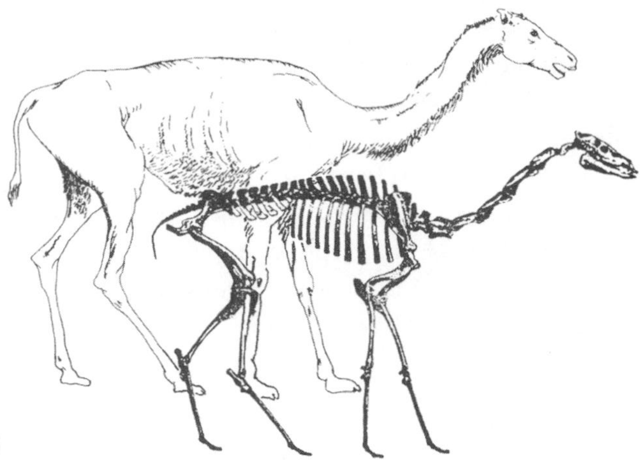

An enormous beaver, seven feet in length, lived in eastern Oregon during the late Tertiary, as did the long-limbed camel. (After Orr and Orr, 2009)

with high percentages of freshwater fish and rodents, were entombed in volcaniclastic sandstones and siltstones of the Deer Butte Formation and the Butte Creek Volcanic Sandstone. Dated at 15 million years ago, the sediments accumulated in structural basins created following the collapse of calderas in the Lake Owyhee field.

The faunal composition and fine preservation of the bones point to a floodplain or lake paleo-setting. Based on the quartz-feldspar (arkosic) make-up of the Deer Butte Formation, Cummings concluded that the major river systems to enter the basin came from the distant highlands of Idaho.

Lake Idaho, one of the most extensive lava-dammed water bodies of the late Tertiary, stretched west from Twin Falls, Idaho, into the Owhyee River canyon as far south as Rome and north toward the Blue Mountains. The lake filled and emptied several times, each episode separated by millions of years. Over 150 miles long and 50 miles wide, it occupied the western Snake River graben during the late Miocene, but ultimately drained some 2 million years ago, when the ancestral Snake River eroded southward to breach the wall. Old lake shorelines can be traced by the alluvial sands and silts of the 8.5-million-year-old Chalk Butte Formation in Oregon and by the Glenns Ferry Formation in Idaho, both of which are known for their rich fossil assemblages. The fragmentary condition of the bones indicates they were carried some distance before being buried.

After Lake Idaho emptied, regional uplift brought renewed down-cutting by the Owyhee River as alluvial fans and gravels on benches above the river were deposited over the lake sediments.

Pahoehoe lava from the rim of Morcom Crater in the Cow Creek area erupted during the Quaternary. These young flows are fresh-looking and free of vegetation, clearly showing surface volcanic features. (Photo from Kittleman, et al., 1965)

Pleistocene Lakes and Floods

The onset of colder weather during the Pleistocene produced glaciers on Steens Mountain and at the higher elevations of Oregon, but evidence of glaciation is lacking on the Owyhee plateau.The area experienced cold temperatures and periods of extreme rainfall, however, which allowed water to accumulate in shallow basins as pluvial lakes. Along the western section of the province, Pleistocene to Holocene lake sediments and alluvial fans at Barren Valley and Big and Little Mud flats record the presence of ephemeral bodies of water, which evaporated when the climate subsequently warmed.

Around 14,500 years ago, floodwaters from Lake Bonneville in Utah discharged across the Snake River Plain and through the Snake River canyon leaving a number of distinct depositional and erosional features. Jim O'Connor of the U.S.G.S. has estimated that the outpouring of floodwater lasted for eight weeks and moved at an average velocity of 30 to 35 miles an hour. Near Vale, silts spread by the Bonneville deluge were redistributed by winds into small sand dunes and a loess covering.

Volcanism continued into the Pleistocene and early Holocene, as faulting created basins that would be buried beneath lavas from vents in the Cow Creek and Jordan Creek drainages. Between 9,000 and 4,000 years ago, six separate basalts and ash-flows from Jordan Craters applied a thin layer over the earlier sediments.

Geologic Hazards

Possible impacts from volcanism, earthquakes, landslides, or flooding on the Owyhee plateau are primarily focused on damage to agricultural land and facilities and secondarily with risks to the sparse human population.

The potential for volcanic activity might come from renewed eruptions in the Jordan Craters field. Movement along Quaternary faults in that area or along the Vale fault zone across the northeast corner of the province could pose a hazard, although the Vale fault has not been active for the past 10,000 years, and historic seismic data is lacking,

Evidence from prehistoric landslides and flooding indicates that such events today could produce unsafe conditions. Pleistocene landslides may have been triggered by excessive rainfall, and most of the recent debris flows took place along the upper Owyhee River. Near the Owyhee reservoir, major soil and rock movements could threaten the dam.

Vale has endured floods from Bully Creek since the early 1900s. Even after placement of a dam in 1915, the community was periodically inundated during periods of very heavy rains and melting snow.

Natural Resources

The presence of precious metals within the McDermitt and Lake Owyhee fields is linked to volcanic layers, which were covered by lake and stream deposits then faulted. The subsequent invasion by mineral laden-thermal fluids along fissures and cracks emplaced gold, uranium, mercury, and other metals.

Mining on federal land in the western United States is regulated by a labyrinth of laws, but the basic comprehensive legislation hasn't been revised since Congress passed the General Mining Act in 1872, responsible for opening the West to settlement. That law allows any person to obtain a title to a claim on public lands for a nominal fee, and the number of claims a company or individual can hold is unlimited. Over 30,000 gold requests have been filed on eastern Oregon alone. The act contains no environmental protection or clean-up requirements, and out-of-state companies are permitted to claim mineral rights on federal lands.

Uranium and Mercury

The Opalite Mining district on the southeast plateau was the richest source of mercury in the western hemisphere. The Opalite, Bretz and Aurora, and Corderzo (Nevada) prospects produced 12,367 flasks, or close to one million pounds of mercury, between discovery of cinnabar in 1914 and mine closure in 1961. The rising price of quicksilver prior to World War II, when large amounts served as ballast in submarines, temporarily stimulated output. Visible waste piles from the abandoned mining facilities are now subject to reclamation.

Gold

Although there are no documented accounts of past mining, Mark Ferns notes old workings on Red Butte dating from the 1880s, and upper Cow Creek and Jordan Creek in the DeLamar district of adjacent Idaho saw notable gold and silver production at that time.

South of Vale, Grassy Mountain is currently being targeted for exploration by Seabridge Gold of Toronto, Canada, which is pushing forward to complete its mining applications. As stated in a 2009 Seabridge report, the "project is believed to represent one of the most studied unpermitted precious metal projects in the United States [and] It is the opinion of [a technical assessment] that if this property were located in Nevada, it would have been mined, milled, and reclaimed." Along with gold prospects at Grassy Mountain, Seabridge is evaluating those at nearby Red Butte, Quartz Mountain, and Mahogany Mountain.

Bentonites and Zeolites

Bentonites (decomposed volcanic ash) and zeolite minerals are produced from localities south of Adrian. Mined from the Miocene Sucker Creek Formation, high-swelling clay-ash bentonites are utilized primarily as sealants in landfills and waste sites and for oil and gas drilling mud. Limited quantities of potassium-rich zeolites from the same formation were first mined in 1975, and commercial uses are now being developed by Teague Mineral Products. Because of their ability to absorb heavy-metals, zeolites are used extensively in industry and agriculture.

Geothermal Resources

Even though temperatures are not high enough to generate electricity, hot springs and near surface thermal waters near Vale and along the Owyhee River canyon reach 200° Fahrenheit. Currently resources near Vale fill the needs for a variety of agricultural projects such as mushroom plants and greenhouses. Vents in the Vale Hot Springs Known Geothermal Resource Area reach the surface along faults cut through late Miocene to Pliocene lake sediments. Where spring waters are 120° in the Owyhee canyon, pools are used by the public for bathing.

Surface and Groundwater

On this arid plateau, surface water is largely intermittent, and only those stream segments augmented by springs or summer downpours are perennial. Because long-term stream gauge records are lacking, data on flow is scant. Most of the stations established prior to 1950 have been discontinued, and of the four remaining, which are operated by the U.S.G.S., two are associated with reservoir outlets. Of the two others, one is on the Malheur

Owyhee Reservoir is enclosed by walls of Owyhee Basalt and rhyolite. The view is looking westward toward Grassy Mountain. (Photo courtesy Condon Collection)

Coffeepot Crater (above, left), on the northwestern edge of Jordan Craters in Malheur County, generated the bulk of basalt, which is characterized by smooth, billowy pahoehoe lava. (Photos courtesy J. Wozniak)

River, and the other on the Owyhee. Readings indicate the Owyhee River discharges at an average rate of 400,000 acre-feet annually.

As with other volcanic regions of eastern Oregon, basalts or stream alluvium are sources for groundwater, but most have only moderate to low yields. Groundwater occurs primarily within fractures and interbeds between flows of Miocene volcanics, but little is known about the chemistry, depth, and movement between units. A similar situation exists with deeper Miocene lakebed sediments. Shallow alluvial aquifers along stream valleys vary from area to area with no documented long-term fluctuation. Recharge comes from rainfall, agricultural sprinklers, and leaking irrigation canals.

Industrial food processors and municipalities rely principally on groundwater wells, while irrigators use surface flows and dam storage. However, even with reservoirs and an intricate system of canals, the supply for agriculture has been inadequate since the 1960s, and in dry years with low water levels, the scarcity becomes acute. In 1959 the Oregon State Engineer declared Cow Creek Valley to be a critical groundwater area, restricting further appropriations. In addition to diminishing in quantity, shallow aquifers are contaminated by coliform bacteria and agricultural chemicals. Mercury, which occurs naturally, is present in Jordan Creek and in the Antelope and Owyhee reservoirs.

Geologic Highlights

Jordan Craters Volcanic Field

Although several areas of Malheur County feature dramatic terrains, the Jordan Craters field, approximately 15 miles northwest of Jordan Valley, is of particular interest because the volcanic surfaces are so well-preserved. The 28-square-mile area has been designated a Natural Resource Area in order to maintain its unique character.

Within the Cow Creek drainage, a ridge of small spatter cones aligned along a fault trending toward Coffeepot Crater first erupted 9,000 years ago. Basalts from Coffeepot cone, one of the largest of the Jordan Craters complex, covered the ground with fluid pahoehoe lava that produced a lumpy, ropy surface of ridges, pits, channels, and fractures in the crust. One wall of Coffeepot Crater was torn away by explosions, after which volcanic debris filled much of the remaining 260-foot-deep pit.

By tracing the Cow Creek channel, the major Coffeepot flow followed a meandering course, which caused a hollow tube system to develop inside the cooled crust once the lava had drained away. Two large pits near the main crater open into a maze of lava tunnels. Immediately south of Jordan Craters, Clarks Butte, Rocky Butte, and Three Mile Hill are small shield cones that erupted about the same time.

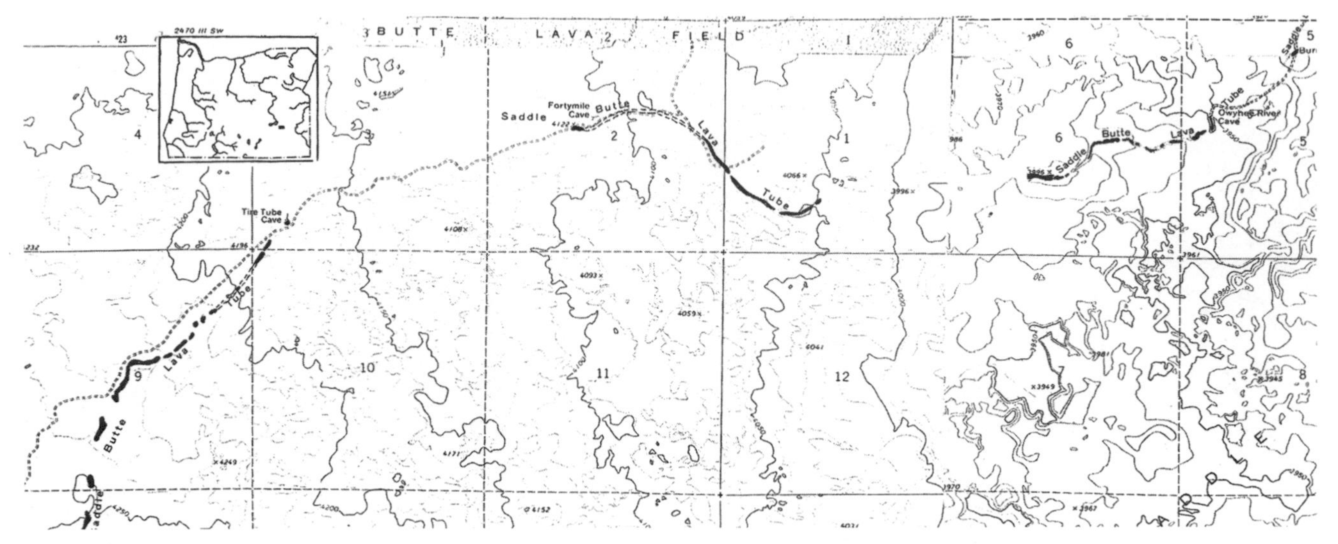

Near Jordan Craters, Saddle Butte caves follow a lengthy sinuous path that has ruptured in places. The caves are particularly hazardous to explore because of the tendency for roof portions to fall. Oversight by the Bureau of Land Management has established federal protection for the Saddle Butte field.

Saddle Butte Volcanic Field

The Miocene Saddle Butte cones, that erupted the Hunter Creek basalt around 16.6 million years ago, produced sinuous lava tube caves that can be traced over eight-and-one-half miles. Before parts of the roof collapsed, the tunnels ran in a continuous meandering chain. Exposed sections of the channel then became winding open trenches, the longest measuring well over one-half mile in length.

Owyhee River and Succor Creek Canyons

Traversing the dry desert plateau of southeast Oregon, canyons of the Owyhee River and Succor Creek dissect some of the state's most scenic topography. Larry Kittleman has related the geologic story particularly well in his guide to the sedimentary and volcanic rock formations. The Owyhee River, dammed in 1932, backed up into a 52-mile-long reservoir.

Ash, lavas, and tuffaceous fluvial and lake sediments form the colorful layers of the Sucker Creek Formation. The fine-grained tuffs are thin and nearly white, while the coarse-grained, massive intervals are green. Weathered exposures of the Sucker Creek are oxidized to yellow, brown, or reddish brown, common in badlands topography. Along lower Succor Creek, the ridges and hills composing the landscape are mostly rocks of the Sucker Creek Formation, but in the upper canyon the Leslie Gulch ash-flow tuffs and the Jump Creek rhyolites have been eroded into cliffs. The red, purple, or grayish-red Jump Creek rhyolites cap the rim of the plateau on both sides of the gorge, with thick Leslie Gulch tuffs exposed below.

Succor Creek winds through the narrow Owyhee gorge, which, over the millennia, was deeply cut into the colorful volcanic and sedimentary rocks of southeast Oregon. (Photo courtesy Oregon State Highway Commission)

The differences between the formational and geographic names came about when the spelling of

The road into Leslie Gulch drops more than 2,000 feet down to the Owyhee Reservoir, a distance of nine miles. The spectacular pinnacles and spires lining the canyon wall are eroded into thicknesses of tuff. (Photo from Kittleman, et al., 1965)

the creek was changed to Succor by the U.S. Board of Geographic Names, but the geologic formation retained the original Sucker Creek designation.

In the Owyhee River canyon, rhyolitic lavas from the McDermitt and Lake Owyhee fields are extensively exposed in the imposing cliffs of pink welded ash that tower over the stream. In the upper stretches, the river cut a deep notch in the rhyolites at Hole-in-the-Ground. Erosion here has exposed a volcanic pipe, which served as a conduit for the molten rhyolite to reach the surface before it cooled and solidified in the cleft.

On the eastern shore of Lake Owyhee, Owyhee Basalt is evident in the 1,000-foot-high cliffs. The dark lower flows break into blocks, while the light-colored upper layers weather to flat slabs. Thin white bands in the basalt are pumice, and the bright red color is where Owyhee lavas baked the sediments of the Sucker Creek below. Forming the prominent Owyhee Ridge, multiple gray flows are interlayered with dark red to purple vesicular basalt (scoria).

North of the reservoir, yellow, orange, and brown iron-stained beds of the Deer Butte Formation are responsible for the resistant knobs of Deer Butte and Mitchell Butte, prominent topographic features.

Leslie Gulch

The narrow steep road from Succor Creek gorge leading to Leslie Gulch is lined with brown, cream, and white minarets and spires of the Leslie Gulch Tuffs. The tuffs are more than 2,000 feet thick in places and cover an area of about 100 square miles. Molten rock and hot gasses that erupted 15.5 million years ago from the Mahogany Mountain caldera, the oldest in the Lake Owyhee volcanic field, cooled to trap gas bubbles (geodes). Locally the distinctive glassy rounded nodules, which filled the cavities, are called Succor Creek eggs.

Pillars of Rome

Approximately four miles northwest of the small community of Rome, cliffs called the Pillars of Rome have been carved into the 400-hundred-foot-thick, nearly horizontal lakebed sediments of the Rome Formation deposited by Lake Idaho. The Miocene lake filled when waters of the Snake River were blocked by a lava dam and backed up into the Owyhee basin. Exposed over an area of 100 square miles, the sediments were sculpted into high vertical pinnacles and pillars that resemble Roman ruins.

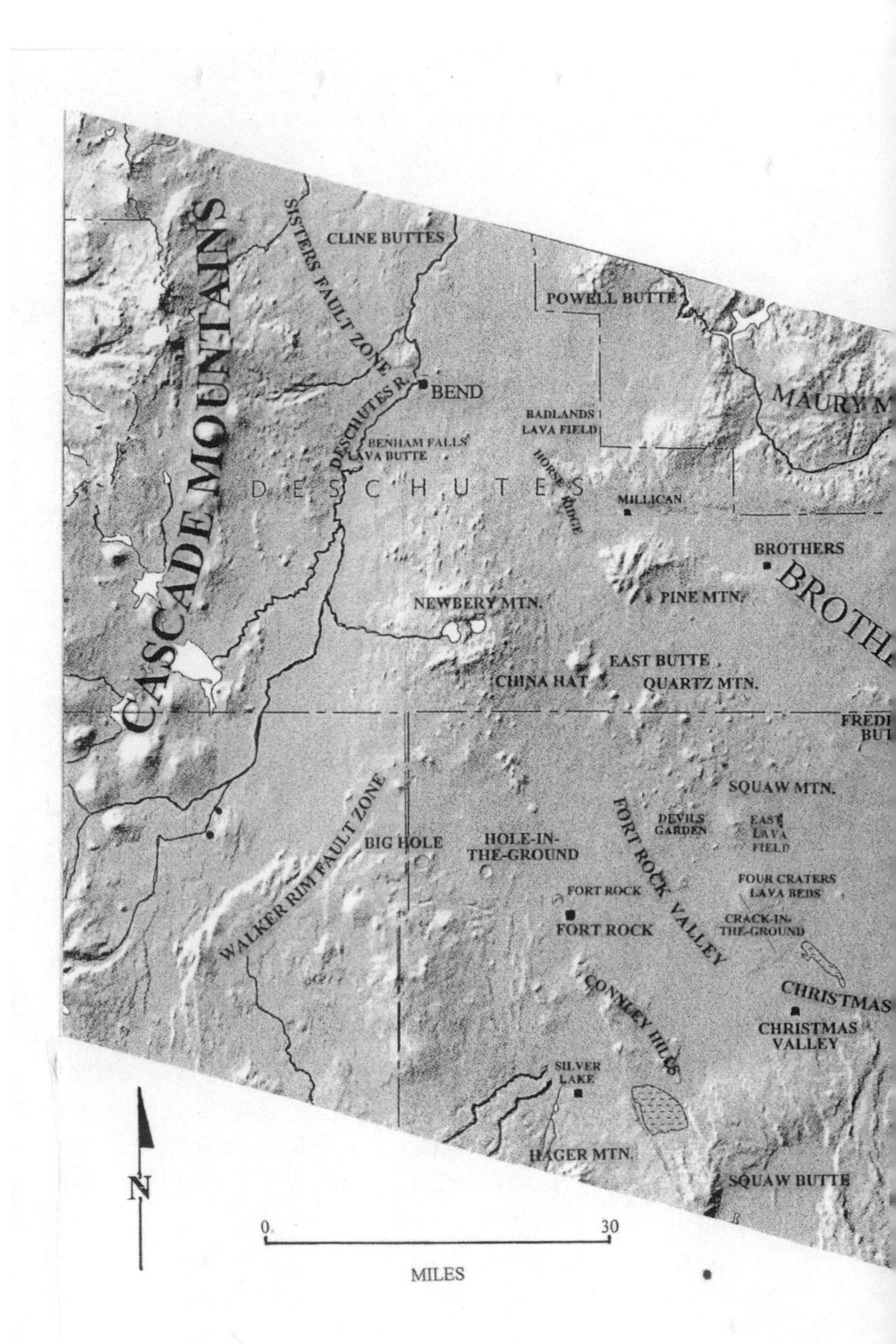

Physiographic map (After Loy, et al., eds., 2001)

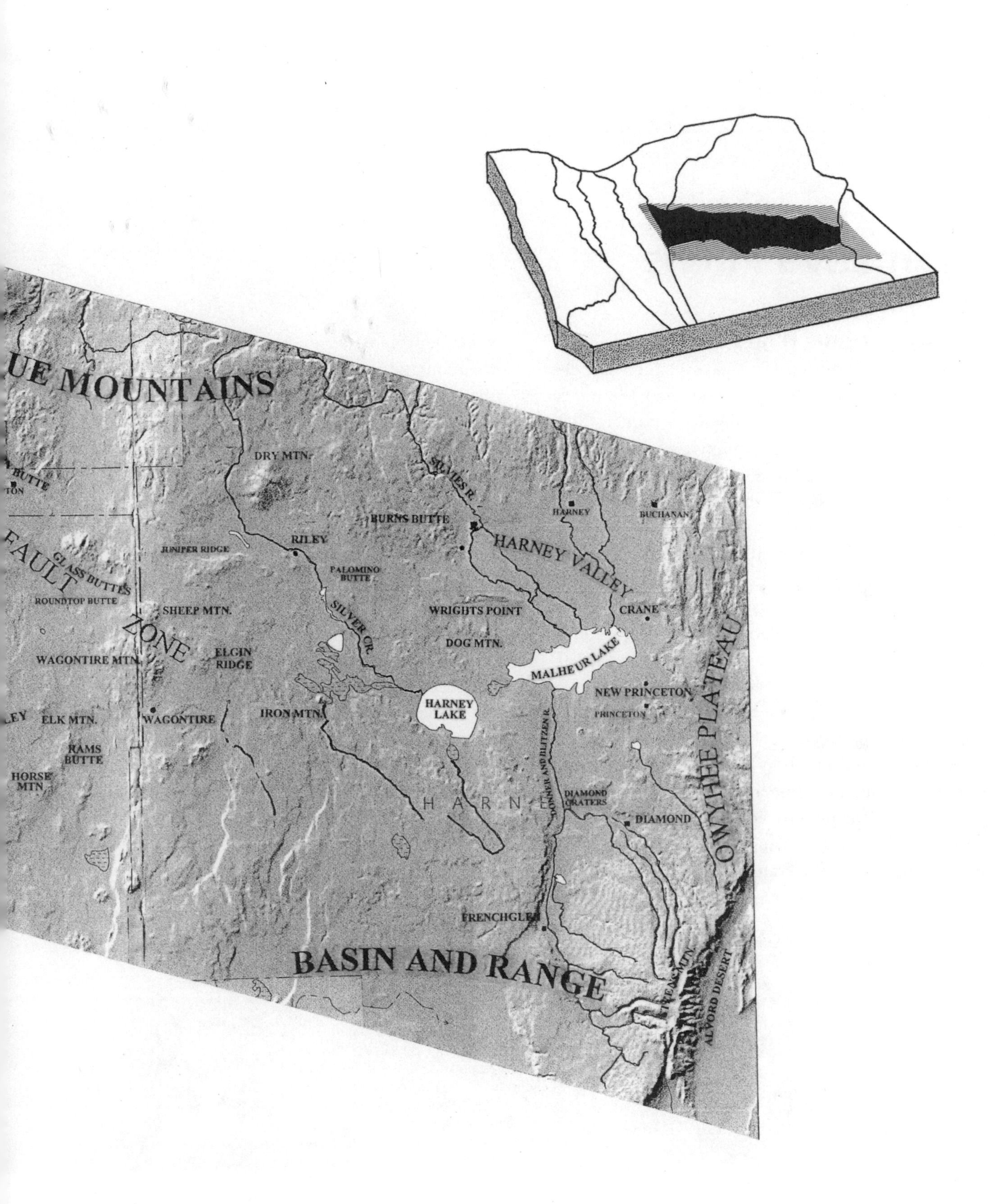

UE MOUNTAINS
DRY MTN.
BUTTE
SILVIES R.
HARNEY
BUCHANAN
BURNS BUTTE
RILEY
HARNEY VALLEY
JUNIPER RIDGE
FAULT
GLASS BUTTES
PALOMINO BUTTE
ROUNDTOP BUTTE
SHEEP MTN.
SILVER CR.
WRIGHTS POINT
CRANE
ZONE
DOG MTN.
MALHEUR LAKE
WAGONTIRE MTN.
ELGIN RIDGE
OWYHEE PLATEAU
NEW PRINCETON
HARNEY LAKE
PRINCETON
ELK MTN.
WAGONTIRE
IRON MTN.
RAMS BUTTE
HORSE MTN.
DONNER AND BLITZEN R.
H A R N E
DIAMOND CRATERS
DIAMOND
FRENCHGLEN
BASIN AND RANGE
ALVORD DESERT

High Lava Plains

Landscape of the High Lava Plains

The geologically young High Lava Plains is an elevated desert plateau 50 miles wide and 150 miles long. Centrally located in Oregon, it borders five other regions. The Blue Mountains lie to the north, the Basin and Range and Owyhee Uplands to the south and southeast, and the Cascade Range is to the west. The northwestern corner shares a small boundary with the Deschutes-Umatilla province.

The overall topography is smooth with moderate relief, averaging just over one mile above sea level. Elevations range from a high of 7,984 feet at Paulina Peak, within the Newberry Volcano center, to 4,080 feet in Harney Basin. Between these two features, the level plain is broken sporadically by low rounded domes and steep-sided, flat-topped ridges.

Annual rainfall amounts are modest, and because of the xeric climate, juniper, sagebrush, and grasses predominate. Deeply eroded topography is lacking, and most tributary streams are seasonal, without well-defined channels. Water in the largest two lakes in Malheur and Harney basins is supplied by precipitation and snowmelt and from the Donner and Blitzen River, Silver Creek, and Silvies River. With its headwaters on Steens Mountain, the Donner and Blitzen drains around 200 square miles. Silver Creek and Silvies River both originate in the Blue Mountains. The watershed of Silver Creek is small, but Silvies River drains close to 950 square miles.

Past and Present

With the exception of Israel Russell's 1905 reconnaissance of central Oregon that included an extensive look at both the geology and water resources, early reports on the High Lava Plains concentrated on Harney Basin or Newberry Volcano. Arthur Piper and others of the U.S.G.S. published on the Harney Basin in 1939, followed by the works of George Walker during the 1960s. Howel Williams reported on Newberry Volcano in 1935 at the same time that Phil Brogan's many articles on the geology began to draw attention to the province.

Few did more to popularize the natural history of central Oregon than Phil Brogan, who contributed hundreds of articles to the *Bend Bulletin* and Portland *Oregonian* during his thirty-year-career. His book *East of the Cascades* remains a standard source for regional information. Born at The Dalles in 1896, Brogan attended school in an old sheep cabin before studying journalism and geology at the University of Oregon. During his years in eastern Oregon, he worked with geologists throughout the northwest. Retiring, Brogan and his wife Louise moved to Denver, where he died in 1983. (Photo courtesy Condon Collection)

Norman Peterson's numerous technical and popular articles and field trip guides on the volcanic terrain make him a major contributor to the investigation of eastern Oregon's geology. Co-authored with private geologist Ed Groh, his papers describe such highlights as Crack-in-the-Ground, Diamond Craters, and Fort Rock. Receiving his degree from the University of Oregon in 1957, Peterson took a position with the Oregon Department of Geology and Mineral Industries (DOGAMI), where his career spanned 25 years. He directed the Grants Pass office until retirement in 1982. In conjunction with his work, Peterson produced mineral evaluations and maps of Harney and Malheur counties and assessments of volcanic hazards, culminating in his editing of the Lunar Geological Field Conference Guidebook. Peterson died in 1994. (Photo courtesy Oregon Department of Geology and Mineral Industries)

Raymond Hatton, who taught at Central Oregon Community College in Bend until retirement in the mid 1990s, has written several books that included the geology and natural history of the high desert. Teaching geology locally for many years, Lawrence Chitwood, with the Deschutes National Forest, also contributed articles on aspects of volcanism. Overview papers on the tectonic and volcanic development of the High Lava Plains by Martin Streck at Portland State University, Anita Grunder at Oregon State University, and Brennan Jordan at the University of South Dakota bring together earlier research and present a current perspective. Robert Jensen's 2006 roadside guide gives a detailed look at the events surrounding Newberry Volcano.

Overview

The High Lava Plains province offers a remarkable variety of exceptionally young lava flows, cones, and buttes which punctuate the otherwise subdued topography. Volcanism that began in the Miocene and continued into the Holocene is often bimodal in composition—lavas vary from rhyolitic to basaltic in composition. Additionally, the locus of rhyolitic activity moved progressively across the province from southeast in the Harney basin to Newberry Crater in the northwest. By contrast, the basaltic lavas were not age progressive, but often spread widely as lava fields, and most vents are located near Bend.

Volcanism didn't cease with the onset of Pleistocene glacial conditions, and the violent interaction of magma and water in pluvial lakes created unusual eruptive features such as circular pits and tuff rings. Similar to those in the Basin and Range, Ice Age lakes in the Fort Rock and Christmas Lake valleys were ephemeral and diminished with post-glacial warming. The broad lacustrine expanses drew mammals and birds to the habitat.

The diversity of eruptive features on the High Lava Plains is its greatest resource. Lava tubes, acres of pahoehoe and aa flows, domes, and the immense Newberry stratocone and its surrounding complex are among the most outstanding sights in the Pacific Northwest.

Norm MacLeod together with George Walker completed the first comprehensive *Geologic Map of Oregon* in 1991. MacLeod was born in Victoria, British Columbia, but moved to the United States at a young age, where he grew up, for the most part, in southern California. His graduate work in the University of California system culminated with a PhD from Santa Barbara in 1970. After 25 years with the U.S.G.S., MacLeod retired in 1986 to live along the Nisqually River in the Washington Cascades. (Photo courtesy N. MacLeod)

Geology

Cenozoic

The oldest rocks in the High Lava Plains are Miocene lavas and tuffs, although Paleozoic and Mesozoic terranes, present in the adjacent Blue Mountains, are presumed to be buried beneath the younger volcanic layers. A veneer of late Miocene to Holocene lavas, obsidian flows, and ash from numerous volcanic vents is widespread.

Bimodal Volcanism

Bimodal volcanism and age progression of the eruptions characterize the geology of the High Lava Plains. With bimodal volcanism, lavas are distinctly different in composition, varying from dark-colored fluid basalt to light-colored viscous rhyolites. Basaltic lavas tend to have a deeper source in the mantle, are extremely hot, and erupt as runny flows. The lower temperature rhyolites originate from shallow depths and erupt with explosive force. The association between basalts and rhyolites is not common and usually occurs where the earth's crust thins due to tension and stretching. Basalts elsewhere appear during an early stage of eruption, whereas rhyolites emerge late in the cycle. But in this province, the rhyolitic eruptions are older, although the younger basaltic lavas are greater in volume.

Age Progression of Volcanic Eruptions

One of the most striking aspects of volcanism here is the geographical age progression of the rhyolitic eruptions, first recognized in the 1970s by Norm MacLeod, George Walker, and Ed McKee. Lavas from a broad belt of approximately 100 centers became

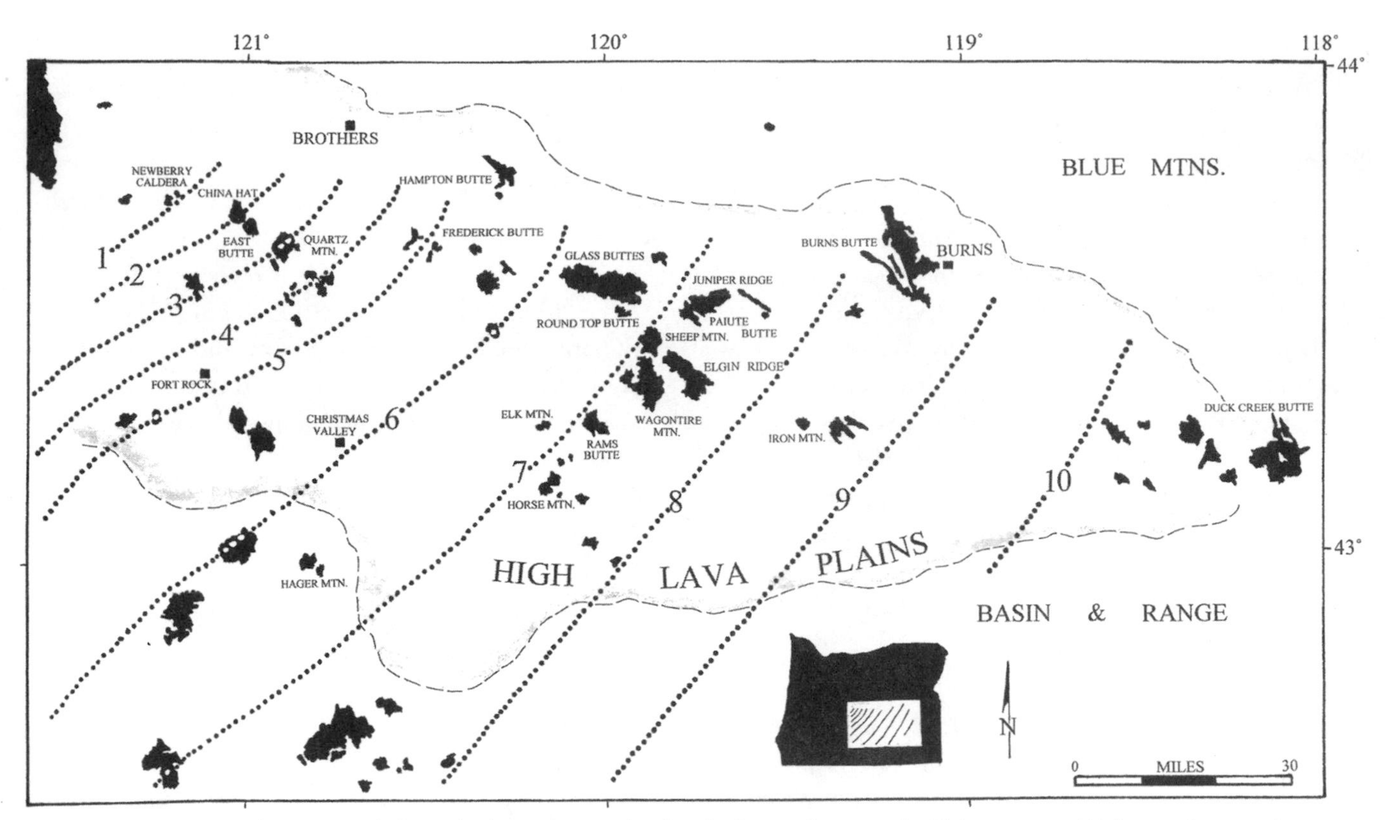

Eruptive centers moved progressively from the late Miocene rhyolites in the southeast to the Pleistocene and Holocene flows and ash in the northwest. Contoured (dotted) lines show the movement of volcanic activity from east to west in millions of years. (After Jordan, Streck, and Grunder, 2002; MacLeod, Walker, and McKee, 1975; Walker and Nolf, 1981)

younger westward from the Owyhee Uplands and across the Lava Plains toward the Cascades. The uniform decrease in age prograded at 20 miles every million years between 10.5 to 5 million years ago, then slowed to 9 to 18 miles after that. The oldest lavas at Duck Creek Butte east of Harney Basin date back over 10 million years, while the most westerly Newberry complex erupted as recently as 1,300 years ago. A wide zone of silica-rich rhyolitic domes in the Basin and Range between Beatys Butte and Yamsay Mountain displays a similar east to west progression.

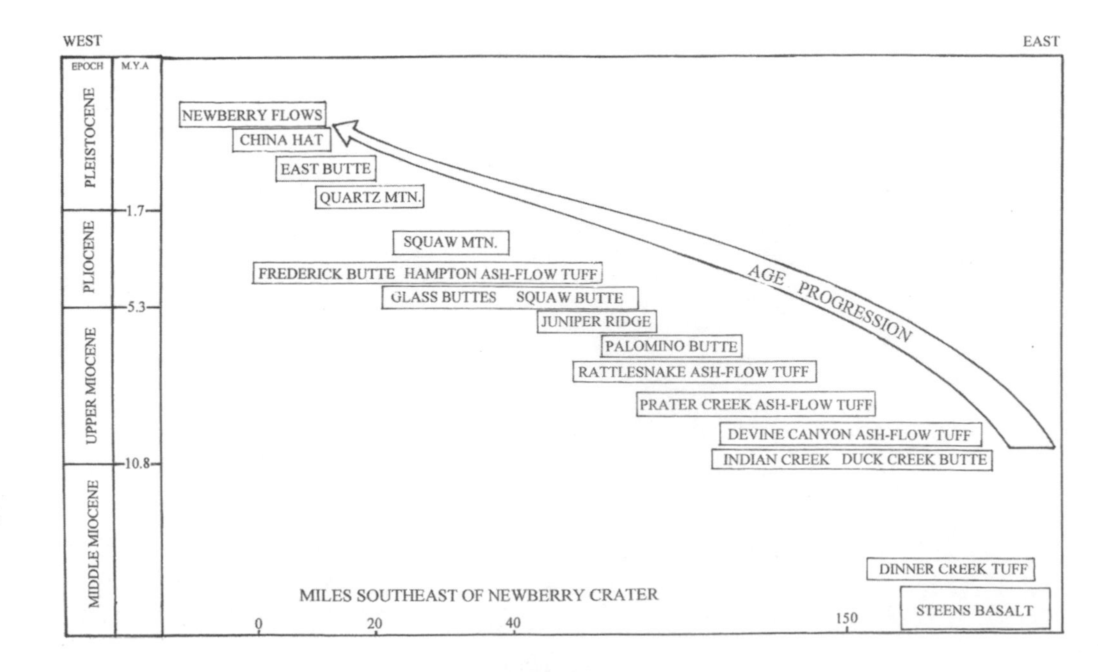

Tertiary stratigraphy of the High Lava Plains. (After Johnson and Grunder, 2000; Jordan, Streck, and Grunder, 2002; MacLeod, Walker, and McKee, 1975; MacLeod, et al., 1995; Sherrod, et al., 2004; Walker, 1979)

A number of theories address the progressive nature of the eruptions. Whereas earlier ideas focused on the movement of tectonic plates, subplate geometry, and crustal thinning, current research favors the activity of a hot spot. Around 17 million years ago, a hot spot or mantle plume (Yellowstone), centered at the junction of Oregon, Nevada, and Idaho, melted crustal rocks which reached the surface as flood basalts. The heat source for mantle plumes is not well understood, but when the balloon-like head of the Yellowstone hot spot encountered the edge of the North American continent, it began to flatten and divide before being sheared off. One of the tongues migrated northward from Steens Mountain into the Columbia plateau, and a second spread westward across central Oregon, triggering the age progression on the High Lava Plains and Basin and Range. In 2004, Camp and Ross proposed that a third tongue projected south across Nevada as a rift zone of volcanics, while the severed stem left a track across southern Idaho as the Snake River Plain.

Rhyolitic Volcanic Centers

Duck Creek Butte and Harney Basin

The oldest rhyolitic eruptions across the High Lava Plains began near the extreme southeastern border at Duck Creek Butte 10.4 million years ago. In contrast to the silica-rich rhyolitic cones which dominate the western segment of the vents, the Duck Creek Butte lavas were more andesitic and dacitic in composition. From the Basin and Range, the north-northeast-striking Steens Mountain fault cuts into Duck Creek Butte as do faults within the Brothers system.

Intersected by the Brothers fault zone, Harney Basin was the next in the progressive sequence to be impacted by eruptions. Volcanic and sedimentary deposits in the 3,000-square-mile basin reflect environmental changes that took place 17 million years ago with the onset of Miocene basalts from Steens Mountain. But as volcanism shifted westward, the region was overwhelmed with outpourings the Devine Canyon, Prater Creek, and

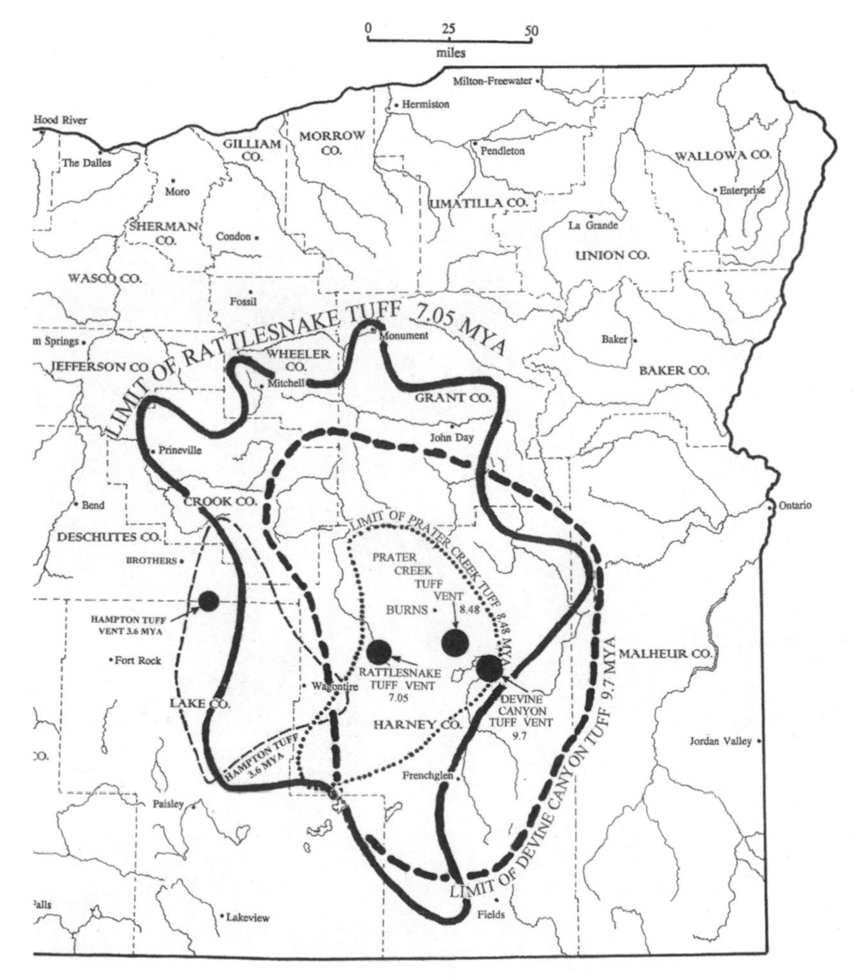

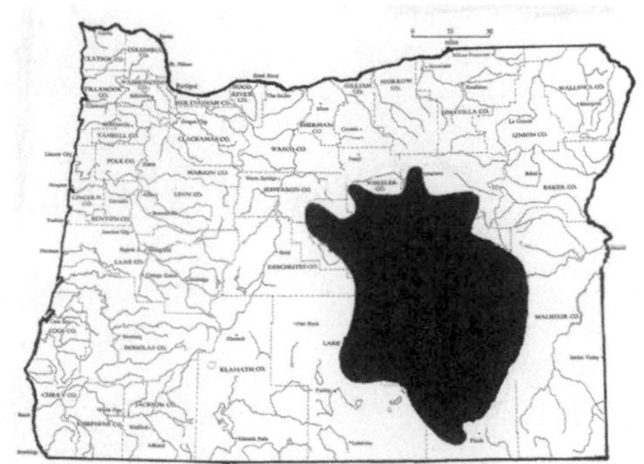

Source vents for the Hampton, Devine Canyon, Prater Creek, and Rattlesnake tuffs are situated in the Harney Basin southwest of Burns. It has been estimated that a huge caldera with a diameter of some 12 miles across would have been necessary for the erupted volume of the Rattlesnake Tuff. For its size, the Rattlesnake is one of the most far-travelled ignimbrites known, reaching 100 miles from the Harney Basin to John Day and Prineville. (After Rytuba, et al., 1990; Streck and Ferns, 2004; Streck and Grunder, 1997; Walker and Robinson, 1990)

Rattlesnake tuffs, representing the greatest volume of rhyolites on the High Lava Plains. Derived from local vents, air-borne pyroclastics and ash-flows of the Devine Canyon Tuff, dated at 9.7 million years ago, the overlying Prater Creek Tuff, and the capping 7-million-year-old Rattlesnake Tuff were originally lumped together as the Danforth Formation. Today, the three distinctive stratigraphic layers are recognized over vast regions of southeast and central Oregon. The greenish-gray Devine Canyon Tuff has been traced from Steens Mountain to Paulina Valley, an area of 7,000 square miles, whereas the Prater Creek ash-flow tuff is limited almost completely to exposures in the Harney Basin.

Palomino Buttes, Juniper Ridge, Squaw Butte, and Glass Buttes

Palomino Buttes at 6.29 million years of age, Juniper Ridge dated around 5.75, and Squaw and Glass buttes at 5 million years ago continue the northwestward-trending line of younger rhyolitic centers. Issuing from clusters of vents, Glass Butte, Little Glass Butte, eastern Glass Buttes, and Roundtop Butte are low domes, aligned in a west-northwest direction for about 12 miles. The domes were constructed during an eruptive phase that may have lasted over a million years. Significant amounts of obsidian at Glass Butte give the complex its name. Besides the common black obsidian, there are mixtures of red and black as well as iridescent silver. Obsidian or volcanic glass has no internal crystalline structure, and its failure to form crystals is due to the rapid cooling of the lavas. Small amounts of quicksilver have been mined from the volcanics of the buttes.

Hampton Butte, Dry Butte, Frederick Butte, Quartz Mountain

Rhyolites from vents near Hampton and Dry buttes erupted close to 8 million years ago. These lie along the northern margin of the High Lava Plains and the most northerly faults in the Brothers zone. A second period of volcanism around 3.8 million years ago produced the Hampton Tuff, an ignimbrite originating from Frederick Butte. Thin layers of this ignimbrite are readily distinguished by the abundance of dark pumice fragments. Frederick Butte is a semi-circular caldera, in which older eruptions may have built a dome before the center collapsed. Subsequently, the pit filled with lava from dikes along Last Chance Ridge.

Situated west of Frederick Butte and approximately 3 million years younger in age, Quartz Mountain consists of two domes. Exposures of the rhyolite flows display streaked layers of dark brown to black obsidian and fresh glassy surfaces.

East of Newberry, the 800,000-year-old China Hat is a rhyolitic dome that erupted near the western end of the province. (Photo courtesy J. Mooney)

At the westward termination of the age progression, Newberry caldera is a stratovolcano with an eruptive life that began with voluminous flows 300,000 years ago. Successions of silica-rich rhyolitic lavas interspersed with andesitic flows and tuffs erected the summit 80,000 years ago, after which collapse produced a caldera similar to that at Crater Lake. More than a dozen postglacial eruptions are interlayered with the distinctive 7,700-year-old pumice and ash from Mt. Mazama to the southwest. (Shaded relief image from 10 meter digital elevation model; Jensen and Donnelly-Nolan, 2009)

Newberry Volcano

Located on the western edge of the High Lava Plains, eruptions from Newberry Volcano are the youngest in the age-progressive cycle. The long outer slopes of Newberry are sprinkled with over 400 parasitic cinder cones, aligned in a 20-mile–long northwest rift zone toward Bend. The most northerly is Lava Butte, whose flows 7,000 years ago entered the Deschutes River channel, damming and diverting the water. A two-mile-long east rim system of 19 rhyolitic domes, flows, and small explosion craters, arranged in a radial pattern around the central Newberry Crater, are blanketed by a thin veneer of Mt. Mazama ash.

Because of Newberry's position east of the Cascade volcanic arc and because of its history of both rhyolitic and basaltic lavas, the nature of its eruptive cycle has been difficult to characterize. In comparing Newberry with the similar Medicine Lake volcano in northern California, Julie Donnelly-Nolan at the U.S.G.S. has concluded that both are rear-arc Cascade volcanoes related to subduction in a region of extensional tectonics.

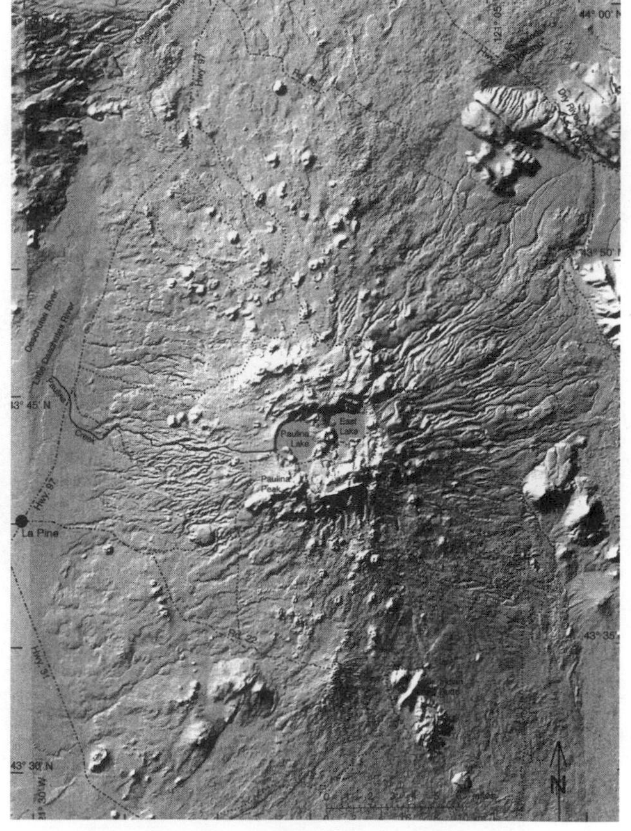

Basalt Volcanic Centers

Miocene to Holocene basalts punctuate the relatively flat surface of the High Lava Plains with distinctive structures and landforms. Eruptions of the fluid basalts are not age progressive like the rhyolites, and most originate from a central vent or from an aligned vent system. The older exposures are found along escarpments, whereas the Quaternary flows, with thicknesses up to 15 feet, cover wide areas that range in size from the Devils Garden at 45 square miles, the Potholes and Badlands close to 25 square miles, and the smaller Four Craters at 12 square miles.

Basalt fields share a number of features such as the alignment of fissures, the age of the eruptions, and the composition of the lavas. Four Craters, East Lava field, Devils Garden, and the Potholes originated from northwest-trending fissures. Vents in the Four Craters field also extend northwestward toward craters in the Lava Mountain shield volcano complex, which produced the East Lava field. Devils Garden and the Potholes are both similar in composition and age, erupting around 50,000 years ago, whereas the Badlands, built by Lava Top Butte, is slightly older at 74,000 years. The 1,000 year date for the Four Craters eruption has been inferred by assessing the vegetation cover and surface erosion.

Some distance from the basalt layers at Bend, Diamond Craters lies at the east edge of Harney Basin. Dated at 17,000 years old, the complex encompasses over 30 cinder cones within a six-mile circular

On the western High Lava Plains, the majority of Miocene to Holocene basaltic centers are clustered southeastward from Bend at the Badlands, the Potholes, the Devils Garden, the East Lava field, and the Four Craters. (After Jensen, et al., 2009; Jordan, Streck, and Grunder, 2002; Meigs, et al., 2009)

The phenomenon of inflated lava is well-illustrated by the basalt fields near Bend. As described by Larry Chitwood, inflation is an upward swelling or doming, known as a tumulus, that results after the initial thin lava flow, some eight to ten inches thick, has spread out, then slowed beneath a hardened outer crust. When molten material continues to move through the structure, the surface rises or inflates to thicknesses from 5 to 60 feet before cracking to produce a very rough topography. Ridges, pits, or raised, broken and tilted basaltic slabs, and caves are signs of internal swelling and inflation. (Photo courtesy Condon Collection)

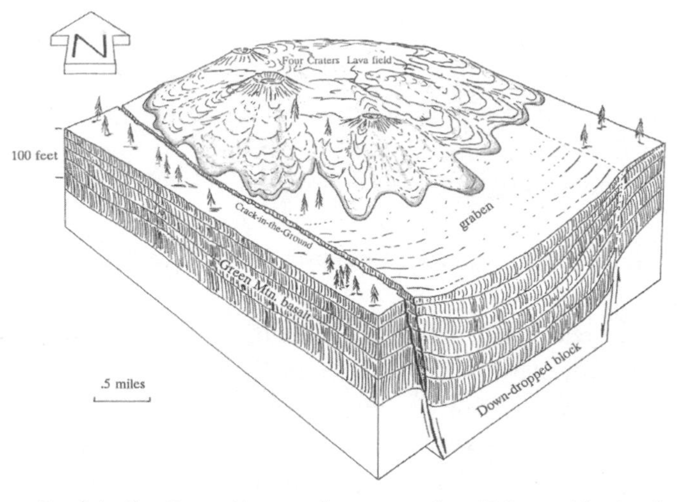

Crack-in-the Ground is a northwest trending 15-foot-wide, 2-mile-long rift that opened in 740,000-year-old basalts from the Green Mountain shield volcano. The crack extends to the Four Craters field where the flow entered the north end. Later subsidence of the Green Mountain basalts produced a lengthening of the crack toward Fort Rock basin. Silvio Pezzopane and Ray Weldon suggest that the fracture coincides with the normal northwest fault pattern in the Basin and Range. (Diagram after Peterson and Groh, 1964; photo courtesy Oregon Department of Geology and Mineral Industries)

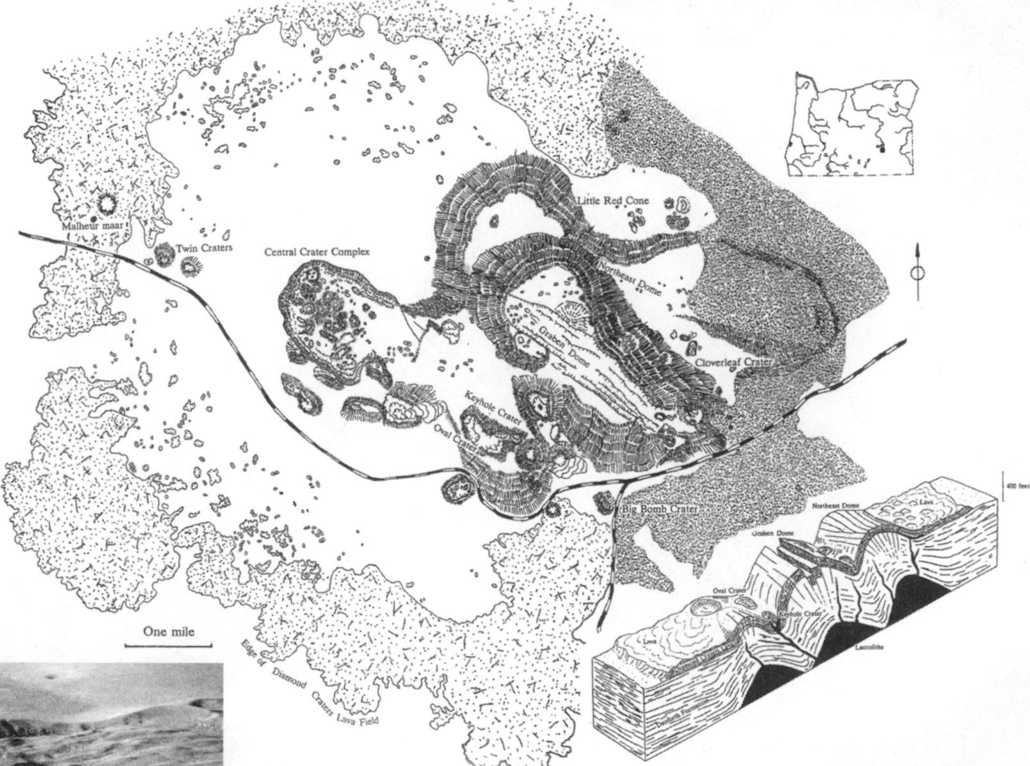

At Diamond Craters, the middle and most notable of the large cones is the Central Crater field (above), a moonscape of undulating pahoehoe and pits. (After Peterson and Groh, 1964; Photo courtesy Oregon Department of Geology and Mineral Industries)

area, where magma welled up from an alignment of vents, and intrusions bowed the ground into three low domes. After the doming, violent eruptions expelled showers of football-shaped bombs that range from pea-sized to as much as two feet in diameter. The debris was briefly airborne, then fell, encircling the crater, which fractured and collapsed.

Structure and Faulting

Three discontinuous fault zones—the Brothers, the Sisters, and the Walker Rim—intersect the High Lava Plains. The east end of the lengthy Brothers fault system emerges from the Steens Mountain escarpment and terminates near Newberry Crater. As with the Sisters and Walker Rim faults, the westerly end point is buried beneath Pleistocene lavas.

The Brothers strike-slip faults have been interpreted as the surface expression of the boundary between the thinner crust of the Basin and Range and the thicker crust of the Blue Mountains. Individual faults are irregularly spaced at quarter-to-two-mile intervals with displacements of less than 50 feet.

Projecting from Black Butte in the Cascades toward Bend, almost 50 separate faults occur within the Sisters zone. This system is around 10 miles wide and 40 miles long, and the most extensive thread is the 30-mile-long Tumalo fault.

Pleistocene Precipitation and Pluvial Lakes

The Pleistocene Epoch, which ended some 10,000 years ago, was a time of continuing volcanism as the climate cooled and continental ice masses moved southward out of Canada. Evidence of glacial erosion in the High Lava Plains was thought to be lacking; however, Robert Jensen, retired from the U.S. Forest Service, and Julie Donnelly-Nolan document glacial features within the Newberry Volcano system. Erratics some distance from the caldera rim, meltwater channels on the east and west slopes, and lava flows interlayered with glacial debris suggest the crater underwent modifications during the Ice Ages.

Increased Pleistocene rainfall filled pluvial lakes at Fort Rock, Christmas Lake, and Malheur Lake. The largest of these, prehistoric Fort Rock Lake, occupied

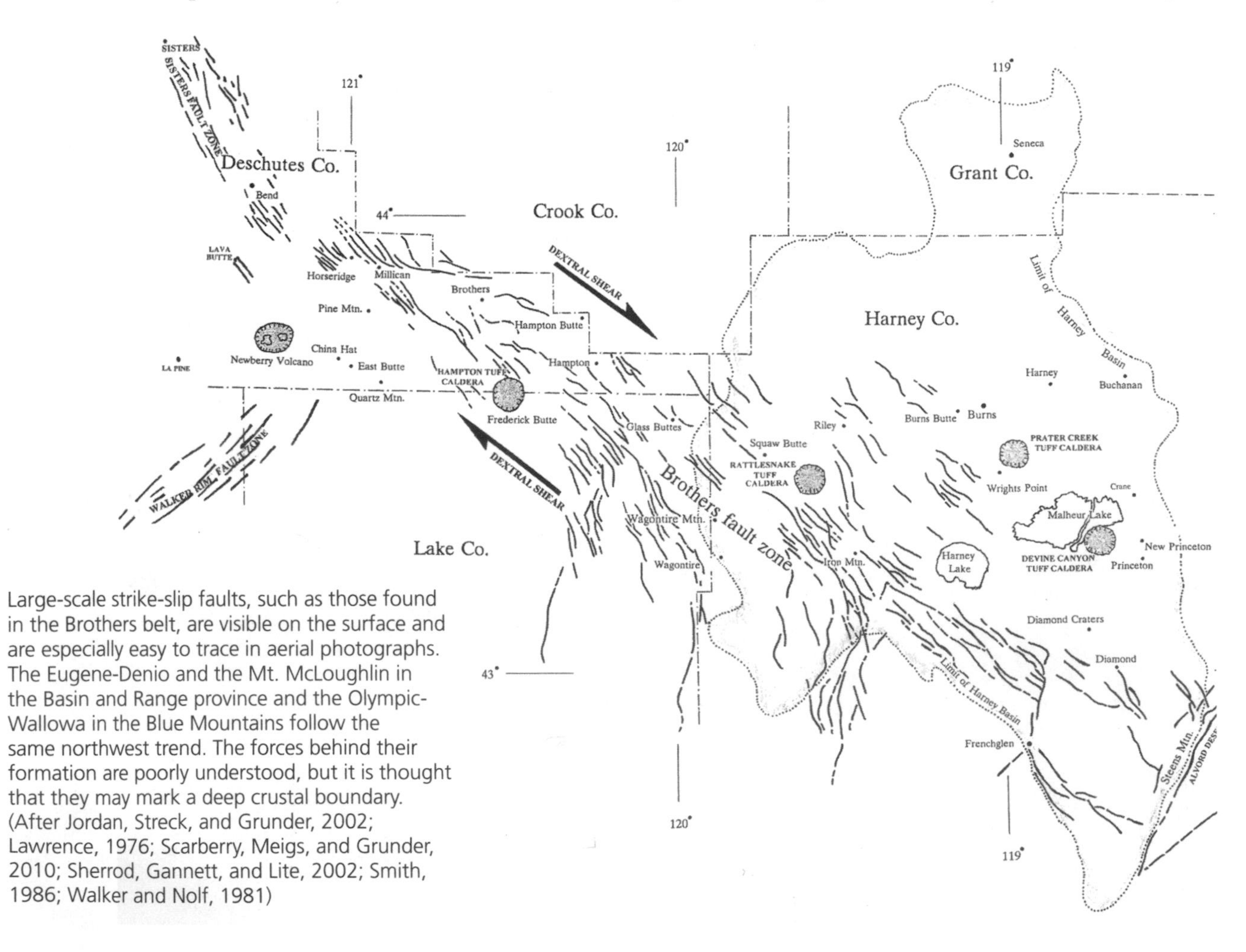

Large-scale strike-slip faults, such as those found in the Brothers belt, are visible on the surface and are especially easy to trace in aerial photographs. The Eugene-Denio and the Mt. McLoughlin in the Basin and Range province and the Olympic-Wallowa in the Blue Mountains follow the same northwest trend. The forces behind their formation are poorly understood, but it is thought that they may mark a deep crustal boundary. (After Jordan, Streck, and Grunder, 2002; Lawrence, 1976; Scarberry, Meigs, and Grunder, 2010; Sherrod, Gannett, and Lite, 2002; Smith, 1986; Walker and Nolf, 1981)

In Lake County, tuff rings and maars are striking volcanic features, formed when magma and groundwater interacted. The Fort Rock tuff ring lies within a shallow Pleistocene lake basin. The wide breach in the south rim is due to wave action breaking the thin volcanic walls. (After Peterson and Groh, 1963; Photo courtesy Condon Collection)

an irregular basin that was 40 miles wide, and only Malheur Lake rivaled it in size. During times of high water, the separate lakes became interconnected, often reaching depths in excess of 350 feet. During dry intervals, rocks, projecting as islands, became peninsulas, and arms of the lake became bays, until only isolated marshes and swamps remained. Sediments of the paleo-lakes preserve an outstanding record of Ice Age mammals, birds, and fish, as well as yielding artifacts by early man. Fossil bones in the lake sands are a distinctive shiny black color.

Except where obscured by younger lavas in the vicinity of the Devils Garden, the 30,000-year-old shorelines of Fort Rock Lake are well marked by gravel bars and beach deposits. At one point the water body included the now dry Silver and Fossil lakes. Bones of the anadromous salmon and the presence of the fresh-water snail, *Limnaea*, found at Fossil Lake, are known only in the Columbia River drainage and are evidence that the lake may have drained through a northern route to the sea.

Prehistoric Malheur Lake occupied an area from the present-day basin southwestward across Mud Lake and into Harney Lake, where the oldest fluvial sediments, which covered the ignimbrites, have been dated at 70,000 years. Paleo-shorelines of the pluvial lake trace Quaternary variations in the water levels that reached a maximum elevation around 4,000 feet. Lake Malheur began to dry by 7,500 years ago, draining into the South Fork of the Malheur River and from there to the Snake River, before the outlets near Princeton and Crane were dammed by lava. Ultimately, the lake shrank in size to small playas, marshes, and saline puddles.

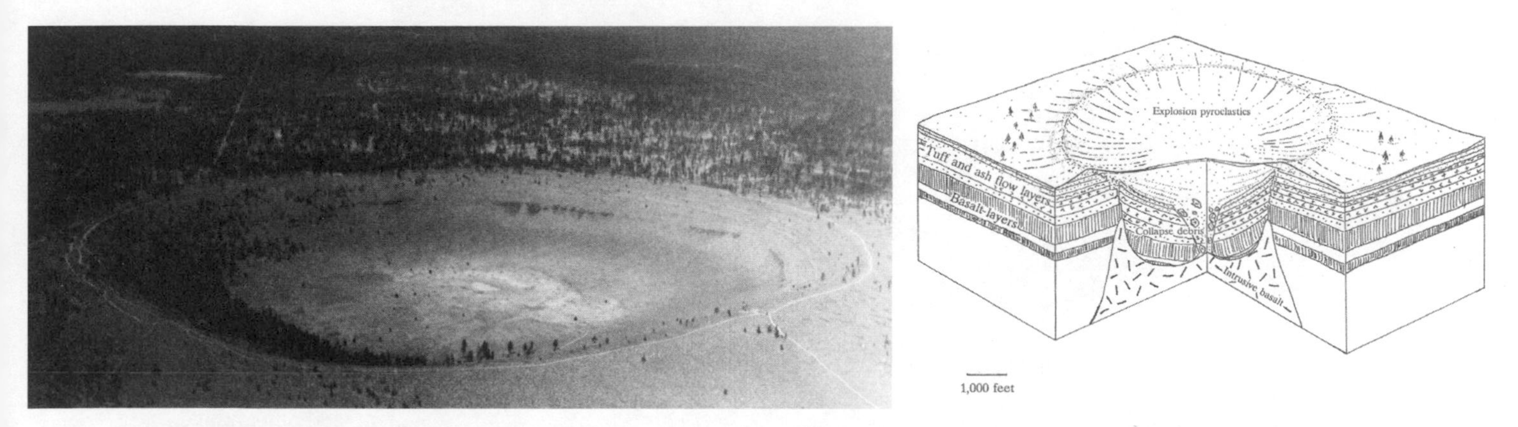

Hole-in-the-Ground is a maar of remarkable symmetry, which exploded between 100,000 to 50,000 years ago when rising magma from Newberry Volcano encountered groundwater. The pit is close to a mile in diameter, and the floor is over 300 feet below the regional ground surface. Massive collapse blocks have slumped from the rim and are overlain by layers of ash and tuff. (After Peterson and Groh, 1961; Photo courtesy Oregon Department of Geology and Mineral Industries)

Pleistocene Volcanism and Groundwater

The presence of tuff rings, maars, and other hydrovolcanic features, clustered in the Fort Rock basin, demonstrate the catastrophic relationship between magma and Ice Age conditions at a time when fluctuating lakes covered hundreds of square miles. Symmetrical craters, maars and tuff rings result when rising magma encounters groundwater (phreatomagmatic). The resulting steam explosion throws rocks and ash high into the air. Settling close to the crater, the debris builds a rim or tuff ring. Saucer-shaped maars are formed by the same process but in shallower water. The craters are often fed from below by a diatreme, or funnel-like vent, which ultimately fills or is blocked with angular pieces of debris. These landforms are frequently obscured by vegetation or are badly eroded. Four tuff rings and maars located in Lake County are Fort Rock, Hole-in-the-Ground, Big Hole, and Table Mountain.

Resembling a castle, Fort Rock is an outstanding example of a crescent-shaped tuff ring, three-quarters of a mile in diameter, projecting 180 feet above the floor of the paleo-lake. Shattered rock, ash, and tuffs that make up the crater walls are a testimony to the violence of the event. After the initial eruptive stage, much of the ash fell back within the ring where the yellowish and brown tuffs dip inward toward the center.

An elongate oval at the edge of the Fort Rock basin, the Table Rock complex includes two overlapping tuff rings, one filled with rubble and the other capped by basalt. Several dikes exposed on the flanks between the two fed the flows. The large tuff cone was the first to erupt during a deep-water interval in contrast to the second surge when the magma encountered groundwater. Dates of the eruptions are uncertain.

A little over a mile in diameter, Big Hole is a circular maar created by a catastrophic phreatic (steam) explosion. A wide ledge within the basin may be the top of a large block that dropped into the depression during the eruption. Unlike Fort Rock and Hole-in-the-Ground, Big Hole is heavily forested. The trees and its immense size make this volcanic feature difficult to visualize from ground level.

Geologic Hazards

The potential for volcanic eruptions and flooding in this province has not been examined in great detail, but a case has been made by Daniele Mckay at the University of Oregon for future activity along the northwest fracture system, which extends from Newberry Crater toward Bend. Renewed volcanism from this structure might produce columns

The floor of Fort Rock Lake can be seen through a huge cleft in the wall. A cave in the wall here is famous as the locality where a 9,500-year-old sandal and other artifacts were discovered. (Photo courtesy Oregon Highway Department)

of tephra and ash. Earlier eruptions blocked the Deschutes River, and such an occurrence today would find the community of Bend especially vulnerable.

In spite of the low rainfall, there are wet intervals when flood waters cover agricultural lands near the shallow Malheur, Harney, and Mud lakes. During El Niño years, the three lakes double or triple in size, even merging into one body. The water level in Malheur Lake is controlled by the Narrows. When it reaches 4093 feet in elevation, the water flows through the Narrows into Mud Lake, which, in turn, may discharge through Sand Gap into Harney Lake. In the early Holocene, Malheur Lake topped the ridge at The Narrows into Harney Lake several times, and record rainfall between 1981 to 1986 caused the three lakes, normally covering 100 square miles, to more than double. During the dry 1990s, only Malheur Lake had a residue of water in the low spots.

The alternating size of the lakes during dry and wet intervals can have serious legal consequences. In 1897, after years of drought, both grain growers and cattle ranchers claimed ownership of the exposed marshy bed of what had been Malheur Lake. The conflict ended in court after the murder of a prominent rancher, French Pete.

In 1934, Malheur, Harney, and Mud lakes were designated as the federal Malheur National Wildlife Refuge, a home for thousands of birds, native plants, and mammals.

Natural Resources

Geothermal Energy

Percolating from deep within the crust through cracks and fractures, heated water reaches the surface as warm or even hot springs. In the eastern High Lava Plains, thermal waters issue from perennial springs near Hines, in the Warm Springs valley, and around Harney and Mud Lakes. Some of the highest temperatures occur southwest of Harney Lake, where the 90° to 154° Fahrenheit waters indicate the proximity of a thermal source. Popular since the late 1800s, commercial facilities have been developed at many of the sites.

Because it is rated as Oregon's top potential geothermal site, Newberry Volcano has been the focus of intensive investigations over the past decade, but the establishment of the Newberry Volcanic National Monument in 1990 limited exploration to the flanks. Drilling encountered both near-surface and deeper hydrothermal systems that recorded temperatures at high as 500° Fahrenheit, however, water in the wells was very limited.

Surface and Groundwater

The High Lava Plains is a desert environment in which the available water supply varies widely. Evaporation is high and rainfall low. Precipitation, which ranges annually from less than eight inches in the valleys to 30 inches at the higher elevations can temporarily fill the streams, pond in shallow basins, or percolate into the ground. During dry years, the streambeds can be farmed, while in wet intervals thousands of acres of land might be flooded for months at a time.

Shallow groundwater levels exist in both alluvial and lakebed deposits and in volcanic rocks. The near-surface water table makes the aquifers susceptible to degradation from saline water at or beneath the playas, as well as from sewage and agricultural chemicals. Prior to the 1970s, less than 20,000-acre-feet was pumped annually with no long-term effects, but water levels have decreased steadily since the Oregon Water Resources Department began to issue 60,000 permits on low-interest federal and state-sponsored farm loans in the 1960s. The water table fell when withdrawal reached 80,000-acre-feet annually. It wasn't until 1986 that new appropriations were restricted, but then only after the aquifer had become critically depleted. Today irrigators vie with households for the resource.

Settlement, ranching, and farming in southcentral Oregon are dependent on adequate water, and at one time it was thought that a bountiful lake lay beneath the ground, but the perception did not match the reality. In 1921 the state legislature funded the drilling of four test wells in the Fort Rock basin to assess the situation. As the precursor to the state department of geology, the Oregon Bureau of Mines and Geology, founded in 1911, was to conduct the study. The Bureau was the logical choice because at that time it was based in the School of Mines at Oregon Agricultural College (later Oregon State University) in Corvallis, which offered courses to students interested in geology. It

Henry M. Parks was appointed Dean of the School of Mines as well as head of the Oregon Bureau of Mines. Parks, who had taken a degree in mining engineering from Iowa State College in 1909, moved to Oregon three years later. The legislature dissolved the Bureau in 1923, at which time Parks purchased land in the Fort Rock valley, where exploratory water wells were to be sited. The drilling program was declared a success, with wells producing 700 to 1,000 gallons a minute. But farmers still had to contend with leaky wooden flumes and water draining away into the lavas or evaporating. For the next 20 years Parks advocated pump irrigation, but it still had not become a reality when he died in 1945. (In the 1910 photo of the faculty at the School of Engineering and Mechanical Arts, Oregon State College, the mustached Parks is 3rd from the right in the back row.; courtesy Oregon State University, Archives)

wasn't until 1932 that the School of Science and a Geology Department were established.

Geologic Highlights

A drive eastward through the High Lava Plains gives an exceptional perspective on Cenozoic volcanism from the imposing Newberry Volcano to Harney basin. Sparsely vegetated lava fields, rounded cones, shrinking lakes, and expansive plains give way to flat-topped ridges and sheets of ash flow tuffs and lavas.

Volcanic Cones, Craters, and Flows

Quaternary volcanism and faulting on the High Lava Plains have erected some of Oregon's most extraordinary landmarks within the past 100,000 years. Preserved in the dry eastern Oregon climate, these features would have been eroded or obscured

South of Bend, Newberry Volcano is a cluster of lava flows, domes, lakes, and craters. Newberry National Volcanic Monument, dedicated in 1990 by President George Bush, includes the caldera, Lava Butte, Lava River Cave, and the Lava Cast Forest. Looking west in the photo, the central crater and East and Paulina lakes are enclosed by the five-mile-wide rim. Paulina Creek flows westward from Paulina Lake, dropping 100 feet as Paulina Falls to flow into the Deschutes River. Toward the south rim of the caldera, the summit of Paulina Peak at 7,984 feet above sea level affords an outstanding view. (Photo courtesy Condon Collection)

Spreading one and one-half miles from its vent, the Big Obsidian flow was Newberry's final and largest eruption 1,300 years ago. (Photo courtesy Condon Collection)

by vegetation in an environment with a higher rainfall.

Within the city limits of Bend, Pilot Butte served as a visible landmark for pioneers traveling to the Willamette Valley. At 600 feet in height, the lone cinder cone offers an impressive panorama of the Cascades and High Lava Plains with glimpses of the winding Deschutes River canyon. Discharged from a fissure 7,000 years ago, cinders and ash settled to build the symmetrical cone. Pilot Butte has been a state park since 1928.

Named for John Newberry, geologist with the Corps of Topographic Engineers in 1855, Newberry Volcano is the most exceptional volcanic feature of the province, and it is among the largest Quaternary stratovolcanoes in the United States. The caldera is an oval 30 miles long and 20 miles wide with a surrounding lava field encompassing 500 square miles. The oldest rhyolitic lavas formed the summit caldera 80,000 years ago, while obsidian flows and basalt cones that line the flanks and floor of the edifice are dated as recently as 1,300 years.

Placed at the northernmost end of a rift zone extending from Newberry Crater, Lava Butte cinder cone erupted close to 6,700 years ago. The event built the 5020-foot-high summit and sent lava over an area of 95 square miles, blocking the Deschutes River. Atop the cone, the newly remodeled Lava Lands Visitor Center has added the Lawrence Chitwood Exhibit Hall in honor of the popular lecturer, forester, and geologist who died unexpectedly in 2008 while hiking Pilot Butte. (Photo courtesy Condon Collection).

At the Lava Cast Forest, flows engulfed pine forests, much like those living in this area today. Some trees remained upright while others fell to be buried by the molten material. Once enclosed, the trees burned inside the hardening crust, leaving a mold the size and shape of the trunk. As the flow receded, rough cylinders in black lava were left standing as the Lava Cast Forest. (Photo courtesy Oregon State Highway Department)

Part of the Newberry Volcano National Monument, Lava River Cave, formerly known as Dillman Cave, is one of the longest unbroken tubes in Oregon. Near the entrance, the roof has collapsed, and the main passageway is blocked by sand. Delicate formations, vertically stacked tubes, ledges marking the levels of lava flow, and the unusual sand castles of a miniature badlands carved by water make the cave of geologic interest. (Photo of the entrance from Williams, 1923)

Caves, Tubes, and Tunnels

Caves, intricate cavern systems, and tubes may develop when the outer crust of a lava flow cools and hardens over a still-moving basaltic stream. As the lava continues to flow within the crust, it drains away leaving an open cave or cavern. Lava tubes are most common in pahoehoe lava, and when the basalt fills a narrow valley a single tunnel or a series of stacked tubes result. Straight caves may follow faults and fractures, but more often the tunnels mark the meandering pathway of a prehistoric stream channel. Interconnected tubes may form when thick, syrupy lava moves in sheets. Once the roof of caves and tubes are attacked by weathering processes, they collapse into openings.

Typically, benches or ledges lining the wall give the cave a keyhole or skull shape in cross-section to show that the lava drained in stages. In one of these, Derrick Cave, the entrance is through an opening in the roof, which rises as high as 50 feet in places. Lava River Cave has a similar keyhole shape. At other times, the molten lava left a coating on the ceiling, which dripped to the floor as pinnacles or formed hanging lavacicles. At Lavacicle

Historically lava caves were used by early inhabitants for shelter, for chilling food, or for supplying ice. The Arnold tube runs northeasterly from Paulina Peak through basalt from fissures on the flanks of Newberry Volcano. In a number of places the roof has fallen in, breaking the tunnel into smaller sections. As part of this complex, Arnold Ice Cave maintains a steady temperature of 40° Fahrenheit. Where the cave intersects water, it cascades down and freezes into silent waterfalls. The entrance has been badly damaged and is unstable. Because of rising groundwater, ice now covers the stairway once in use, so that special climbing equipment is needed to enter safely. (In the photograph, ice is being harvested from Arnold ice cave; courtesy Condon Collection).

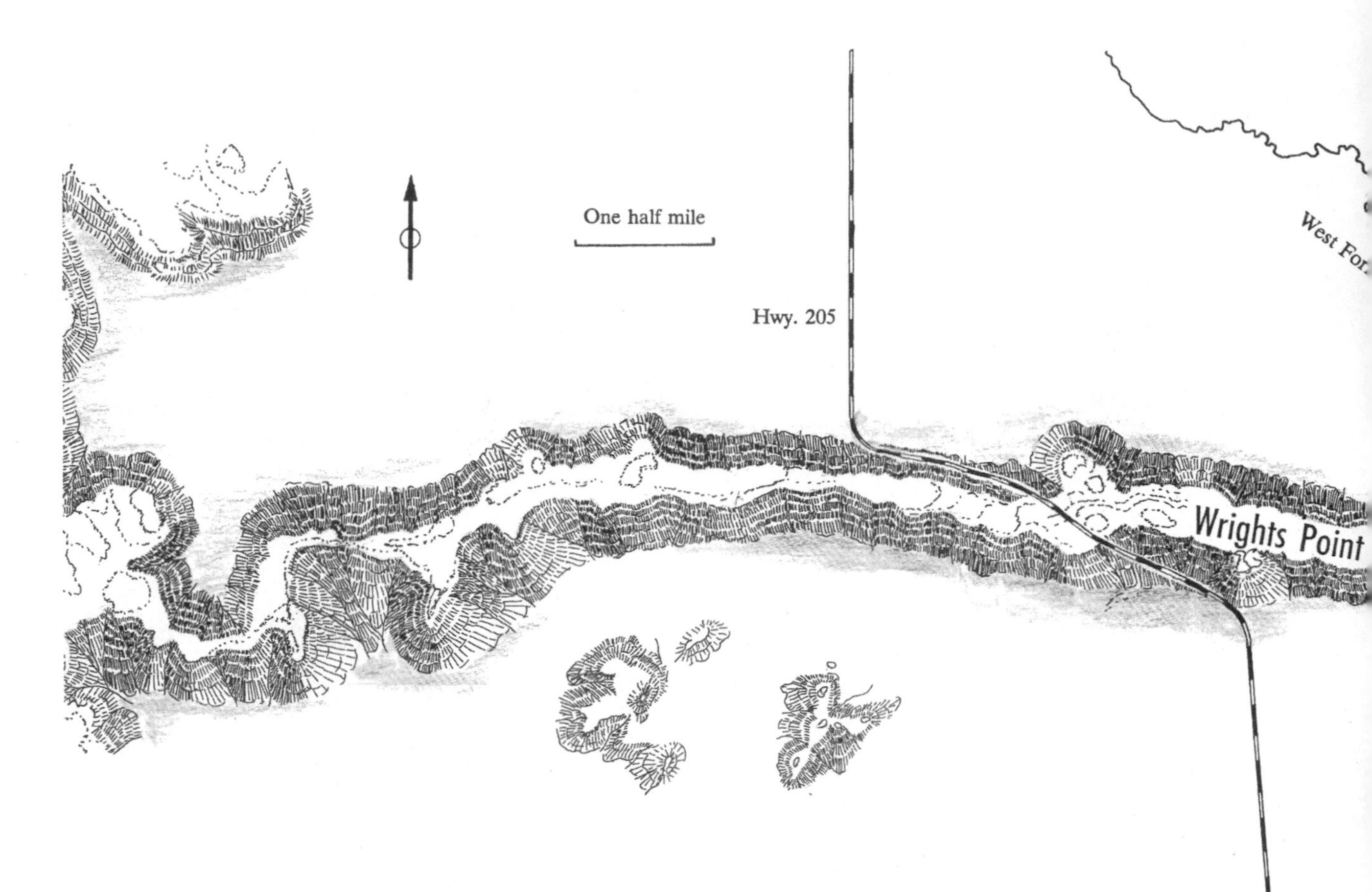

Wrights Point in Harney County traces the ancient channel of a meandering stream that had been filled in by basalt. Projecting 250 feet above the plain, the flat-topped sinuous ridge is six miles in length and averages 400 yards in width before it merges with a broad mesa at its western end. Fossilized fish bones and freshwater clams and snails, collected in the lower strata of Wrights Point, are evidence of the older stream environment.

Cave south of Millican, the lavacicles are only a few inches long, but those rising from the floor are two to six feet high. Because of the fragile nature of the formations, the cave entrance is barred and permission to enter must be obtained from the U.S. Forest Service.

Lava caves in basalt terrains often served as conduits to carry the fluid material great distances underground. Robert Jensen notes that the basalt from Lava Top Butte spread northward some 10 miles through the Arnold Lava Cave system until it was blocked. At that point the lava emerged to cover the surface as the Badlands topography. What has been called the Badlands volcano is in actuality a rootless vent that was fed through the Arnold Cave passageway.

On the extreme eastern edge of the Lava Plains, the privately owned Malheur Cave is both obscure and difficult to explore. At least one-half of the cave is filled with water, creating a long underground lake where it intersects the groundwater table. The cave can only be traversed by canoe or raft, although the lake and configuration of the tunnel prevent passage after a certain point. Permission must be obtained to enter.

Wrights Point

As the softer layers of Pleistocene fluvial sands, conglomerates, and ignimbrites, which had invaded the Wrights Point paleo-valley, were stripped away with erosion, the streambed was left cast in lava as an excellent example of inverted topography. Because of the orientation of the lavas and stream deposits, Alan Niem suggests that they flowed from the west and that Palomino Butte was the possible source.

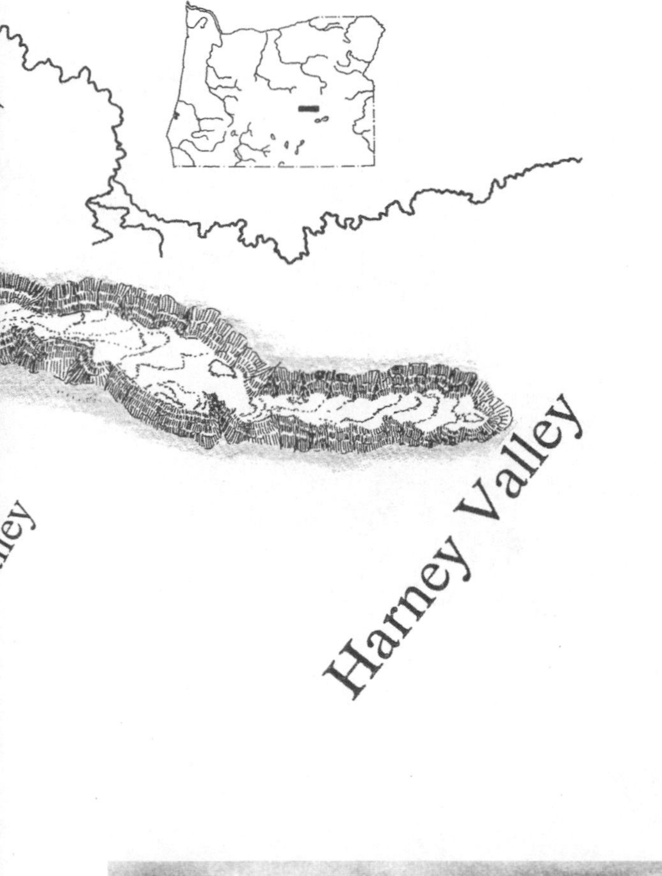

In this oblique aerial view looking westward, Wrights Point appears to wind along the ground. ((From Niem, 1974; photo courtesy Oregon Department of Geology and Mineral Industries)

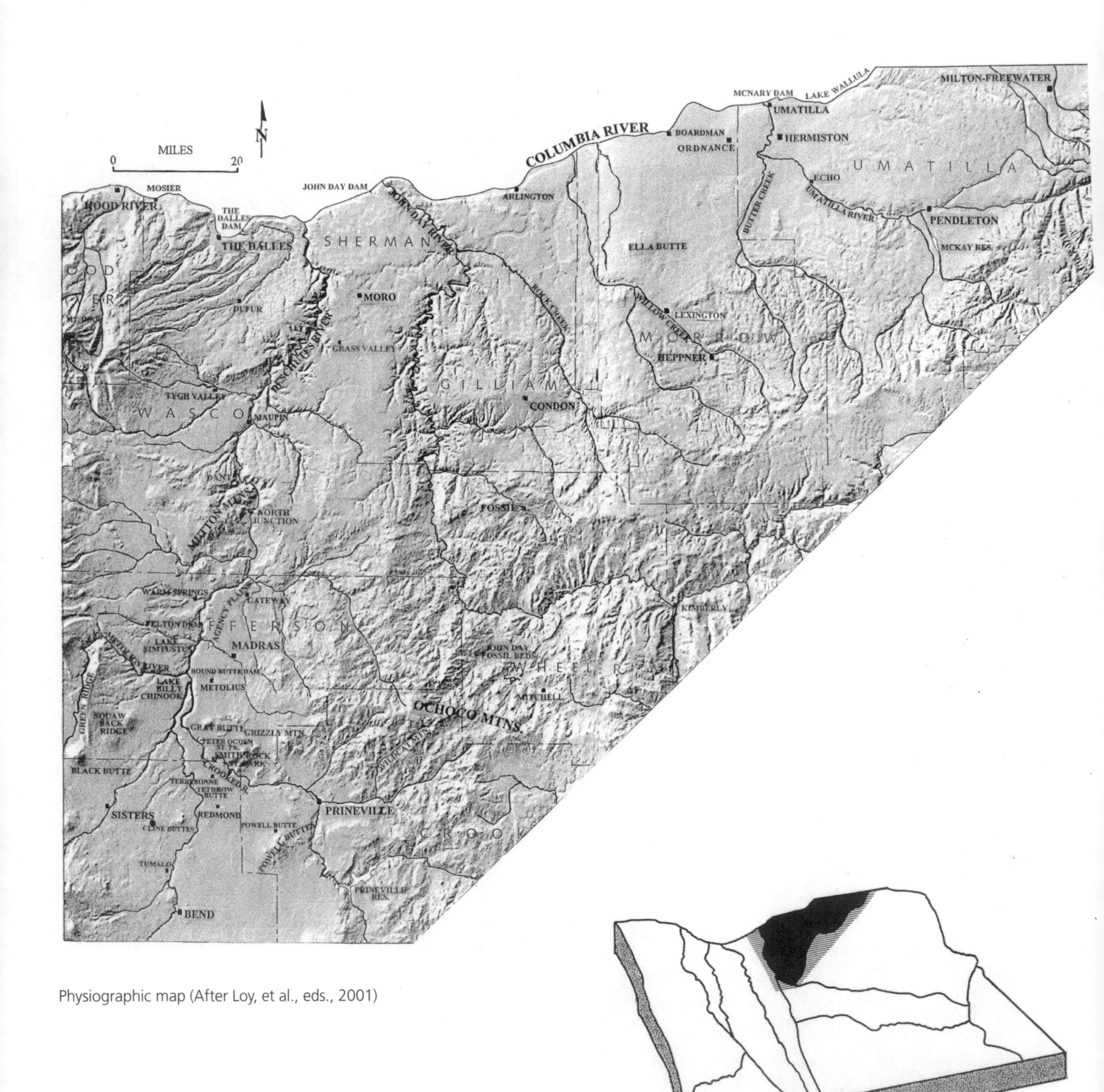

Physiographic map (After Loy, et al., eds., 2001)

Deschutes-Umatilla Plateau

Landscape of the Deschutes-Umatilla Plateau

The Deschutes-Umatilla province extends nearly 200 miles westward from the Blue Mountains to the Cascade Range and is 100 miles at its widest. The region slopes gently northward from elevations of 4,000 feet toward the Blue Mountains down to a few hundred feet above sea level near the Columbia River.

The surface of the province is interspersed with deeply incised canyons and broad uplands, all of which contribute to its characterization as a plateau. The Deschutes-Umatilla tableland lies at the southern edge of the larger Columbia River flood-basalt plateau that extends into the Blue Mountains, central Washington, and western Idaho. Few basalt provinces worldwide have been studied as thoroughly as this one.

Stream flows are steady as they make their way northward to enter the Columbia. The third largest river in North America, with a watershed covering 260,000 square miles, the Columbia has its headwaters on the west slopes of the Canadian Rockies and runs through Washington to Wallula Gateway where it makes a sharp bend toward the west. Along its westerly course, the river defines the boundary between Oregon and Washington before ultimately reaching the Pacific Ocean.

Originating in the Blue Mountains, the Umatilla River, Butter Creek, and Willow Creek are entrenched in their higher reaches, but in the lower stretches they are shallow and less dissected. At 90 miles in length, the Umatilla is considerably smaller than the Deschutes and John Day systems, which dominate the central and western portions of the province. From its beginning as small streams near Mount Bachelor, the Deschutes drains over 10,000 square miles and has a length of 252 miles. Supplemented by numerous springs, rainfall, and Cascade snowmelt, the Deschutes is joined by its principal tributaries, the Metolius and Crooked rivers near Madras. Because it originates in the drier Ochoco Mountains, Crooked River has a more moderate flow. From its source in the Blue Mountains, the John Day is Oregon's longest at 284 miles.

Irrigation has made the rolling topography ideal for raising wheat and other grasses.

Past and Present

Israel Russell, of the U.S. Geological Survey, named and described the Columbia River basalts in 1893 as the Columbia lava "in the precipitous walls of the coulees or canyons . . . and in the remarkable gates eroded by Yakima River through ridges" in the layers. Russell's report included the Eocene to Recent rocks of Washington, Oregon, Idaho, and northeastern California. In 1901 John Merriam of

Aaron Waters was a leading volcanologist of the twentieth century who pioneered studies of basalts in the Pacific Northwest. He was best-known for his knowledge of the Columbia River basalts, and his research, along with that of his many students, ultimately led to the subdivision and classification of the group. Waters came by his interest as a young man growing up in Waterville, Washington, where his family's ranch bordered the Columbia River plateau. His PhD in 1930 from Yale University focused on the geology of the Chelan Quadrangle, but his interests covered not only regional volcanology but geomorphology and tectonics as well. Waters' career led to teaching at several universities. He died in 1991 at age 86 in Tacoma. (Photo courtesy Condon Collection)

Recognized for his expertise on the geology of the Columbia River Basalts, Marvin Beeson's interests also included the geology of the northern Willamette Valley and especially the Portland area. Born in 1937 at LaGrande, Oregon, he completed a PhD from the University of California, San Diego, in geology and geochemistry. Accepting a position at Portland State University, Beeson remained at that institution until retirement. His recognition and demonstration that the Miocene lavas of coastal Oregon originated from vents on the Columbia plateau will be one of his most lasting contributions. He consulted in the private sector until his unexpected death in 2004. (Photo courtesy T. Tolan)

Terry Tolan grew up in Portland, Oregon, and attended Portland State University where his research emphasized the Columbia River Basalt Group and its relationship to the Troutdale Formation. After graduation, Tolan joined the Hanford nuclear waste facility, where he focused on the regional geology. Authoring numerous papers on the geology, structure, and hydrogeology of the Columbia Plateau, the northern Willamette Valley, and the Pacific Northwest, Tolan is in private practice in Kennewick, Washington. (Photo courtesy T. Tolan)

While spending 1968 to 1969 at Washington State University on sabbatical from the University of Wales, Peter Hooper developed an interest in the Columbia River basalts. He returned two years later to serve as chairman of the Department of Geology until retirement to England in 2009. As theories about the origins and composition of the Columbia plateau basalts evolved, Hooper argued in favor of their complexity. Hooper's Raiders, who combined their efforts on the basalts, are (from left to right) Steve Reidel, Peter Hooper, Victor Camp, and Marty Ross. Presently at Washington State University, Reidel has worked with geology of the Northwest for over 30 years. Vic Camp, originally from West Virginia, took his degree from Hooper, then worked in Africa and the Middle East, before accepting an appointment at San Diego State University in 1993. At Northeastern University in Boston, Marty Ross began as a graduate student with Hooper on the CRBGs before completing his PhD from the University of Idaho in 1978. (Photo courtesy P. Hooper)

the University of California, Berkeley, limited the flows to the Miocene layer most prominent along the Columbia River and in the John Day basin. During the 1930s, U.S.G.S. Director George Smith introduced the term *Yakima Basalt* for all of the flows. This nomenclature was in use until Aaron Waters' stratigraphic revision.

Overview

The physiographic boundaries for many regions are difficult to define because much of the underlying geology and structure is obscured or has been modified by later events. This is especially true for the Deschutes-Umatilla province, where the margins are covered by thick layers of middle to late Miocene lavas. Immense outpourings Columbia River flood basalts from numerous north-trending dikes created a tableland of stacked sheet flows in Oregon, Washington, and Idaho. Other volcanic plateaus, such as the Deccan in India or the Siberian Traps in Russia, are larger in volume and areal extent, but the superb exposures, accessibility, and distinct chemical and mineral composition of the Columbia River group make them ideal for the study of flood basalt provinces. Characterized by a rapid eruptive rate rather than by fluidity, the sheet-like lavas are thought to have been triggered by interaction between a hot spot and the earth's crust.

Throughout the late Miocene and into the Pliocene, basins atop the basalt surface near Madras, The Dalles, Arlington, and McKay accumulated volcanic and fluvial sediments that preserve an assortment of fossil plant and animal remains. With the gradual subsidence of the High Cascades into a deep regional graben, the sediment source was blocked, and deposition ceased.

Pliocene and Pleistocene lavas from Newberry Crater and vents near Bend obstructed the Crooked, Metolius, and Deschutes channels before uplift brought renewed downcutting that allowed the rivers to establish the deep gorges seen today. The late Pleistocene brought stupendous floods from glacial Lake Missoula in Montana when ice dams on the Clark Fork River broke, releasing calamitous deluges that reached to the Pacific Ocean. In the course of these floods, water and sediment backed up from the constriction at The Dalles, across the Umatilla plateau, and into the Deschutes canyon as far as Maupin.

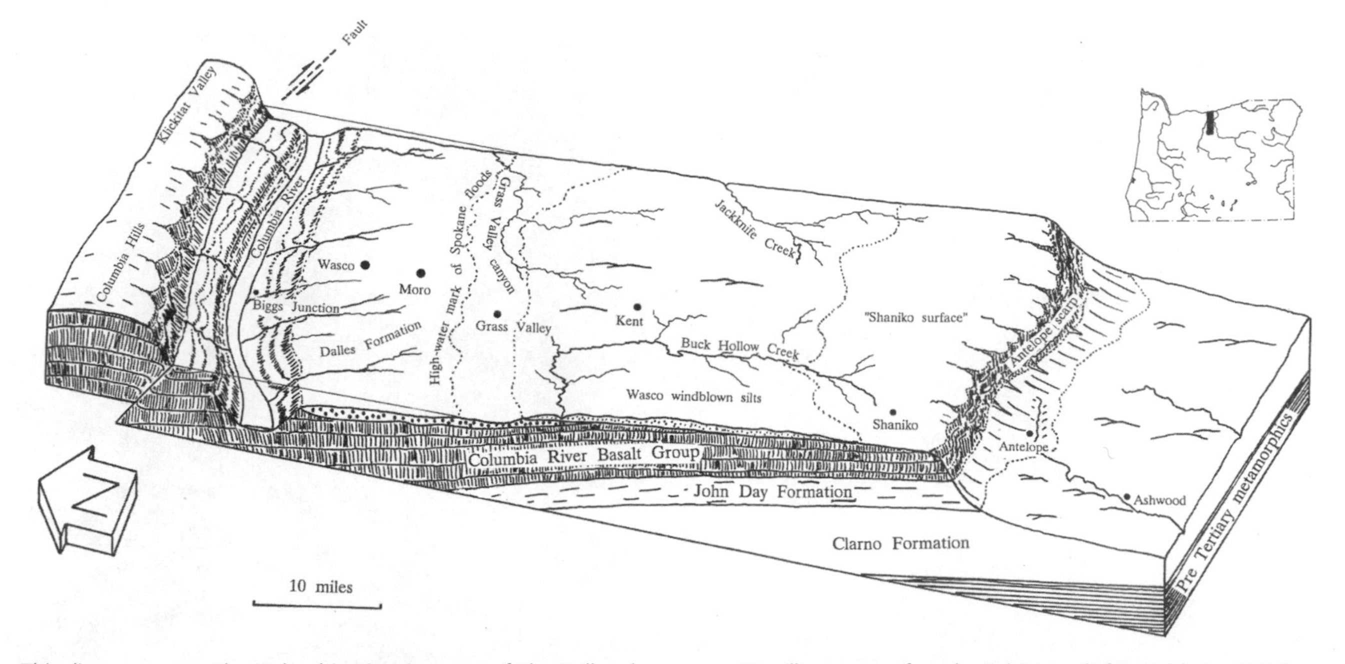

This diagram across the Columbia River just east of The Dalles shows over 50 million years of geologic history. (After Baldwin, 1981)

Overall, the province is well-watered by rivers and tributaries, however, the growth of urbanization along with demands for irrigation have brought a significance decline surface flows and a consequent drop in the groundwater table.

Geology

Paleozoic-Mesozoic

The pre-Tertiary rocks of the Deschutes-Umatilla plateau are largely covered by Miocene Columbia River basalts, although Paleozoic, Mesozoic, and early Cenozoic exposures along the margins of the province hint at older accreted terranes below. In the 1980s, four exploratory petroleum wells were drilled in the vicinity of Yakima, Washington, by Shell Oil Company. Targeted were Cretaceous and early Tertiary strata lying within anticlines, which were judged to have sufficient organic content and porosity to generate natural gas. Although the potential for hydrocarbons proved to be negligible, Eocene to early Miocene sediments were encountered.

Cenozoic

Extraordinary Flood Basalts

The dark gray, fine-grained, tholeiitic (olivine deficient) Columbia River basalts that spread over adjoining sections of Oregon, Washington, and Idaho are the geologic centerpiece of this province. During the middle to late Miocene 17 to 6 million years ago, an extraordinary succession of lava flows created a broad, nearly level plateau that covered 77,000 square miles, an area only slightly smaller than the state of Washington. Over 300 separate basalt sheets were emplaced during an 11 million year period for a total volume of 56,000 cubic miles. The individual flows vary, but locally the combined layers reach 8,000 to 9,000 feet thick.

Instead of forming a central cone, the basalts moved as horizontal sheets from north-northwest-trending fractures or dike swarms. The sheet-like nature is characteristic of flood basalts and is due to the exceptionally rapid eruptive rate and not to the low viscosity of the lavas. Because of the rapidity, large volumes of material expanded over vast distances. Steve Reidel calculated that emplacement rates fluctuated from just a few weeks or months to several years and that the older flows in the central plateau were invaded and inflated by younger extrusions.

The final cooling and hardening phases produced a number of characteristic layers, each distinguished by a top, core, and bottom. The top is a ropy pahoehoe with brecciated angular fragments, and the interior most often consisted of columnar blocky jointing. The middle and base are arranged into two distinct parts, the colonnade and the

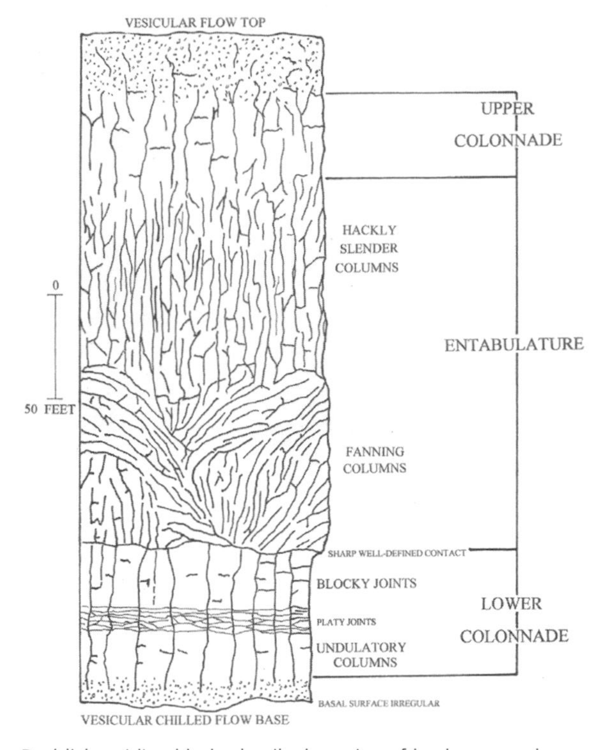

Reddish oxidized baked soils, breccias of broken angular basalts, or pillow structures where the magma interacted with water differentiate individual lava flows. In some, upper columnar intervals mimic the lowermost colonnade. (After Tolan, Beeson, and Vogt, 1984; Tolan, et al., 2009)

Dikes of the Grande Ronde Basalt display horizontal columns like stacked cordwood. (Photo courtesy Oregon Department of Geology and Mineral Industries)

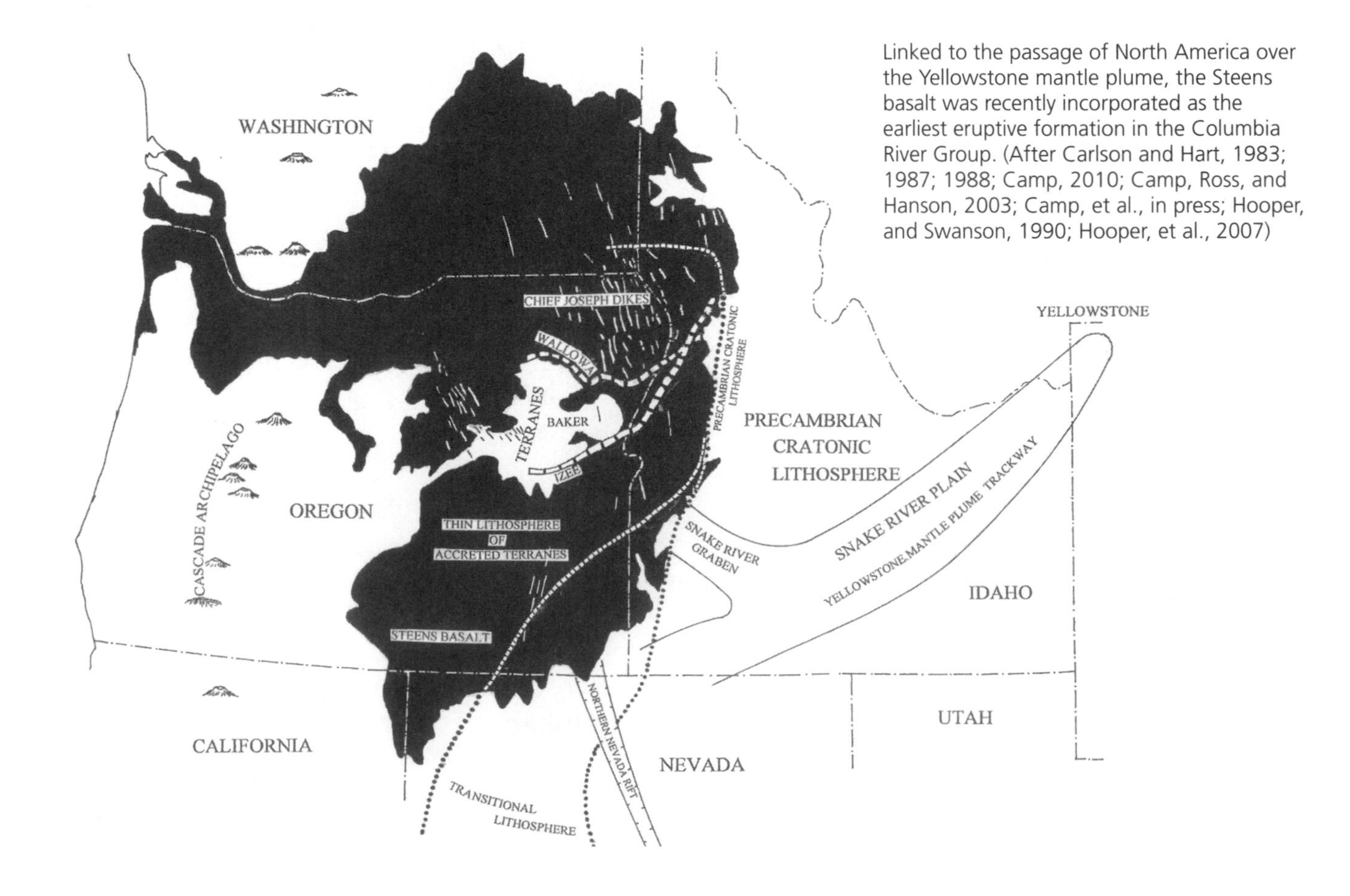

Linked to the passage of North America over the Yellowstone mantle plume, the Steens basalt was recently incorporated as the earliest eruptive formation in the Columbia River Group. (After Carlson and Hart, 1983; 1987; 1988; Camp, 2010; Camp, Ross, and Hanson, 2003; Camp, et al., in press; Hooper, and Swanson, 1990; Hooper, et al., 2007)

entablature. Named after Greek temple architecture, the lower portion or colonnade is so-called because the basalt cooled and contracted into large, well-defined columns perpendicular to the surface below. The remaining upper entablature, with smaller poorly defined columns, constitutes up to four-fifths of the entire volume. Near the top of the entablature, vesicular basalt formed where gas bubbles accumulated from the molten mass. Often long, vertical tubes or pipes mark the pathways.

Individual Flows of Columbia River Basalts

Initially lumped together as the Yakima Basalts, the individual flows of the Columbia River Basalt Group are now distinguished by subtle variations in geochemistry, mineralogy, and magnetic polarity. Of the five main formations, the oldest is the Steens basalt, then the Imnaha, the Grande Ronde, the Wanapum, and the youngest Saddle Mountains. In 1988 Richard Carlson of the Carnegie Institute and William Hart at Miami University suggested that the Steens might be an early manifestation of the Columbia River series. With a flow-by-flow analysis, Peter Hooper distinguished the properties of the individual sheets, and in 2010 Vic Camp formally recognized the Steens Basalt as the oldest formation in this group.

The extent of Steens eruptions was confined to the Basin and Range and Owyhee provinces and Camp now considers them to be the equivalent of the Imnaha and Grande Ronde basalts. The widespread Imnaha issued from feeder dikes of the Chief Joseph swarm and occurs from Pullman, Washington, east to the Clearwater embayment in Idaho, and west to the Pasco basin. The 17.5-to-16.5-million-year-old Imnaha filled deep-canyons in the eroded pre-Tertiary surface to smooth out the topography.

The 120 individual Grande Ronde lavas, which were significantly younger and more silica-rich than the Imnaha, built a flat tableland. Erupting from fissures as much as 100 miles long in the Chief Joseph dike system of central Oregon, the extrusions were so rapid that the lavas did not mound up but instead spread across the uneven ground. Even though the Grande Ronde episode lasted almost 1 million years, over 95 percent of the material poured out

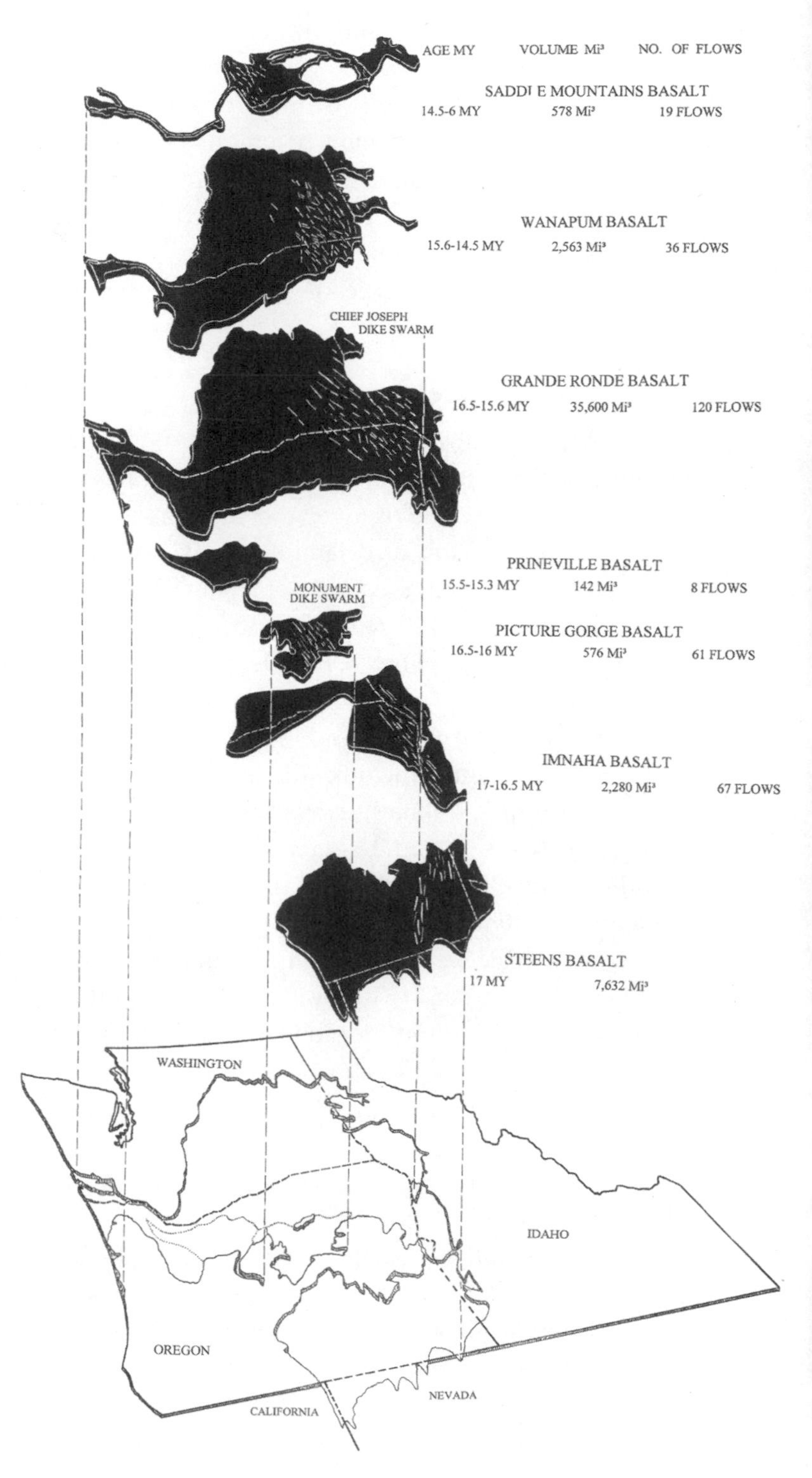

The distribution of formations within the Columbia River Basalt Group (After Beeson and Moran, 1979; Beeson, Tolan, and Anderson, 1989; Hooper, et al., 2007; Reidel, et al., 1989; Tolan, et al., 1989; 2009)

during an interval of less than 250,000 years. The flows reached between 250 to 1,200 cubic miles in volume, and the Grande Ronde comprises more than 85 percent of the entire output of the Columbia River basalts.

In frequency, the Grande Ronde averaged one flow every 8,000 years, but the eruptive intervals slowed during the later Wanapum and Saddle Mountains times. The Saddle Mountains members were smaller and erupted within an 8-million-year interval. They were mainly confined to the central part of the plateau, where they occupied river valleys eroded into the earlier layers as intracanyon flows.

Yellowstone Hot Spot

A consensus regarding the evolution of the Columbia River Basalt Group has yet to be reached, but the various scenarios involve extension, mantle plume emplacement, or delamination of the lithosphere. Advocates for backarc spreading contend that the eruptions are related to crustal thinning, whereas others support a mantle plume or hot spot. Extensional processes in the Basin and Range stretched the crust (lithosphere) in a backarc setting to generate mantle melting and eruptions. Even though thinning ordinarily accompanies extension, the issue is whether the stretching was the cause of the massive eruptions or the consequence of the magmatic activity itself.

Present evidence favors a mantle plume (Yellowstone) accompanied by delamination of the lithosphere. As the North American plate migrated westward, it passed over a stationary hot spot, which, in turn, flattened (pancaked) when encountering the thickened Precambrian edge of the North American craton. The plume head broadened along the base of the crust, opening it to the extrusion of basalts.

At comparatively shallow depths the plume generated the vast volumes of the Grande Ronde basalts before impact with the thickened edge of the North American craton altered the eruptive style to produce the smaller sporadic Wanapum and Saddle Mountains flows. Because of the large-scale features associated with the track, Kenneth Pierce and Lisa Morgan of the U.S.G.S. estimate that, at present, the Yellowstone plume projects at least 600 miles into the mantle.

Series		Group		Formation	Member	Age MY
Miocene	Upper	Columbia River Basalt Group	Yakima Basalt Subgroup	Saddle Mountains Basalt	Lower Monumental Member	6
					Ice Harbor Member	8.5
					Basalt of Goose Island	
					Basalt of Martindale	
					Basalt of Basin City	
					Buford Member	
					Elephant Mountain Member	10.5
	Middle				Pomona Member	12
					Esquatzel Member	
					Weissenfels Member	
					Basalt of Slippery Creek	
					Basalt of Tenmile Creek	
					Basalt of Lewiston Orchards	
					Basalt of Cloverland	
					Asotin Member	13
					Basalt of Huntzinger	
					Wilbur Creek Member	
					Basalt of Lapwai	
					Basalt of Wahluke	
					Umatilla Member	
					Basalt of Sillusi	
					Basalt of Umatilla	
				Wanapum Basalt	Priest Rapids Member	14.5
					Basalt of Lolo	
					Basalt of Rosalia	
					Roza Member	
					Shumaker Creek Member	
					Frenchman Springs Member	
					Basalt of Lyons Ferry	
					Basalt of Sentinel Gap	
					Basalt of Sand Hollow	15.3
					Basalt of Silver Falls	
					Basalt of Ginkgo	15.6
					Basalt of Palouse Falls	
					Eckler Mountain Member	
					Basalt of Dodge	
					Basalt of Robinette Mountain	
					Vantage Horizon	
				Grande Ronde Basalt	Sentinel Bluffs Member	15.6
					Slack Canyon member	
					Fields Springs member	
					Winter Water member	
				Grande Ronde Basalt / Prineville Basalt	Umtanum member	
					Ortley member	
	Lower				Armstrong Canyon member	
					Meyer Ridge member	
					Grouse Creek member	
				Grande Ronde Basalt	Wapshilla Ridge member	
				Grande Ronde Basalt / Picture Gorge Basalt	Mt. Horrible member	
					China Creek member	
					Downy Gulch member	
				Grande Ronde Basalt	Center Creek member	
					Rogersburg member	
					Teepee Butte Member	
					Buckhorn Springs member	16.5
				Imnaha Basalt		17.5

Stratigraphy of the Columbia River Basalt Group (After Beeson and Moran, 1979; Beeson, Tolan, and Anderson, 1989; Reidel, et al., 1989; Swanson, et al., 1979; Tolan, et al., 1989, 2009)

Faulting and Structure

In conjunction with volcanic episodes, a north-south compression combined with east-west extension progressively distorted, sheared, and tilted the rocks of the Deschutes-Umatilla plateau. Consequently, it exhibits a suite of fractures, joints, and folds such as the Yakima fold belt, which includes the Horse Heaven anticline, the Ortley and

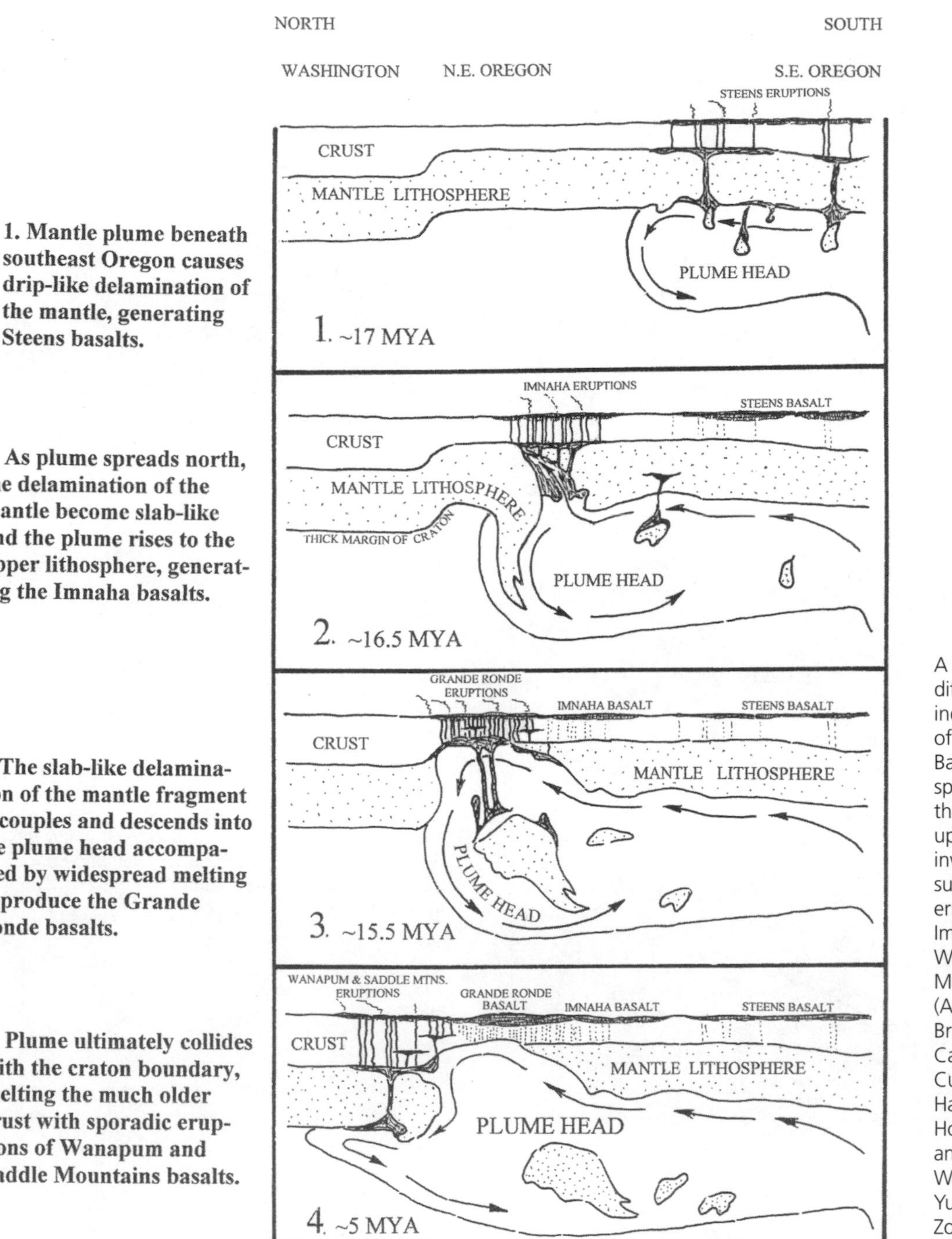

A model to explain the differences between individual formations of the Columbia River Basalt Group features a spreading plume head that peeled back the upper mantle to allow invasion of the crust with subsequent northward eruptions of the Steens, Imnaha, Grande Ronde, Wanapum, and Saddle Mountains basalts. (After Baksi, 2010; Brueseke, et al., 2007; Camp and Hanan, 2008; Cummings, et al., 2000; Hart and Carlson, 1987; Hooper, 2007; Pierce and Morgan, 2009; Waite, et al., 2006; Yuan and Dueker, 2005; Zoback, et al., 1984)

Bingen anticlines, the Mosier-Bull Run syncline, and the Columbia Hills anticline. Camp and Ross in 2004 interpreted the Yakima fold belt as thin crust, which was deformed as a tongue of the Yellowstone mantle plume spread beneath northcentral Oregon.

Emplacement of the individual flows was guided in part by the regional structure and subsidence. As new vents opened, the entire basalt platform was tilted gently toward the northwest by gradual uplift along the Idaho batholith. Tilting allowed the lavas to spread into central Washington, where they ponded in the Pasco basin. From that point, they moved westward through the trough-like conduit of the Columbia trans-arc lowland. Continuing eruptions were also accompanied by subsidence of the crust. Since the rate of subsidence was equal to the extrusion rate of the flood basalts, Reidel has shown that it was most pronounced during eruptions of the voluminous Grande Ronde but diminished thereafter and ceased around 3 million years ago.

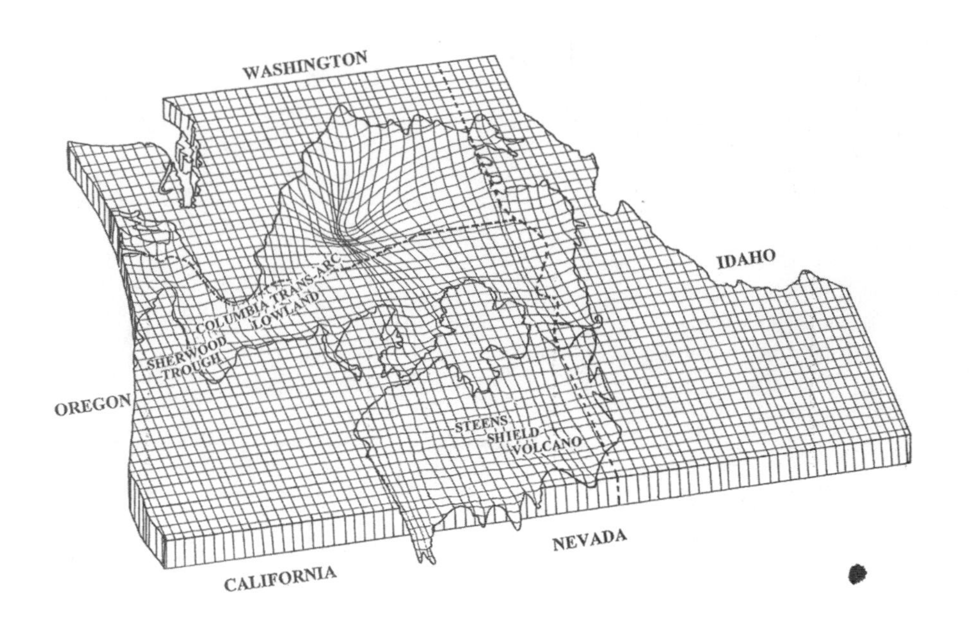

A spoon-shaped mass of the combined Columbia River basalts is nearly three miles thick where it is centered in the basin beneath Yakima and Pasco, Washington, but the layers thin to a mile along the Columbia River and to a feather-edge near the Blue Mountains. This suggests that the lava filled a shallow depression, which subsided steadily. (After Beeson and Tolan, 1990; Reidel and Hooper, 1989; Reidel, et al., 1989; Tolan, et al., 2009)

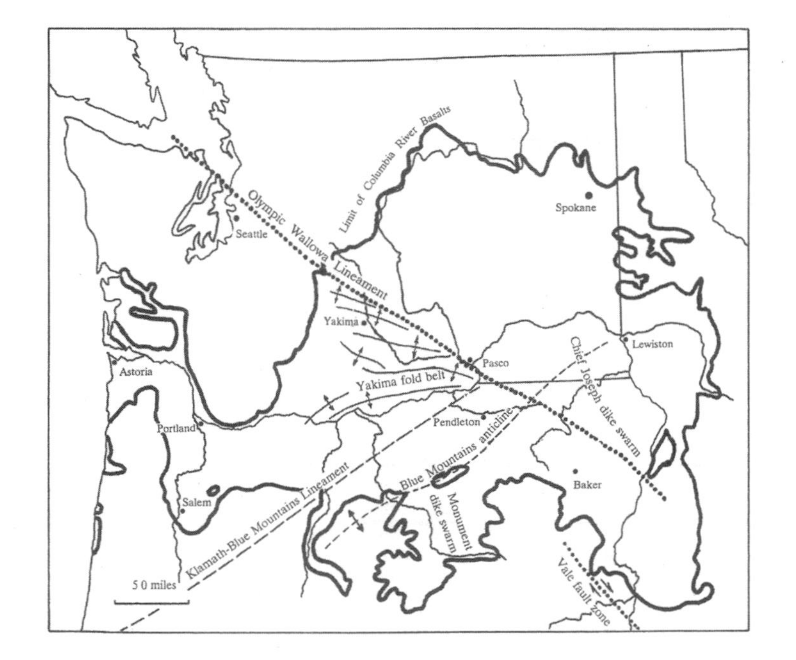

Lineaments, faults and folds, and the trans-arc lowland are among the major structural features that cut the Deschutes-Umatilla province. (After Glen and Ponce, 2002; Hooper and Conrey, 1989; Tolan, Beeson, and Vogt, 1984; Tolan, et al., 2009)

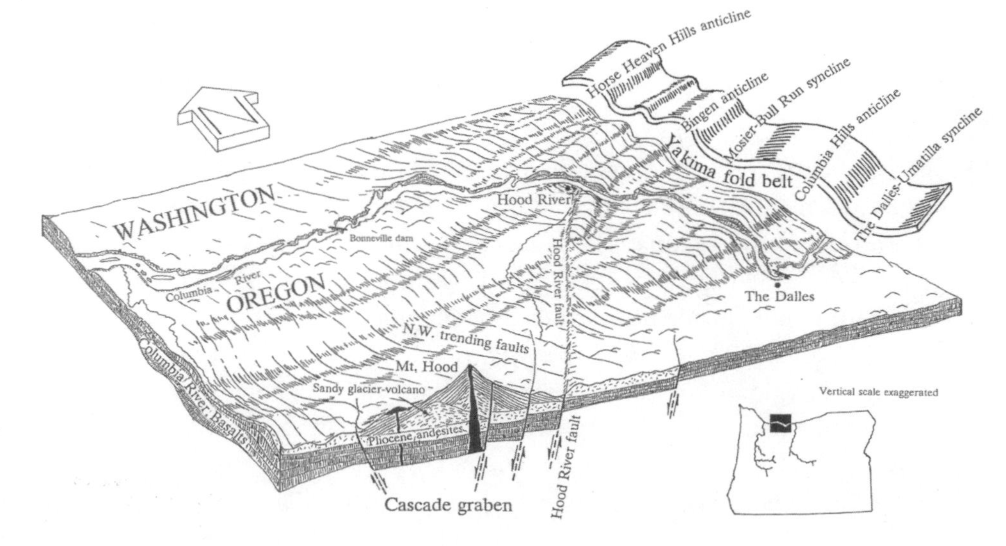

Running southwesterly into the Cascades from central Washington, the gentle folds of the late Miocene Yakima belt form ridges separated by broad valleys. West of The Dalles to Bonneville Dam, the folds are exposed in cross-section where the Columbia river cuts through. (After Camp and Ross, 2004; Tolan, Beeson, and Vogt, 1984; Williams, et al., 1982)

Southwest of the Yakima basin, the Columbia trans-arc lowland was the main conduit by which the Columbia River basalts traversed the Cascades to the Pacific Ocean, a distance of 450 miles. Subsiding more than 16 million years ago, this broad southwest-northeast trending route served as a channel for the lavas as well as for the ancestral Columbia River itself. An extension of the lowland in the Willamette Valley and Coast Range, the Sherwood trough further directed the molten flows westward.

Presently at the University of New Mexico, Gary Smith grew up outside Cincinnati, Ohio, where his fascination with volcanoes was rare for a Midwesterner. That interest led to Oregon State University and on a field trip to the Cascade Mountains and Deschutes basin he saw an opportunity to work out the geologic problems. Basing his conclusions on the lithology and distribution of sediments in the Deschutes region, Smith revised the stratigraphy and named the Miocene Simtustus Formation for his PhD. Further research refined Miocene and Pliocene terrestrial sedimentation on the Columbia plateau in Washington, Oregon, and Idaho. Smith now combines teaching and administration. (Photo taken in 2010; courtesy G. Smith)

Sedimentary Basins

Although eruptions of the Columbia River basalts were an elemental force in shaping the Deschutes-Umatilla province, they are only part of the geologic picture. During the middle to late Miocene, enormous quantities of pyroclastic debris and tuffaceous sediments filled depositional basins, which subsided along the western and northern margins of the uplands. Richard Conrey concluded that volcanism began just prior to intra-arc rifting along the Cascades and associated each basin with a structural segment on the arc. Some of the basins were ephemeral, while others persisted, and many lie within the adjacent Blue Mountains. A network of rivers, draining the surrounding mountains, distributed volcanic detritus into the Deschutes, The Dalles, Tygh, Umatilla (Arlington), and Agency depressions where the preservation of animals and plants provides a picture of changing paleo-environments and climates. The onset of regional uplift 3 to 2 million years ago brought sedimentation to a close.

The Deschutes Basin

The Deschutes (Madras) Basin extends from Redmond northward to Madras and Gateway, east toward the Ochoco Mountains, and west to the Cascade Range. Even though most of the older strata are concealed beneath the Columbia River basalts, the John Day Formation is exposed in small buttes south and west of the Ochoco Mountains, along the eastern flank of the Mutton Mountains, and in the Deschutes canyon. The Columbia River basalts invaded between the mid to late Miocene and interfinger or are overlain by younger sediments and volcanics erupted from regional cones.

Deposited from 15.5 to 12 million years ago, thin brown and white tuffaceous silts, muds, and pyroclastics (volcanic fragments) of the Simtustus Formation represent floodplain and fluvial conditions. Near Madras, mudflows of the Simtustus entomb leaf fragments of *Populus* (cottonwood) and *Salix* (willow), which grew in a temperate climate similar to that of today. Originally included as part of the coarse conglomerates of the Deschutes Formation, sediments of the Simtustus are lithologically distinct, and the two are separated by a substantial unconformity—a gap in the rock sequence.

Following Simtustus activity, lava flows, volcaniclastic sediments (fragmented eroded volcanic rocks), and incandescent clouds of ash of the 7.5 to 3.9 million-year-old Deschutes Formation fanned out for 30 miles from vents on the eastern margin of the High Cascades. In addition to the Cascades, Cline Buttes, Tethrow Butte, and Round Butte also contributed basalts to the formation. In the central basin combined thicknesses reach 2,000 feet, thinning to 50 feet toward the Ochoco Mountains. The volcanic sources Mount Jefferson and the Three Sisters were shut off 3.5 million years ago when the 2,000-foot-high Green Ridge escarpment on the eastern flank of the High Cascade graben rose. For a time, erosion from the Ochocos continued to provide debris to the eastern portion of the Deschutes basin, but that supply also eventually ended. With the fluvial systems no longer overwhelmed by volcanic debris, streams began to incise and establish their channels.

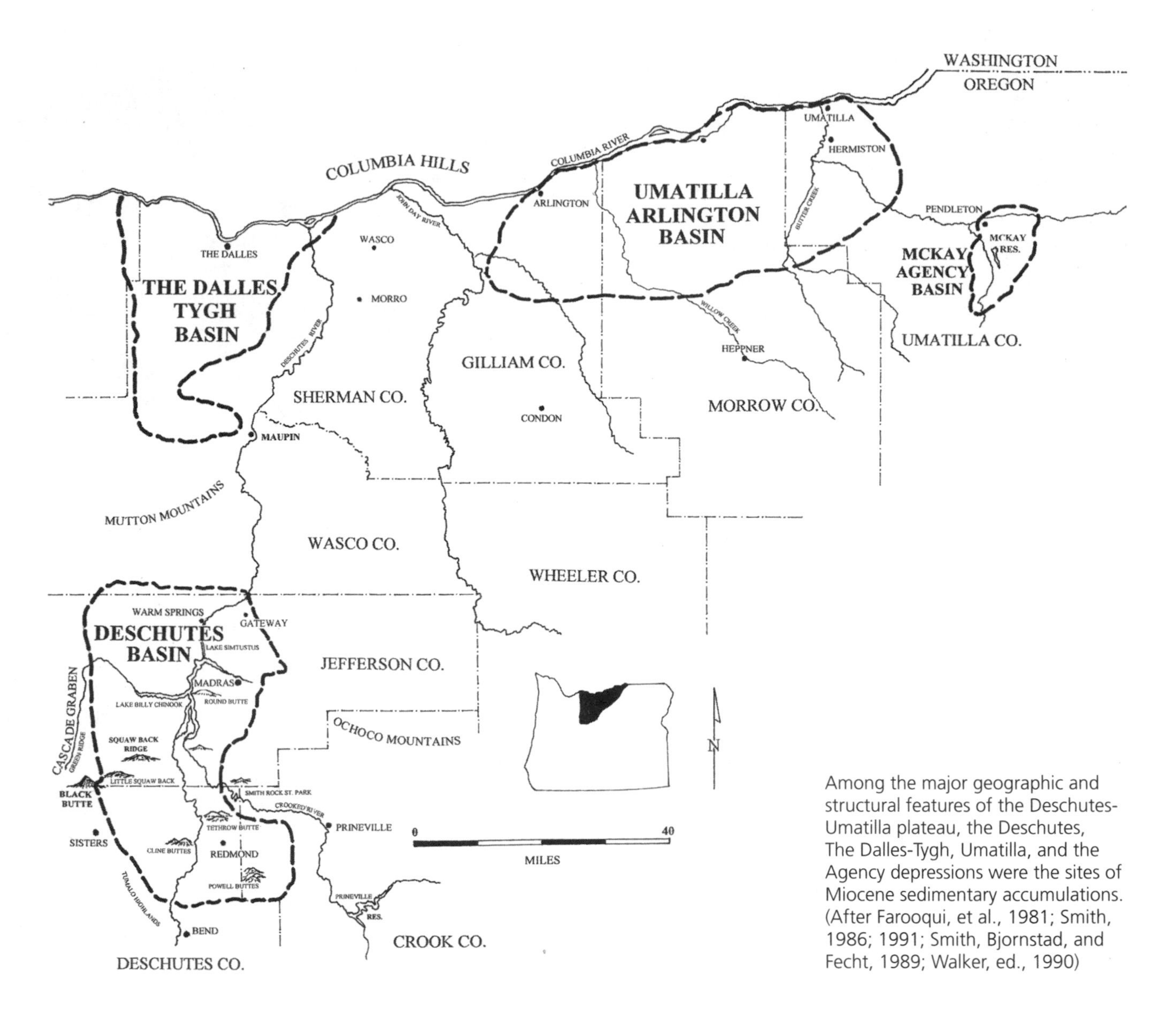

Among the major geographic and structural features of the Deschutes-Umatilla plateau, the Deschutes, The Dalles-Tygh, Umatilla, and the Agency depressions were the sites of Miocene sedimentary accumulations. (After Farooqui, et al., 1981; Smith, 1986; 1991; Smith, Bjornstad, and Fecht, 1989; Walker, ed., 1990)

The Dalles-Tygh Basins

The Tygh and The Dalles basins accumulated late Tertiary lahars and volcaniclastic rocks that spread eastward as alluvial fans from the Cascade volcanoes. Mapped as The Dalles Formation, over 500 feet of sediments in the Tygh Valley are confined by the Mutton Mountains and Tygh Ridge. Primarily fluvial, the sediments overlie the Columbia River basalt, but their variable lithology makes a specific formational assignment difficult.

In the hiatus between intervals of the Simtustus and the Deschutes formations, tuffaceous sediments, lahars, and basalts of the late Miocene Dalles Formation can be traced eastward from the Cascades and northward from the Mutton Mountains to the Columbia River. Floodplain, stream channel, and alluvial fan deposits of The Dalles overlie the Priest Rapids Member of the Columbia River basalts and are dated as upper Miocene. Only a handful of vertebrate fossils have been recovered from rocks of The Dalles Formation exposed in creeks along the Columbia Gorge.

Umatilla (Arlington) and Agency Basins

Overlying the Saddle Mountains basalt (Elephant Mountain Member) on the Umatilla uplands, late Miocene Alkali Canyon sediments were concurrent with those of the McKay Formation in the Agency basin. Bordered to the south by the Blue Mountains in Oregon and to the north by the Columbia Hills in Washington, the Umatilla depression is drained by the steep-walled canyons of

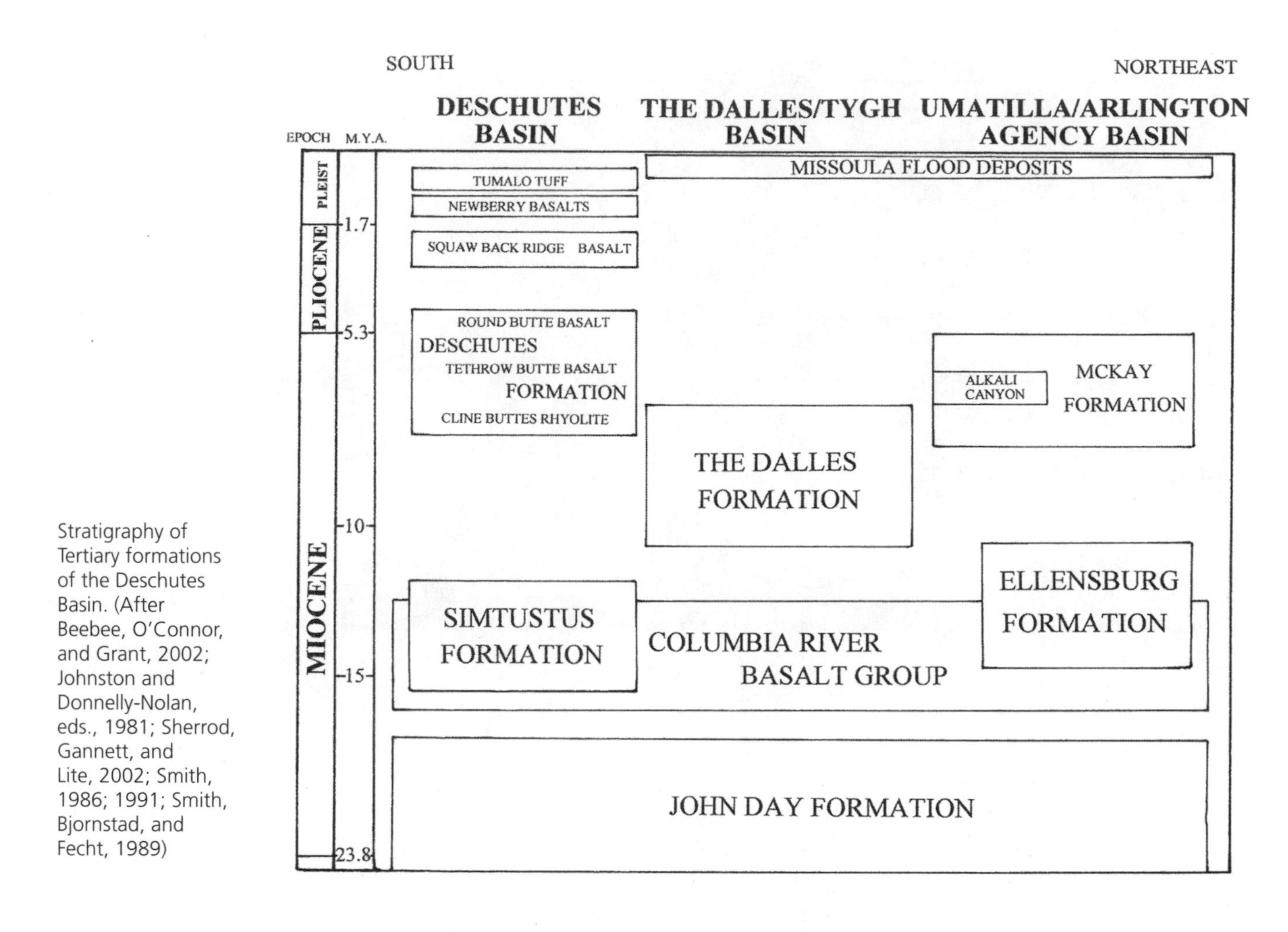

Stratigraphy of Tertiary formations of the Deschutes Basin. (After Beebee, O'Connor, and Grant, 2002; Johnston and Donnelly-Nolan, eds., 1981; Sherrod, Gannett, and Lite, 2002; Smith, 1986; 1991; Smith, Bjornstad, and Fecht, 1989)

Willow and Butter creeks and the Umatilla River. Basaltic gravels and tuffaceous sediments of the Alkali Canyon Formation, deposited in braided streams and alluvial fans, were transported by the ancestral Umatilla River system from the Blue Mountains. The 150-foot-thick rocks indicate a drainage pattern that was much more extensive than at present.

High in the Umatilla watershed, the Agency basin is delineated by the Blue Mountains and the Horse Heaven and Reith anticlines. Confined to this area, the McKay Formation is composed of basaltic gravels interbedded with fine tuffaceous sand and silt. Up to 160 feet thick along McKay Creek, it thins considerably westward. McKay deposits are situated above the Grande Ronde Basalt and the Frenchman Springs Member of the Wanapum Basalt, and they, in turn, are covered by Quaternary wind-blown loess that often forms dunes. Fossils are common in the loose sands.

Pliocene-Pleistocene Volcanic Eruptions

A number of Pliocene shield volcanoes and cinder cones dot the Deschutes highlands at Tethrow Butte, Squaw Back Ridge, Little Squaw Back, and Round Butte. Widespread basalts from Tethrow Butte built a 165-foot-thick flatlands from Redmond to Agency Plains and Gateway. Distinctive red and black scoria from the Tethrow cinder cones also rims Cove Palisades State Park. After an initial eruptive phase, the 3.9-million-year-old Round Butte was reduced to two small summit cones. With a remarkably symmetrical profile and offering a spectacular view, Round Butte was the site of Indian ceremonies and is still regarded as sacred.

Hot ash clouds and pyroclastic fragments from sources on Tumalo Highland enveloped the region. The oldest Desert Spring Tuff, dated at 600,000 years, covered terraces in the Deschutes canyon with ash and pumice and blanketed the Bend area to depths of 60 feet. Successive 300,000-year-old showers of ash clouds (air-fall), violent surges of

The Columbia Hills to the left are the up-thrown block of a large east-west reverse (compressional) fault. Looking east up the Columbia River, Wishram, Washington, is left center and the mouth of the Deschutes River is on the upper right. (Photo courtesy Oregon Department of Transportation)

Bend Pumice, and the pink to salmon-colored strata of Tumalo Tuff serve as distinctive regional marker beds. The younger Shevlin Park and Century Drive episodes similarly inundated the Tumalo upland with volumes of lava, cinders, and ash. The source vents are no longer evident, but the build-up of debris produced the current topography.

Glaciation and Floods

In addition to a final volcanic covering, much of the Pacific Northwest experienced intervals of Ice Age flooding between 15,000 and 13,000 years ago. As continental glaciers advanced southward from Canada, ice plugged and backed up the Clark Fork River in western Montana, impounding the vast reservoir of Lake Missoula. The sudden release of waters when the barrier broke sent catastrophic floods across vast stretches of the Columbia River plateau.

Because Wallula Gap, a narrows at the big bend on the Columbia River, was less than one mile wide, it was unable to handle the 15 to 18 cubic miles of floodwaters exiting Pasco Basin hourly. A hydraulic dam, which impounded water at that point, sent the overflow into Lake Lewis north of the Horse Heaven Hills. In his readable, well-illustrated 2008 book, Robert Carson at Walla Walla University notes that this was the biggest hydraulic dam in the history of the earth. As the rushing water from Lake Missoula slowed, a five-foot-thick layer of sediments was strewn across the Umatilla basin.

Blocks of ice, stones, sand, and gravel jammed into the constriction at The Dalles to create the temporary Lake Condon to the south in the Deschutes River channel toward Maupin. Terraces along the river are composed of massive gravel layers that coincide with the water levels. Enormous rocks (erratics), that were rafted by icebergs, are common southwest of Arlington. Many mark shorelines of the ancient lake. In the Umatilla area, such stones form huge circles, having been pushed aside by farmers operating rotating irrigation sprayers.

Redirecting the Rivers

Even major rivers like the Columbia and Deschutes can be forced to change their direction with uplift of the land, faulting, and when encountering ice or lava. On the Deschutes-Umatilla plateau, volcanic eruptions played a leading role in rerouting the waterways. When moving water is impeded by lava, it works to remove the barrier, but if the dam proves to be permanent, the river bypasses the blockage along a new route. Frequently, paleo-drainages can be reconstructed by mapping individual basalt flows that invaded and occupied the channel or by tracing the presence and location of fluvial sediments and erosional patterns.

The Columbia River changed its channel many times during the past 15 million years when deflected by the Columbia River basalts, by eruptions from the Cascades, by regional subsidence, or by uplift and tilting of the plateau. Terry Tolan, Marvin Beeson, Beverly Vogt, and Karl Fecht have written particularly detailed accounts of the evolution of the drainage patterns.

Prior to 16 million years ago, the ancestral Columbia River crossed Oregon and Washington through a broad low plain, flowed near Mount Hood, and entered the Willamette lowland south at Salem. Here it turned northwest, reaching the Pacific Ocean somewhere in Lincoln County, a route that can be traced by sediments as well as by the location of the basalts.

Alterations to the Columbia River pathway close to Mount Hood began around 15.6 million years ago, when the channel was impeded by Frenchman Springs flows. This member of the Wanapum Basalt forced the river toward the northwest where it followed the Mosier-Bull Run syncline. At the end of the Wanapum episode, Priest Rapids basalts filled and destroyed the route along the Mosier-Bull Run syncline, moving the Columbia River farther north to follow the Columbia trans-arc lowland. This trough allowed the Saddle Mountains basalts to enter western Washington and Oregon. For more than 10 million years the river remained in this broad route, depositing sands and gravels of the Troutdale Formation. Later Saddle Mountains eruptions failed to move it, but regional folding, uplift, and subsidence of the plateau combined with lavas from the Boring cones and High Cascade volcanoes brought a reorganization of the drainage. By 2 million years ago, the river was again forced northward close to its present channel.

Although obscured by later flows, two previous pathways of the Columbia River in the gorge can seen where they are intersected by the Columbia River today. Looking upriver from the Women's Forum State Park near Corbett, one of these is exposed in cross-section at Crown Point. In the sheer walls of the canyon below this point, a single flow of Priest Rapids can be seen. The blocky jointed 500-foot-thick Priest Rapids basalt covers a 200-foot layer of glassy, volcanic sediment emplaced before the lavas.

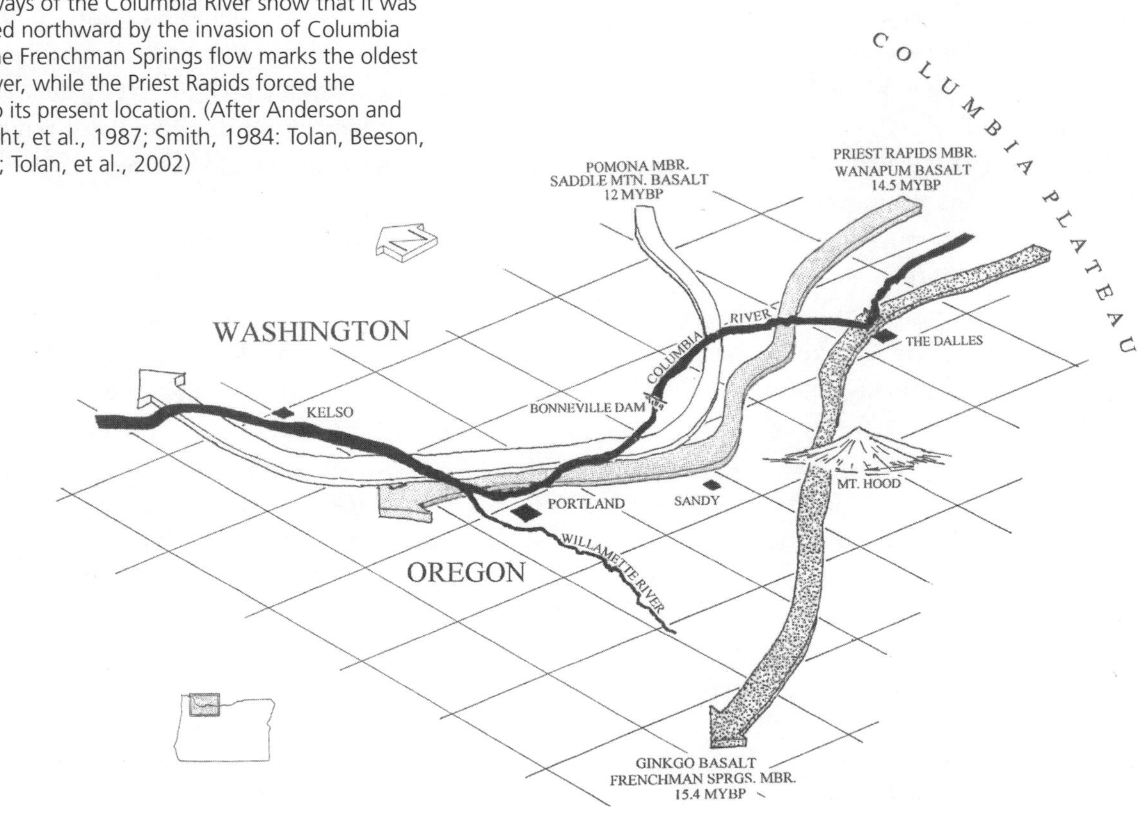

Ancestral pathways of the Columbia River show that it was repeatedly forced northward by the invasion of Columbia River basalts. The Frenchman Springs flow marks the oldest course of the river, while the Priest Rapids forced the channel close to its present location. (After Anderson and Vogt, 1987; Fecht, et al., 1987; Smith, 1984: Tolan, Beeson, and Vogt, 1984; Tolan, et al., 2002)

The feature known as The Island is an erosional remnant of Pleistocene intracanyon basalt flows, which projects northward into Lake Billy Chinook at the confluence of the Crooked River on the right and the Deschutes River on the left. (Photo courtesy Oregon State Highway Department)

Between the Miocene and through the Pleistocene epochs, basalt dams and fluvial debris routinely modified the drainages of the ancestral Deschutes and Crooked rivers. From 7.4 to 4 million years ago, the Deschutes River was spread over a wide alluvial plain in contrast to the narrow canyon it occupies today. Sheets of volcanics, derived from the Cascades, repeatedly pushed the channel toward the Ochoco Mountains where it followed a route similar to that taken by the Crooked River. Once Cascade volcanism had diminished, both rivers began to cut downward before Pleistocene intracanyon flows from Newberry Volcano 700,000 years ago again clogged the channels as far as Lake Billy Chinook. The Crooked River skirted the margins, while the Deschutes cut through the barricade. But it was well into the Pleistocene Epoch before the canyons were deepened to the current levels.

In a final episode about 6,000 years ago, damming of the Deschutes River by extrusions from Lava Butte impounded the water into Lake Benham. Diatoms and other plant material from sediments suggest the lake persisted until 2,000 years ago, at which time the river cut through, and the water drained, leaving the falls over the basalt. In addition to the location of Benham Falls, rapids in the river bed mark the position of the lava flows.

Geologic Hazards

As with much of the Pacific Northwest, the Deschutes-Umatilla plateau experiences its share of earthquakes, landslides, and flooding. In general, efforts at reducing risks are greatest in areas of high population and growth. Consequently, with substantial urbanization, the Deschutes River valley has seen a more detailed examination of potentially hazardous situations than has taken place near Umatilla, Arlington, or Pendleton.

Earthquakes

Along the Columbia River between Umatilla and The Dalles, periodic earthquakes have been noted as far back as the 1800s, but local seismological data is limited. In the Umatilla basin, earthquakes are concentrated between Hermiston and Milton-Freewater along the Olympia-Wallowa lineament (OWL), but in the Deschutes River valley they may be related to the eastern California shear zone (Walker Lane).

The earliest historic earthquakes were recorded at The Dalles in 1866, at Umatilla in 1893, and at Milton-Freewater in 1936. Oregon's strongest event, with a magnitude of 6.1, struck just west of Milton-Freewater. Chimneys and houses were damaged, shelved items scattered, and 75-foot-wide, 150-foot-long cracks opened. Water rose in wells and emerged from the smaller fractures in what was interpreted as liquefaction. No deaths were recorded, but costs amounted to $100,000 in 1936 dollars. With the U.S.G.S., Gary Mann and Charles Meyer attributed the quake to recent motion along the Wallula fault zone. The Wallula system is part of the northwest-southeast-trending Olympic-Wallowa lineament at the point where it emerges from the Blue Mountains anticline. Traversing the Deschutes-Umatilla plateau from northwest Washington to the Snake River Plain, the OWL or megashear is readily visible on aerial photographs but is of uncertain origin. Mann's examination of the fresh fault scarps led him to conclude that Holocene seismicity could pose a threat to local communities and infrastructure.

On the lower Deschutes River, the community of Madras shook slightly and doors rattled during November, 1942, and a year later Bend experienced a similar incident. The Maupin quake of magnitude 4.8 in April, 1976, was so strong that houses swayed and creaked, and a roaring noise was heard. Since 2006, Maupin has averaged weekly tremors registering less than 3.0 magnitude. Geologists initially suspected that the release of stress may have been from changes in water levels deep below the surface, but a more recent explanation presented by Oregon State University researcher Jochen Braunmiller and coauthors at the 2008 American Geophysical Union conference connects them to the Eastern California shear zone, which reaches from southern California into Nevada and central Oregon.

Landslides

Steep gradients, a large number of tributary streams, sediments, and human impact all can contribute to landslides, and in the Deschutes Basin a combination of sediments of the John Day Formation, overlain by lavas of the Columbia River Basalt Group, and interfingering with the Deschutes and Dalles formations lead to deep bedrock slumping, shallow soil and debris flow, creep, and rockfalls.

Slumping rocks of the John Day Formation triggered several impressive Pleistocene landslides in a 20 square mile area of the Deschutes canyon between Round Butte Dam and North Junction. Dated from 40,000 to 10,000 years ago, enormous blocks and debris, which temporarily dammed the river, can be seen at Whitehorse rapids (The Pot), Wapinitia, Boxcar, Trout Creek rapids, and near Dant. The Whitehorse landslide is the most recent of these. At Lake Billy Chinook, an ancient slide is evident in the hummocky topography at the north end of The Peninsula, and it also underlies the flat areas where the campground, boat launch, and headquarters were placed.

In 1985, John Beaulieu mapped historic slides at The Dalles including one located along the southern limits of the city. A slowly creeping slope in the clay-rich Dalles Formation became apparent after the city allowed construction that blocked springs draining from the underlying Columbia River basalts. The water began to infiltrate the unstable rocks of The Dalles, which initiated movement over a broad area despite a low regional gradient.

Flooding

Considering the low rainfall, flooding is surprisingly frequent in this region. However, precipitous valleys, blocked drainages, channel modifications, and poor building practices play a role. If sudden summer storms generate more water than a channel can carry, a flash flood over-tops the bank.

Evidence of Pleistocene flooding in the Deschutes River at Dant is readily visible. Sandy bars of coarse cobbles and scoured bedrock are an indication of the remarkable Outhouse flood, so-named because the Bureau of Land Management constructed its toilets on boulders in the channel. The powerful flood some 3,000 to 4,000 years ago took place after an exceptional rainfall, when the build-up of water was discharged at a rate two to three times higher than any on record. Similarly, the 1861 flood, which left sediments along the lower Deschutes, demonstrated an intensity far greater than those cause by the 1964

The momentous Heppner flood of 1903 devastated the small community of Heppner in Morrow County. The photo was taken near Main Street. (Courtesy Morrow County Historic Society)

and 1996 storms, when heavy precipitation and snowmelt combined to overwhelm stream channels.

In June, 1894, what is called the greatest flood along the Columbia River, resulted from snowmelt high in the watershed. Pouring past The Dalles at 1.2 million cubic feet per second, the water overtopped the Willamette River channel, which rose over 34 feet at Portland. Undaunted residents took to boats and rafts to celebrate the Carnival of Waters.

Modern day flash floods on the Umatilla plateau vary in magnitude. The destructive flood in June, 1903, at Heppner in Morrow County followed excessive rainfall in the upper Willow Creek drainage. While flash floods are often anticipated, this one came as a surprise to residents because precipitation had been light in the community itself. The deep waters killed 247 Heppner residents out of a population of 1,500. This same region experienced one of the largest flash floods in the history of the United States. In July, 1964, rains in Speare Canyon, a tributary of the Umatilla River, pushed a wall of water, mud, rock, and debris that rose over 200 feet wide and 10 feet deep. Highways and homes were destroyed and one person killed above the community of Echo.

Natural Resources

Industrial Rock

With the exception of limited quantities of diatomite and pumice, the province is not rich in mineral resources. Near Terrebonne, diatomite was mined by Ori-Dri Company from the late 1950s until the deposit was exhausted in 1992. Created when lavas from Newberry Volcano blocked the Deschutes River, a Pleistocene freshwater lake accumulated a 60-foot-thick layer of diatom-rich ash. Today little remains of the beds except for waste piles from the strip mine operation. Highly absorbent,

The ease with which the young person lifts the chunk of pumice is proof of its light weight. (Photo courtesy Oregon Department of Geology and Mineral Industries)

light-weight, and porous, diatomite is marketed as an absorbent to clean up spills, as a non-chemical insecticide, and as cat litter.

Near Bend 15-to 40-foot-thick tephra beds are mined by the Cascade Pumice and Central Oregon Pumice companies. Once extracted, the pyroclastic material is air dried, crushed, then screened to various sizes. Oregon is the leading national producer of pumice, which is used as a low density concrete aggregate in landscaping, roofing, and soil mixes. Reclamation of the site was undertaken by the company in 1982.

Geothermal Resources

The geothermal potential on the Deschutes-Umatilla plateau is moderate to low. North of Cove Palisades State Park, the Confederated Tribes of the Warm Springs Reservation operate a spa built around mineral springs arising from Clarno basalts. The waters reach 140° Fahrenheit and have a mild sulfur odor.

Surface and Groundwater

Surface water in the Deschutes-Umatilla province is controlled by the lengthy Columbia River system and its tributaries the Deschutes, John Day, Willow Creek, and the Umatilla. Regional groundwater sources are supplied by interbeds within the Columbia River basalts, by Miocene basin sediments, and by Pleistocene lake and stream deposits. Recharge of both surface and groundwater comes from snowmelt or rainfall or through leakage from unlined irrigation canals and sprinklers.

The basalts may store ample amounts of groundwater, but vertical transfer can be limited by faulting or clay seals. Consequently, the separate layers act as discrete aquifers with slow percolation and recharge. Papers by Washington consultants Terry Tolan, Kevin Lindsey, and co-authors address the regional hydrogeology of the Columbia River Basalt Group with attention to the relationship between aquifers and the flow sheets.

After monitoring water levels on the Columbia plateau for 40 years, the Washington Department of Ecology found that the aquifers began to show early signs of decline. The shallowest groundwater levels were the first to drop, but in the last 15 years drastic losses began to appear in the deeper zones. In 2010 geologists with the U.S.G.S. confirmed an 83 percent drop in water levels across the entire Columbia plateau since 1984, although they had earlier predicted there would be no long-term problems. Aquifers supplying Bend and Redmond experienced a 20-foot decline.

Attempting to rectify the shrinking water situation and rescue the local farm economy, the 2009 Oregon legislature financed a plan to pump water from the Columbia River into the aquifers, regardless of possible contamination and a limited storage capacity.

Aquifers supplying Bend and Redmond experienced a 20-foot decline. In areas where surface water appropriations have been closed for years, groundwater permits are still being granted by the Oregon Water Resources Department, and the state

In the dry climate of the Deschutes plateau, a combination of low precipitation, intensive agriculture, population expansion, and the perception that the water supply was boundless have proven to be calamitous. To sustain growth, water diversion canals, groundwater pumping, and numerous creative schemes aim to adjust water use. Near Bend, present-day irrigation canals (seen in photo) divert over 60 percent of the river's annual flow. The definitive *Waters of Oregon*, a source book by Rick Bastasch, lists 116,000 irrigated acres in Umatilla County and 37,160 in Deschutes. (Photo courtesy J. Mooney)

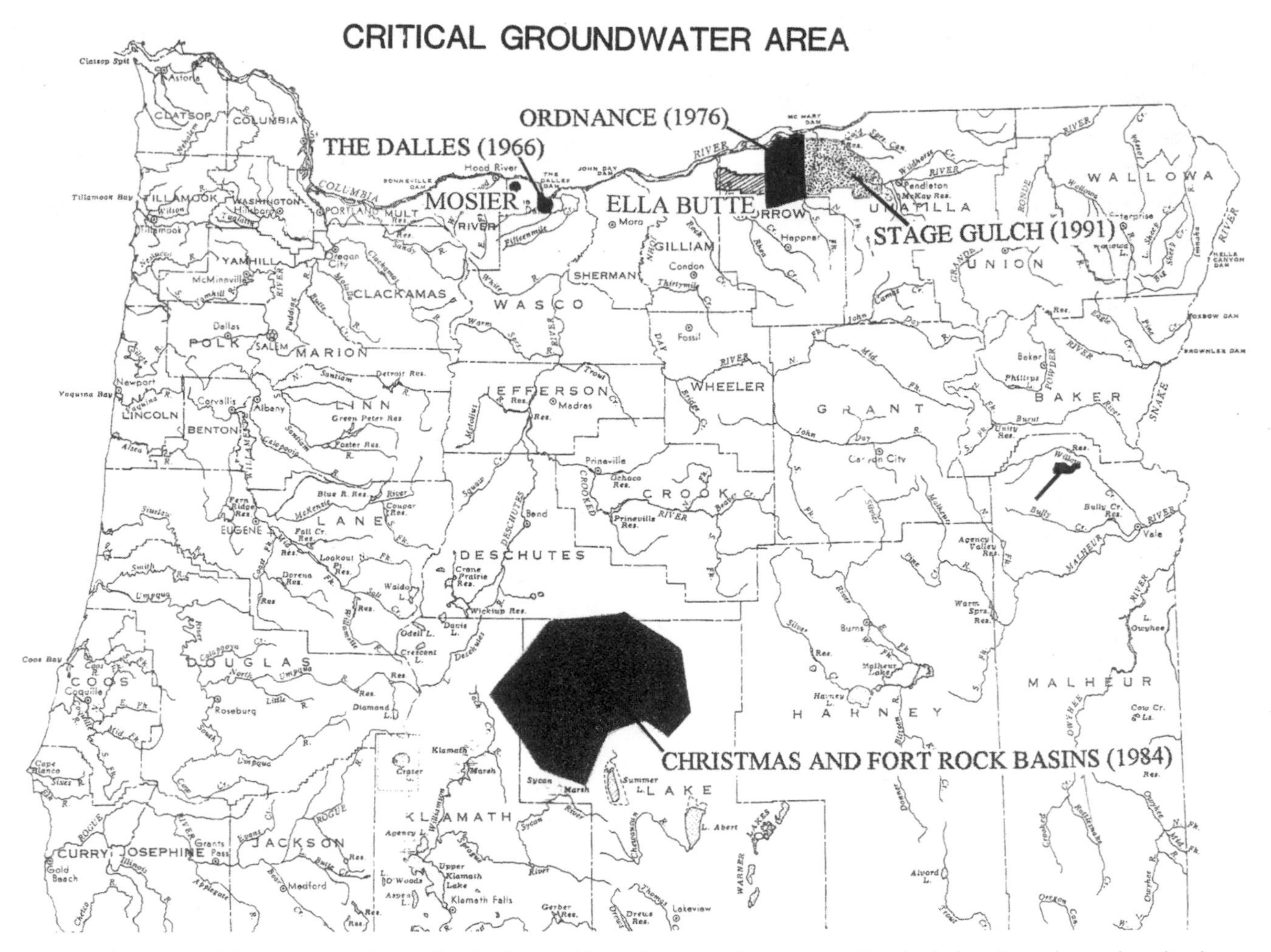

Designated as areas of depleted groundwater by the Oregon Water Resources Department, the shaded sections show where levels on the Columbia plateau have been dropping since the 1960s. (After Bastasch, 1997; Oregon Water Resources Department, 1988; Orr and Orr, 2005; Zwart, 1990)

legislature extended the permissible pumping period in the Deschutes basin until 2013.

Deschutes Basin. The hydrology of the Deschutes River is distinguished by two characteristics, the flow and the erosion. The flow at the mouth of the river remains relatively consistent throughout the year, with little variation in volume between the lowest in late summer and the highest in mid-winter. This is because an extensive aquifer complex receives a continuous year-round supply from the High Cascades. While there is little difference in the discharge at the mouth, the natural flows in the upper section have been altered substantially by dam construction and irrigation. Here the normal high winter-low summer situation found elsewhere has been reversed.

Another notable trait of the Deschutes River is that very little erosion occurs and only a small amount of sediment is being carried and deposited in the channel. Because the bed is primarily through hard, unweathered volcanic terrain, it is not easily broken down by fluvial processes. Behind the Round Butte and Pelton dams the build-up of debris since construction 45 years ago is minimal.

Groundwater in the Deschutes basin moves eastward from the Cascades and north from Newberry Volcano. Near Madras, the low permeability of Clarno and John Day strata forces an enormous volume of water into the more porous overlying Deschutes Formation, where it contributes heavily to streams and appears as springs. The Metolius and Opal springs are the largest two in the watershed. At a chilly 48° Fahrenheit, springs at the base of Black

Butte are the source for the Metolius River, which discharges at a remarkably constant rate of 1,500 cubic feet per second toward Lake Billy Chinook. However, in times of heavy rains such as fell during the winter of 1996, the flow soared to a record high of 8,430 cubic feet per second.

Along the Crooked River Gorge, springs emerge intermittently within a seven-mile stretch between The Cove at Lake Billy Chinook and Smith Rock. Opal Springs exits from the east bank at a temperature of 53° and at a rate of 80 million gallons a day. Opalized pebbles in the basin gave the springs its name.

Umatilla Basin. As elsewhere on the plateau, surface and groundwater in the Umatilla basin display wide seasonal variations. The arid climate, a slow rate of recharge for both alluvial and basalt aquifers, and agriculture usage have steadily reduced the levels. Discharge peaks in the spring but diminishes to low periods in August or September. Sudden storms can temporarily elevate the amount of water in the channel, and diversions below Pendleton additionally alter the flow.

Declines in groundwater near Ordnance, Hermiston, and Stanfield, and in the Willow Creek drainage led to the Oregon Water Resources Department imposing legal limitations on withdrawal. Designation as a critical groundwater area means that the amount being drawn from the aquifer exceeds the estimated natural long-term recharge.

Geologic Highlights

Cove Palisades State Park

A secluded spot, just above the juncture of the Deschutes, Crooked, and Metolious rivers, was known as The Cove by settlers in the early 1900s. The site and surrounding 7,000 acres were purchased and officially designated as Cove Palisades State Park in 1940. The Deschutes River has both a State Scenic Waterway and a National Wild and Scenic River designation, and, although some sections of the banks are privately owned, most is public land.

The Ship at Cove Palisades State Park is composed of tuffaceous Deschutes Formation capped by rimrock basalt. (Photo courtesy Oregon State Highway Department)

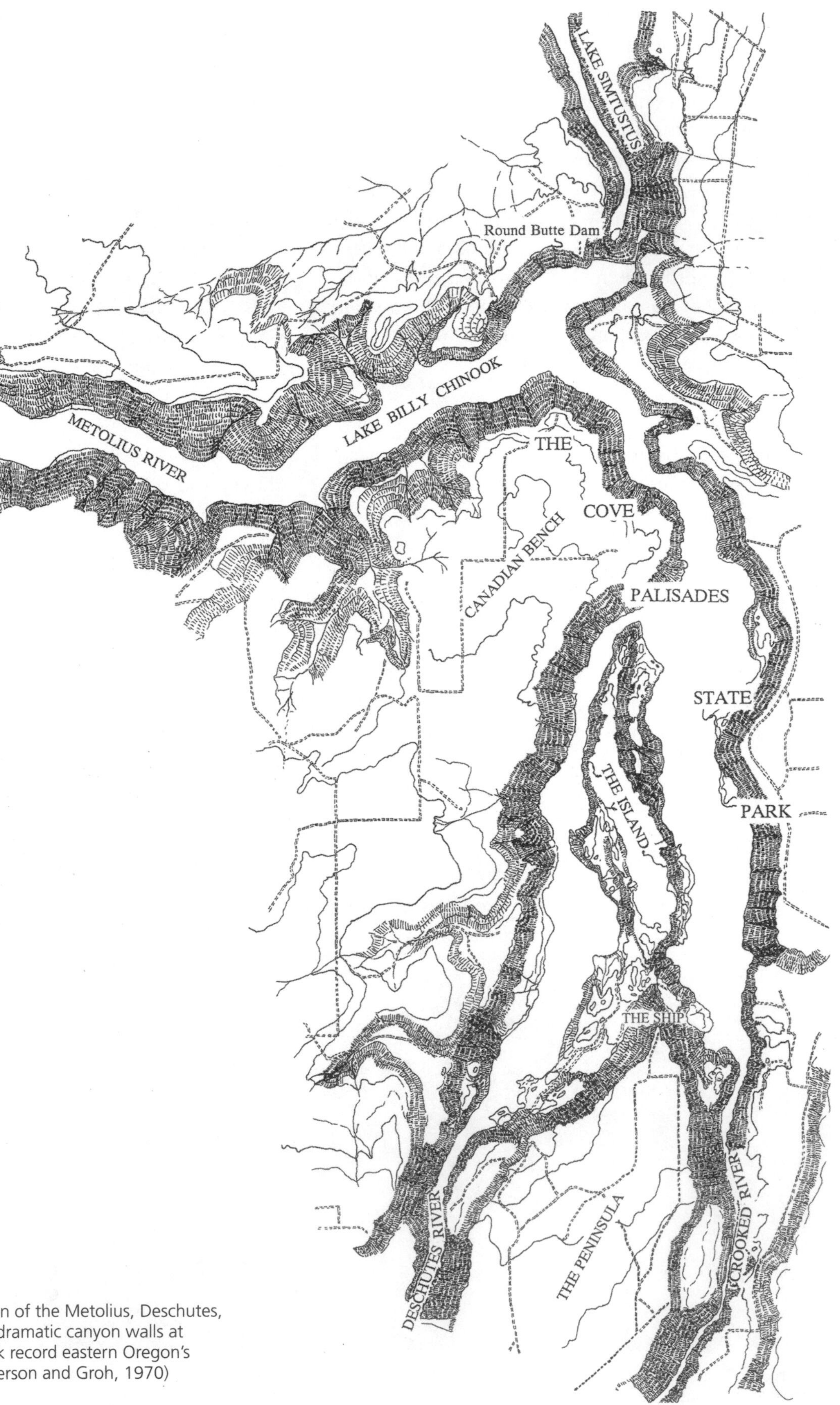

Situated near the junction of the Metolius, Deschutes, and Crooked rivers, the dramatic canyon walls at Cove Palisades State Park record eastern Oregon's geologic past. (After Peterson and Groh, 1970)

The park includes Lake Billy Chinook and Lake Simtustus. Impounded by Portland General Electric, the Pelton Dam in 1958 backed up Lake Simtustus, and in 1964 Round Butte Dam filled the three arms of Lake Billy Chinook. These altered the complexion of the river considerably, arresting the pace of the water and decimating the migrating steelhead population. Only in 2004, when federal re-licensing came due, did PGE agree to improve fish passageways.

The geology of Cove Palisades State Park was reviewed in 1970 by Norm Peterson and Ed Groh, and updated by Ellen Bishop and Gary Smith in 1990. The emplacement of mid to late Miocene lava flows and thick fluvial sediments is recorded by colorful rocks and in scenic features such as The Ship, The Island, and The Peninsula. Contrasting yellowish-brown sands and black basalts in the canyon walls are from High Cascade eruptions, whereas the distinctive light-colored red to lighter pink and white intervals are ignimbrites of the Deschutes Formation. Ignimbrites are incandescent air-borne ash that falls to the ground to cool as a glassy layer. At Cove Palisades, an ignimbrite armors the distinctive white prow of The Ship, a high promontory at the end of the Peninsula. Dark sandstones and conglomerates of the Deschutes Formation lie below that, and ash is visible at the base. The Peninsula is made up of lavas extruded 5.4 million years ago from the Tethrow Butte cinder cones.

Distinctive iron-stained brown lavas, which erupted during the Pleistocene from Newberry Volcano, invaded and plugged the canyons as far as Round Butte dam, a distance of 40 miles from source vents. These layers can be seen in the cliffs around the park. Even though the basalts are considered to be the source for The Island, some geologists favor a yet-to-be-found fissure toward the south. Known to pioneers as the Plains of Abraham, The Island is an isolated flat mesa, which rises to a spectacular 450-foot-height above the lake shore.

Balanced Rocks

John Newberry, who accompanied a railroad exploring expedition to Oregon in 1855, examined the Deschutes basin in great detail. One of his discoveries was a cluster of precariously situated rocks on the north-facing slope of the Metolious River, where it enters Lake Billy Chinook.

Balanced rocks, each weighing over a ton, are perched atop tapering pedestals that reach to 30 feet in height. Resembling the Easter Island statues, the pedestal and top knot are Deschutes Formation ignimbrites. Both the columns and pedestals are the product of differential weathering, in which lower strata have been removed to leave the resistant cap. (Photo courtesy Oregon State Highway Department)

Peter Skene Ogden State Park

Named after one of the Northwest's first fur traders, Peter Ogden, the state park was dedicated in 1927 at the point where U.S. Highway 97 crosses the Crooked River gorge. A little over 100 acres in size, the park includes the bridge, which is 290 feet above the Crooked River at this point. The original structure over the chasm was constructed in the same year, but in 2000 a new concrete arch was opened to the east. The older bridge is now limited to foot and bicycle traffic. Pleistocene lavas from Newberry Volcano form the canyon walls in the park.

Both perspectives show the Crooked River Bridge on Highway 97 where it is 290 feet above the river. (Courtesy Oregon State System of Higher Education and Condon Collection)

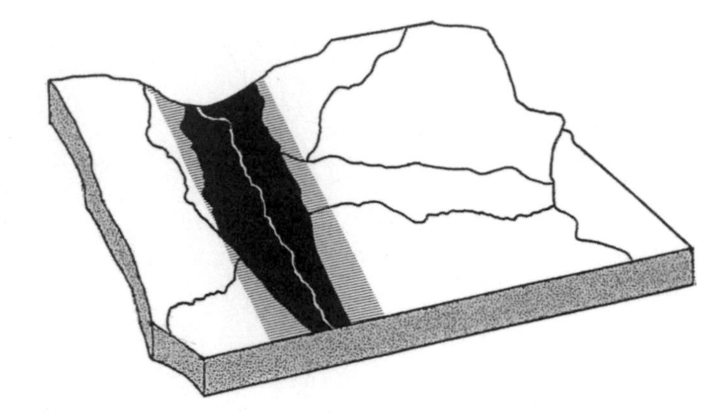

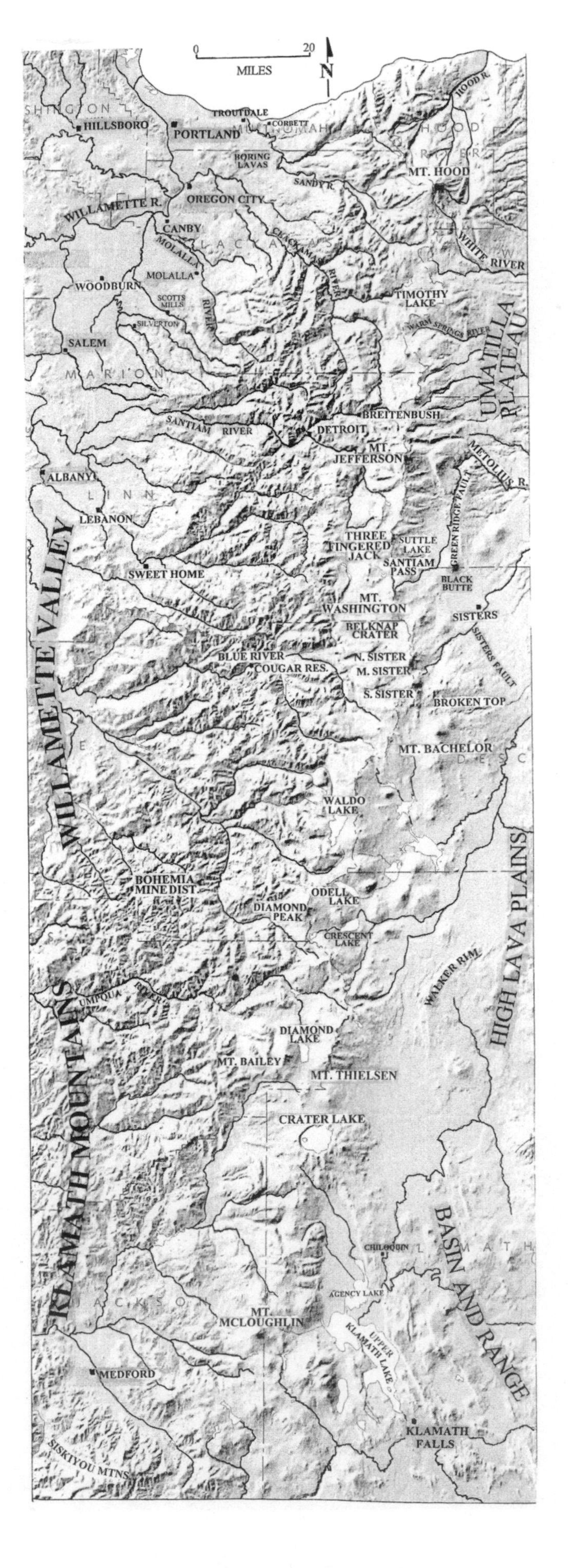

Physiographic map (After Loy, et al., eds., 2001)

Cascade Mountains

Landscape of the Cascade Mountains

The 600-mile-long Cascade Mountains from northern California through Oregon, Washington, and British Columbia provide some of the Pacific Northwest's most dramatic scenery. Touching all other provinces in Oregon except for the Blue Mountains and the Owyhee plateau, the Cascades lie east of the Willamette Valley. The Deschutes-Umatilla plateau is toward the northeast, the High Lava Plains is directly eastward, with the Basin and Range and the Klamath Mountains to the south. A small boundary is shared with the Coast Range.

In Oregon, the range is divided into the older, deeply eroded Western Cascades and the sharply contrasting snow-covered High Cascade peaks. With heights from 1,700 to 5,800 feet, the Western Cascades are only half the elevation of the younger summits, which reach in excess of 11,000 feet. This north-south volcanic chain divides the state into two distinct climate zones. Mild temperatures, high rainfall, thick soils, and heavy vegetation characterize the western slope, but on the east side the climate is warmer and drier, and the vegetation more sparse.

The major waterways drain westward. The Clackamas, Molalla, Santiam, and McKenzie rivers all wind their way from headwaters in the Cascade range to merge with the Willamette River system. Draining from the Calapooya Mountains of the Western Cascades, the Umpqua crosses the Coast Range to the Pacific Ocean at Reedsport, a distance of 112 miles. By flowing west then turning toward the north at Roseburg, the entire route of the Umpqua lies within Douglas County.

Eastward-flowing streams originating in the Cascades are comparatively small. With an average length of 30 to 35 miles, the White, the Warm Springs, and the Metolius all enter the Deschutes waterway. The Columbia, which is the demarcation between Oregon and Washington, is the only river that completely traverses the mountains from east to west.

Past and Present

In the 1930s Edwin Hodge from Oregon State University, Howel Williams from the University of California, and others addressed specific aspects of Cascade geology. Twenty years later a cooperative effort between Dallas Peck of the U.S.G.S. and the Oregon Department of Geology and Mineral Industries (DOGAMI) produced the first large-scale reconnaissance of the stratigraphy, structure, and lithology of 7,500 square miles of the central and northern Western Cascades. During the 1960s into the 1980s, papers by Paul Hammond from Portland State University and Edward Taylor at Oregon State University filled in details of the geology.

Born in 1898 in Liverpool, England, Howel Williams' undergraduate emphasis was on geography, archaeology, and geology, but he began to concentrate more on geology after his twin brother entered that field. Williams' PhD in 1928 from the University of Liverpool was awarded after his appointment to the University of California, Berkeley. In addition to providing some of the first geologic overviews of selected High Cascade peaks, Williams is, perhaps, best known for his recognition of Crater Lake as a collapse caldera, and for his compilations of volcanic data worldwide. Meticulous pen and ink drawings and field sketches illustrate his many publications. Williams died in 1980. (Photo courtesy Geology Department, University of Oregon)

A professor of Geology at Oregon State University, Edward Taylor encountered many students in his 33-year career of teaching mineralogy, crystallography, and field geology. Growing up in Corvallis, Taylor completed a PhD from Washington State University in 1967. His early research focused on the ages and evolution of volcanic activity in the central High Cascades with publications on individual peaks. He named and developed a stratigraphic sequence for the many ash-flow sequences west of Bend. An emeritus professor, Taylor lives in Corvallis. (In the photo, Collier Cone and North Sister are in the background; courtesy E. Taylor)

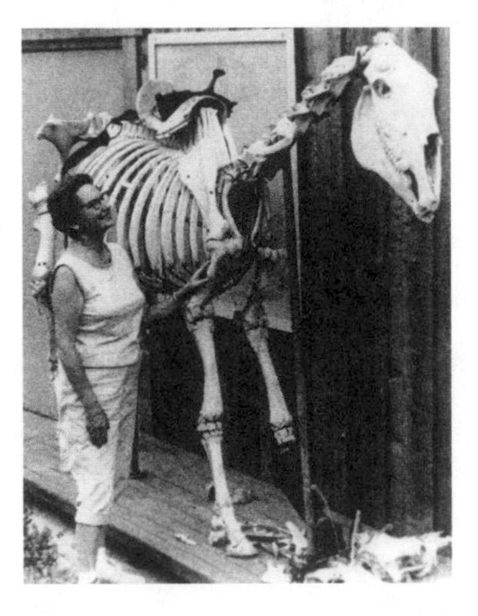

The careers of both Margaret Steere and Beverly Vogt at DOGAMI involved the roles of geologist, editor, and consultant. Originally from Michigan, Steere moved to Oregon to become the resident state paleontologist, identifying fossils, leading field trips, and giving talks. Editing articles for the book *Fossils in Oregon*, Steere retired in 1977 and died unexpectedly in 1995.

Beverly Vogt joined DOGAMI in 1977 where she produced *Oregon Geology* magazine, set up conferences, and conducted public outreach. Also from the Midwest, Vogt finished her graduate degree at Portland State University while still accomplishing the many aspects of her work in the department. After twenty years of service, Vogt retired in 1997 and currently lives in Portland. (Photos courtesy Oregon Department of Geology and Mineral Industries; and B. Vogt)

In light of plate tectonics, current researchers are attempting to clarify different aspects of Cascade volcanism. Among these, George Priest at DOGAMI and Richard Conrey at Washington State University are proposing several ideas to explain the complexities of arc evolution. Maps and reports by William Scott, David Sherrod, and James Smith at the U.S.G.S. refine the geology of individual stratovolcanoes.

Overview

It is curious that the Cascade Range is regarded as the backbone of the state, yet it is one of Oregon's youngest geologic provinces. The Cascades began during the Eocene Epoch with outpourings of ash and lava from a line of cones parallel and adjacent to the Pacific Ocean. Fed by the subduction of the Farallon oceanic plate beneath North America, the lengthy volcanic arc erupted thick deposits that built the foundation of the Western Cascades. An interval of tilting and folding 5 million years ago brought a cessation to the activity and accelerated stream erosion.

The eruptive centers migrated steadily eastward between the late Miocene and Quaternary to erect the High Cascade peaks. Interspersed between the High Cascade stratovolcanoes and shield cones, black cindery fields erupted from hundreds of sites just a few thousand years in the past. An overprint of faults and the development of a High Cascade trough or graben, which extends most of the length of the state, represent crustal failure on a grand scale.

Cascade lavas exhibit a diversity of timing, spacing, and composition, and several models have been proposed to explain the differences. Most are related to subduction and crustal extension. Additionally, the volcanic arc has been divided into segments based on the location of the vents, and the individual centers have been grouped by the eruptive frequency or chemistry of the lavas.

During the late Pleistocene, less than 100,000 years ago, ice modified the Cascade landscape even as lava erupted above and beneath glacial sheets. The size of lahars, fast moving muddy slurries of ash, water, and rock, was increased by the addition of glacial melt-water and rubble, which choked channels downstream. On the mountain tops, glaciers enlarged valleys, filled lakes, and eroded sharp crests.

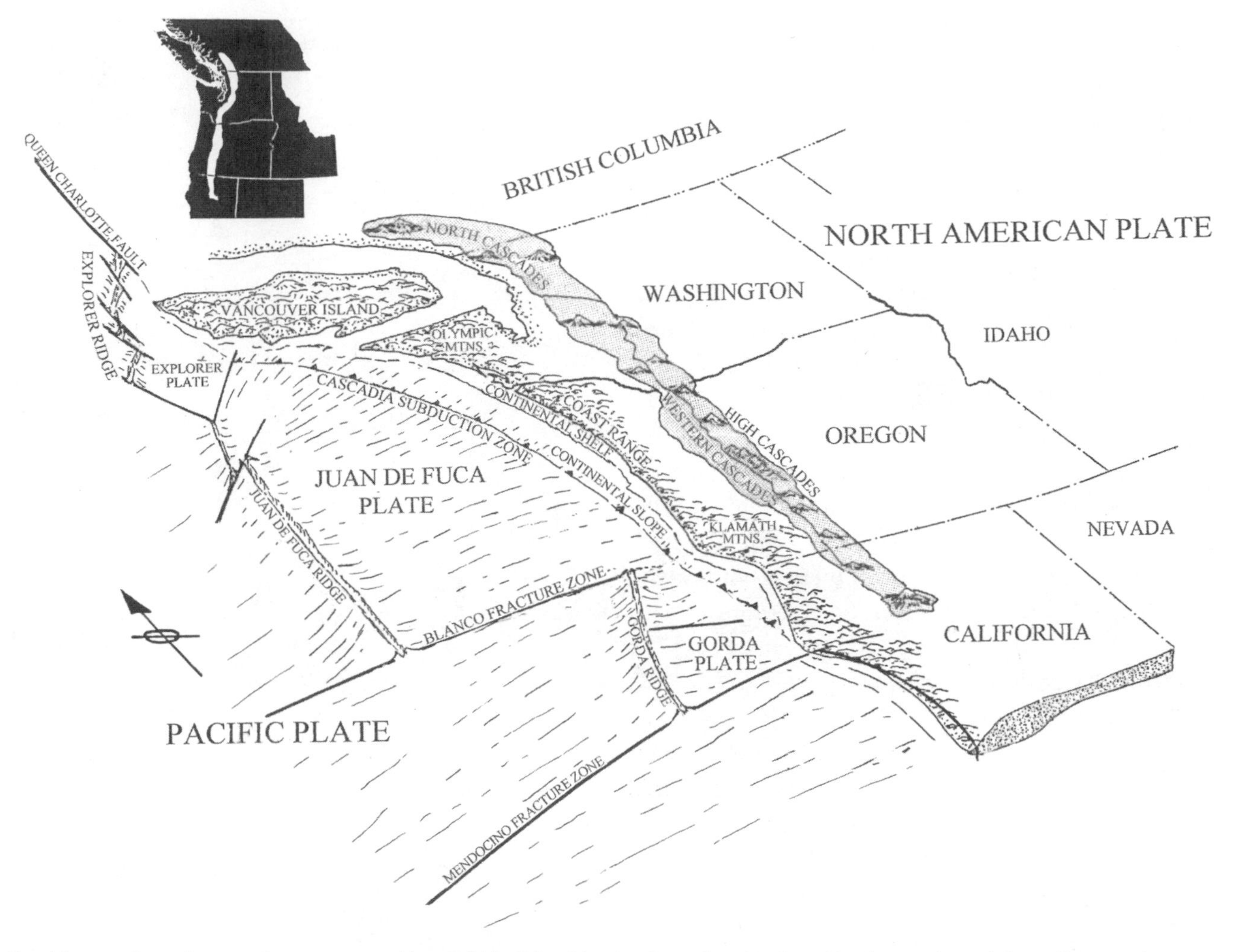

Reaching northward across three states and into British Columbia, the Cascade volcanic arc is a by-product of the collision and subduction of three tectonic plates and the North America continent.

Since the onset of High Cascade volcanism just over 7 million years ago, stratocones have continued to be active into historic times. European settlers witnessed smoke from Mount Hood, and most recently the Pacific Northwest saw the eruption of Mount St. Helens and the slow bulge in the Three Sisters area. Increased public awareness and government monitoring programs for potential volcanic activity may help to mitigate potential disasters.

Geology

Cenozoic

Beginning in the Eocene, around 40 million years ago, a volcanic arc was constructed parallel to the continental margin and well to the east of the Pacific Ocean shoreline. Volcanism was generated by convergence of the Farallon and North American plates. As the Farallon plate plunged beneath North America, partial melting of the mantle at depths of 60 to 75 miles created magma chambers that rose through the crust to feed the volcanoes.

Lavas and ash from numerous volcanoes in the chain built the Western Cascade range, but tectonic adjustments brought a profound reduction in the amount of eruptive material, a narrowing of the arc, and limited periods of activity in the late Miocene Epoch around 7.5 million years ago. As the volcanic front shifted eastward, eruptions of the High Cascades commenced. Differences in the nature of the High Cascade regime relate to distance between the subduction trench and the volcanic arc, to the temperature, buoyancy, and rate of plate movement, and to the angle of the subducting slab.

The Cascade Range can be divided into a northern segment from British Columbia to Snoqualmie Pass, Washington, and a southern portion from Snoqualmie Pass into California. The southern section has been broken into the older dissected Western Cascades, which consists of Eocene to Miocene volcanic debris, intrusives, and marine sediments, and the High Cascade peaks that include late Miocene to Holocene ash and lava. Because of deep erosion and burial by younger lavas, traces of the volcanoes and landforms of the Western Cascades were obscured, and in Washington exposures of batholiths are the only evidence of early activity. High Cascade volcanism differs in Washington, where stratovolcanoes are isolated edifices, in contrast to Oregon and California. where large composite peaks are surrounded by fields of cinder cones, shield volcanoes, and recent flows.

Western Cascade Volcanism and Sedimentation

Because the 10,000-to-20,000-foot-thick sedimentary and volcanic underpinnings of the Western Cascades are limited in areal extent and not well defined, the individual layers are arranged by stratigraphic position and composition. Andesites, basaltic andesites, and dacites of the early Western Cascades from 35 to 17 million years ago were followed by a second late Western Cascade episode of flows and ash between 17 and 7.5 years ago. Volcanic output was highest during the first period, but declined thereafter.

Deposited along the ocean shoreline, which paralleled the eastern border of what is now the Willamette Valley, marine and nearshore sands and volcanic tuffs of the Eugene and Fisher formations denote the earliest sediments of the central Western Cascades. Outcrops near Eugene preserve molluscs and leaf imprints. Overlying the Fisher, the Little Butte flows and tuffs erupted from some 30 vents irregularly spaced from Roseburg in Douglas County to the North Santiam River in Marion County. Near Molalla, the Little Butte is adjacent to or interfingers with the late Oligocene Scotts Mills Formation along the eastern marine strand. To the north, it is overlain by the Columbia River basalts and the Sardine lavas. Mudflows and ash of the Little Butte preserve wood, leaves, and pollen in

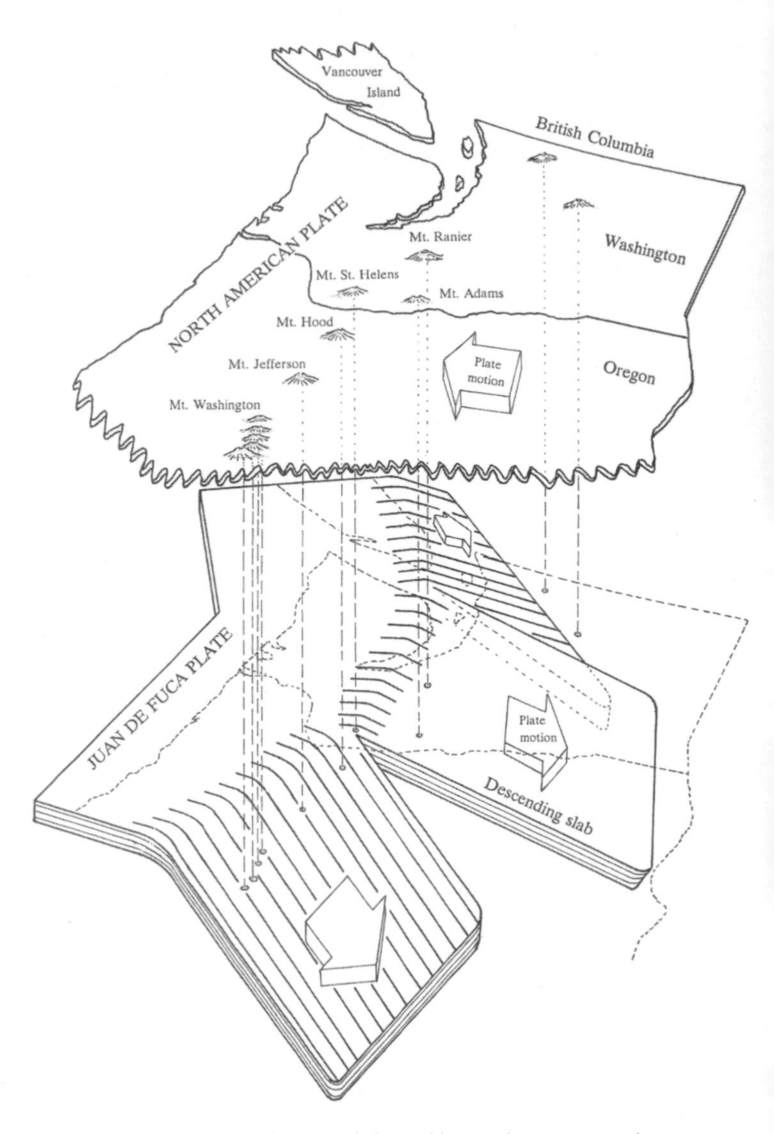

As the Juan de Fuca plate is subducted beneath Oregon and Washington, it splits into smaller tongues. Because the leading edge of the plate is tattered and uneven, the volcanic peaks in the Cascade chain are not aligned. (After Michaelson and Weaver, 1986; Duncan and Kulm, 1989)

Clackamas County and near Eagle Creek in Multnomah County.

At Siskiyou Pass in southern Oregon, Little Butte eruptives cover the Payne Cliffs and Colestin formations. Lavas, pyroclastic flows, ash, and lahars (mudflows) of the Eocene Payne Cliffs and the Oligocene Colestin formations record some of the earliest volcanic pulses of the Western Cascades. Delineated to the north by the Siskiyou Summit fault, a slowly subsiding graben filled with Colestin detritus carried by streams from Cascade slopes. Sediments of the Colestin, rich with plant remains, reflect mild coastal conditions. Petrified wood of

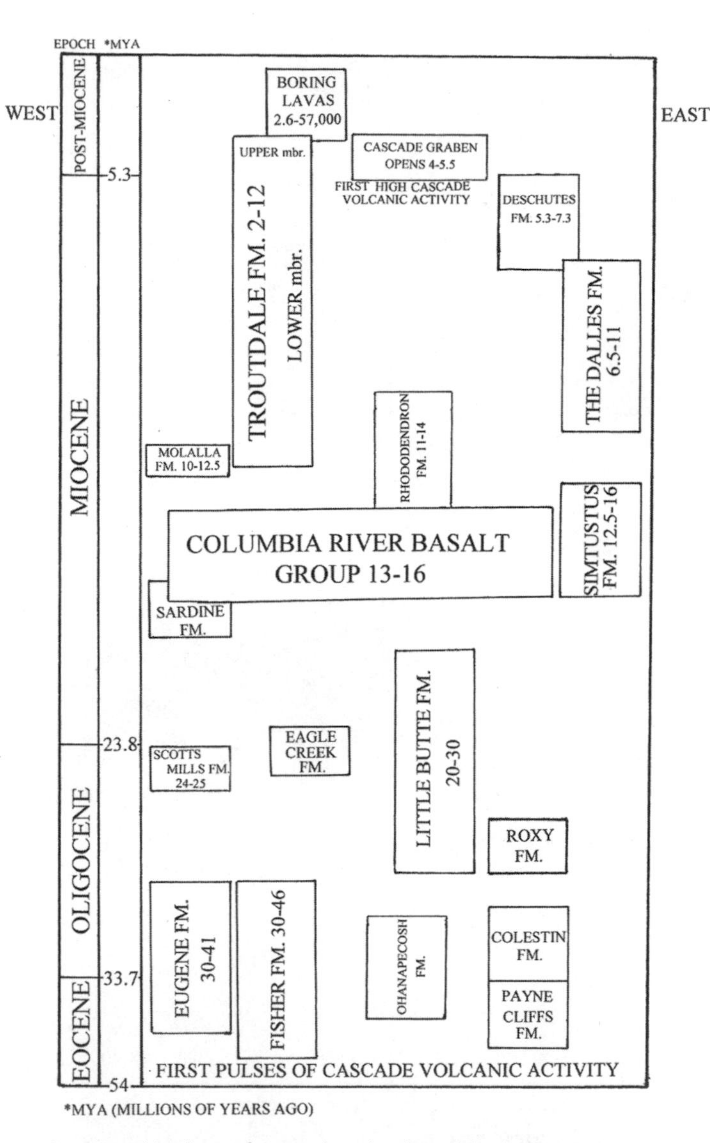

Stratigraphic chart of Cascade volcanic activity in Oregon. (After Bestland, 1987; Hammond, 1989; McKnight, 1984; Peck, et al., 1964; Priest, 1990; Priest and Vogt, 1983; Sherrod and Smith, 2000; Smith, 1993)

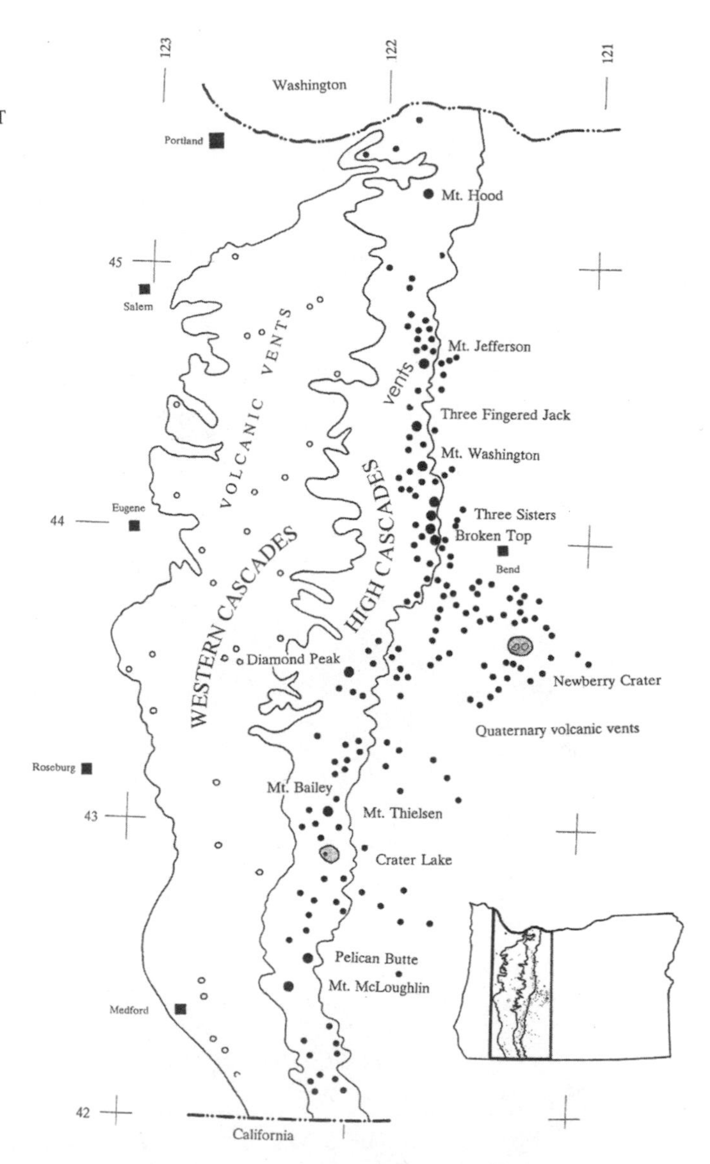

This map shows the contrast between the frequency and spacing between the older Western Cascade vents and the High Cascade volcanic cones. (After Peck, et al., 1964)

tropical tree ferns (*Cibotium*) and palm (*Palmoxylon*) mark a profound climatic shift from tropical to temperate by the early Miocene, a change also reflected by fossil plants in central Oregon.

A decline in central Western Cascade activity coincided with the onset of the Columbia River basalt eruptions from vents further east. But following this decrease, a late Western Cascade episode began around 17 million years ago with andesitic, basaltic andesite, and dacitic lavas of the Sardine Formation. Mapped by Dallas Peck, about a dozen vents of the Sardine are spaced from the McKenzie River in Lane County to Breitenbush Hot Springs in Marion County. Successions of Sardine Formation lavas cover much of the northern Cascades with thicknesses reaching 3,000 feet.

The Western Cascades were elevated in the early Pliocene, around 5 million years ago, contemporaneous with faulting and formation of High Cascade grabens. Entrenched streams and narrow canyons attest to the rapid erosion initiated by the uplift.

High Cascade Volcanism

Shifting eastward between the late Miocene and Pliocene epochs, the Cascade volcanic arc became increasingly narrow, signaling the onset of High Cascade eruptions and the renewal of volcanism after a period of low output. Composed of basaltic

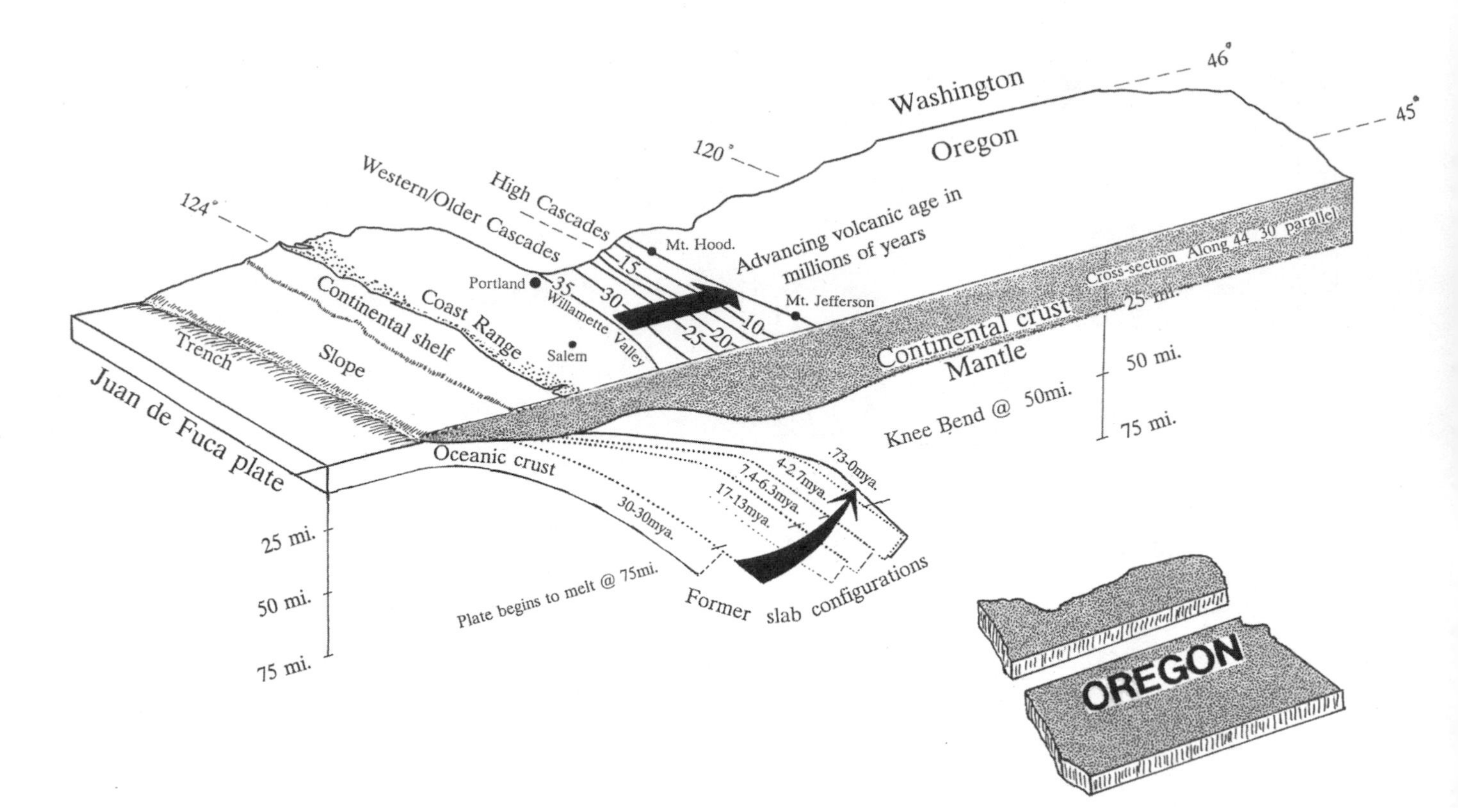

During the late Tertiary, the active Cascade volcanic arc migrated eastward and became increasing narrow, as the volume of lavas diminished. One synthesis for these phenomena calls for a knee-bend in the subducting slab as it swung upward. (After Duncan and Kulm, 1989; Priest, 1990; Verplanck and Duncan, 1987)

Volcano	Eruptive periods (years ago)
MOUNT HOOD	500,000 Hood event …. 424,000 Cloud Cap event …. 50,000-30,000 Hood River event
MOUNT JEFFERSON	1 MILLION …. 300,000 …. 150,000 …. 100,000-35,000 Main Cone ….
THREE-FINGERED JACK	…. 150,000 100,000
MOUNT WASHINGTON	?700,000 …. 300,000-200,000
SAND MOUNTAIN-NASH CRATER	….
BELKNAP CRATER	….
YAPOAH CONE 4-IN-ONE CONE	….
COLLIER CONE	….
NORTH SISTER	…. 400,000 Lower Shield Stage …. 182,000-99,000 Glacial Stage …. 80,000 Upper Shield Stage …. 70,000-55,000 Stratocone Stag
MIDDLE SISTER	…. 100,000 …. 37,000-14,0
SOUTH SISTER	…. 178,000 …. 30,00
BROKEN TOP	?700,000-100,000
MOUNT BACHELOR	….
MOUNT THIELSEN	…. >300,000
CRATER LAKE MOUNT MAZAMA	…. 420,000 Mt. Scott eruption. 400,000 Phantom Cone eruption …. 300,000-200,000 Cloudcap Bay eruption .. 185,000-110,000 Llao eruption …. 35,000 Dacite dome …
MOUNT MCLOUGHLIN	…. >300,000

Chart of Cascade eruptive periods in Oregon during the past 700,000 years. (After Bacon and Lanphere, 2006; Conrey, et al., 2002; Harris, 2005; Schmidt and Grunder, 2009; Scott, 1990; Sherrod, et al., 1996; Wood and Kienle, 1990) and basaltic andesite, lavas from both strato-and-shield volcanoes built today's familiar snow-capped High Cascades.

The High Cascade volcanoes differ from each other to some degree, and for that reason there are various parameters for classification. Based on age, George Priest broke the eruptions into an early interval from 7.4 to 4 million years ago and a late period from 3.9 million years ago to the present. Additionally, the High Cascade regime has been separated into steep-sided composite or stratovolcanoes that are long-lived and active over hundreds of thousands of years and shield cones that are broad, gently sloping structures of ephemeral but persistent activity. From north to south, eight of the Oregon composite peaks are Mount Hood, Mount Jefferson, Mount Washington, Three-Fingered Jack, the South and Middle Sisters, Broken Top, and Crater Lake. Belknap Crater, the North Sister, Mount Bachelor, Mount Thielsen, and Mount McLoughlin are shield volcanoes.

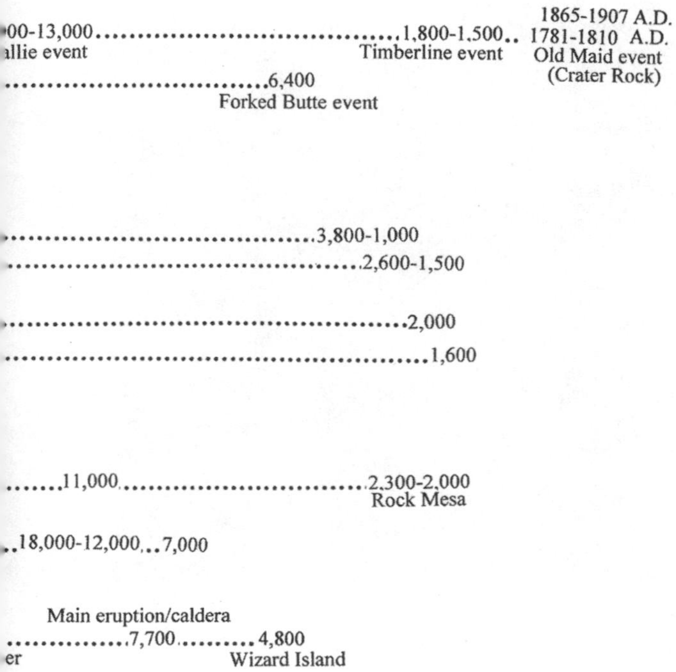

00-13,000..................................1,800-1,500.. 1781-1810 A.D. 1865-1907 A.D.
allie event Timberline event Old Maid event (Crater Rock)
..............................6,400
Forked Butte event
..................................3,800-1,000
....2,600-1,500
....2,000
....1,600
.......11,000..............................2,300-2,000
Rock Mesa
..18,000-12,000...7,000
Main eruption/caldera
..............7,700.........4,800
er Wizard Island

Of the 13 stratocones within the entire Cascade Range, few have erupted during the past 5,000 years, and in Oregon only Mount Hood, the South Sister, and Crater Lake fall within that time frame. Mount Hood and the South Sister are identified today as having the highest potential for activity. Along the length of the volcanic regime, the last recorded flows from Mount Lassen were in 1917, and Mount Baker produced steam in 1975. The explosive episodes that began in 1980 at Mount St. Helens saw dome construction and as late as 2005.

Evolution of Individual High Cascade Peaks

Northern Cascades At the northern end of the province in Oregon, Mount Hood and Mount Jefferson are both andesitic in composition, but they have different histories. Lavas erected a platform at Mount Hood 1.5 million years ago, but construction of the central cone began at 500,000 years. Although a volcanic field has existed in the vicinity of Mount Jefferson as far back as 4 million years, the stratocone dates back only 1 million years. After a dome-building phase, Mount Hood experienced several eruptive periods and continued to emit gas, smoke, and pumice as late as 1907, attesting to the proximity of an active magma source below the summit. Even today steam plumes rise near Crater Rock. By contrast, activity at Mount Jefferson ended around 20,000 years ago. The total output of lava and ash for the long-lived Mount Hood edifice is estimated at over 12 cubic miles, whereas the volume from Mount Jefferson was considerably smaller, even though a large eruption caused ash to fall in southeast Idaho.

Attempts to decipher Mount Hood's past began with observations in the early 1900s by Harry F. Reid, professor at Johns Hopkins University. From the 1940s to the 1970s, the geology and petrology were examined by William Wise of the University of California, whereas U.S.G.S. geologist Dwight Crandell recounted the eruptive history and potential future hazards. One of the best sources is the overview and fieldtrip guidebook by William Scott and others.

While the older vents at Mount Hood are obscured by younger lavas, they probably lay close to the main edifice, which was built from extensive Hood River and Cloud Cap lavas. The eruptive

Lava flows and dome construction during the last 20,000 years account for many of Mount Hood's surface features. Looking north in this photograph, Crater Rock and Illumination Ridge can be seen on the upper left skyline (west). Illumination Rock was eroded from a thick, 121,000-year-old andesitic flow, and Crater Rock is the remnant of the central dome emplaced in the Old Maid period. During the Polallie interval, Steel Cliff, on the upper right (east) near the summit, sent pyroclastics and lahars toward White River and Mount Hood Meadows. Forests of pines and fir blanket the lower slopes. (Photo courtesy Oregon State Highway Department)

events began with the dome-building Polallie from 20,000 to 13,000 years ago, the Timberline from 1,800 to 1,500 in the past, the Zigzag from 600 to 400, and the Old Maid from 1781 to 1810 A.D. The domes emplaced during the Timberline and Old Maid periods subsided into avalanches of ash, lavas, and snowmelt. The partial collapse of Crater Rock that deposited pyroclastics along the White River took place as recently as 200 years ago.

Central Cascades Edwin Hodge summarized volcanic events in the central Cascades after surveying Broken Top, the Husband, the Three Sisters, and other peaks in 1924, concluding that they were once part of a single large volcano, which had exploded and collapsed. He named the vanished structure Mount Multnomah, and his hypothesis gained wide acceptance until a reexamination by Howel Williams demonstrated that such an edifice never existed.

The South and Middle Sisters and Broken Top are major composite volcanoes that experienced long eruptive periods, in contrast to North Sister and Belknap shield cones, which have had shorter cycles. Within this group, the North Sister is the oldest at almost half a million years, and Belknap Crater is the most recent at 1,500 years ago. The Three Sisters history included at least four early pyroclastic flows, but most episodes, which built the Middle and South peaks, occurred from single

Sitting in a deep graben, Mount Jefferson has had a long eruptive history that includes a cluster of numerous domes and small shield-like volcanoes. Richard Conrey showed that the early wide expanse of eruptive sites was the consequence of crustal extension which allowed magma to spread over a broad area, whereas late Quaternary volcanism focused on a single stratocone. Lying in a deep glacially-scoured basin, Marion Lake is in the foreground. (Photo courtesy Oregon State Highway Department)

Viewed from the southwest, Mount Bachelor is a remarkably symmetrical shield cone. Broken Top is on the left, Lucky Lake is in the foreground, Lava Lake is directly behind, and Little Lava Lake is to the right. These are typical of the mountain lakes dammed by basalt. Cascade Lakes Highway (Highway 46) is the white line between the lakes. (Photo courtesy Oregon State Highway Department)

conduits during the past 100,000 years. On the northwest margin of North Sister, Collier Cone covered an extensive region with lava, cinders, and bombs (blobs of airborne lava) around 1,600 years ago, blocking and backing up the McKenzie River drainage into Spring Lake and Linton Lake.

Projecting as vertical walls above the surrounding lava fields, parallel dike swarms are characteristic along the flanks of the Three Sisters. The seven-mile-long Matthieu Lakes fissure that extends northwest from North Sister is the largest in the Cascades. Erupting through ice fields about 75,000 years ago, lava from the Matthieu Lakes vents ponded into thick platforms. This was the final overprinting after which the magma supply shifted away from the main center.

South of the Three Sisters, the Mount Bachelor cone has seen little glaciation, whereas the sharply eroded spines of Mount Thielsen, and Mount McLoughlin are all that remain of the once larger volcanoes. The 15-mile-long Mount Bachelor complex is composed of cones, flows, and three shield volcanoes, Sheridan Mountain, Kwolh Butte, and Mount Bachelor. Between 18,000 and 7,000 years ago, volcanic activity moved northward from the oldest lavas at Sheridan Mountain, to those at Kwolh Butte, to Mount Bachelor. Papers by William Scott and Cynthia Gardner discuss the eruptive history of Mount Bachelor and in particular the interaction between the lava and Pleistocene glaciers.

Southern Cascades Mount Thielsen and Mount McLoughlin are shield volcanoes that were probably active over 300,000 years in the past. Studies of Mount Thielsen began with Joseph Diller in 1902 and continued with Howel Williams in 1933. The summit of Mount Thielsen was built of pyroclastic debris and thin lava flows before two volcanic plugs intruded the central core.

Little was known of the geology of Mount McLoughlin until Leroy Maynard's field mapping for his Masters degree from the University of Oregon in 1974. Severe glacial erosion, which has carved out a distinct semi-circular basin or cirque on the northeast slope, suggests this mountain is probably one of the older High Cascade peaks, although the exact periods and sequence of activity have yet to be clarified. Multiple eruptions built the cone and surrounding blocky lavas, and in the final Pleistocene stages andesites oozed from fissures low on the western side to construct North and South Squaw Tips.

The geology of Crater Lake was first described in detail by Joseph Diller and Horace Patton of the U.S.G.S. in 1902. Diller questioned whether the caldera resulted from volcanic subsidence or from explosive eruption, an idea favored by many geologists. Some 40 years later, Howel Williams wrote a dramatic account of the growth and eruption of the volcano, and from the 1980s to the present Charles Bacon of the U.S.G.S. has employed several

In recent explorations of the floor of Crater Lake, Charles Bacon used submersibles and underwater cameras to reveal many bathymetric features such as lava flows and volcanic domes, landslides, and older shorelines from varying water levels. In this perspective looking toward the west, the slopes of Wizard Island cone and the central platform edifice record past lake levels. (U.S.G.S. digital image; courtesy Bacon, et al., 2002).

techniques to reveal additional details of the eruptive cycle and features of the crater wall and floor.

The construction of Mount Mazama began over 400,000 years ago with overlapping shield and stratovolcanos from successive andesite and dacite lavas that ultimately erected the foundation. The earliest eruptions from Mount Scott, Phantom Cone, and Cloudcap Bay continued until about 170,000 years ago, after which new vents at Llao Bay sent out sheets of andesites. Immediately before the main eruption, The Watchman, Redcloud Cliff, and Steel Bay domes were emplaced, and flows from Llao Bay shield cone and Cleetwood Cove were still fresh when the entire mountain exploded.

Erection of a central dome and large magma chamber around 30,000 years ago was followed by an intermediate dormant period, which ended with climactic violence from a single vent on the northeast side of Mount Mazama dated at 7,700 years. The slopes and surrounding valleys were covered with avalanches of pyroclastics to depths of 300 feet, while an incandescent ash cloud, over 30 miles high, darkened much of the Pacific Northwest. Bacon determined that the eruption took place over three days during a warm dry interval when ice would have been restricted to the highest regions. Once the immense quantities of rhyodacite magma had been evacuated from the chamber, the roof collapsed.

The caldera filled rapidly with water to its present levels within the first few hundred years, even as vents on the central and western crater floor produced lavas beneath the rising lake. It was during this period that Wizard Island and Merriam Cone erupted. Only the crest of Wizard Island rises above the water, and Merriam Cone, submerged close to the northern rim, is a mile across at the base and projects 1,320 feet from the caldera floor. The last known lavas from vents at the base of Wizard Island constructed a central platform around 4,800 years before the present.

Future violent explosions are unlikely, but smaller eruptions on the floor could propel rocks or ash beyond the rim, and submarine landslides might produce small waves.

The cold bright blue lake waters that occupy the caldera are remarkably clear because algal growth is limited to thick belts of moss on the walls where light penetrates. By dating lake sediments with pollen and diatoms from core and dredge samples, Bacon established a post-collapse chronology for the crater.

Variations in Cascade Volcanism

Patterns in the evolution of High Cascade volcanoes differ considerably, and several models account for the contradictory amounts of extruded material, for the spacing between volcanic vents, for the composition of lavas, and for the style (composite vs. shield, vs. cinder cone). Volcanologists have attributed most of these to interaction between tectonic plates and the effects of crustal extension.

With modifications of the subducting slab, the eruptive volumes dropped, and the Cascade arc diminished to a strip only eight miles wider than at present. By averaging mapped thicknesses and quantities of rocks in Oregon, George Priest was able to demonstrate that the total volume in the early High Cascades was 1,200 cubic miles as opposed to 960 in the late High Cascades. Greater rates characterize the Three Sisters area, but obvious decreases are evident south of Crater Lake and north of Mount Jefferson. A more specific approach by David Sherrod and James Smith calculated the rate of output per edifice for the entire Cascade chain. The relatively brief but explosive Crater Lake event had substantially greater production than is the case for the range in general.

Recent theories to explain the spacing and eruptive style of the Cascades have subdivided the arc into structural segments, ranging from three to six

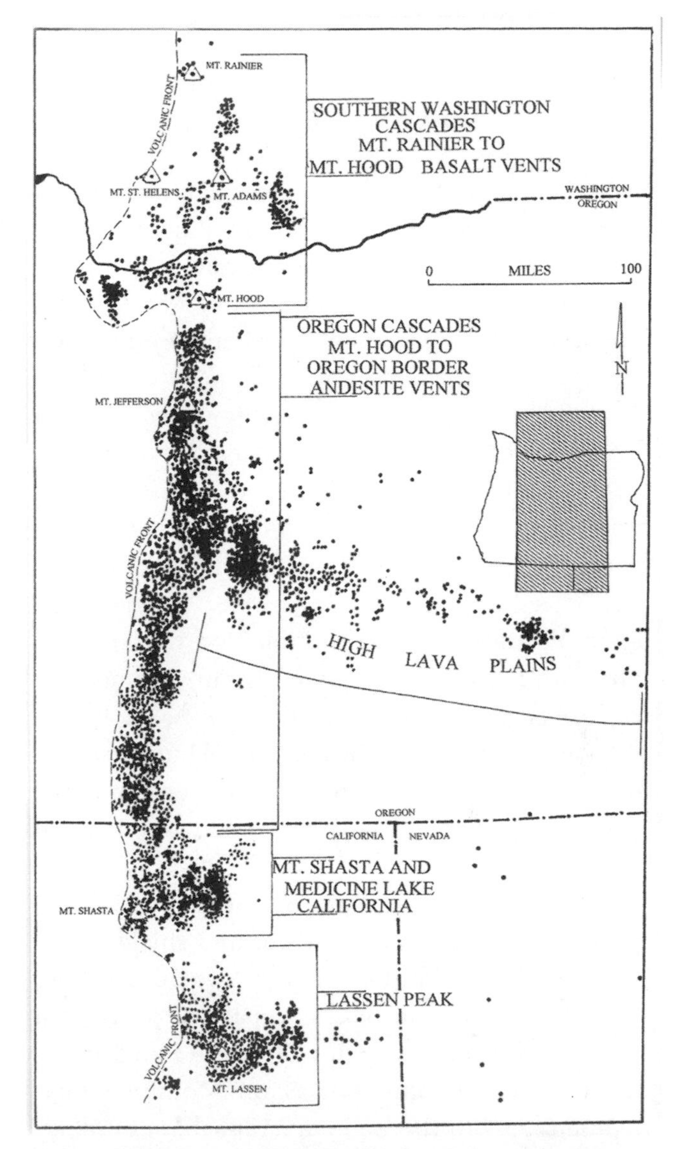

In their 1988 paper, Marianne Guffanti and Craig Weaver of the U.S.G.S. turned their attention to the late Tertiary in the Pacific Northwest. Studying the spacing of 2,821 separate vents, active since the Pliocene, they divided the Cascade range into six regions, five of which are in the Cascades and the sixth toward the High Lava Plains. The most northerly segment is from British Columbia to Mount Rainier. From Mount Rainier to Mount Hood the vents are primarily basaltic, but segment three from Mount Hood to the California border includes a dense cluster of andesitic eruptive centers. The fourth and fifth segments are Mount Shasta, Medicine Lake, and Lassen Peak in California. (After Blakely and Jachens, 1990; Christiansen and Yeats, 1992; Guffanti and Weaver, 1988; Hughes, Stoiber, and Carr, 1980; Peck, et al., 1964; Riddihough, 1984)

increments, each with a set of unique characteristics. There is presently little agreement on the various parameters, but it appears that no one section is representative of the entire range. Ed Taylor summarized the problems related to spacing and organization of High Cascade composite volcanoes noting that they are not widely distributed but instead are clustered in a few areas, they appear to be randomly placed, and they vary considerably in age.

By compiling data on 3,000 volcanic centers in the High Cascades, Paul Hammond showed that the areal distribution and concentration of centers increase between Mount Rainier and Mount Hood but diminish in numbers between Mount Hood and Mount Jefferson and in northern California. Examining the spacing of basaltic cones, Richard Conrey concluded that they average 12 miles apart while stratocones are consistently almost four times that. Since stratocones and shield volcanoes evolve separately (one is not parent to the other), two styles of volcanism in the north-central Oregon Cascades are responsible for the spacing. The incidence of long-lived stratovolcanoes is controlled by processes in the mantle or in the lithosphere (crust), while ephemeral but persistent basaltic cones are the product of extension along the arc or from bending of the subducting Juan de Fuca plate.

In the central section of the High Cascade province, the majority of cones and shield volcanoes produced basaltic (mafic) lava. During the late Quaternary, the volcanic pattern progressed southward from the oldest 3,850-year-old eruptions at Nash Crater and Sand Mountain near Santiam Pass to the Belknap, Yapoah, and Collier cones along the McKenzie Highway dated at 1,500 years, reflecting an underlying fracture zone.

Impinging on the Sand Mountain field, eruptions from Belknap Crater, Yapoah, and the Four-in-One complex were the most recent and widespread. The youngest shield volcano in the Cascades, Belknap has two summit craters, the largest of which is 250 feet deep and 1,000 feet wide. Belknap's repeated eruptions impacted 40 square miles, immersing trees and filling the McKenzie River channel. Cylindrical molds from one to five feet in diameter, as well as 35-foot-long trenches, are the remains of tree trunks consumed by lavas, which cooled and hardened around them. Where the McKenzie disappears into the porous Belknap field, it percolates underground to reemerge at Tamolitch Falls.

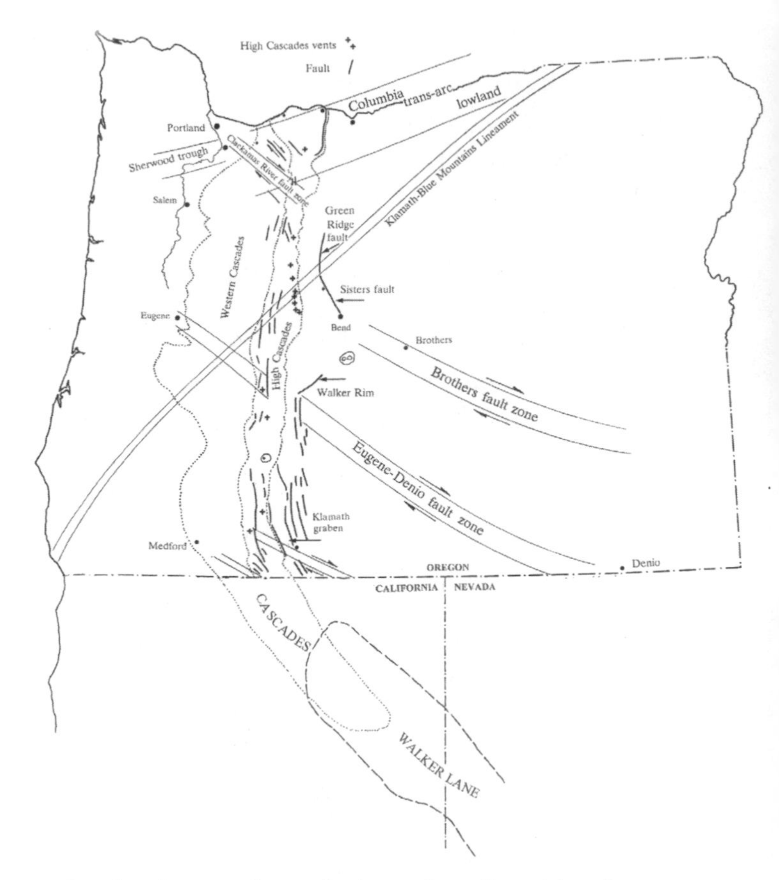

Crossing the state diagonally, the north-northwest-trending McLoughlin, Eugene-Denio, and Brothers fault zones extend from the Basin and Range and High Lava Plains into and through the Cascades. The Clackamas River belt of faults, which projects northwesterly to merge with the Portland Hills, may be an extension of the Brothers fault system, and Walker Lane, at the northern reach of the eastern California shear zone, may extend into the Cascades. (After Dokka and Travis, 1990; Faulds, Henry, and Hinz, 2005; Priest and Vogt, 1983; Venkatakrishnan, Bond, and Kauffman, 1980)

Faulting, Volcanics, and Sediments

The late Tertiary shield volcanoes and subsidence of the High Cascades into a deep intra-arc graben are associated with extensional tectonics. The word *graben*, from the German *grave*, refers to the sunken trough above a collapsed block. Structurally a graben is a depression between two faults.

As early as 1938, Eugene Callaghan and Harold Buddenhagen of DOGAMI suggested that a series of faults ran the length of the Oregon High Cascades, and considerable documentation since the 1980s has confirmed the presence of an intra-arc graben. Paralleling the axis of the Cascades between the Three Sisters and Mount Jefferson, the east side of the graben is well-defined by the steep Green Ridge scarp and the west by the Horse Creek fault. In the central section, the graben lies along the upper Clackamas River, and in the north it follows the Hood River fault.

After field mapping in 2004, Richard Conrey at Washington State University and Ed Taylor proposed a segmented time-propagating rift beginning as far back as 8 million years. Opening like a zipper from the south, a lengthy intra-arc graben began to develop, subsiding as deep as two miles in the south to less than a mile in the center. Lavas, emerging with the rifting, closely resemble mid-ocean ridge basalts (MORB).

Thick layers of lava, pyroclastics and ash from High Cascade volcanoes were discharged both east and west along the intra-arc rift. In basins to the west, explosive eruptions of breccias, tuffs, and flows of the Rhododendron Formation are intermixed with or overlain by basaltic and andesitic cobbles of the Troutdale Formation. The Troutdale is composed of sediments transported along the ancestral Columbia River canyon and spread laterally into tributary stream valleys. Above the Troutdale, the Pliocene–Pleistocene Boring lavas erupted from the many buttes and smaller hills in and around Portland and Vancouver, Washington.

Eastward, the late Western Cascades and High Cascades contributed muds, silts, and volcanic debris to the Simtustus, The Dalles, and the Deschutes formations in basins atop the Columbia River basalts. Conrey concluded that each structural portion

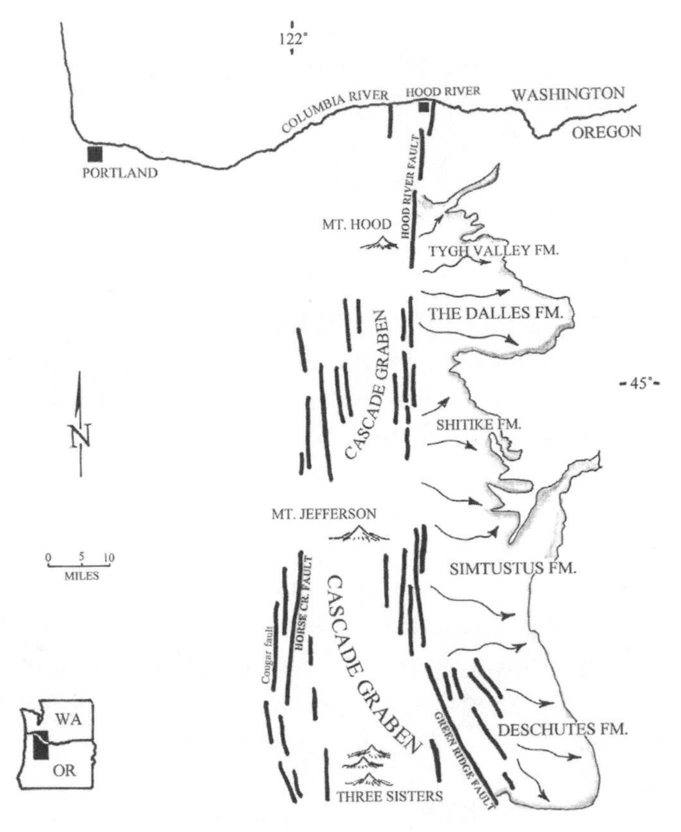

A northward propagating rift along the Cascade arc has been divided into three segments, whose boundaries line up with the stratovolcanoes. The youngest portion from the Columbia River to Hood River valley and Mount Hood is a half graben with a broken western hinge. The central segment between Mount Hood and Mount Jefferson lies along the upper Clackamas River and is a complete graben bounded by faults, while the oldest southern part from Mount Jefferson to the Three Sisters is between the Green Ridge escarpment and Horse Creek fault. (After Allen, 1966; Callaghan and Buddington, 1938; Conrey, Grunder, and Schmidt, 2004; Conrey, et al., 2002)

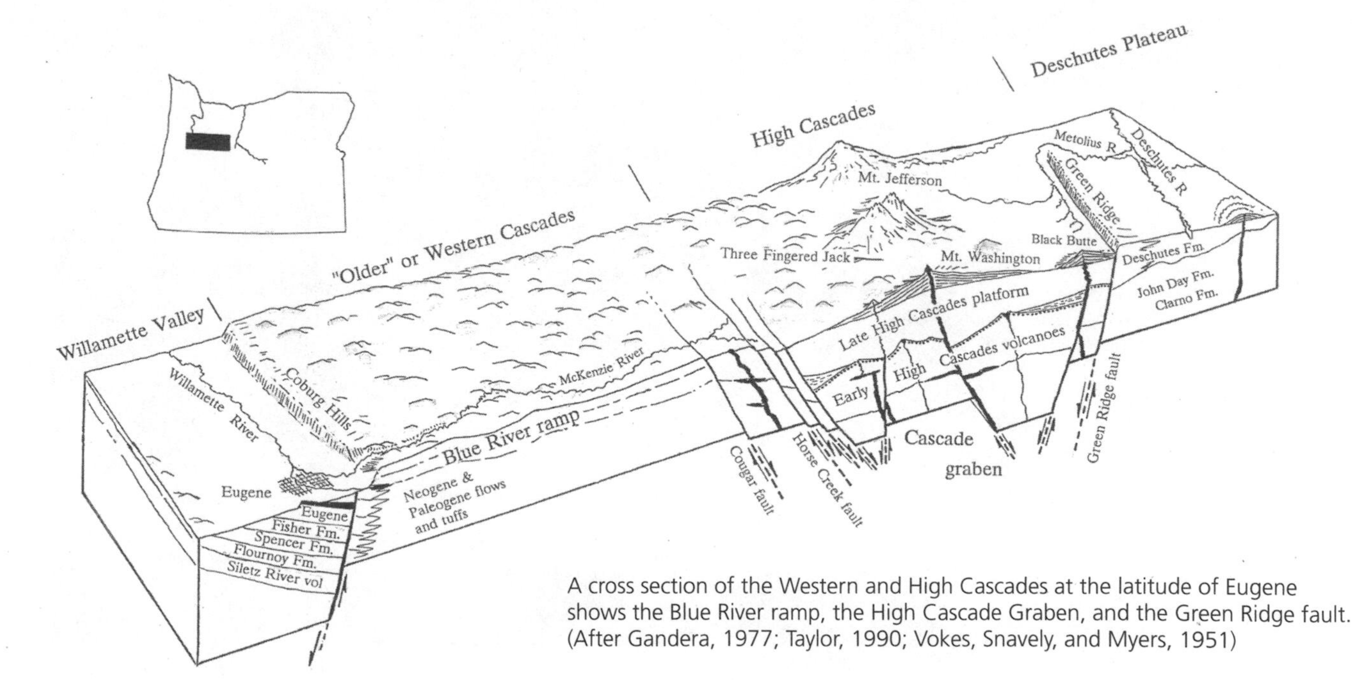

A cross section of the Western and High Cascades at the latitude of Eugene shows the Blue River ramp, the High Cascade Graben, and the Green Ridge fault. (After Gandera, 1977; Taylor, 1990; Vokes, Snavely, and Myers, 1951)

of the intra-arc rift is associated with the three eastside basins, and the eruption of late Miocene to Pliocene lavas and tuffs began immediately prior to rifting During the early Pliocene, sedimentation ceased when the source was cut off by the wall of the Green Ridge scarp as the High Cascades subsided into a graben.

Pleistocene—Ice and Lava

In the High Cascades 2 million years ago, snow and advancing ice sheets, interacting with lava and ash, sent lahars (mudflows) down the mountain sides. Subglacial lava flows were sometimes flattened beneath the ice (tuya).

While northern Washington experienced continental glaciers, a nearly continuous ice field capped the Oregon Cascades, and sizeable ice masses in the lower canyons advanced and retreated. Many of the naturally occurring lakes in the Cascades owe their origin to moraines of gravel, sand, silt, and clay which blocked streams. Semi-circular basins (cirques) hold smaller lakes (tarns).

Although the final cold phase ended around 11,000 years ago, glaciers persist on Mount Hood today. The total volume of 12 billion cubic feet of ice and snow on the slopes, if melted, would provide enough water to enable the Columbia River to sustain its normal flow for 18 hours. The amount of ice above Crater Rock at Mount Hood is estimated at 47 million cubic feet.

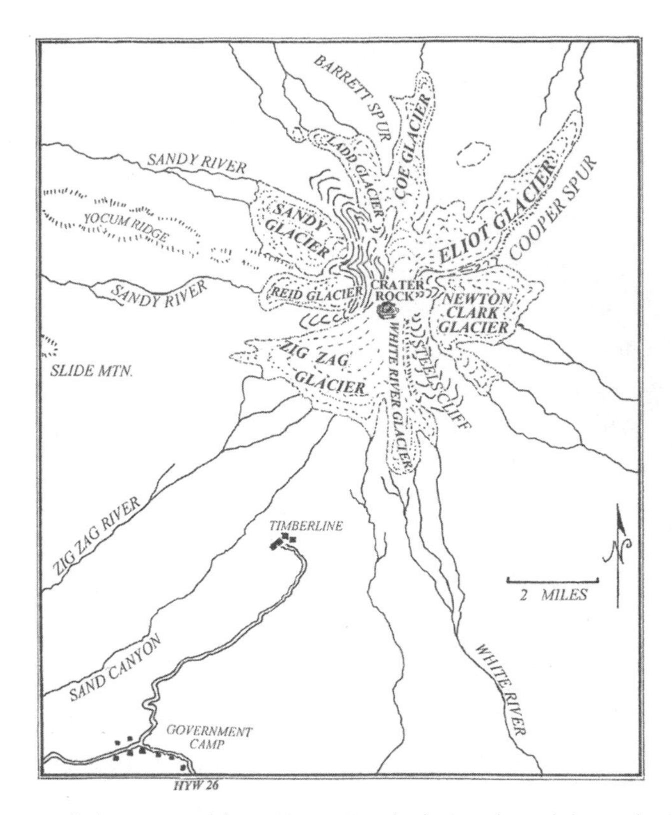

Radiating outward from Mount Hood, glaciers shaped the peak, cutting deep crevices to leave sharp spurs and ridges projecting between ice fields. Meltwater, rocks, and mud combined as lahars moving down the river valleys and depositing broad fans of volcanic debris. (After Crandell, 1980; Williams, 1912)

Once central Oregon's lava fields cooled, fluvial and glacial erosion began to wear away at the tall Cascade peaks. The older volcanoes, which ceased activity over 100,000 years ago, were reduced by Pleistocene glaciation to craggy amphitheaters and ragged ridges as seen at Broken Top (above). In the foreground, a snow-covered lava field appears to be flowing. (Photo courtesy Oregon State Highway Department)

The steep eastern face of the Middle Sister is covered by the Hayden and Diller glaciers, while Collier Glacier extends from the north slope of the Middle Sister across to the base of the North Sister. The largest of the five ice fields on the Three Sisters today, Collier Glacier covers 200 acres, a decrease of over 50 percent during the past 80 years. Since the last ice advance in 1700 A.D., all of Oregon's glaciers have been shrinking at a rapid rate. (In this view looking south, the Three Sisters are in the foreground, Broken Top is back center, and Mount Bachelor is behind right; Photo courtesy Delano Photographics)

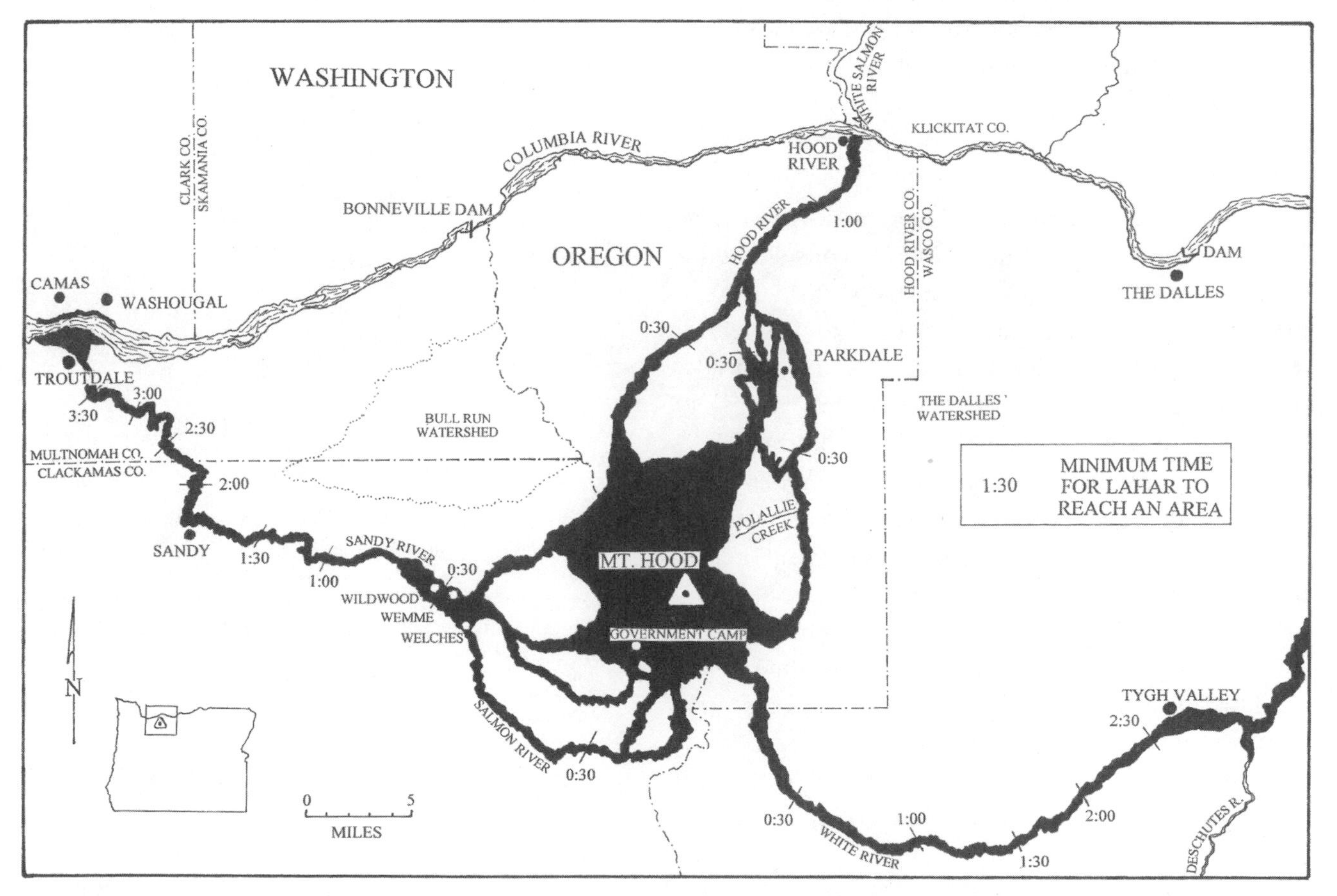

Hazard zonation maps, such as this one of Mount Hood's historic eruptions, are drawn to depict possible routes, intensity, and times of future events. (After Cameron and Pringle, 1986; Pierson, et al., 2009; Scott, et al., 1997; Scott et al., 1977a. Photo courtesy U.S. Geological Survey)

Geologic Hazards

Its tectonic position, topography, and stratigraphy make the Cascade province especially vulnerable to earthquakes, landslides, and volcanism. Vastly improved imagery and mapping is now available through the use of LIDAR (Light Detection and Ranging), a technology that literally sees through the vegetation cover to provide details of the bare land surface and expose ancient and modern features beneath. This tool enables geologists to identify potentially risky conditions.

Earthquakes

The U.S.G.S., the University of Washington's Pacific Northwest Seismograph Network, and other cooperating groups have been monitoring the Cascades for close to 20 years. Mount Hood experiences small swarms annually, each involving over one hundred individual events. In 2011, by the use of LIDAR, Ian Madin discovered a fault scarp that stretched for several miles on the north flank of Mount Hood. Madin estimates that the fault was generated by a 6 to 9 magnitude quake. Because of its location and potential for causing damage, Mount Hood is closely watched.

Landslides

When the Cascades and Columbia River gorge experience prolonged heavy rains, loosely consolidated rocks and soils move in mass on steep hillsides. An understanding of the intricate relationship between supersaturated soils, denuded slopes, and stratigraphy is crucial to assessing the potential for hazards.

No area of the Cascades has been more heavily studied than the H.J. Andrews Experimental Forest in the upper Blue River and Lookout Creek watersheds, where ecosystem research by the U.S. Forest Service and Oregon State University has been ongoing since 1948. In conjunction with other forest service personnel, Fred Swanson has documented local streamflow, landslides, and slope processes, relating them to the dynamics of clearcutting. Not surprisingly, he found that the removal of trees on slopes composed of the Sardine Formation and Little Butte lavas increased the annual streamflow by 40 percent, a condition that persisted longer at the upper elevations. Sediment load, flooding, and stream transport were dramatically higher where the terrain was cut over or where the placement of roads created instability. When combined with severe storms, as in the winters of 1964 and 1996, slumping, rock avalanches, and degradation of water quality and stream habitat ensued. Additionally, damage to campgrounds, roads, hiking trails, and boat launching ramps was extensive.

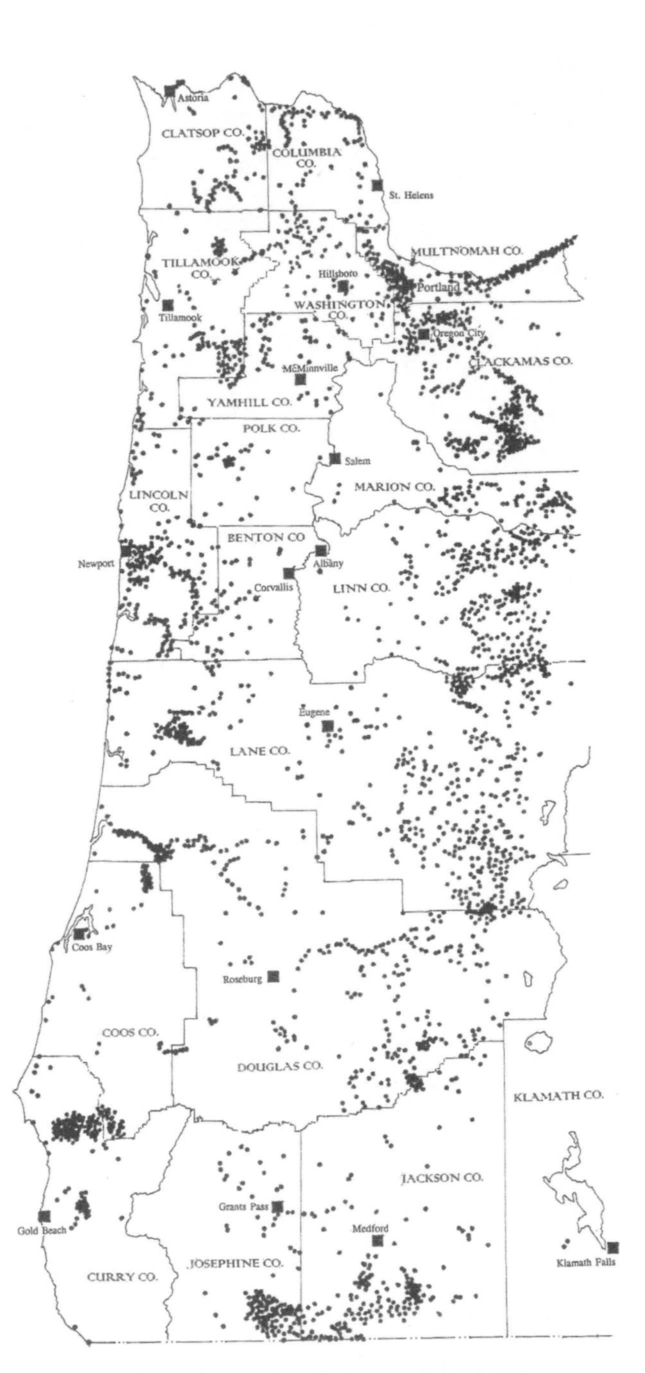

Oregon experienced exceptionally heavy rainfall in February, 1996, a landmark time for flooding and slides, which DOGAMI designated as an 100-year event. The 1996 to 1997 landslide occurrences in western Oregon are indicated by the dots on this map. (Map courtesy University of Oregon; after Hofmeister, 2002)

Similar conclusions were reached in studies of the Mount Hood National Forest following the floods of November, 1995, and February, 1996. Surveys by the U.S. Forest Service in the upper Clackamas River drainage near Estacada revealed over 250 landslides in logged over areas or where roadways had been sited. A more focused look at conditions on Fish Creek and Roaring River confirmed these findings. Because of steep slopes and soft, easily eroded rocks of the Rhododendron Formation, the intensely managed Fish Creek watershed experienced a tenfold increase of landsliding in areas of timber harvest. Roughly 75 percent of the slides in Fish Creek were associated with clear-cutting, in contrast to the Roaring River area where minimal logging activated only two percent of the slides.

The slopes of the High Cascade peaks have often been subjected to free-flowing mixtures of mud, ice, water, and trees after winter and spring rains. Triggered by precipitation around Christmas, 1980, the Polallie Creek avalanche was one of the largest on record. After the slumping of a headwall along the creek on the northeast flank of Mount Hood, the debris traveled up to 35 miles an hour to enter the East Fork of Hood River. Over 100,000 cubic yards of material temporarily dammed the water to form a lake, which, in turn, failed sending a second wave carrying whole trees, enormous boulders, sections of bridges, and pavement slabs across Highway 35 just south of the community of Hood River. One person was killed, and property damage was $13 million.

In September, 1998, and again in November, 2006, debris flows, resembling a watery cement-like mixture, pushed large boulders and gravel down the White River channel on the south side of Mount Hood. Triggered by melting ice and ponding water, the surge is estimated to have traveled at 10 to 15 miles an hour. Since then, high gradient streams, easily eroded deposits, and melting ice continue to deliver sediments to the fan in the lower channel at roughly five-to-ten-year intervals.

The bridge on Oregon Route 35, crossing the White River below Mount Hood, was blocked by debris in November, 2006 (lower photo). The White River channel is frequently impacted by flow surges, and 2 to 14 feet of deposits layer the floor. New channels have been scoured by glacial melt water (upper photo). (Photos courtesy T. DeRoo, C. Hedeen, D.Jones)

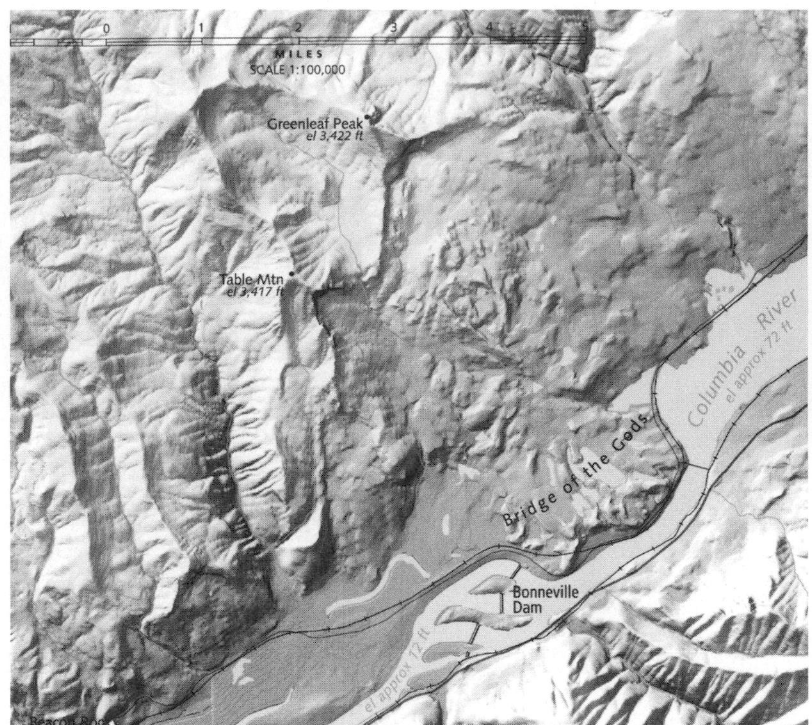

The map illustrates the location of landslides and lahars on the Columbia River between The Dalles and Portland, and the digital images pinpoint the strip near Bonneville dam and Cascade Locks where innumerable slides have taken place. (After O'Connor and Burns, 2009; Loy, et al., eds., 2001; Wang, et al., 2002; photo courtesy University of Oregon)

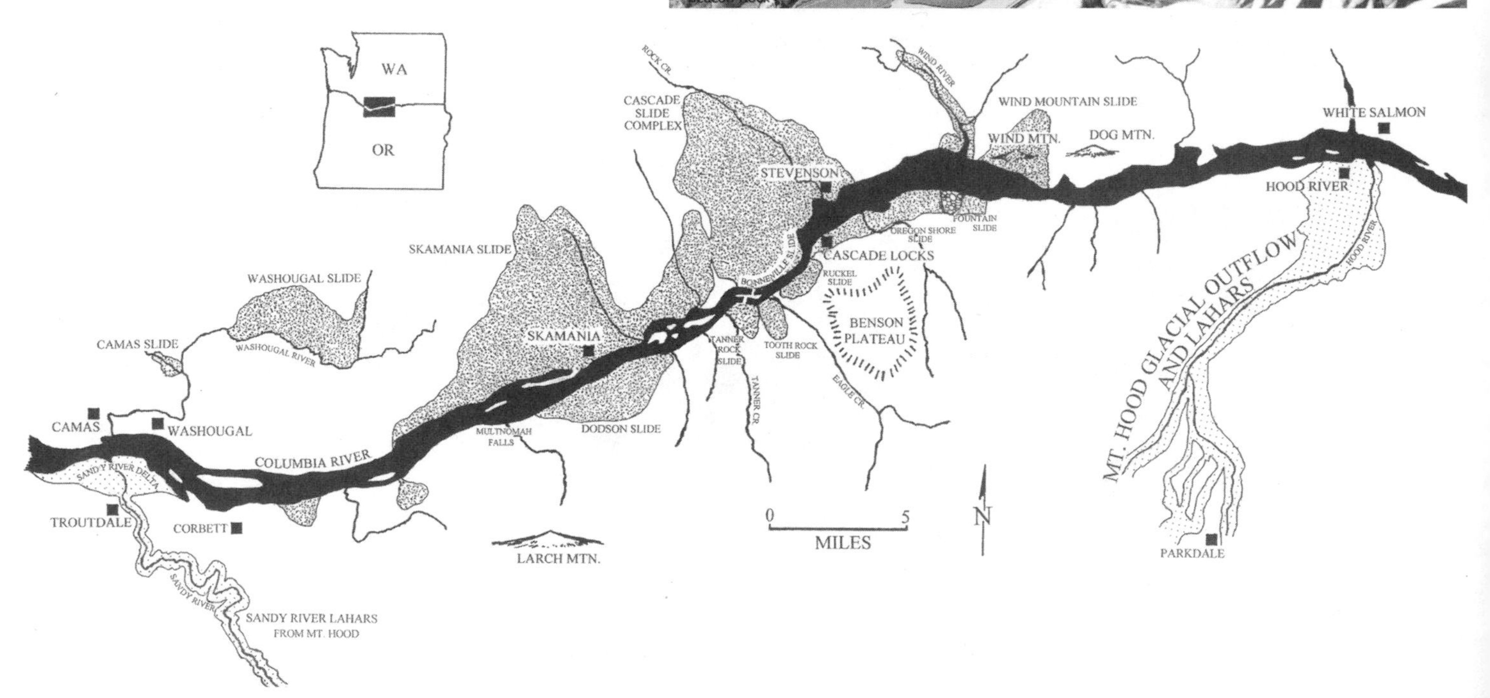

In the Columbia River gorge, large slides, up to 20 square miles in extent, originate predominantly from the Washington side, because the rocks along the river are sloping (dip) toward the south. For the same reason, most of the waterfalls are on the Oregon side. Typically, the slides are fostered by thick clays derived from the decomposed Ohanapecosh Formation lying beneath the Eagle Creek Formation and the Columbia River Basalts. As Aaron Waters observed in 1973, once they become saturated with water, the clays act as a "well greased skidboard."

The Cascade group, incorporating multiple separate slides near Cascade Locks, extends over five miles from Table Mountain, Greenleaf Peak, and Red Bluffs in Washington to the Oregon shore, an area of 14 square miles. Rapids in the channel here, which constitute evidence of mass movements, have been the subject of speculation since first

Heavy rain and a slurry of mud washing through Dodson were harbingers of the 1996 debris flow that would destroy the Royse home. On February 7, 1996, Carol Royse heard a loud rumbling and felt the ground vibrating just before she looked out to see a wall of mud, trees, and rocks moving in her direction. The Royses, their horses, and cat escaped, but their two-story home was lifted off its foundation before being immersed in 15 feet of rocky debris. The flow continued across the highway and railroad to the Columbia River. (Photos courtesy C. Royse)

observed by Indians and explorers. In 1887 Reverend Gustavus Hines wrote that the river dropped "in continued rapids for three miles, not less than fifty feet." A few years later, government geologists such as Clarence Dutton and G. K. Gilbert also took notice of the blockage.

The most recent of the Cascade landslides, the Bonneville rubble, which temporarily dammed the Columbia channel around 500 years ago, may be the origin of the legend of the Bridge of the Gods. As related by Indians, the story describes a bridge arching over the river that crashed down to form the rapids. Rising waters behind the natural dam submerged trees upstream as far as The Dalles until the river cut through the rubble to create the whitewater in the channel. Using radiocarbon to date the drowned trees, Jim O'Connor with the U.S.G.S. and Scott Burns place the Bonneville slide between 1425 to 1450 A.D.

To enhance ship navigation at Bonneville, the U.S. Army Corps of Engineers built a system of locks and canals around the rapids in 1896, but their 1937 dam submerged most of those features. The footings of the dam itself are anchored into the toe of the Bonneville slide.

Land movements originating on the Oregon side of the Columbia River tend to be of limited extent. West of Bonneville Dam, recurring debris flows near Dodson are of impressive size, while east of the dam the Tooth Rock, Ruckel, and Fountain slides only cover one square mile each. An article by Yumei Wang DOGAMI, as well as a 2009 guide by O'Connor and Burns, provide an excellent overview of gorge landslides.

The Dodson-Royse-Warrendale-Tumalt Creek slides all fall within the informally-designated Dodson fan that has been accumulating fluvial sediments since the late Pleistocene. The Columbia River basalt, the Troutdale Formation, and the Boring lavas supply most of the debris for the fan, and when sections of the rock upstream slump off, as happened in 1996 and again in 2001, housing, Interstate 84, and the railroad tracks were covered. During the same day in February, 1996, boulders and gravel of Boring lavas came loose along Tumalt Creek east of the Royse property to pond water in the upper reaches. When the obstruction broke, the mud, rock, and water moved at 30 miles an hour, hitting trucks and train cars, already halted because of the Royse slide, and pushing them into the Columbia River channel.

Adjacent to the Bonneville Dam, Tooth Rock is an ancient slide that was studied extensively in the 1980s by the Army Corps of Engineers before they pinned a Bonneville navigation lock into the slump block. The large block had been rotated 40°, however, engineers concluded that further movement was unlikely because the slab had been stabilized by bedrock that projected into the toe. No motion has been reported since construction.

Even during emplacement of a railroad portage at Ruckel Creek in 1924, the crew described the moving slope as a "glacier." Although dormant at present, the landslide involves the Eagle Creek Formation lying beneath the Columbia River basalt. Similar in composition, the Fountain is an older slide that was mobilized by highway construction in 1952. Heavy rainfall along with the widening of Interstate 84 in 1966 has contributed to the present-day movement.

Volcanic Eruptions

A close examination of historic and current volcanic activity is crucial to anticipating future risks from the potentially explosive Cascades. While many of the stratocones from British Columbia to northern California appear to be dormant, earthquakes, evidence of hydrothermal fluids, steam eruptions, and fumaroles are clear signs that they are still active. Risks from an eruption involve more than flowing lavas, which rarely extend over ten miles from their source. Rapidly moving clouds of incandescent ash, cinders, and chunks of rock can cover great distances at speeds well over 100 miles an hour, and lahars can travel up to 30 miles an hour, stretching tens of miles from their sources.

Typically, the volcanic history of Mount Hood has been one in which lavas, ash, and water combine in a soupy mixture to move as destructive lahars. In his guidebook, William Scott documents some of the most notable. The oldest lahar, dated at 38,000 years ago, swept into the Hood River valley. Of impressive size, it overwhelmed the site of present day Hood River before crossing the Columbia River and moving up the White Salmon River into Washington, where the deposits piled 350 feet high. Starting 500 years ago, numerous lahars spread down the Zigzag and Sandy rivers toward the Columbia and down the White River to the Deschutes, depositing coarse rubble. As recently as the 1800s, during Mount Hood's Old Maid eruptive period, mud and ash flooded the White and Sandy river beds to reach the Columbia.

Over the past 2,000 years, coniferous forests almost 50 miles from Mount Hood have been overwhelmed by lahars, then later exhumed by erosion. Of these, the best known is the Stadter forest on the south side of Illumination Ridge below Zigzag Glacier. Named by Edwin Hodge in 1931, the Stadter was described in 1991 by Kenneth Cameron, formerly of the U.S.G.S., and Patrick Pringle of the Washington State Department of Natural Resources. They compiled the glacial and volcanic history of the Stadter and six similar sites from the Timberline and Old Maid periods.

Eruptions of ash and lava are not unknown along the central Cascades, and around 2,300 years ago Rock Mesa was built on the flanks of South Sister. Historically, one of the Three Sisters was viewed belching forth dense volumes of smoke from its summit. In 1996, calculations by Charles Wicks of the U.S.G.S. revealed an area three miles west of the South Sister that was doming upward. The cause was thought to be an accumulation of magma at a depth of about four miles. Three years later, when an earthquake swarm was centered near the uplift, geologists began to monitor the bulge closely, but by 2007 it had diminished in size. Studying what came to be called the South Sister bubble, Wicks suggested that major subduction earthquakes might even trigger an eruptive cycle in the Cascade arc. The idea that a quake of such magnitude might accompany volcanic activity brings to mind an almost unimaginable calamity.

Natural Resources

Erosion removed much of the overlying Western Cascade rocks during the past 5 million years, exposing mineralized zones along valley floors. Mineralization is associated with granodiorite intrusions that developed during separate phases of volcanism, and thus the ore bodies are not interrelated. Where veins of shattered breccias were invaded by hydrothermal fluids at temperatures of 250° to 350° Fahrenheit, minerals precipitated in the host rocks.

Gold and Silver

Bohemia Mountain south of Cottage Grove has been the most productive and versatile in this province, although mineralization is confined to only nine square miles. Discovery of placer gold on Sharps Creek in 1858 came shortly before that of lode veins by James Johnson, who had traveled north to the "extreemly wild and untenanted wilderness." Since Johnson was from Bohemia in eastern Europe, the region was named for him. The Champion, Helena,

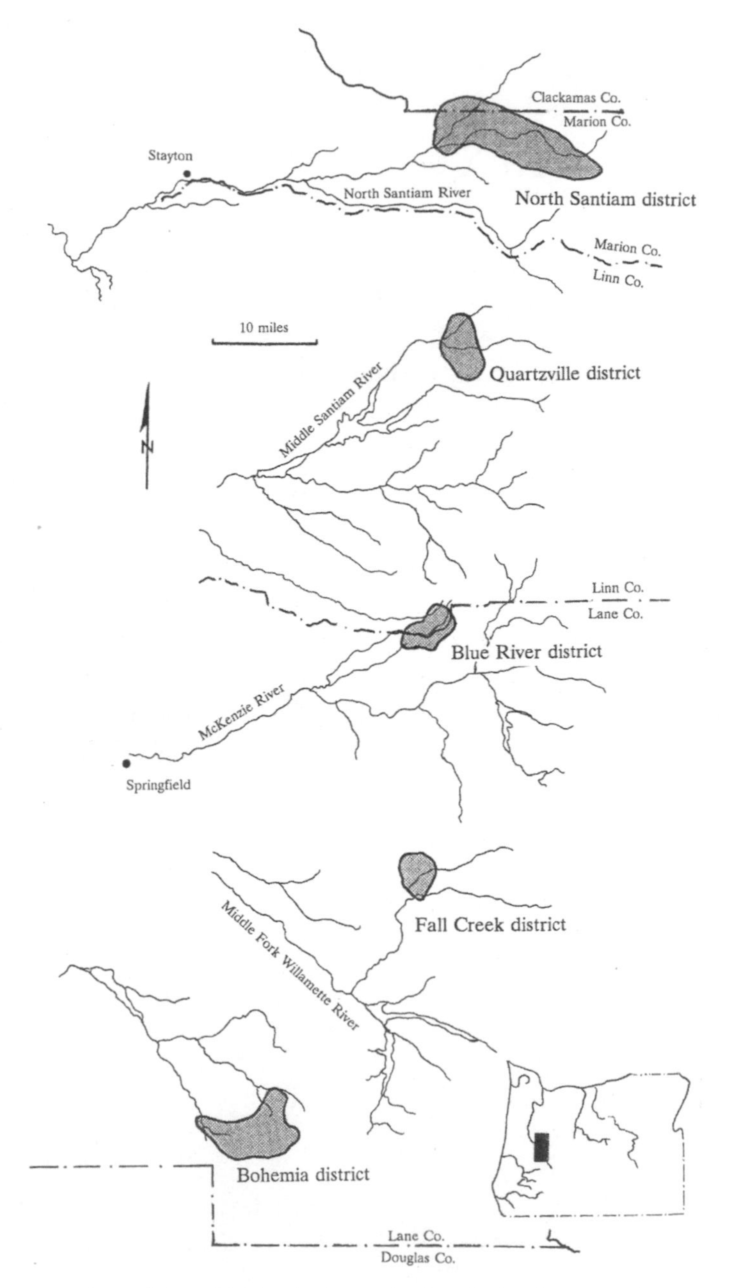

Five mining districts, the North Santiam in Clackamas and Marion counties, the Quartzville, the Blue River, the Fall Creek, and the Bohemia in Linn and Lane counties all lie within a 25-to-30-mile-wide belt of the Western Cascades. Even though yields were small in comparison to other areas of Oregon, there was a surprising variety of gold, silver, and other ores. (After Brooks and Ramp, 1968)

Musick, and Noonday mines dominated the Bohemia district, generating a combined total of about $1 million from both gold and silver.

Dorena Lake, a U.S. Army Corps of Engineers flood control reservoir, is being contaminated by high levels of mercury leaching from the Bohemia Mine. A mercury amalgamation process was used to extract the gold.

Revenue from the Quartzville and Blue River districts was marginal. At Quartzville, gold-bearing veins were located in 1864, and the complete output amounted to $181,000. Somewhat less was removed in later years, when the mines only operated intermittently. In 1984 the Quartzville Recreation Corridor was set aside for panning.

About 45 miles east of Eugene in Lane County, the Blue River district was largely confined to the Lucky Boy Mine, located in 1887. Practically all of its operations ceased by 1913, when the vein was exhausted. Production figures for gold and silver have been estimated at $175,000. Gold ores at the Fall Creek mines in Lane County are of comparatively low grade, and mining has been limited.

Along with the Bohemia Mountain area, the North Santiam district yielded a variety of copper, zinc, lead, silver, and gold since it opened in the 1860s. Most activity focused on the Ruth vein where ores of zinc and lead have been mined periodically. The quantity of gold and silver is low in comparison with that of other minerals, and the total for all ores was $25,000. The most recent exploration involved surveying and drilling for copper by the Shiny Rock Mining Company, which holds a large block of claims. The claims were subsequently leased in 1980 to Amoco Minerals Company.

While production figures for all Cascade minerals can only be estimated because the records are lacking, the revenue since 1858 is $1.5 million, keeping in mind that the price of gold was well below $35 an ounce before 1960.

At the sophisticated Champion Mill, a 100 ton selective flotation plant processed ore in the Bohemia area. (Photo courtesy Oregon Department of Geology and Mineral Industries)

Geothermal Energy

Between the High and Western Cascade mountains, an irregular north-south belt of hot springs, narrowing to less than 12 miles in width, marks a major thermal boundary or heat flow change. Low temperature gradients in the Willamette Valley and adjacent Western Cascades increase toward the High Cascades. Many of the thermal springs are located along faults, where superheated waters migrate through fractures to reach the surface. A 1983 monograph edited by George Priest and Beverly Vogt summarizes the geology of the central Cascades and reviews heat flow patterns.

A combination of warm groundwater and partially molten material beneath the younger volcanic rocks is responsible for the thermal springs. A decade of examining gravity anomalies (variations) and heat flow near the Three Sisters led David Blackwell at Southern Methodist University to conclude that the source was a large magma chamber or wide thermal zone at a depth of six miles (midcrust).

However, William Stanley and Steven Ingebritsen at the U.S.G.S. reached an alternate hypothesis. They surmised that a narrow zone of high heat flow along the axis of the High Cascades is caused by groundwater moving laterally though porous Quaternary rocks. Based on drill holes reaching depths of several hundred feet, they considered the area of crustal melt to be smaller than that postulated by Blackwell, and, because the volcanic rocks were permeable, the heat flow was shallow.

Thermal waters discharge at an average rate of 80 gallons per second from the Cascades in California, Oregon, and Washington in comparison to hot springs at the Yellowstone caldera, which release 800 gallons per second. Austin Hot Springs in Clackamas County has the highest thermal discharge of any in Oregon at 30 gallons a second.

Even though the geothermal potential is only modest, the Cascades has been rated as a Known Geothermal Resource Area by the U.S.G.S. At present, DOGAMI is working with the Oregon Department of Energy to inventory Oregon's geothermal systems, most of which are located on federal lands. Since the 1970s, these agencies have drilled exploratory wells to pinpoint heat flow on Mount Hood. The volcanic history of the mountain implies significant hydrothermal systems, but the results were

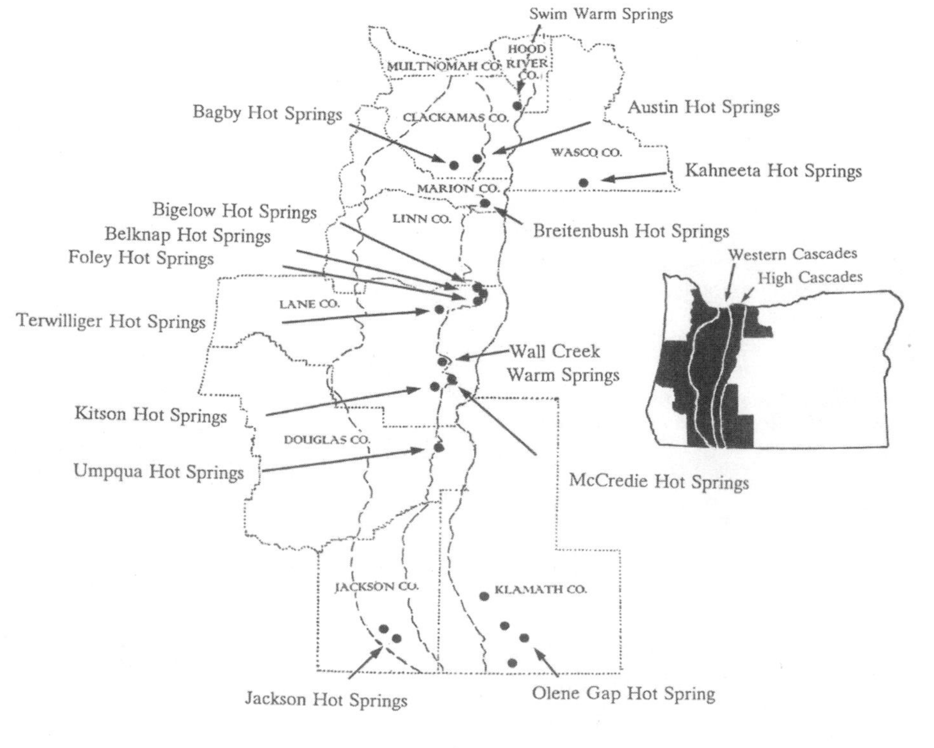

A north-south belt of hot springs marks a thermal boundary between the Western and High Cascades. Springs in the northcentral portion of the Western Cascades have higher water temperatures than those at either end. Jackson Hot Springs is the most southerly in the Oregon system, while the warm springs at the old town of Swim in Clackamas County are at the northern end. Temperatures average between 90° to 190° Fahrenheit. (after Black, Blackwell, and Steele, 1983; Blackwell, et al., 1978; Blakely, 1994; Blakely and Jachens, 1990; Ingebritsen, Mariner, and Sherrod, 1994; Mariner, et al., 1990; Priest and Vogt, 1983)

inconclusive, and there was little evidence of high temperature waters circulating near the surface or even in deep-seated magmas. As with Newberry Volcano and Crater Lake, the recreational value must be considered.

Private companies, seeking to explore Crater Lake for geothermal resources, have been opposed by environmental groups, fearing that drilling, even a distance from the lake, might interfere with plumbing conduits. Ongoing assessments by Charles Bacon confirmed the presence of discharging thermal vents on the floor of Crater Lake as well as on the south and east flanks in the Wood River Valley. Evidence from drill holes led him to conclude that the springs are associated with heat from the final eruption. However, their flow has diminished since early in the lake history. Temperatures in the Crater Lake hydrothermal system reach a maximum of over 200° with an average of 55° Fahrenheit.

Surface and Groundwater

When moisture-laden air masses from the Pacific Ocean encounter the Coast Range and Cascades, rain soaks Oregon's western region, but much of it is blocked by the mountains from reaching the drier eastern part of the state. As a result, annual precipitation in the Western Cascades averages from 70 to over 150 inches, contrasting to the approximately 12 inches on the east slopes of the High Cascades, most of which falls in the wintertime. Thus the north-south Cascade physiographic division has shaped the climate of the state since the Miocene.

A mature soil cover and high gradient streams in the Western Cascades absorb rainfall rapidly after a dry fall, quickly become saturated, and have little storage capacity. These characteristics lead to winter and spring flooding and a diminished summer flow. In the High Cascades, a 2004 investigation examined the relationship between the aquifers, lavas, and topography. Anne Jefferson and coauthors from Oregon State University found that along the upper McKenzie River the movement of shallow groundwater is controlled by the geographic limits of the volcanic fields, by the high permeability of the lavas, and by the rate of recharge. As rainfall percolates through young High Cascades lavas it supplements the groundwater and emerges downslope as springs.

There have been periodic attempts to divert what is perceived as Oregon's abundance of water. The Columbia River has been a prime target for schemes to pipe water out of state to the dry southwest, and the 2008 "discovery" of reservoirs in volcanic rocks of the Cascades was characterized in the *Oregonian* newspaper as "a secret stockpile of water." U.S. Forest Service and Oregon State University staff viewed the aquifers as unused and a potential supply for future needs, focusing on the economic value. This notion played to the popular but falsely held belief in an unlimited underground reservoir waiting to be discovered and tapped. Hydrogeologists were quick to point out the importance of this already well-known supply to basins on both sides of the range, where it is has been utilized for decades. They noted that the Cascade aquifers periodically suffer from drought and do not constitute a hidden resource.

Geologic Highlights

Columbia River Gorge

The best perspective on the inner workings of the Cascade Range can be found in the Columbia River gorge, which cuts through the mountains. Laid bare and swept clean by Ice Age floods, the strata and the multiple ancient channels are revealed.

The gorge comprises 75 miles of geologically spectacular scenery along the Columbia River from the narrows at The Dalles to Portland. Following this route, the Columbia River Highway opened in July, 1915, and was dedicated one year later. The deterioration and abandonment of many portions of the route by the 1940s were rectified in 1986 with the acknowledgment of its historic and picturesque value. Through the efforts of a number of people, the 292,615-acre Columbia River Gorge National Scenic Area was dedicated. One stipulation of the partnership between government agencies was to preserve and restore parts of the original highway for public use.

An early account of the geology by Ira Williams in 1916, was not updated until John Allen's 1984 guidebook *The Magnificent Gateway*. Terry Tolan, Marvin Beeson, and Beverly Vogt offer a Tertiary overview, while 2009 articles by Jim O'Connor and Scott Burns integrate the geology of many sites with impacts of the Missoula floods.

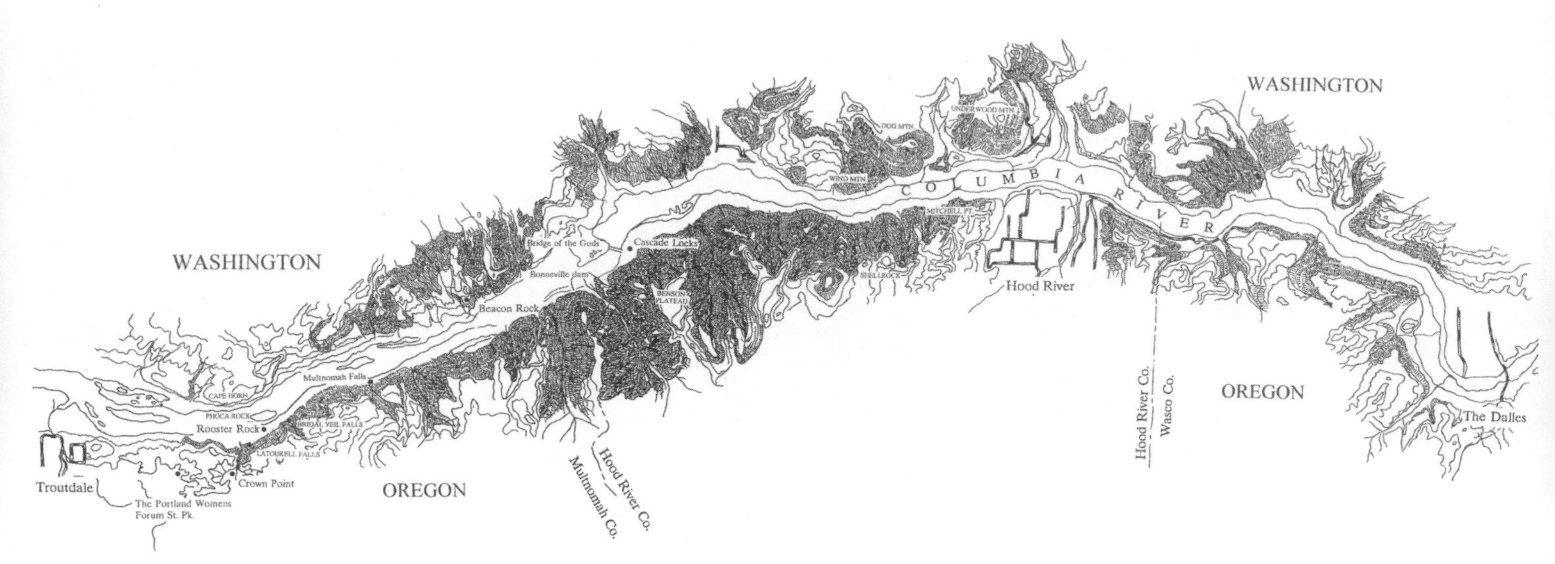

From The Dalles to Portland, the Columbia River is entrenched in a 75-mile-long gorge, famous for its exceptional array of geologic highlights. (After Allen, 1984; Williams, 1916)

Viewpoints Overlooking the River

Formerly named Chanticleer, the promontory at Womens Forum State Park near Corbett rises 925 feet above the river and gives a view of adjacent Rooster Rock and Crown Point to the east, of Mount Zion and Cape Horn directly across the river, and of Beacon Rock a distance upriver.

Observations from Vista House and the cliff at 725-foot-high Crown Point are even more impressive. On the windswept height once known as Thor's Crown, Vista House was the site of the opening ceremony dedicating the Columbia River Highway. The English Tudor-style stone building rests directly on the Priest Rapids intra-canyon flow, the youngest member of the Wanapum Formation (Columbia River basalts), which invaded the gorge some 14 million years ago. The blocky jointed basalt that makes up the cliff is 500 feet thick, covering over 200 feet of older volcanic layers that had filled the channel. Floodwaters from Pleistocene Lake Missoula, cresting at Crown Point, would have enveloped Vista House.

Exposures of the Priest Rapids flow are also prominent at Womens Forum State Park and Shepperds Dell. Pepper Mountain, Larch Mountain, Mount Pleasant, and Mount Zion, cinder cones and volcanic plugs of the Pliocene Boring lavas, are visible for many miles in both directions. Larch Mountain is a low-profile shield volcano.

Born in Seattle, Washington, in 1908, John E. Allen grew up in Eugene where he attended the University of Oregon as a journalism major. Influenced by Edwin Hodge and Warren D. Smith, he completed a Masters on the geology of the Columbia River Gorge before taking a PhD at the University of California, Berkeley, in 1944. His initial job involved field mapping with DOGAMI in the Baker City office. Years of professorships in geology at several institutions led him to Portland State University in the 1950s, where he remained until retirement. His two books on the Columbia gorge are now classics. Allen died in 1996. (Photo taken in 1942; courtesy Geologic Society of the Oregon Country)

Waterfalls

The Columbia River is lined with 71 waterfalls, 11 of which drop over 100 feet. Of these, Multnomah Falls, which is the fourth highest in the United States, is actually two falls, precipitating a distance of 620 feet from a projecting ledge of the Grande Ronde basalts. Elowah Falls at 289 feet, Latourell at 250, Wahkeena at 242, and Oneonta Falls at 221 feet are the next highest.

Waterfalls tend to be temporary, and most of those in the gorge originated during glacial floods between 18,000 to 15,000 years ago. As surging waters stripped away hundreds of feet of basalt from the canyon walls, the streams were left to plunge over the sheer face. Falls most often occur when the lavas are flat-lying, and, after erosion removes

Oneonta Falls, typical of those along the the Oregon side of the Columbia River, is fed by snowfields of the High Cascades. (Photo courtesy Oregon State Highway Department)

the less resistant material between the basalt, the lip gradually retreats. Amphitheater-shaped grottos below form when mist and splashing water penetrate the blocky jointing to freeze and expand, splitting the rock.

Four separate layers of the Columbia River basalts, each with differing erosional properties, have created Multnomah Falls. The steep slopes are not without incident. A fire in 1991 removed the vegetation above the bridge, and a few years later a bus-sized chunk of rock fell, breaking loose a shower of pieces, and injuring 20 people. The U.S. Forest Service has since erected cable nets to protect the public.

Pinnacles and Promontories

Upriver, sharp pinnacles and cliffs project at Rocky Butte, Rooster Rock, and Phoca Rock. Now within the city limits of Portland, Rocky Butte is the core of a late Pleistocene volcano that discharged Boring lavas. The Federal Works Progress Administration built a park on top of Rocky Butte in the 1930s, and the Multnomah County jail was constructed at its base some 10 years later with rock quarried from the east side.

On the north bank near Skamania, Washington, the vertical 850-foot-high column of Beacon Rock is the andesitic basalt plug of an ancient volcano that erupted the Boring lavas only 57,000 years ago.

The layers surrounding the plug were carried away by Missoula floods, exposing the pillar midstream. When Bonneville slide debris engulfed the monolith, the shoreline shifted, and Beacon Rock was effectively transferred to Washington. During the construction of jetties at the entrance to the Columbia River in the early 1900s, the U.S. Army Corps of Engineers proposed quarrying Beacon Rock for use. To that end, they drilled holes into the monolith preparing to dynamite it for the needed stone, before a private investor purchased the landmark.

The proposal to demolish Beacon Rock was not unusual, and many of the pinnacles and projections have been removed or modified to make way for highways and railroads. The Pillars of Hercules, two sharp upright columns of basalt between Bridal Veil and Latourell falls, were cut away when the highway was constructed. On the Washington side, Cape Horn, a steep promontory near Mount Zion, was destroyed to make way for the highway and railroad tunnel. Tunnels were bored through the Tooth Rock landslide block at Eagle Creek as well as through the overhanging Mitchell Point basalt cliff west of Hood River.

Mountain Peaks

The High Cascades offer a variety of majestic peaks, dazzling glaciers and snowfields, forested slopes, cold streams, lakes, and waterfalls. While the crest of the range averages just over 5,000 feet in altitude, a number of conical summits, which rise considerably higher, are more conspicuous than the older eroded spires.

Stephen Harris's *Fire and Ice* provides an invaluable overview of all Cascade volcanoes from California to Canada, while Raymond Hatton's *Oregon's Sisters Country* gives a very readable account of the history and geology. Fieldtrip guides by Edward Taylor, Richard Conrey, and University of Oregon geologist Katharine Cashman examine different aspects of volcanic activity including the widespread Quaternary cover.

Most areas of the High Cascades attract climbers, but Mount Hood, with its proximity to Portland, has long presented a challenge. Joel Palmer, Territorial Commissioner of Indian Affairs, was the first reported to have scaled the south face in October, 1845, reaching around 9,000 feet before turning back. Three years later, newspaperman

The monoliths of Rooster Rock (left) and Phoca Rock are believed to have been moved by a large landslide activated when Missoula flood waters undercut cliffs along the Columbia River. Rising 200 hundred feet, Rooster Rock and the surrounding acreage were purchased for $10,000 as a state park in 1938. Lying nearby in the middle of the channel, Phoca Rock was so named because seals (phocids) rested here on their way upriver. (Photo courtesy Oregon State Highway Department)

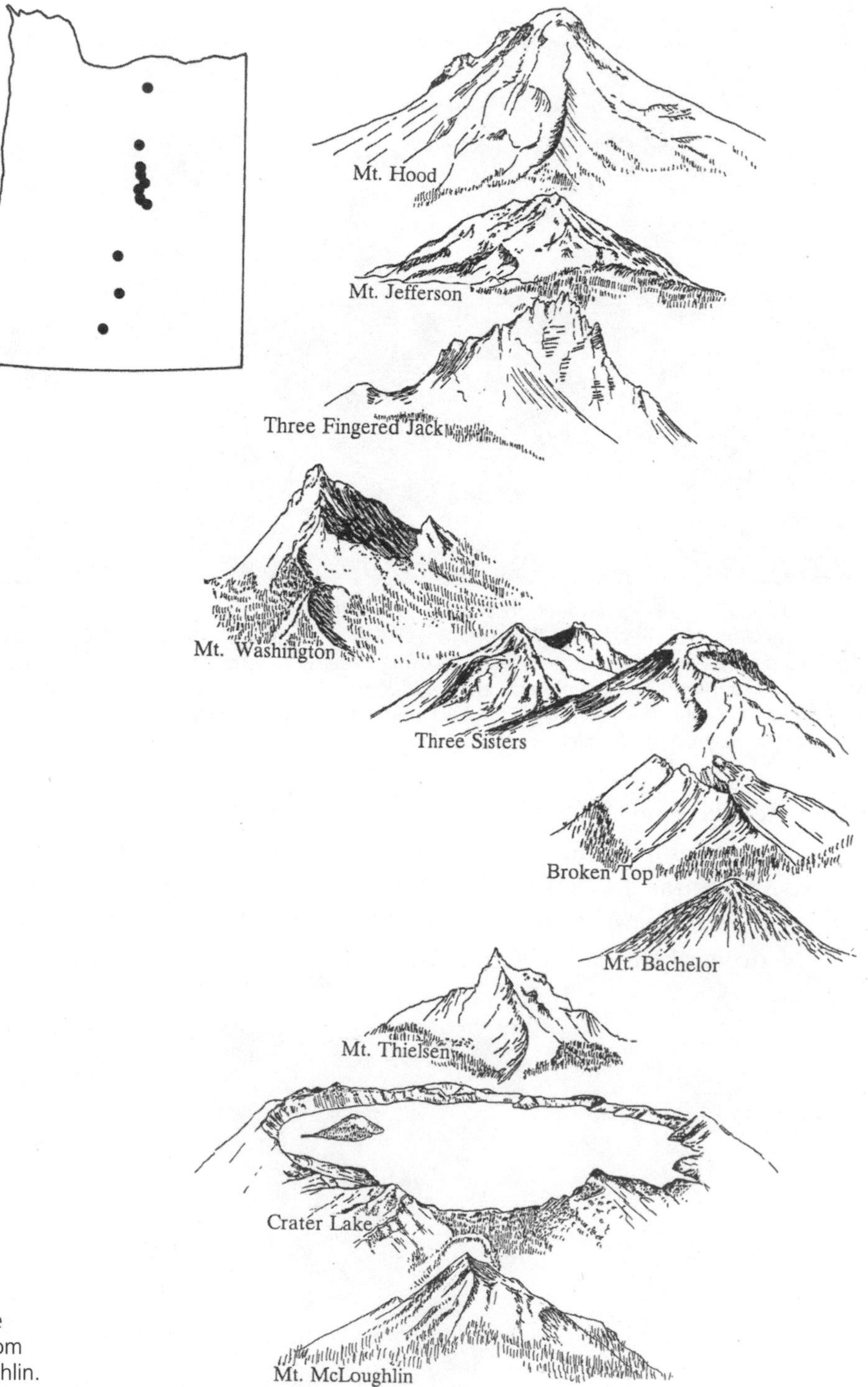

Significant volcanic peaks of the Oregon High Cascade Range from Mount Hood to Mount McLoughlin.

Henry Pittock with a party of six placed a flag on the top. Many who ascend Mount Hood fail to heed weather conditions or take safety measures, and 20 lives have been lost over the last decade, although many others have been rescued.

Facing Mount Adams and Mount St. Helens across the Columbia River, Mount Hood was named for Samuel Hood, an officer in the British Royal Navy, who was stationed off the Northwest coast during the American Revolution. At 11,235 feet in elevation, it is the highest of Oregon's Cascade mountains, the prominent summit serving as a reminder to travelers that their epic journey to the Willamette Valley was coming to an end.

Named Faith, Hope, and Charity in the 1940s, the Three Sisters are pitted with small glacially cut

Mount Washington and Three-Fingered Jack both reach close to the same elevation at 7,800 feet, both are composite volcanoes, and both have been glacially carved into horns. Built on thin lava flows, the central conduit of Mount Washington was intruded by a large resistant plug that projects as a spire on the horizon. Looking toward the south in the photo, the saw-toothed ridgeline of Three-Fingered Jack, which consists of loose tephra covering a dike, is in the foreground with Mount Washington and the Three Sisters behind. Broken Top and Mount Bachelor are in the back on the left, and Mount Thielsen is in the background on the right. (Photo taken by D. Rohr; courtesy Condon Collection)

ravines, crevasses, and ridges. Within the Three Sisters volcanic cluster, the South cone is the highest at 10,354 feet, but the 10,182-foot-high North Sister is the oldest and most deeply eroded with no evidence of a central crater. The South Sister is the best preserved with a circular summit crater, in which snow-melt forms a small lake during the summer. Ice has been persistently cutting away at the Three Sisters, leaving sharpened crests and cirque valleys. Established in 1957, the Three Sisters

A series of flows and domes around the rim of Crater Lake reveal the different eruptive stages of Mount Mazama. Prior to the final explosion, the cone is estimated to have been between 10,000 and 12,000 feet above sea level, but today the caldera rim averages 8,000 feet in elevation. At 8,929 feet, Mount Scott is the highest point along the rim. (Acoustic backscatter map of Crater Lake, U.S.G.S., 10 m DEM, 2002)

Overshadowing Diamond Lake, Mount Thielsen is distinguished by its needle-like plug, evidence of the erosive force of ice that scalloped its north side. In this view looking to the east, Summit Rock is to the right, and the spire is to the left. Diamond Lake was constructed when lava from Mount Thielsen blocked a basin already deepened by Pleistocene ice. (Photo courtesy Oregon State Highway Department).

Wilderness covers 287,000 acres and is under federal protection.

Closest to Oregon's southern border, Mount Thielsen and Mount McLoughlin both average just over 9,000 feet in elevation. Mount McLoughlin, which is the highest between the Three Sisters and Mount Shasta, has had many designations in its past. Once called Snowy Butte or Big Butte, the peak became Mount Pit in 1842, a name derived from the many pits dug by Indians to trap game. It was subsequently named for John McLoughlin, director of the Hudson's Bay Company.

Mountain Lakes

The wealth of lakes in the Cascade Mountains is the product of several geologic processes. Landslides, lava, or glacial till block and alter stream flows, while rainfall and snow melt may fill a basin or caldera. Lakes dammed by lava or glaciers are the most numerous in the Oregon Cascades, where volcanic eruptions and ice interacted during the Pleistocene.

Oregon's most celebrated is Crater Lake, a caldera that filled after the volcanic explosion and collapse of Mount Mazama 7,700 years ago. The symmetrical basin containing the lake today is five miles in diameter with a depth of one mile, making it the deepest fresh-water body in the United States and the second deepest in North America. The crater, lake, and surrounding 183,180 acres in Klamath County were dedicated as Crater Lake National Park in 1902.

North of Crater Lake, Diamond, Crescent, Odell, and Waldo lakes are among the many that lie in glacially scoured valleys dammed by moraines. With a surface area exceeding 3,582 acres, Odell Lake is one of the largest in the Cascades. When a terminal moraine from layers of High Cascade ice blocked Odell Creek, the basin filled with melt-water. The formation of Crescent Lake was similarly due to glacial activity. Lying in the same valley as Suttle Lake but of volcanic origin, Blue Lake is relatively small but one of Oregon's deepest at over 300 feet. The depth gives the water its unusual blue color. The lake resulted from a violent explosion when magma encountered groundwater around 3,500 years ago.

In the central portion of the range, the spread of basalt from the Sand Mountain-Nash Crater chain altered the flow of streams. This linear field of 23 cinder cones and 41 separate fissures and vents discharged nearly a cubic mile of lava and ash, blocking the McKenzie River to impound Lava Lake, Fish Lake, and Clear Lake. Sahalie and Koosah falls drop

Near Highway 20, the popular Suttle Lake had its beginnings when glacial moraines of ice and rock impounded Lake Creek. The long, narrow outlines and depth of the lake reflect its origins in a glacially carved valley. The composite cone of Black Butte is in the background. (Photo courtesy Oregon State Highway Department)

In this view toward the west from Dee Wright Observatory, Belknap shield is on the horizon, and rubble from the Yapoah eruption is in the foreground. The stone observatory was built by the Civilian Conservation Corps in 1935 and named for the person who spent many years working there. (Photo courtesy Oregon State Highway Department)

over a resistant lobe of the basalt. At Clear Lake, an entire forest was submerged. Radiocarbon dates from upright *ghost* trees, visible in the depths of the unusually clear water, have pinpointed the time at 2,750 years in the past. Successions of diatoms living in the lake deposited a snow-white layer on the bottom. Fed by cold springs, Clear Lake is the source for the McKenzie River.

Recent Volcanic Eruptions

The central High Cascade region is essentially a Quaternary volcanic platform of overlapping cones, flows, and volcanoes. Here an interval of very recent volcanism emplaced dark cindery fields between 3,850 to 1,500 years ago from the Sand Mountain-Nash Craters, Yapoah Crater, Collier Cone, Four-in-One vents, and Belknap shield cone.

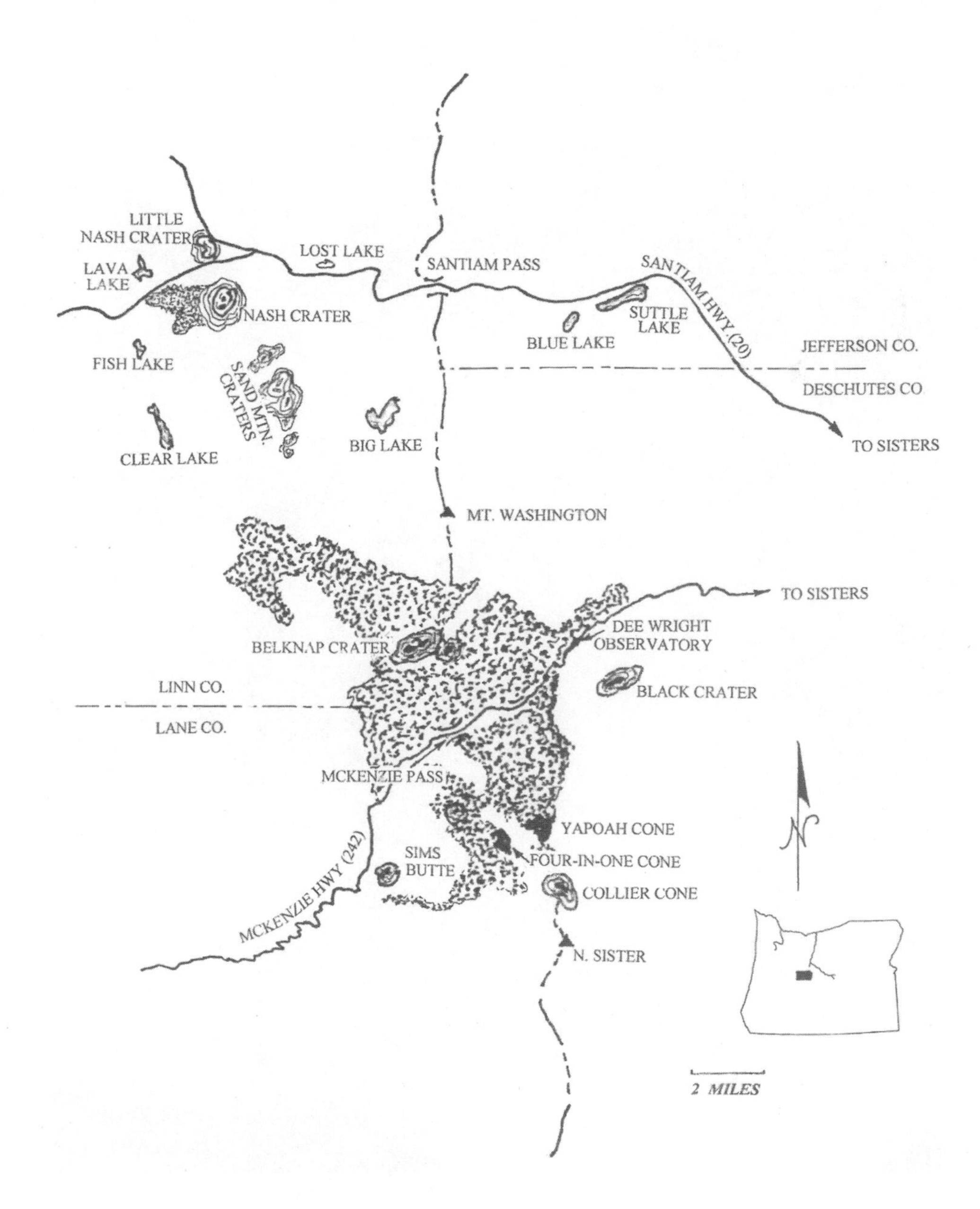

Lava fields along the Santiam and McKenzie highways provide one of the best places in western Oregon to see a fresh volcanic cover. From the summit of McKenzie Pass at Dee Wright Observatory, the desolate landscape has been created by blocky Yapoah basalts, Belknap flows, and the Four-in-One vents. Reaching 500 feet above the surroundings, Yapoah Crater is mantled by red cinders. Nearby, the Four-in-One cones, aligned in a northwest-southeast direction, were initially active along a one-half mile fissure but subsequently concentrated at four conduits.

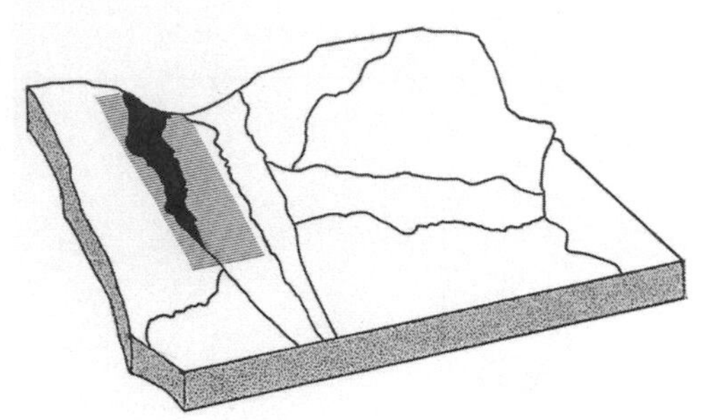

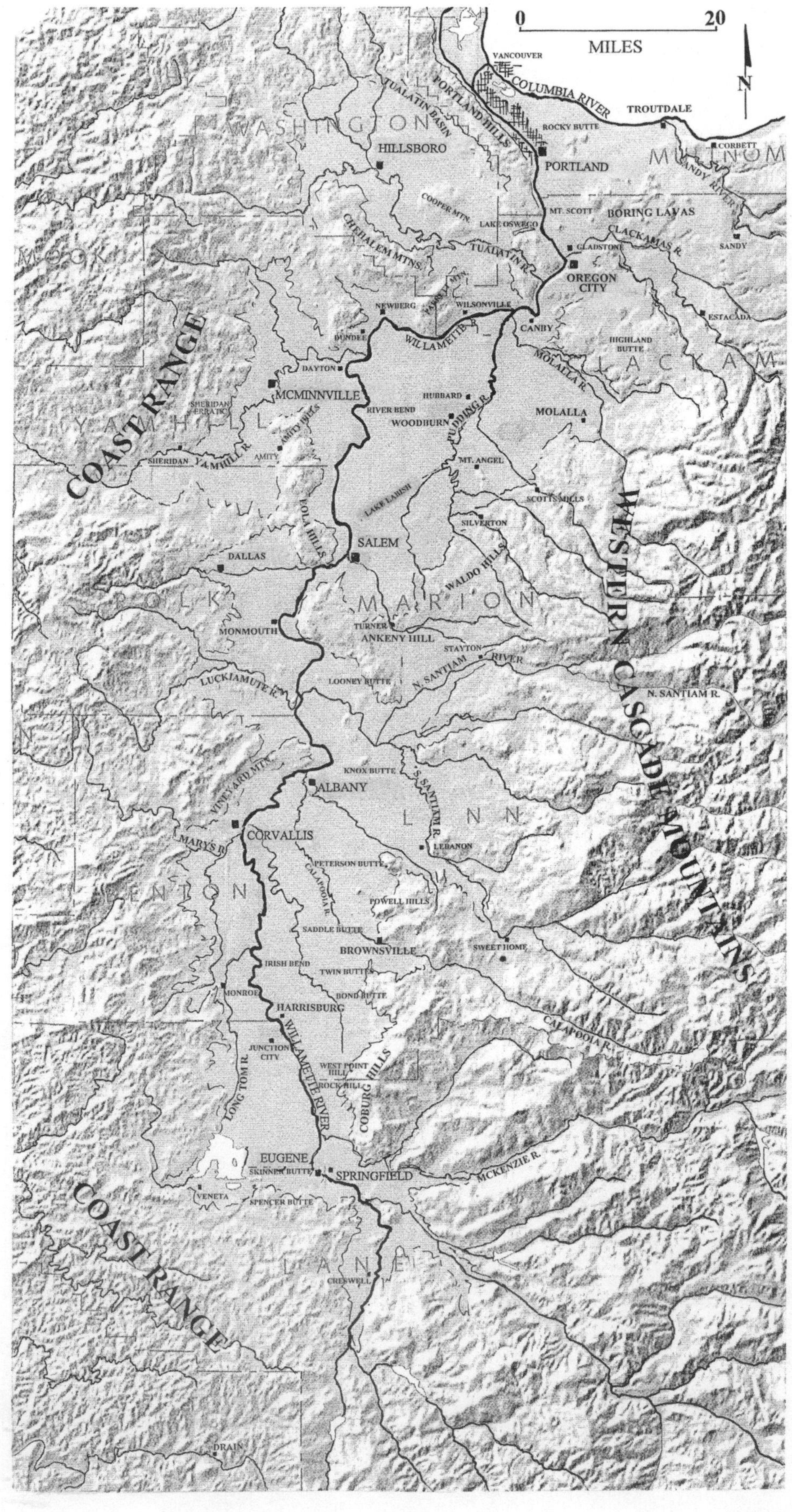

Physiographic map (After Loy, et al., eds., 2001)

Willamette Valley

Landscape of the Willamette Valley

The Willamette Valley and Puget Sound lowland stretches from Cottage Grove, Oregon, to British Columbia in Canada. In Oregon, the Willamette Valley is an elongate basin which narrows at either end before pinching out. Enclosed on the west by the Coast Range, on the east by the Cascade Mountains, and to the north by the Columbia River, the basin is approximately 130 miles long and up to 40 miles wide. From an elevation of 400 feet at Eugene, the surface drops close to sea level at Portland, an average gradient of only four feet per mile.

Contrasting topography divides the Willamette Valley into northern and southern regions at Salem. For the most part, the southern portion is nearly featureless, distinct from the hilly terrain in the north. Near Salem, the Eola Hills lie to the west, the Ankeny Hills are to the south, and the Waldo Hills to the east. The Tualatin basin is separated from Portland by the Portland Hills and from the Willamette Valley by the Chehalem Mountains. Smaller topographic projections on the valley floor include volcanic cones and buttes around Portland and Oregon City and from Brownsville to Eugene. Near the center of the province, the 45th parallel, halfway between the equator and the North Pole, passes close to Salem.

The Willamette Valley is a lengthy alluvial plain of the river, whose watershed drains 11,200 square miles. Originating at the junction of its Coast and Middle forks near Eugene, the Willamette River meanders toward the north to enter the larger Columbia system at Portland. The overall gradient is northward and not from the margins toward the center, making it one of the few rivers in the United States to flow in that direction.

South of Albany, tributary streams parallel the main Willamette channel for a distance before merging. The McKenzie, Calapooia, North and South Santiam, Pudding, Molalla, and Clackamas rivers have their headwaters in the Cascade Mountains, while the Long Tom, Marys, Luckiamute, Yamhill, and Tualatin rivers drain the Coast Range.

Although comparatively small, the Willamette Valley is the economic and cultural heart of Oregon. Supporting 70 percent of Oregon's population, a preponderance of the industry, and a varied agriculture, the province boasts fertile soils, generous rainfall, and a mild climate.

Past and Present

In conjunction with westward explorations, an examination of Pacific Northwest geology began indirectly with the Lewis and Clark expedition, commissioned by President Thomas Jefferson in 1803. With the purpose of furthering that knowledge, fifty years later the U.S. Congress authorized railroad surveying parties that included mineralogists and geologists. After the Civil War, the newly created U.S. Geological Survey initiated the westward search for mineral and fuel resources.

In the mid 1800s, Thomas Condon began an informal study of the regional geology and paleontology while stationed at The Dalles as a missionary. His interest and collections from the John Day fossil beds of central Oregon caught the attention of nationally known paleontologists. After his appointment as state geologist and later as professor at the University of Oregon in 1876, along with his book *The Two Islands*, Condon led the development of the science in the state. He died in 1907 at Eugene.

Chairman of the Geology Department at the University of Oregon from 1914 until 1947, Warren D. Smith participated in the early growth of geology in the state for 33 years. He saw the department through the emotional transfer of the sciences to Oregon State College (University) at Corvallis in the 1930s and experienced the turmoil that World War II brought to education. Because much of Oregon's geology was largely unknown when he arrived, Smith was called on to participate in many aspects of land-use, mineral identification, and paleontology. In addition to popularizing the state's scenic geology, he felt that one of his primary jobs was to provide the public with accurate information on geologic issues which surfaced. Politically active, Smith assisted in creation of the Oregon Department of Geology and Mineral Industries. He died in Eugene in 1950, three years after retirement. (Photo taken around 1930; courtesy Condon Collection)

Edwin T. Hodge also had to deal with much of Oregon that was still unmapped territory in the 1920s and 1930s. In that role, he was a consultant for government agencies on regional projects such as placement of the Bonneville Dam. His diverse interests included mineral resources and stratigraphy of the lower Columbia River and Cascades. Teaching at all three major institutions in the state at various times, Hodge, along with Ira Allison and William Wilkinson, was one of the initial faculty members of the new Geology Department at Oregon State College (later University) in 1923. As an outgrowth of his classes, he organized the Geological Society of the Oregon Country in 1935. Hodge died in 1970 at 81 years of age. (Photo taken in 1942; courtesy Geological Society of the Oregon Country)

Overview

The Willamette Valley and Coast Range had their beginnings as part of a wide continental shelf along the western margin of North America, where the two provinces shared closely related environments and sediments. The older foundation rocks are part of a volcanic island terrane, Siletzia, that developed in the Pacific Ocean basin. Colliding and accreting with the continental landmass, Siletzia subsequently subsided and was blanketed with layers of fossil-rich sediments from the late Eocene through the Oligocene. Uplift and tilting of the Coast Range in the late Cenozoic accompanied subsidence of the Willamette Valley into a lengthy structural trough, in which ocean waters shallowed and gradually retreated northward. By the middle Miocene, the Columbia River lavas, which spilled across from eastern Oregon, invaded the valley as far south as Salem.

The face of the valley was thoroughly reworked by events of the Pleistocene Epoch. A setting of tranquil lakes, meandering streams, and swamplands shifted to a more turbulent environment as torrential melt waters of the ancestral Columbia River and Cascade streams built great alluvial fans and gravel terraces atop the older bedrock. In the Willamette basin, thick silts were left by catastrophic floods, which originated from ice-dammed lakes in Montana. Once the ice blockage was breached, rushing flood waters cascaded across Idaho, southeastern Washington, and through the Columbia gorge, backing up into the Willamette Valley and forming the temporary Lake Allison. The multiple floods ceased when the climate warmed.

Lacking significant mineral wealth, the Willamette Valley boasts a plentiful supply of ground and surface water as its most valuable asset. This resource, the Willamette aquifer, has fostered the province's rapid growth and population. On the other hand, the basin experiences its share of natural geologic hazards. A surplus of rainfall brings flooding and landslides, while movement along the many faults intersecting the valley results in damaging earthquakes. Though not well understood, seismic activity originating along the offshore Cascadia subduction zone could prove even more destructive.

Geology

Cenozoic

Paleocene to Eocene

The Willamette Valley has played a somewhat passive role against the backdrop of moving tectonic plates. During the Paleocene to early Eocene, 60 million years ago, a chain of volcanic seamounts (Siletzia) carried atop the Farallon plate, were accreted to western North America. After the docking or initial contact, the Siletzia terrane began to slide beneath the continent and rotate into position, subsiding into a marine forearc trough west of the emerging Cascade volcanoes.

From the Eocene to the late Oligocene, copious sediments spread over the Siletzia volcanic plateau, which was to become the Willamette Valley and Coast Range. Consequently, rocks of the valley are continuous with those of the coastal province. The oldest are 15-mile thicknesses of submarine Siletz River Volcanics, which are covered by deep-water mica-rich sandstones and siltstones of the Tyee Formation. Tyee strata form a shallow delta at the southern end of the shelf near Roseburg and a larger deep-sea fan toward Salem. Although the Klamath Mountains supplied much of the sediment, there is evidence that an additional source lay well to the east.

Muds, sands, and silts of the Yamhill Formation, named by Ewart Baldwin, mixed with ash from the Western Cascades and were carried into the shallow seaway to overlap the Tyee. Within the Yamhill, shoals around offshore banks of the Rickreall and Buell limestones in Polk County hosted warm-water mollusks, foraminifera, and plants. These fossils, along with plant fragments, confirm a shelf setting close to a wave-washed beach.

Westward from Benton, Polk, and Lane counties, deep-water contrasts with the shallow marine and shoreline Yamhill conditions. In this environment, mollusks in sandstones of the Spencer Formation trace a flat, open continental shelf some distance from the shore. These strata are covered by nonmarine volcanic tuffs, sands, and conglomerates of the Fisher Formation. Plant imprints, preserved in the Fisher near Cottage Grove, reflect a warm, moist tropical climate where *Cinnamomum* (camphorwood), *Ficus* (fig), *Ocotea* (lancewood), and the broad-leafed *Aralia* grew adjacent to a coastal plain or a lakeside setting. Fisher deposits are overlain by the Little Butte volcanics.

Beneath the community of Eugene and northward toward Salem, almost a mile thickness of upper Eocene to Oligocene shallow marine and nonmarine sandstones and siltstones of the Eugene Formation entomb a varied assemblage of mollusks, crabs, sharks, and plants. The interpretation of this formation has evolved through multiple revisions in age and environments, but in the 2000s Donald

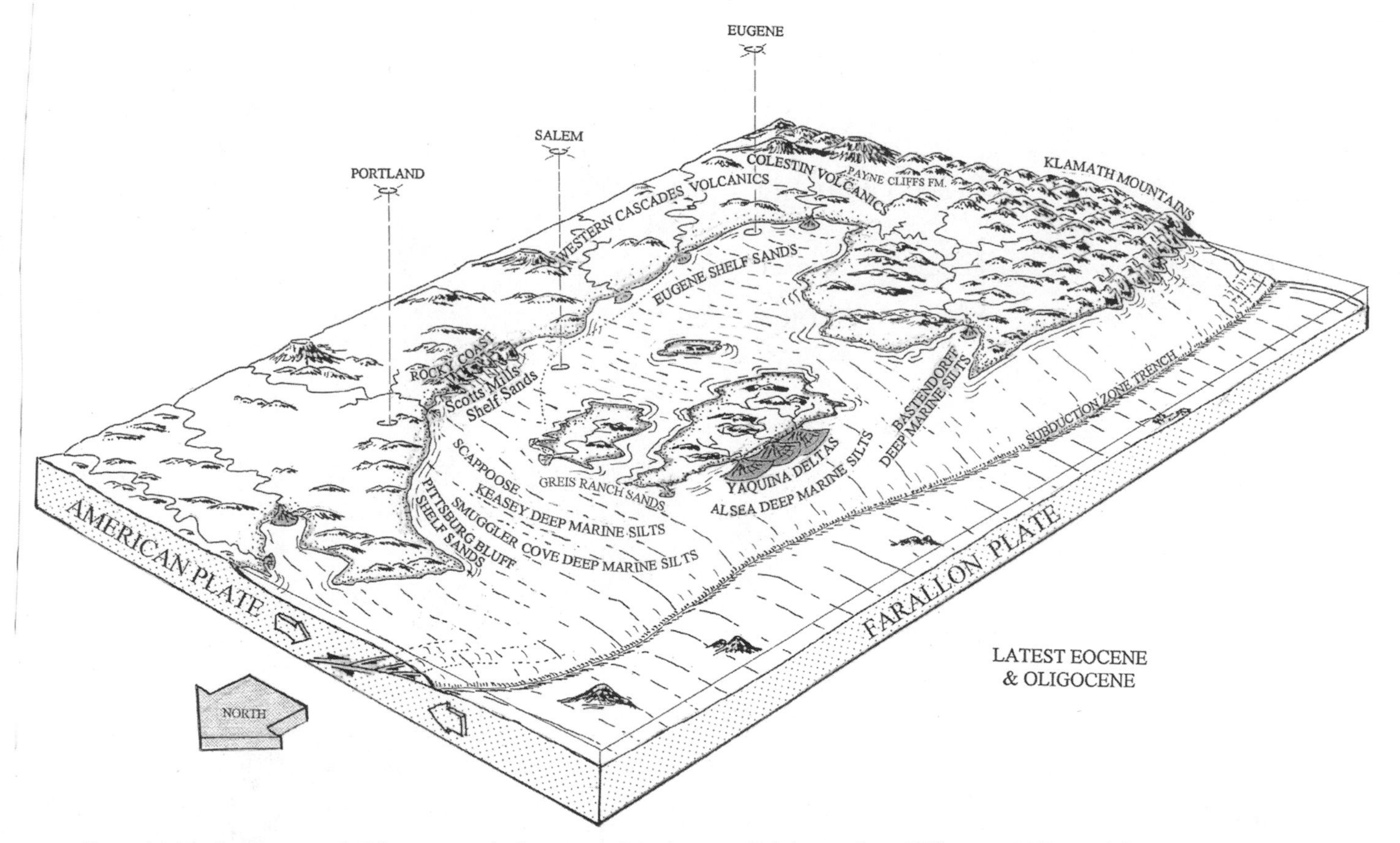

From the late Eocene to early Oligocene, a shallow tropical ocean occupied the southern Willamette Valley and the present Coast Range, before it was replaced by cooler climate conditions of late Oligocene time. (after Orr and Orr, 2009)

Prothero at Occidental College examined and revised many of the Pacific Northwest rocks using magnetic stratigraphy and biostratigraphy. Focusing on the timing of extinctions and the accompanying paleo-climatic changes on land and in the ocean, paleontologists at the 1999 Penrose Conference moved the Eocene-Oligocene boundary upward almost 4 million years from 38 to 33.7 million years ago. Their conclusions were published in the book *From Greenhouse to Icehouse*, edited by Prothero, who continues to adjust ages and rock correlations locally as well as internationally when new techniques emerge.

Oligocene

A gradual shift from a tropical to cool-temperate climate worldwide brought the demise of many species at the end of the Eocene. Locally the extinction of plants reached 60 percent with a corresponding 32 percent for marine invertebrates. Coinciding with cooling, the warm subtropical flora was replaced by temperate Oligocene plants while invertebrates continued to decline as the ocean withdrew from the valley. The faunal changes were noted as far back as 1968 by University of California paleontologist Carole Hickman, and in 2003 she concluded that in Oregon the climate alteration matched the onset of Cascade volcanism. A later overview paper by Greg Retallack at the University of Oregon, which also examines extinction rates and paleoclimates, recorded a similar loss of diversity for invertebrates and plants of the Eugene Formation by the early Oligocene, attributing it to long-term processes such as coastal uplift and shifting ocean temperatures rather than to meteorite impacts or volcanic eruptions.

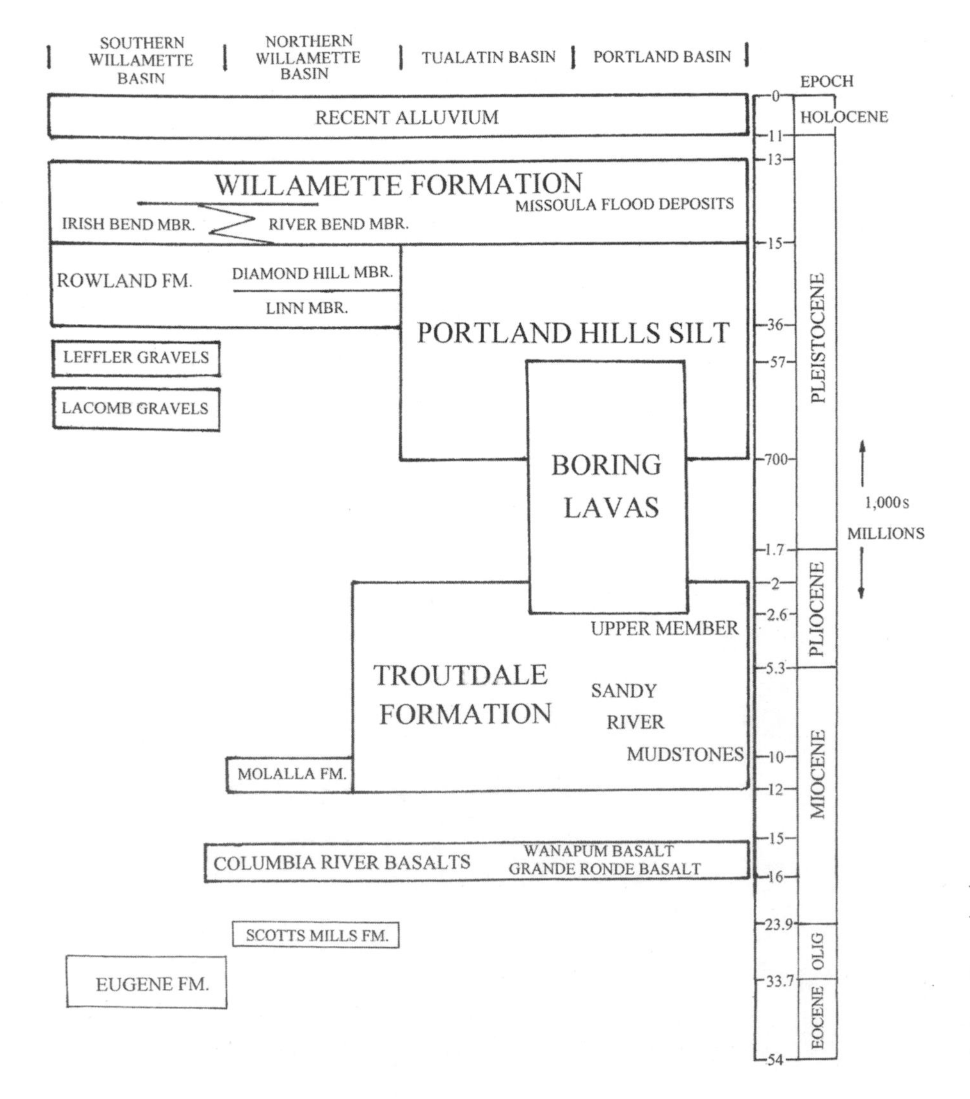

Tertiary to Quaternary stratigraphy of the Willamette Valley. (After Allison, 1953; Armentrout, et al., 1983; Balster and Parsons, 1969; Beaulieu, 1971; Gannett and Caldwell, 1998; Glenn, 1965; Mc Dowell, 1991; O'Connor, et al., 2001; Roberts, 1984)

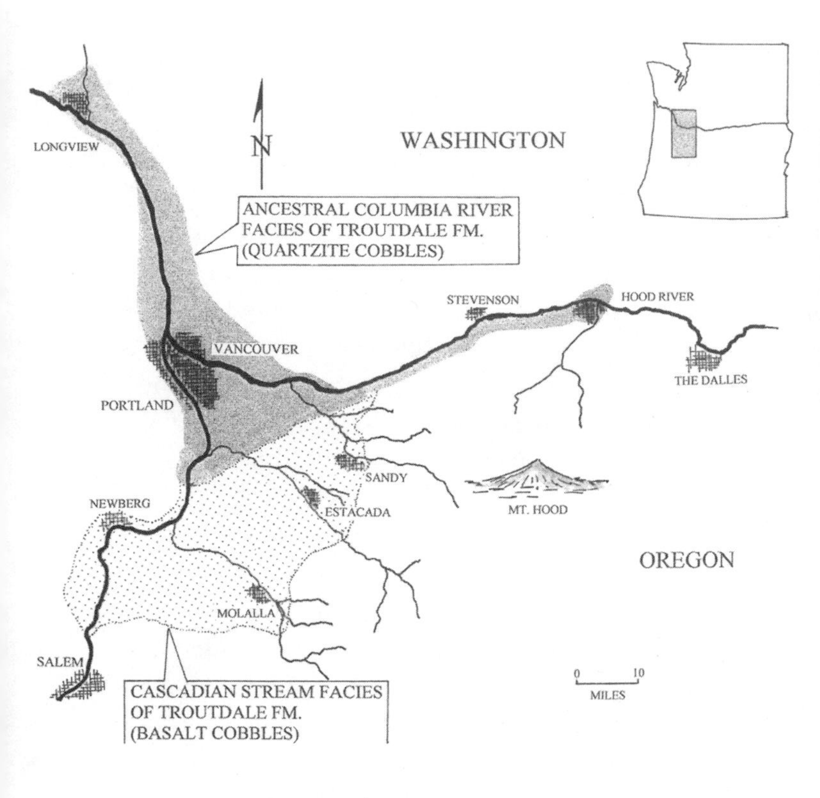

Terry Tolan, Marvin Beeson, and Beverly Vogt recognized two facies of the Troutdale Formation, each reflecting variable sources. Pebbles and cobbles of quartzite, schist, and granite transported by the ancestral Columbia River from the northern Rockies have origins in the Precambrian Belt series, whereas the high-alumina basalt sands and gravels of the Cascadian layer were derived locally from the Cascades and Boring lavas. The photograph is a close-up of the exotic quartzite conglomerate cobbles. (After Tolan, Beeson, and Vogt, 1984; Trimble, 1963; Wilson, 1998; photo courtesy Oregon Department of Geology and Mineral Industries)

Regional uplift by the late Oligocene caused the ocean to retreat. In the vicinity of Silverton in Marion and Clackamas counties, Scotts Mills sediments denote the position of the marine shoreline along the eastern edge of the Willamette Valley. The Scotts Mills Formation initially records a transgressive advancing seaway, followed by storm conditions, shallow water, and coastal swamps as the ocean withdrew.

Miocene

During the Miocene, between 23 and 5 million years ago, elevation of the coast range, subsidence of the Willamette Valley, and the invasion by lavas profoundly changed the topography and climate of western Oregon. The ocean continued to recede until the shoreline was close to its present position.

Sheets of dark Columbia River basalts, pouring from fissures and vents in northeastern Oregon and Idaho, crossed the Cascade Mountains and spread over the valley floor to depths of 600 feet. The basalts underlie most of Portland, the Tualatin Valley, and the northern Willamette basin, as well as mantling the uplifted Waldo, Eola, Amity, and Salem hills, the Red Hills near Dundee, and the Tualatin highlands. Near Portland the layers produced a monotonous flat surface with only the highest peaks projecting above the flows. In western Oregon's wet climate, the basalts began to decompose rapidly, yielding a thoroughly dissected volcanic landscape.

The Columbia River Basalts in northern Marion and Clackamas counties are covered by 1,000 feet of mudflows and volcanic tuffs of the Molalla Formation, the first terrestrial deposits after the retreat of marine waters. Fossil floras in the strata suggest a well-dissected hilly topography dominated by *Liquidambar* (sweet gum), *Platanus* (sycamore), *Carya* (hickory), and swampy lowlands favored by *Taxodium* (cypress).

Late in the Miocene and into the Pleistocene, a series of fault-bounded blocks created depositional basins at Portland and Vancouver, in the Tualatin and northern Willamette valleys, at Stayton, and in the southern Willamette Valley. These depressions are separated from each other by ridges underlain by Columbia River basalt. Because of the barriers, the individual basins have different histories and multiple sediment sources, making correlations between them difficult. Fill in the southern valley may be one continuous unit, but in the northern region sequences of thick nonmarine strata are present. Here the Sandy River Mudstone and gravels of the Troutdale Formation reach depths of 1,000 to 1,500 feet atop the Columbia River Basalts.

The Sandy River Mudstone and Troutdale Formation were thoroughly eroded by streams, before being perforated and covered by Boring lavas and late Pleistocene silts and sands. In 1933 Edwin Hodge named the combined strata as the Troutdale Formation, but in his classic 1963 publication on the geology of Portland Donald Trimble of the U.S.G.S. assigned the Pliocene conglomerates in the upper portion to the Troutdale and the underlying fine-grained Miocene to Pliocene layers to the Sandy River. Exposures of the mudstones along the Sandy and Clackamas rivers have been traced south to Estacada and Oregon City. Originally interpreted as a lake setting, the sediments are now regarded as predominantly fluvial in origin.

Cores drilled at Tualatin in Washington County and near Monroe in Benton County yielded lake sediments dating back to the middle Tertiary. Because of its population density and geographic position, the Tualatin basin has received considerable attention. In 1967 Herb Schlicker published on the sedimentary history, and Doyle Wilson at Portland State University presented an overview of its geologic evolution in 1998. He reported that the surrounding highlands were the main source for the Tualatin Valley fill, informally renaming the previously mapped Troutdale Formation and Sandy River Mudstone as the Hillsboro Formation. Pollen of Cupressaceae (cypress), which make up 45 percent of the Hillsboro flora, with lesser percentages of *Pinus*, *Alnus* (alder), *Salix* (willow), and *Abies* (fir), in conjunction with the diatoms *Aulacosira* and *Melosira* allowed Wilson to assign a late Miocene to Pliocene age to the alkaline lake deposits.

In 1976, Michael Roberts, then at Simon Fraser University, identified a large lake of similar age after he examined pollen from conifers and broad-leafed deciduous trees recovered from cores drilled into a terrace near Monroe. The oldest unit encountered was the Eocene Spencer Formation, overlain by an organic-rich clay that may correlate with the Sandy River Mudstone, and an uppermost Pleistocene gravel.

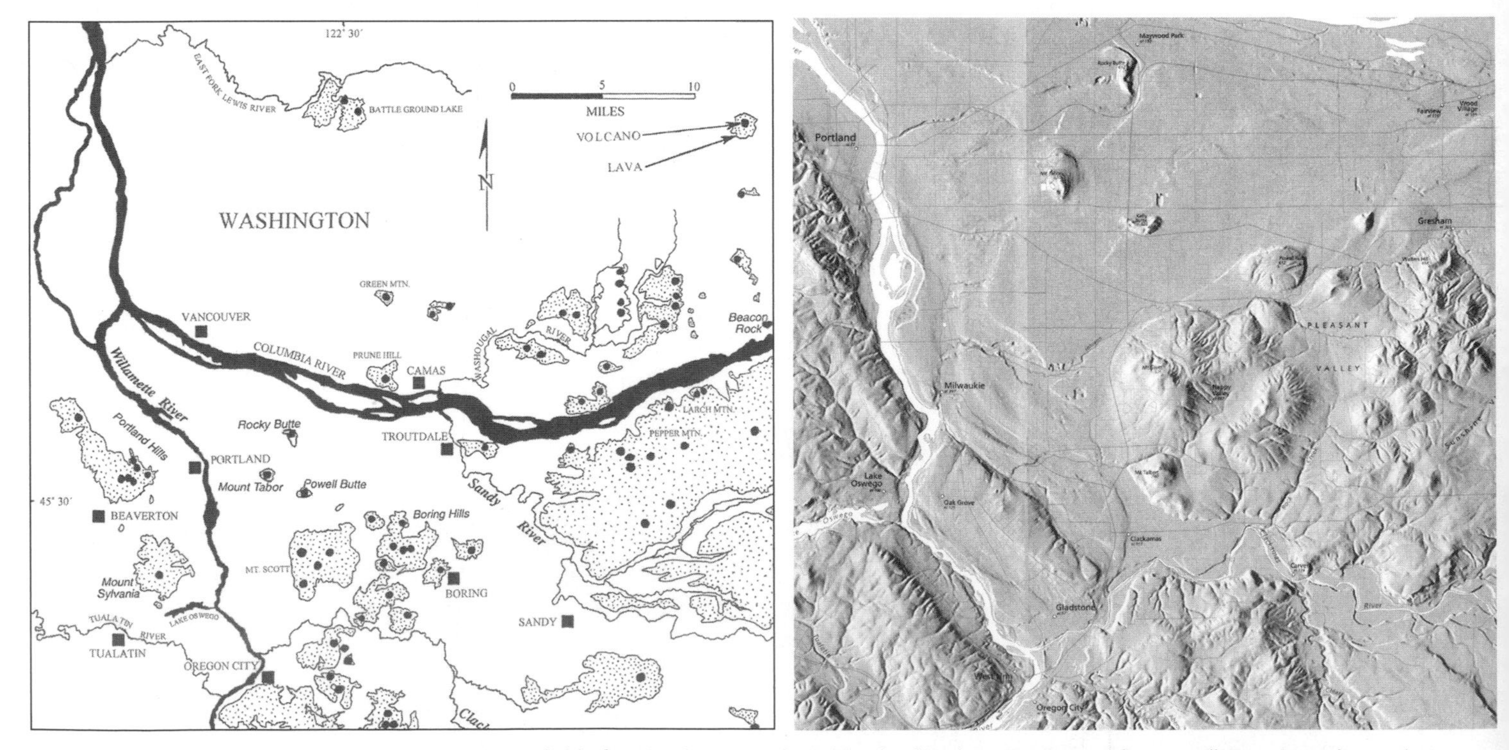

Between 2.6 million and 50,000 years ago, a volcanic field of Boring lavas enveloped Portland and vicinity. Named for a small town in Multnomah County, the flows extend westward to the Portland-Vancouver metropolitan area, south to Highland Butte, and southwest to Mount Sylvania near Beaverton. Among the most prominent are Rocky Butte, Mount Tabor, Powell Butte, and Mount Scott. Battle Ground Lake, filling a pronounced depression north of Vancouver, is an intact maar that formed from a violent explosion when rising magma encountered cooler groundwater. In the map on the right, the buttes are east of Portland (After Allen, 1975; Evarts, et al., 2009; Peck, et al., 1964; Trimble, 1963; photo courtesy University of Oregon)

Pleistocene—Volcanism, Floods, and Sediments

Because of continued subduction of the Juan de Fuca plate, uplift and gentle folding of the coastal mountains, and subsidence in the Willamette Valley during the Pliocene and Pleistocene, the province emerged as a separate physiographic region, no longer part of a wide coastal plain.

Volcanism in the Portland-Vancouver metropolitan area commenced 2.6 million years ago with eruptions from over 80 small cones, vents, and shield volcanoes. Merging to the east with the High Cascades, the Boring lava field is located in the forearc region. Its unique positioning with relation to the Cascades and the origin of the eruptions, while still unclear, indirectly relate to the mechanics of plate subduction. Following a hiatus lasting almost a million years, activity was resumed as the Boring lavas became widely dispersed, were more variable in composition, and were roughly continuous until 50,000 years ago. The youngest cones are in the north part of the field, and the oldest toward the south. The fine-grained texture and fracture pattern of the lavas, which weather into large blocks, readily distinguish them from the much older Miocene Columbia River basalts, in which small columns are common. In a 2009 paper, Russell Evarts of the U.S.G.S. addresses the possibility of future eruptions from the Boring cones, which he considers to be very low.

Portland Hills Silt

Interfingering with the Boring lavas around Portland and Tualatin, the gritty, fine, yellowish-brown Portland Hills silt was deposited from 700,000 to 15,000 years in the past. Commonly 25 to 100 feet thick, the silts cover much of the higher elevations from the Tualatin and Chehalem mountains all the way to Gresham and Boring. Microscopically, the silt is remarkably uniform, and its physical characteristics match the wind-blown (loess) Palouse sediments of southeast Washington. During dry interglacial periods, fine rock flour, ground up by the crushing and milling action of ice, was carried aloft in enormous clouds and distributed across the Columbia River basin by strong winds. Four different loess intervals mark interglacials, while soil horizons separating the silts indicate times of glacial advance.

River Deposits

Shallow-water lake (lacustrine) and stream (fluvial) environments in the Willamette lowlands were heavily modified during the Pleistocene when the rivers and tributaries spread sheets of gravel and sand, constructing fans and terraces throughout the area. The older deposits were subsequently covered by Willamette Formation silts, which were deposited by flood waters late in the epoch.

The stratigraphy and geomorphology of the Willamette Valley were studied extensively by early researchers Ira Allison, Jerry Glenn, and Donald Trimble, while in the 1990s and 2000s Jim O'Connor of the U.S.G.S. revised the stratigraphy and Patricia McDowell at the University of Oregon outlined the problems and complexities of correlating the units.

An examination of the earliest Pleistocene landscape underlying the valley was begun in 1969 by Clifford Balster and Roger Parsons with the U.S. Soil Conservation Service to determine the relationship between the older soil and rocks and the overlying fluvial debris. These authors recognized 15 separate geomorphic units. The oldest surfaces are represented by weathered gravel and mud pediments, signifying various phases in the development of ancestral river systems.

From the mid to late Pleistocene, massive sediment loads carried by the Willamette River and tributaries constructed high glacio-fluvial terraces that today are preserved around the margins of the basin and along river channels. In his landmark 1953 publication on the Albany Quadrangle, Ira Allison produced a definitive stratigraphy of the units, dividing them from oldest (highest) to youngest (lowest)—the Lacomb, the Leffler, and the Linn. Composed of sediments from the Cascades to the east and the Coast Range on the west, gravels and muds of the Lacomb and Leffler terraces stand 100 to 300 feet above the valley floor. Extensively eroded or obscured by later deposits, only small exposures of these terraces remain.

Tectonic subsidence and renewed downcutting by the Willamette River preceded deposition of the 200-foot-thick Linn gravels, which occupy much of the valley floor. Representing glacial outwash, the Linn was flushed from the Cascades and distributed as a complex series of coalescing alluvial

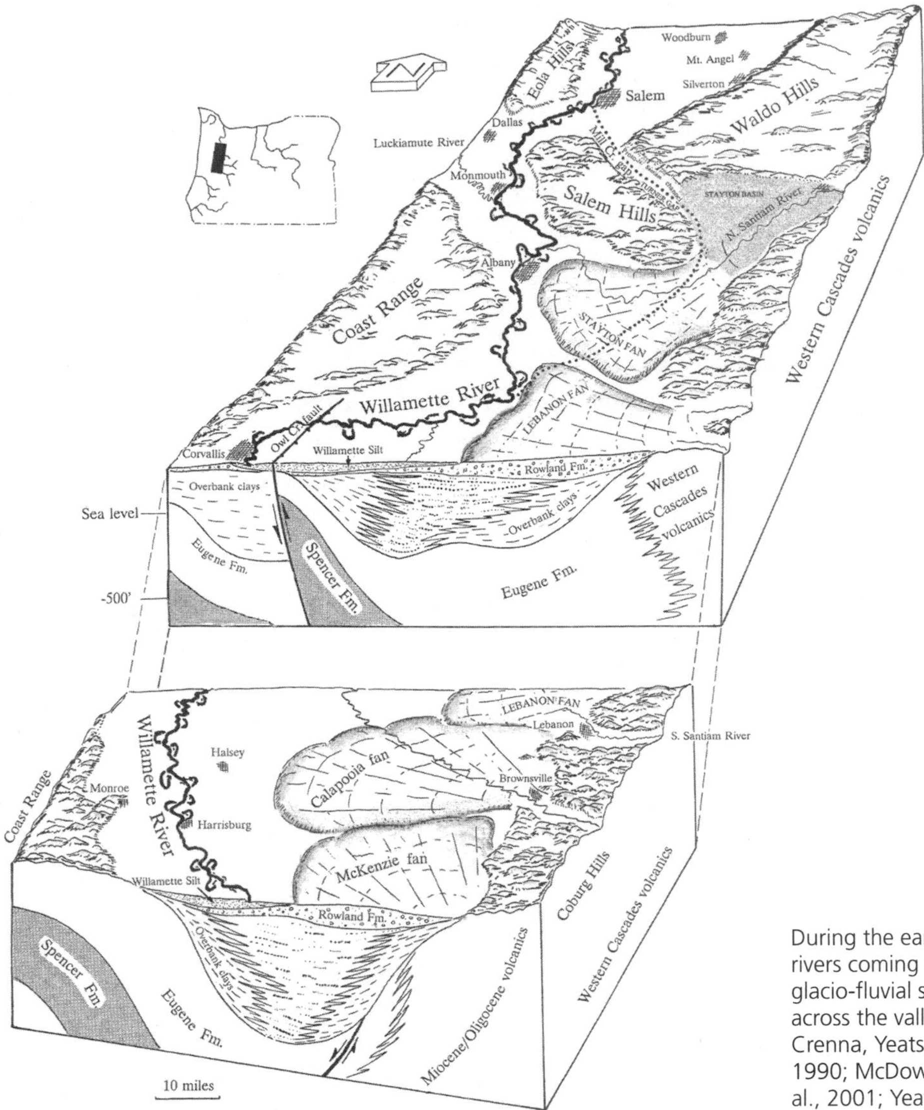

During the early to middle Pleistocene, rivers coming off the Cascades spread glacio-fluvial sand and gravel fans across the valley. (After Allison, 1953; Crenna, Yeats, and Levi, 1994; Graven, 1990; McDowell, 1991; O'Connor, et al., 2001; Yeats et al., 1991)

fans. These, in turn, pushed the Willamette River channel over to the west.

Allison's divisions have since been revised. Balster and Parsons renamed the Linn as the Rowland Formation and subdivided it into the upper Diamond Hill and the lower Linn members. Wood fragments from the Diamond Hill have been dated between 23,000 to 20,000 years old. In 2001, O'Connor delineated four Quaternary stages in the Willamette Valley—the oldest gravels and sands from 2.5 million to 0.5 million years ago correspond to Allison's Lacomb and Leffler; lake and stream fill going back at least 420,000 years is equivalent to Allison's Linn gravel; Missoula flood deposits are dated from 15,000 to 13,000 years in the past, and Willamette River sediments are younger than 12,000 years.

Missoula Floods

Thick, uniformly laminated Willamette silts, over the Linn gravels and older bedrock, originated from episodes of catastrophic floods that swept from Montana, across Idaho and Washington, and through the Columbia River gorge to Oregon. When the notion of colossal flooding was first proposed in the 1920s by J. Harlen Bretz, a professor at the University of Chicago, the evidence was compelling, but the geologic community was slow to accept the idea without a source for such a huge volume of water. Richard Flint at Yale University, Ira Allison, and Edwin Hodge appealed to landslides, ice dams, glaciers, and streams, or a combination of these to substantiate alternate theories and to explain the erosion and deposition. Strong opinions prevailed

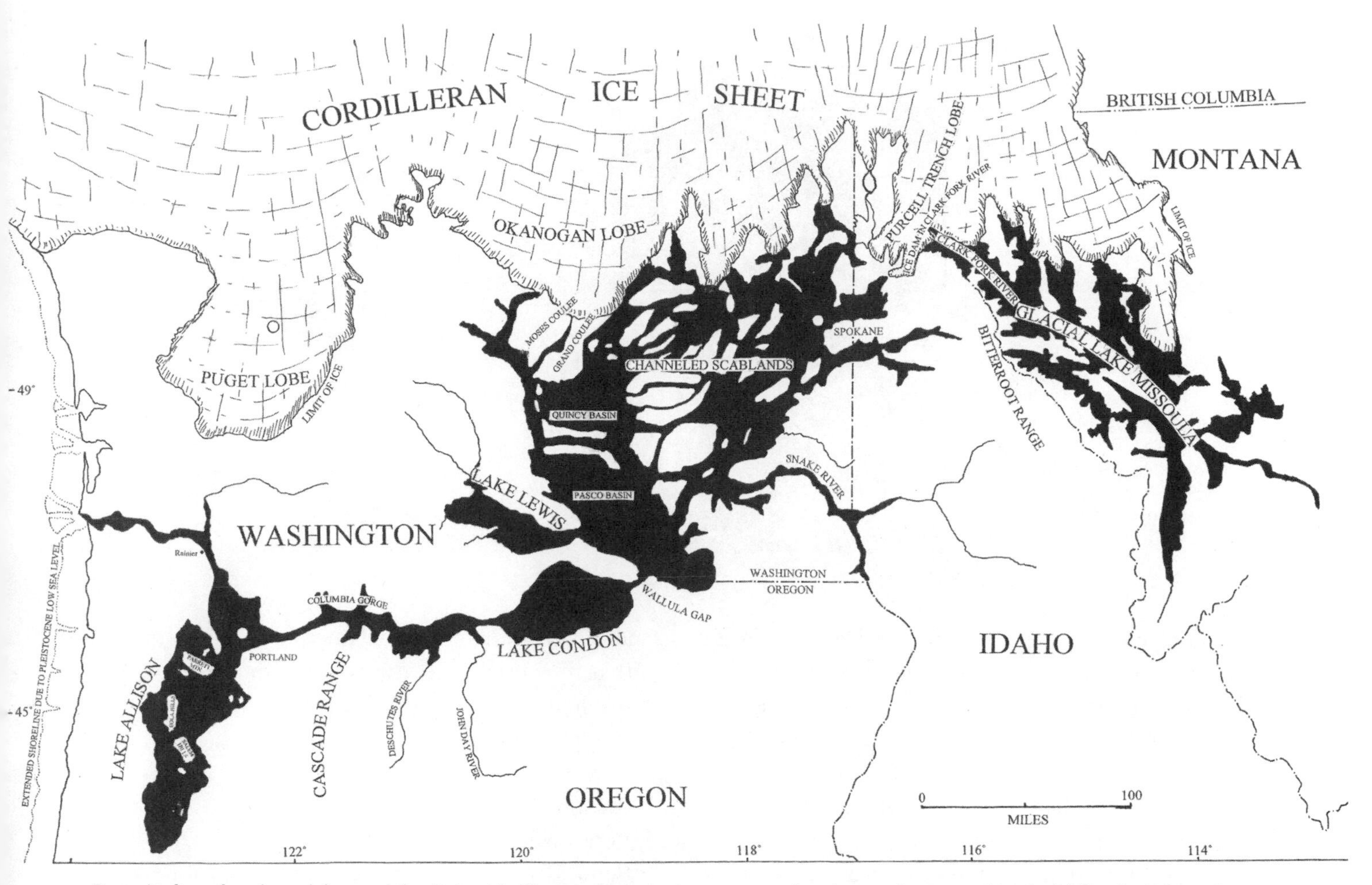

Deposits from four large lakes and the Columbia River gorge trace the source and pathway of catastrophic glacial floods. (After Allen, 1986; Allen, Burns, and Burns, 2009; Allison, 1953; Baker, 1973; O'Connor and Burns, 2009; Waitt, Denlinger, and O'Connor, 2009)

on all sides. The discovery of an ice-dammed Pleistocene lake in Montana by Joseph Pardee at the U.S.G.S. was the final puzzle piece required for the flooding theory to be accepted.

In the 1970s and 1980s, Richard Waitt of the U.S.G.S. and Victor Baker from the University of Arizona examined various aspects of the deposits and chronology of events, supplying new data on their magnitude and frequency. The story of the cataclysms and the controversy surrounding them is related particularly well in John Allen's book, recently updated by Marjorie Burns and Scott Burns, and by Waitt's 2009 field guide that compares early flood hypotheses, summarizing the hydraulics, discharge rates, and geomorphology.

Beginning about 15,000 years ago and continuing for a period of 2,000 to 3,000 years, the lower Columbia River drainage experienced a succession of devastating floods when the Clark Fork River in northern Idaho was dammed by the Purcell lobe of the Canadian ice sheet, backing up enormous Lake Missoula across much of western Montana. As the dam was breached, over 500 cubic miles of glacial ice, water, and debris poured out, traveling from an elevation over 4,200 feet to sea level. Flushing through the Idaho panhandle and scouring the area now known as the channeled scablands of southeast Washington, the deluge ponded briefly into the 1,000 foot deep Lake Lewis at the narrows of Wallula Gap, then a second time at The Dalles to create Lake Condon before coursing westward, scouring the canyon of the Columbia River. Water moving at the rate of 9.5 cubic miles an hour would have taken three days before draining the lake.

Geologists long ago concluded that separate layers along the route suggested many floods, but exposures of the Touchet Beds at Burlingame Canyon near Walla Walla in 1926 provided indisputable

Filling the Willamette Valley, Lake Allison covered Portland with 400 feet of water, and Lake Oswego, Beaverton, Hillsboro, and Forest Grove would have been submerged between 100 to 200 feet. (Image courtesy S. Burns)

evidence after water accidentally released from an irrigation canal revealed numerous distinct beds. Representing individual events, each unit is several inches thick, grading upward into the classic finer flood sequences. Since the 1920s, the number of proposed floods has steadily increased. The figures initially proposed ranged from seven, eight, or even to 35 or 40 deluges, but the consensus at present is there were 90 to 100 occurrences.

When flood debris choked the narrow Columbia River channel downstream from Portland at Kalama, Washington, the muddy water backed up into the Willamette Valley as the temporary Lake Allison. Corresponding in large part to Thomas Condon's designated Willamette Sound, the turbid freshwater body extended to Eugene. Repeated surging and ebbing as additional water entered through the gaps at Oregon City and Lake Oswego kept the level constantly fluctuating. After a brief interval of a week or two, Lake Allison drained back into the Columbia River and the Pacific Ocean, leaving innumerable layers of lacustrine silts, sands, and clays over the older gravels.

Willamette Formation (Silt)

Ira Allison was the first to suggest that late fine-grained Pleistocene sediments in the Willamette Valley were deposited by glacial floods, and after his examination of the nine-to-twelve-foot-thick section at Irish Bend south of Corvallis, he described and named them the Willamette Silt. He concluded that they originated from a backwater lake, which resulted from countless glacial floods, but his theory was modified in 1965 by Jerry Glenn in his PhD from Oregon State University. Glenn worked out the stratigraphy of a section south of Dayton at River Bend, where the facies was similar to that at Irish Bend. His hypothesis fixed the number of floods at 40 with a single much larger climactic event.

In 1969, the Willamette Silts were redesignated as the Willamette Formation and subdivided into four members by Balster and Parsons. The oldest Wyatt layer represents river channel fill, the thick Irish Bend silt and fine sand mantled a wide central area, while the younger Malpass and Greenback covered the lowlands and foothills.

The flood torrents that spilled into the valley carried icebergs laden with erratics, which were dropped over a wide area when the ice melted. There are 300 recorded occurrences, but thousands more are still unrecognized. Whereas over 40 boulders exceeding three feet in diameter have been located, numerous smaller stones have been uncovered in fields, along roadcuts, and in old river terraces. Varying in composition, the erratics are clearly foreign to the Willamette Valley and more typical of rocks in western Montana. Their exotic nature makes it possible to trace the flood pathways down the Columbia River channel and through Wallula Gap to The Dalles, where they are found up

Large stones, called erratics, brought into the Willamette Valley atop ice floes, are scattered from Eugene to Portland. In this map the localities of over 300 erratics are based on work by Ira Allison. Of these, the largest known is the Sheridan erratic (photo) between McMinnville and Sheridan in Yamhill County. Composed of the metamorphic rock argillite, the stone originally weighed about 160 tons, but almost half has been removed by collectors. The site has been designated as a state park to protect and preserve the specimen. (After Allison, 1935; Photo courtesy Oregon State Highway Department)

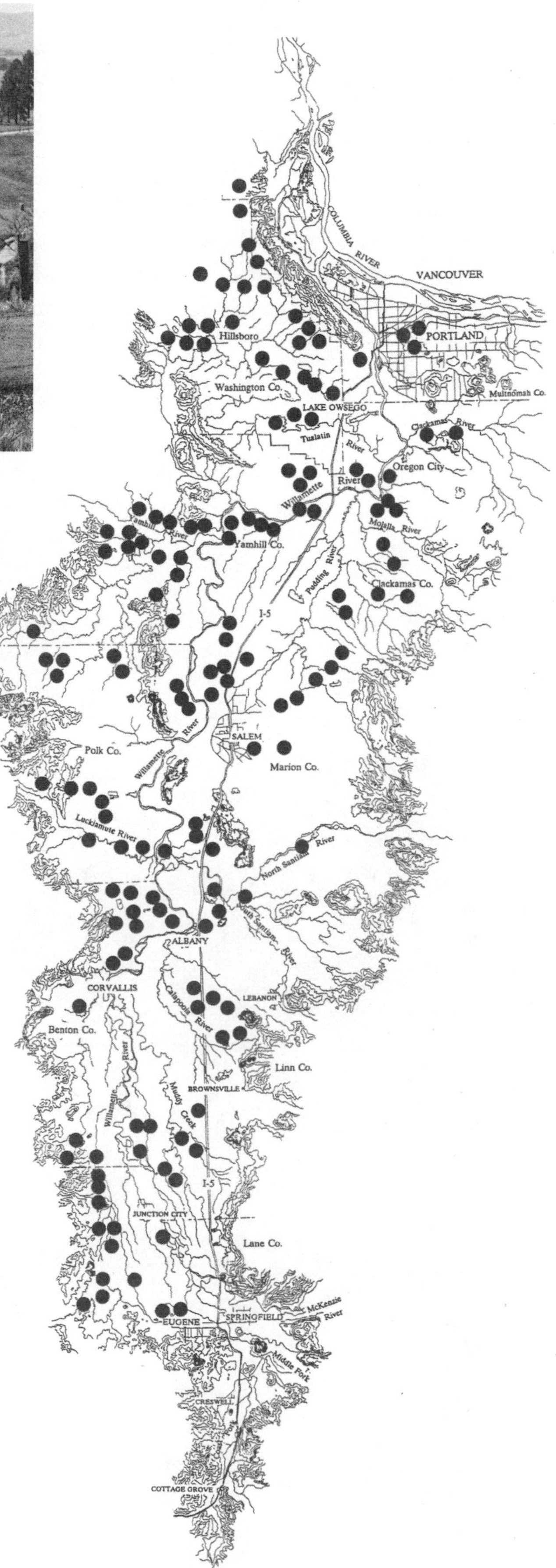

Perhaps the most famous Oregon erratic is the Willamette meteorite (photo), which was apparently rafted by ice from Montana to a site later acquired by the Oregon Iron and Steel Company at Lake Oswego. Surreptitiously but strenuously moved by Ellis Hughes to his own property, the 31,107 pound behemoth became the focus of a court battle about ownership. The company won the judgment, and the rock was ultimately purchased in 1902 and donated to the American Museum of Natural History in New York. (Photo courtesy Oregon Department of Geology and Mineral Industries)

to 1,000 feet above sea level. In the valley, erratic fragments occur at elevations as high as 400 feet in the upper layers of the Willamette Silt.

Rearranging the Rivers

After the final Missoula flooding episode around 13,000 years ago, the Willamette River and Cascade streams were reestablished on the valley floor. Braided river networks, abandoned channels, and oxbow lakes characterized the lowlands during this period. Post-Missoula flood sands and gravels up to 60 feet thick fanned out in broad swaths where the Cascade rivers entered the valley, displacing and redirecting the meandering pathways.

During the late Pleistocene, the North Fork of the Santiam River repeatedly changed its confluence point with the Willamette. Transporting fine Cascade glacial deposits of the Missoula flood, it followed a course northward from the Stayton basin through the narrow pass at Turner (Mill Creek) Gap to join the Willamette River at Salem instead of merging at its present-day junction north of Albany.

In the northern Willamette basin, the channel of the Willamette River may have been pushed westward by gravels of the Troutdale Formation, by heavy Cascade glacio-fluvial sediments, or even by lavas from the Boring field. Establishing its present route some 11,000 years ago, the river cut the prominent cliffs and falls through the Columbia River basalts at Oregon City.

Lake Oswego occupies a channel scoured by the Missoula floods. The Cascade Mountains and Mount Hood can be seen in the distance across the valley. (Photo by Delano Photographics; courtesy Condon Collection)

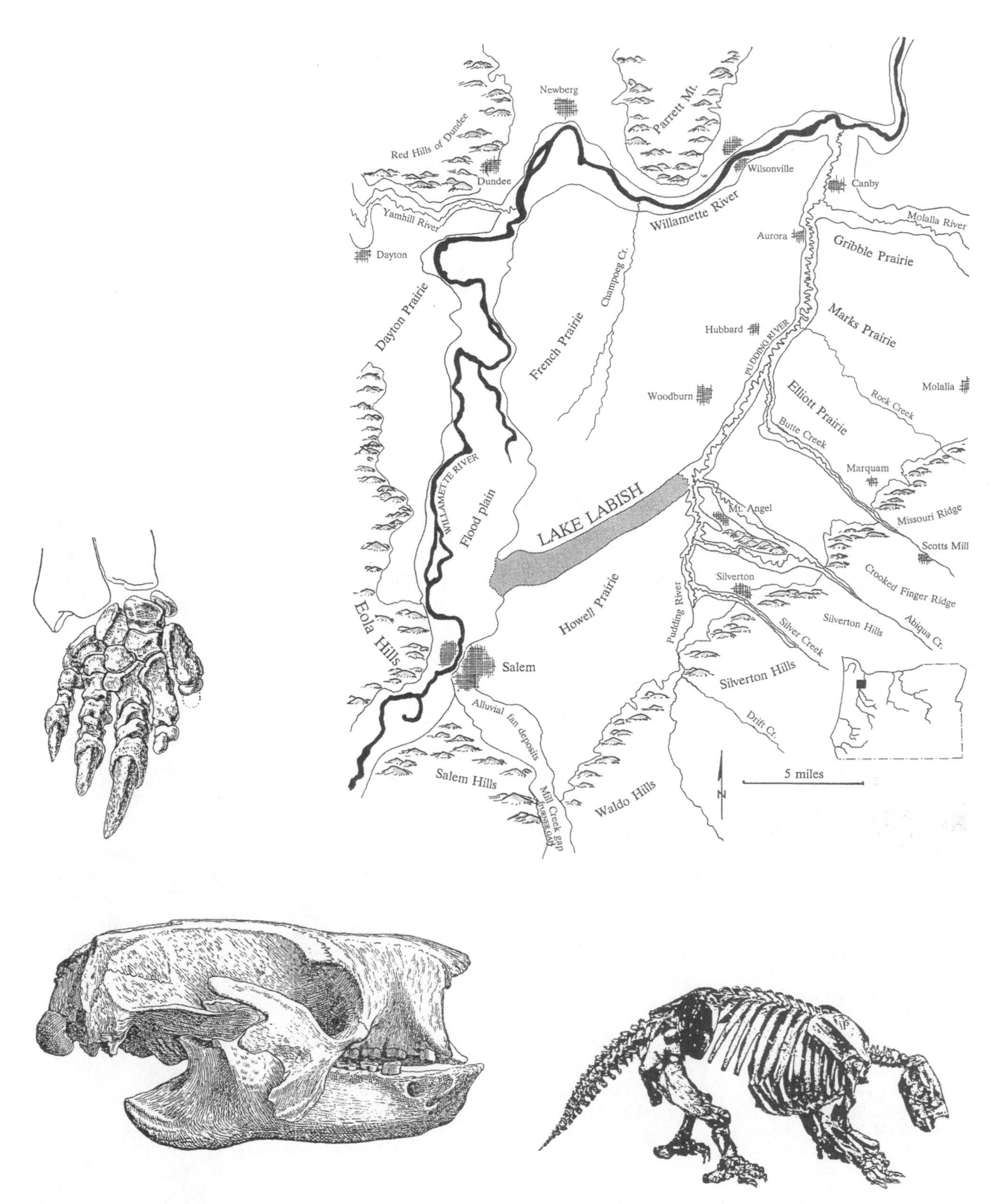

What may be a former route of the Willamette River, Lake Labish extends in a straight strip for almost 10 miles northeast from Salem. The river was blocked by a flux of sediments from Silver, Abiqua, Drift, and Butte creeks, and the resulting shallow lake slowly converted into a peat-rich marsh. Dated around 14,000 years ago, the bones of Ice Age mammals such as mammoths, mastodon, giant sloth (claw, skull, and skeleton), and bison are frequently exposed during farming operations. Unlike the LaBrea tar pits of Los Angeles, the animals were probably not mired in the bog, but the carcasses were washed in and covered, allowing the remains to be preserved in the oxygen-poor setting away from the attention of scavengers. Today the fertile lakebed soils support a thriving vegetable industry. (After Glenn, 1962; Orr and Orr, 2009; Schlicker and Deacon, 1967)

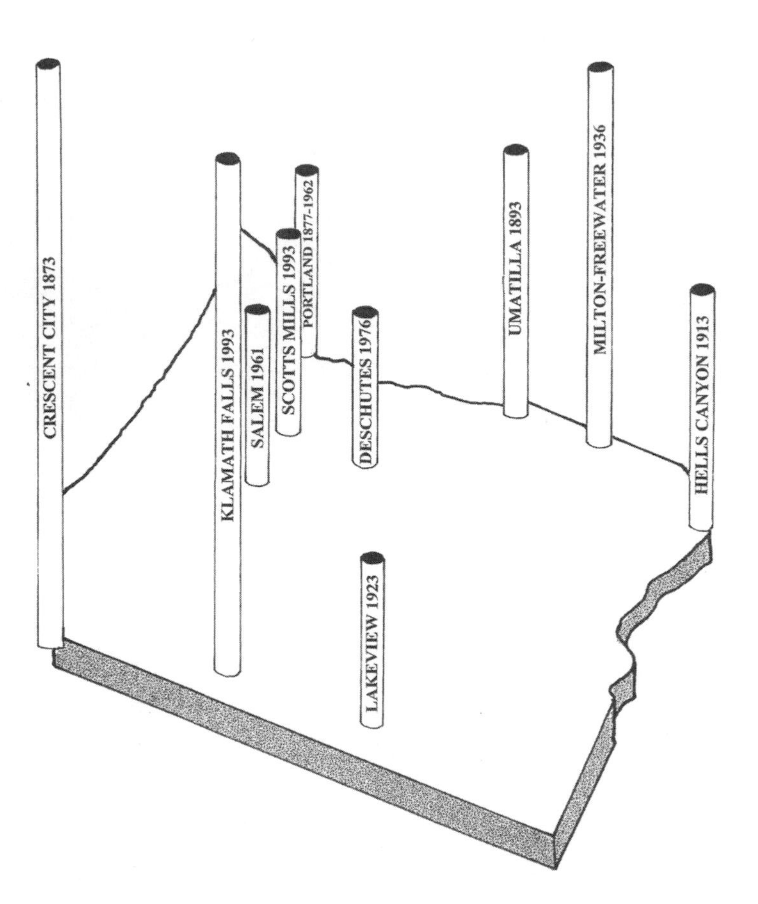

A look at large earthquakes in Oregon shows most measured VI on the Mercalli scale. Only the Crescent City (California), Klamath Falls, Milton-Freewater, and Portland quakes were stronger at VII. (After Berg and Baker, 1963; Byerly, 1952; Madin and Mabey, 1996; Niewendorp and Neuhaus, 2003; Townley and Allen, 1939; Wong and Bott, 1995)

Geologic Hazards

In western Oregon, and especially in the Willamette Valley, the increase in population, the decrease in available land, and inadequate land-use evaluations have led inhabitants to situate themselves, their structures, or their activities on floodplains, atop unstable soils, or near seismically active regions. As a consequence, work on geohazards has become of primary importance, with concerted efforts by state and federal agencies to make information on risks available through seminars, newspapers, maps, and reports.

Earthquakes

Situated inland from the Cascadia subduction zone and adjacent to the Cascade volcanic arc, the Willamette Valley is vulnerable to seismicity from crustal faults, from the subduction zone (megathrusts), or from events within the subducting slab itself (intraplate). In contrast to seismicity related to subduction processes, movement along crustal faults in the Willamette is shallow, at 6 to 12 miles in depth, and more frequent than activity connected with subduction.

A seismograph station at Corvallis has been part of a worldwide system since 1963, but improved monitoring was established for Portland in 1980 when the University of Washington expanded the Pacific Northwest Seismograph Network into northern Oregon. Improved record keeping now shows that there is a high rate of seismicity in

Soils engineer and geologist Herbert Schlicker was one of the earliest proponents to regard geohazardous and environment conditions as restraints on development. Schlicker was born in 1920 and grew up on a farm near Salem. Graduating from Oregon State University, he joined DOGAMI in 1955, where his first study on the Tualatin Valley and his last on the geology and hazards of northwestern Clackamas County bracket 25 years with the state. Providing expertise for many boards and government agencies, Schlicker also was instrumental in initiating registration for geologists in Oregon, and, as the first, he was assigned the number One. He died in Clackamas in 1992. (Photo courtesy Oregon Department of Geology and Mineral Industries)

During his career, John Beaulieu also emphasized the need for land-use planning that would take geohazards into account. Born during World War II to parents who worked on the Manhattan project, Beaulieu and his twin brother grew up near the Hanford Nuclear facility in Washington. Finding mastodon bones at Hanford led Beaulieu to later field work on the site. Finishing a PhD from Stanford in 1969, he began with DOGAMI shortly afterward, where, as director he frequently worked with the legislature to create environmental awareness and goals. Beaulieu retired in 2003 and currently lives in Portland, where he works on selected geologic projects. (Photo courtesy J. Beaulieu)

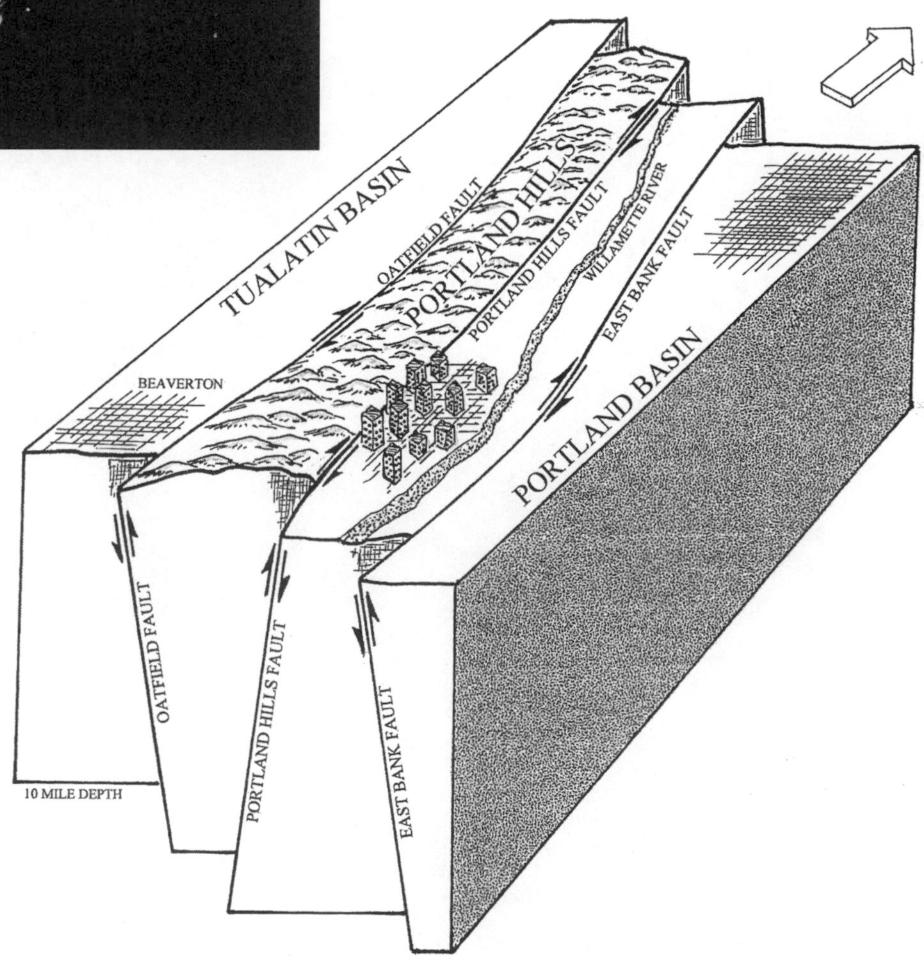

Swarms of small to medium tremors during July, 1991, centered at the north end of the Portland Hills fault, were noticed in many communities of western Oregon, and the 2001 discovery of up to six feet of displaced flood deposits beneath Rowe Middle School in Milwaukie shows that activity is continuing. On April 24, 2003, a 3.6 earthquake near Kelly Butte Park shook Portland and Vancouver, nine miles distant. There were no damages or injuries, but this was one of the largest quakes locally within the past 35 years. In the photograph, Portland is located on the down-thrown (northeast) side of the Portland Hills fault, which follows the Tualatin Mountains (upper left). (After Blakely, et al., 1995, 2000; Wang, et al., 2001; Photo courtesy Delano Photographics)

the Willamette Valley. Ground shaking from a 6.8 magnitude rupture of the Portland Hills crustal fault could exceed even the damage produced by a Cascadia subduction zone event. Ground motion maps also measure the distance from the epicenter and the susceptibility of the soils to liquefaction. When saturated with heavy rainfall, the soft loose Willamette Silts, that fill the Portland basin, might liquefy if shaken. Somewhat like Jell-O, they would continue to shake after the event had stopped, then begin to flow, thus accentuating the destructive effects.

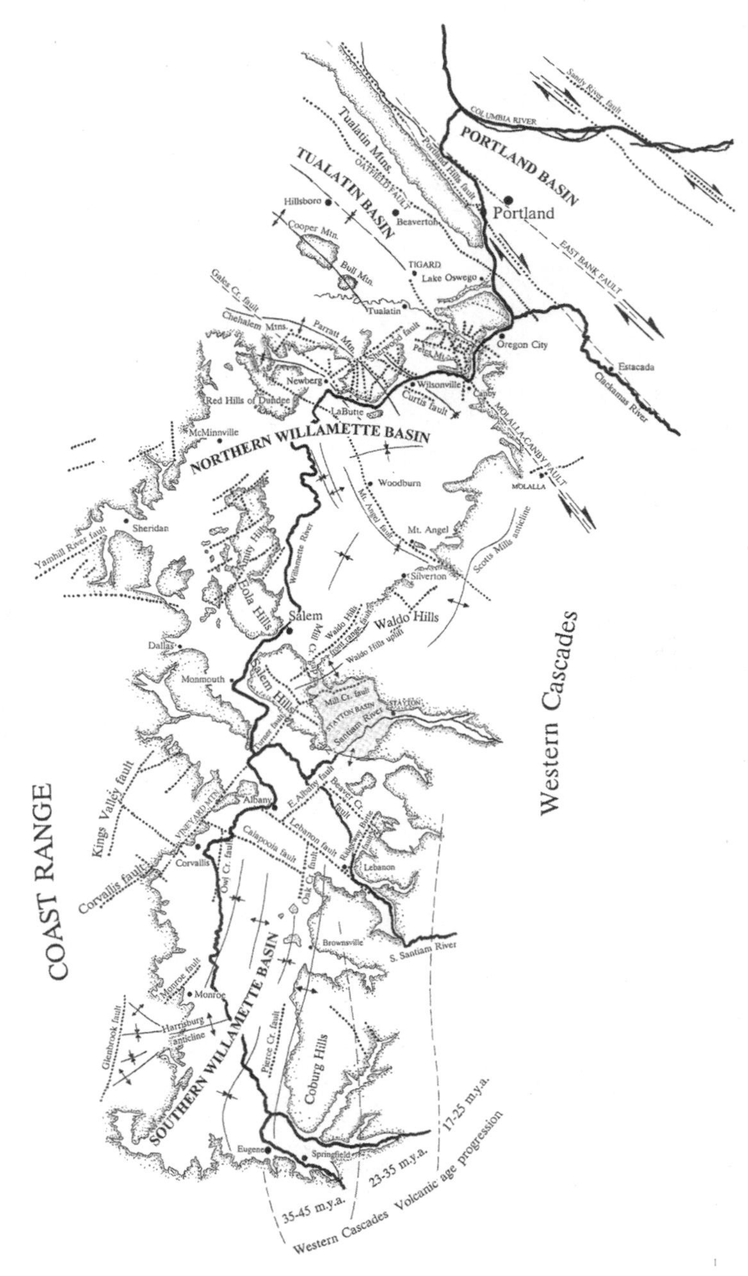

Over the past 20 years, detailed mapping has allowed geologists to connect disparate faults in the Willamette Valley that were originally regarded as isolated features. The faults of greatest concern in the north Willamette Valley follow the Portland Hills anticline that date to the late Miocene and Pliocene. The Portland Hills system, over 40 miles long, trends northwest by southeast and includes the Portland Hills, the Oatfield, the East Bank, and the Frontal (Sandy River) faults. Aeromagnetic surveys flown in 1992 show that Portland Hills faults, which parallel the Willamette River in the downtown area, are concealed beneath Quaternary sediments and may join the Clackamas River fault belt to the south. This alignment of surface fractures traverses all the way to the Steens Mountains as part of the Brothers Fault zone. (After Blakely, et al., 1995, 2000; Crenna, Yeats, and Levi, 1994; Gannett and Caldwell, 1998; Graven, 1990; Werner, 1990)

Accounts of historic earthquakes in the Pacific Northwest have been extensively re-evaluated in order to anticipate future high-magnitude episodes. In 1993, private consultants Jacqueline Bott and Ivan Wong chronicled the crustal earthquake record for the Portland area, revealing at least 17 shocks of magnitude 4.0 or larger on the Richter Scale since the late 1800s. Brick buildings swayed, windows rattled, and inhabitants rushed into the streets during the tremor of February 3, 1892, although there was minimal damage. In 1961 and 1962 and again in 1968 magnitude 4.7 and 5.2 quakes struck Portland from epicenters on the eastern edge of the city, causing some structural problems. Richard Blakely with the U.S.G.S. notes that the strongest 1962 quake occurred between the Portland Hills and Frontal (Sandy River) fault zones.

Elsewhere in the Willamette Valley seismic records are scant, although there have been a number of shallow crustal quakes. On November 16, 1957, residents near Salem described a shaking that registered an intensity of 4.5 but with no damage, although a similar occurrence at Albany and Lebanon in 1961 toppled chimneys, broke windows, and knocked over signs. The more recent Scotts Mills 1993 quake, centered 25 miles northeast of Salem,

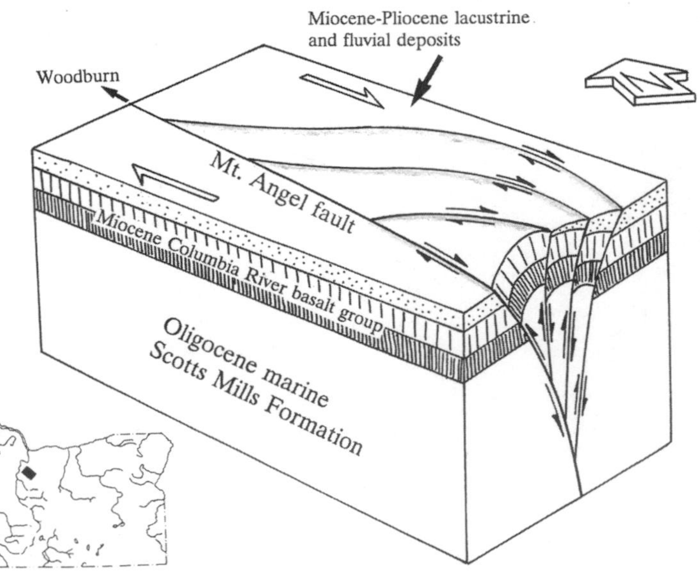

The butte at Mt. Angel, where the Abbey sits, forms what is called a fault "pop-up" in which rocks, caught and squeezed between a series of intersecting fault blocks, are slowly extruded to the surface. (After Werner, et al., 1992)

Eastward from Corvallis, a criss-crossing network of normal and thrust faults extends to the Western Cascades. The 30-mile-long Corvallis fault has been identified as the source of three to four quakes of 3.0 to 5.0 intensity. In the photograph looking north, the fault passes diagonally between the city and the distant low Vineyard Mountain at the edge of the McDonald State Forest. Marys River is in the foreground. (After Goldfinger, 1990; Graven, 1990; Kienle, Nelson, and Lawrence, 1981; Werner, 1990; Yeats, et al., 199l; Photo by Western Ways of Corvallis; courtesy Condon Collection)

caused considerable structural damage to the State Capitol, as well as to buildings in Molalla and Mt. Angel at a cost of $28 million. Rated at 5.6 magnitude, it was triggered by motion along the 35-mile-long Mt. Angel fault. The Mt. Angel structure is on the southeast end of a fault complex that trends northwestward beneath Woodburn to project into the Gales Creek fault zone from the Coast Range and southeastward into the Waldo Hills front range system.

In 2000, Richard Blakely described the previously unrecognized Canby-Molalla lineament, which runs 40 miles through the communities of Canby and Molalla in Clackamas County. He regards this fault as similar to the parallel Mt. Angel zone in that both are extensive strike-slip features displacing significant sections of the upper crust.

Subduction—Intraplate Earthquakes

Marking the boundary between the Juan de Fuca and North American plates, the Cascadia subduction zone lies roughly 100 miles west of the Willamette basin. Subduction earthquakes are generated along the interface between the two plates, but historically

The chance of a subduction quake in the Pacific Northwest is more probable than for an intraplate episode, which is associated with deformation, temperature, and pressure changes within the Juan de Fuca plate itself. However, there are exceptions. A 1949 mid-plate quake in Olympia, Washington, registered 7.1 magnitude, one near Tacoma-Seattle in 1965 recorded 6.5, and in 2001 Nisqually, Washington, experienced a 6.8 magnitude shaking. (After Hofmeister, Wang, and Keefer, 2000)

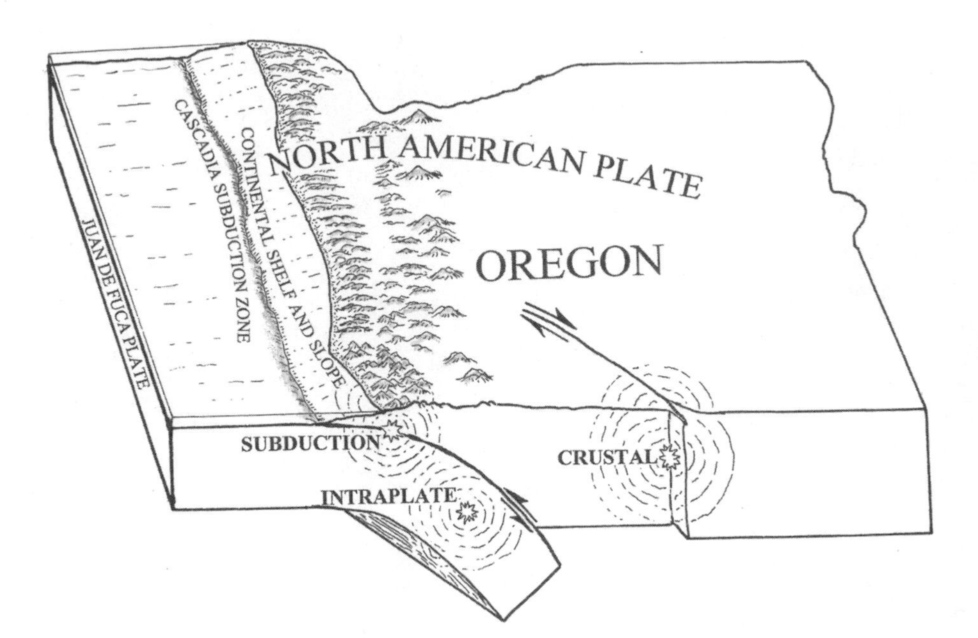

the Willamette region has not experienced this type of massive earthquakes since establishment of a Pacific Northwest monitoring system. Worldwide, there have been eight strong subduction quakes, which exceeded 9.0 in magnitude, during the last 100 years. The Chile event of 1960 was the highest with a 9.5 magnitude, and the 2011 quake in Japan is the most recent.

Many estimates for the timing and damage from earthquakes in western Oregon are based on the worst-case scenarios, where ruptures are postulated as taking place along the entire length of a fault or where activity in the Cascadia subduction zone is a factor. Since a great deal of uncertainty exists about the size and location of earthquakes, crustal faults, and subduction plate motion, the calculations vary widely from a 5.5 magnitude quake every 100 to 150 years to a 7.0 or an 8.5 event every 475 years. Blakely, Yeats, and other seismologists urge caution before drawing detailed conclusions for future predictions.

Landslides

Landslides involve the downslope movement of rock, soil, or related debris in response to gravity. In this province, slides are rare on the valley floor but common in hilly topography, where weathering and

In recent years, Scott Burns has worked to bring the issues and problems connected with landslides to the public's attention. His book *Environmental, Groundwater, and Engineering Geology* provides invaluable information on geohazards and is a critical tool for responsible planning. Burns' varied research includes engineering and environmental geology, soils, and geomorphology. Completing degrees at Stanford and the University of Colorado, he began a teaching career that took him abroad as well as elsewhere in the United States. A sixth-generation Oregon native who grew up in Beaverton, Burns returned to the state in 1970, when he joined the staff at Portland State University. (In the 1996 photograph Burns is inspecting the Newell Creek Canyon slide at Oregon City; Photo courtesy S. Burns).

Structural geologist Bob Yeats received his PhD from the University of Washington in 1958, then worked a time for Shell Oil Company before teaching at Ohio University. His research focused on mapping earthquake faults in three dimensions using subsurface data. In 1977 he transferred to Oregon State University to take over the chairmanship of the geology department. Through his numerous papers, including *Tectonics of the Willamette Valley* and his several books—notably *Living with Earthquakes in the Pacific Northwest*—and his many students, Bob has been a major player in unraveling Oregon's geology. Living in Corvallis, he is currently working on an international project to create a worldwide active fault database. (Photo courtesy A. Yeats)

The February, 1996, storm, now known as a 100-year event, fostered landslides and flooding throughout the state. The Portland metropolitan area alone suffered $10 million in losses. The highest cost was to private residences, since few homeowners were prepared or had insurance. Inspecting over 400 sites, Burns observed that Portland became the "City of Plastic" as owners futilely attempted to prevent further sliding with slope coverings. Most of the moderate to small earthflows and slumps were in Portland Hills loess that mantles the West Hills. (This photo was taken near S.W. Montgomery Drive and Elm Street; photo courtesy S. Burns)

erosion have over-steepened the gradients. Heavy rainfall, removal of the vegetation cover, excavating, poor drainage, and loading all destabilize the slope. In the northern Willamette basin, winter mass movements at multiple locations happen almost annually.

While predicting volcanic eruptions and earthquakes with any precision is not yet possible, forecasting the occurrences of landslides and flooding is more certain. For some time geologists such as William Burns and Ian Madin at DOGAMI have been working with LIDAR (Light Detection and Ranging), which effectively sees through vegetation, to map and define unstable slopes with precision. Burns estimates that there are close to one-fifth of a million landslides in Oregon and that about one-tenth of these have been mapped.

In the Portland, Vancouver, and Tualatin basins, landslides develop where the Boring lavas cover the Troutdale Formation and Sandy River Mudstone or within layers of the Portland Hills silt. Over half of all identified landslides in the Willamette Valley are comparatively moderate in size, and geologists surmise that the larger prehistoric ones may have been induced by Cascadia subduction zone earthquakes.

In 2002 Jon Hofmeister, formerly at DOGAMI, completed a number of site specific maps for landslide-prone areas in 19 western Oregon counties. Politicians and government officials, however, felt that the maps labeled too much of the area as hazardous, and since economic growth might be affected, the maps were withdrawn.

The 125-acre Highlands landslide is perhaps Portland's most notorious, involving Washington Park, the zoo, residential housing, the World Forestry Center, and the former Oregon Museum of Science and Industry (OMSI). Up to 90 feet deep, the block-and-earth flow of Portland Hills silt above a shear zone of decomposed Columbia River basalts may be traced as far back as 700,000 years. Excavations for city reservoirs in Washington Park in 1894 initiated movement, which has since been reactivated by periodic construction. After years of difficulty with an unstable building foundation, OMSI abandoned the site for a new facility on the east bank of the Willamette River.

Close to Oregon City, the Troutdale Formation has been responsible for the great concentration of ongoing slides within the 600 acres of Newell Canyon. In 1993 Scott Burns and his students mapped the canyon, which is bisected by Highway 213, recording 53 landslides. Of the seventeen new earthflows and slumps discovered in 1996, the largest was approximately 200 cubic yards in size. However, it was enlarged to a 15,000-cubic-yard debris flow by the rains of 1996.

Landslides fostered by the 1996 storm were not restricted to Portland. Lane County had the largest number with 2,264. Douglas County was second with 1,084, followed by Linn County with 913 and Clackamas with 880. A state of emergency was declared, and the effects were compared to the high waters of 1964. Destruction was most pronounced on federal forest lands, where logging roads were washed out and blanketed with rocks and mud. The Eugene district of the Bureau of Land Management suffered some $2.5 million in losses as over 20 miles of roadways, along with campgrounds, bridges, and boat ramps were destroyed.

Calling them *giant* landslides in the Coburg Hills, Ian Madin and Robert Murray of DOGAMI mapped the quadrangle north of Eugene in 2004, recognizing a series of major slides along 2,500-foot-high escarpments. They identified decomposed Little Butte basalts covering the Eugene and Fisher formations as responsible for a prehistoric debris avalanche, which continues to move. Although the Coburg Hills offer spectacular views of the valley and coastal mountains, such existing hazardous conditions would necessitate extensive study before any development on the steep slopes.

The use of LIDAR by the Oregon Department of Geology has revealed hundreds of broad, shallow depressions, approximately 100 feet across and 5 to 10 feet deep, on the Willamette Valley floor. One suggestion as to the origin of these intriguing "pits of mystery" is that they may be sites where huge Missoula flood icebergs grounded and slowly melted to create a divot in the soft silt layers.

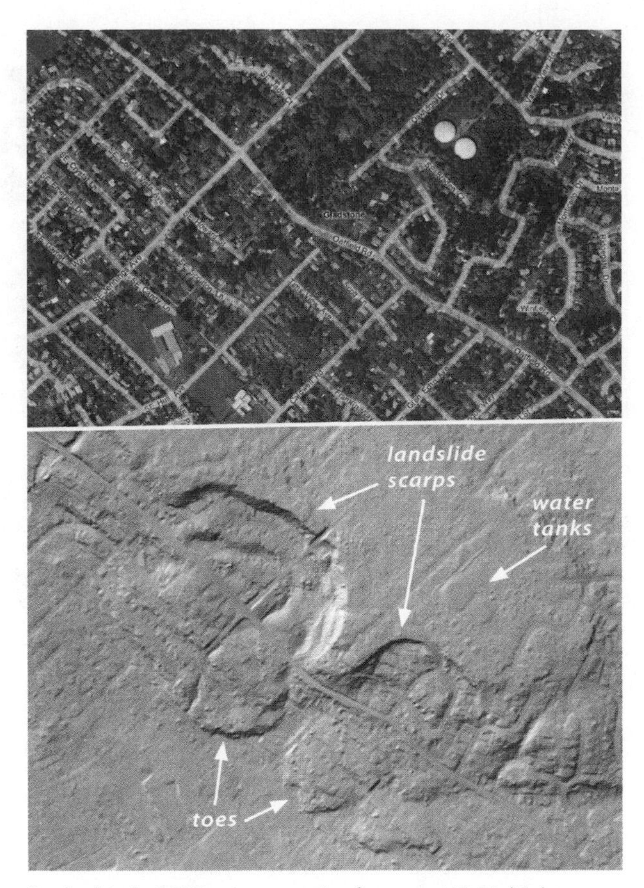

Beginning in 2005, a cooperative five-year LIDAR (Light Detection and Ranging) project between DOGAMI, Portland State University, and the U.S.G.S. hoped to identify areas at risk from landslides along the immediate coastline and in the Willamette Valley. LIDAR provides clear images of surface geomorphology to show features associated with slides. (In this 2005 digital image, the toes and scarps of active slides along Oatfield Road in Gladstone are clearly visible; courtesy Oregon Department of Geology and Mineral Industries)

Flooding

Flooding on low-lying lands throughout Oregon was recognized by the first Europeans, who attempted to control high waters with drainage ditches, dams, or levees. Since these early endeavors in the 19th Century, agriculture and towns have steadily expanded onto floodplains and wetlands, where they are vulnerable to inundation by waters rising from a combination of melting snow and heavy rainfall. Floods can occur at any season of the year, but they are most frequent during late winter or early spring, and today they have become almost yearly occurrences.

Over one million acres of recognizable floodplains in Oregon—areas where river bars, banks, and terraces were overtopped—had been mapped by the 1970s; however, as noted by consultant Frank Reckendorf, even with this knowledge, hazard prone regions continued to be developed. Reliance on inadequate federal and state maps, guidelines, and monitoring, which failed to account for channel constrictions and fill, for the encroachment of buildings, or for the natural stream geomorphology, contributed to the $200 million damage caused by the 1996 floods. Throughout the valley, cities became islands, while individual houses and businesses were isolated in the middle of lakes. Minimal oversight by Oregon state as well as federal agencies still exacerbates flooding problems.

Oregon's most disastrous flood overwhelmed the community of Vanport on May 30, 1948, in the area now occupied by Portland's Delta Park. Many accounts detail the events, in which the

In what has become known as the first "great flood" of December, 1861, a prolonged warm rain and snow melt inundated many towns the length of the Willamette Valley. Intermittent episodes over the next century were punctuated by the record-breaking storms when intense rains caused destructive landslides and overflowing rivers. In February, 1890, the second largest flood on the Willamette River took out virtually every large bridge in the valley, shifted the river channel, and inundated Salem (above). (Photo courtesy Oregon State Archives)

In the Willamette Valley, the construction of large dams for flood control began in the 1940s, and today there are 13 Army Corps of Engineers storage projects. The five earliest facilities are in the southern part of the valley near Cottage Grove and Eugene, and the other eight are aligned toward the north along the east side. The final one was placed across Blue River in Lane County in 1968. Because they are aging, the dams are facing repairs. Anticipating a high chance of valley flooding in the winter of 2011, the Army Corps announced that about 42 spillways urgently needed attention.

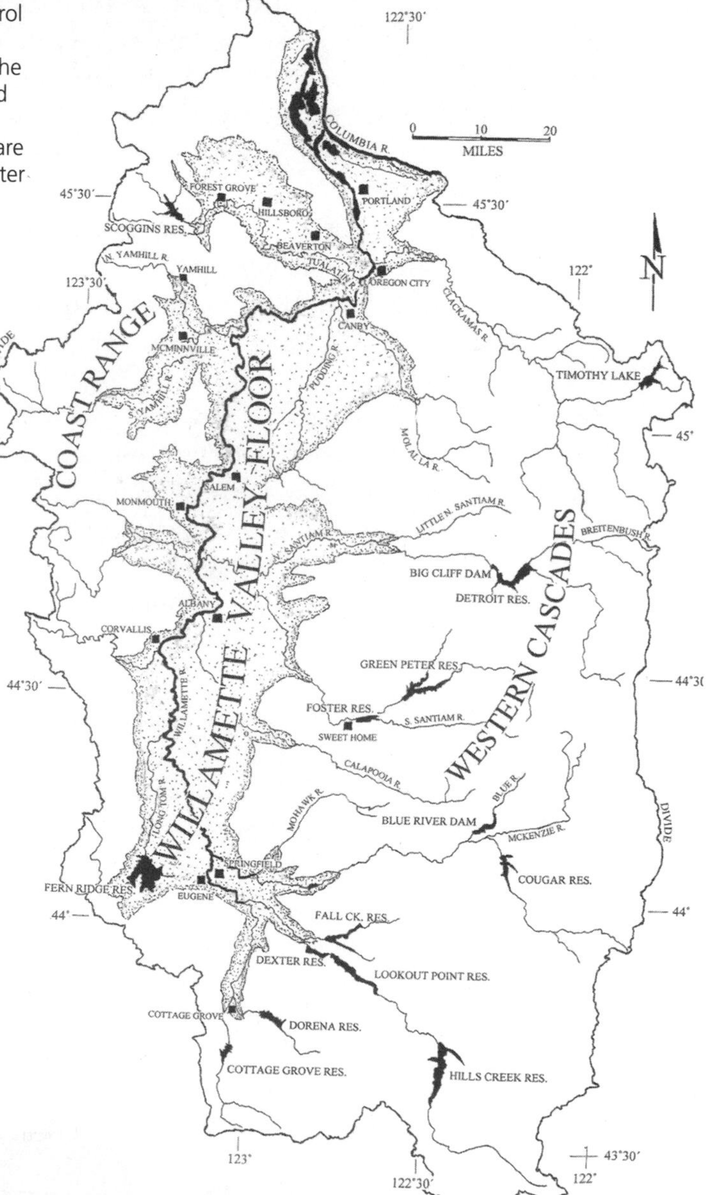

ill-conceived federal housing project for workers at the Kaiser Company shipyards was destroyed and 15 people killed. Situated on the floodplain of the Columbia River, the two square miles of Vanport village were protected from the river by narrow dikes maintained by the Army Corps of Engineers. Even as the water reached the top of the levees, 13 feet over river flood stage, the populace was assured there was no danger. Rainfall and runoff from late snow-melt over-topped the dike and opened a six-foot-wide break, allowing the surge to enter the housing area.

As noted by John Beaulieu, one of the unexpected consequences of dam construction is that changes in the meandering patterns of larger rivers lengthen the period of flooding. Prior to dam construction, floods peaked over short periods of time and inundated wider areas, whereas, with dams in place, the duration of high water, channel changes, and erosion can be more extensive. During the 1996 flooding, high water persisted in Portland for two weeks.

Natural Resources

Bauxite and Iron

Where surfaces of the Columbia River basalts are deeply weathered, residual ores of aluminum and iron develop in swampy areas. With rock decomposition, soluble chemicals are leached out leaving a clay soil with the insoluble bauxite enriched with aluminum and iron. In the northern Willamette Valley, the thickest deposits are in Washington and Columbia counties.

The process of extracting aluminum from bauxite ore requires considerable electric power, and Oregon's readily available inexpensive hydroelectric resources made the state attractive to the industry even before the demands of World War II. A global increase in the use of aluminum in 1987 was reflected in expanded production by the state's two smelters, Reynolds Metals Company at Troutdale and Northwest Aluminum at The Dalles. After the Reynolds operation closed in 2002, the structure was demolished, and the site was placed on the National Priorities List for remedial cleanup because of contaminated soils.

Attempts were made to mine and process the iron oxide mineral, limonite, near Scappoose and Lake Oswego from 1867 to 1894. Furnaces of the Oregon Iron and Steel Company produced 83,400 tons of pig iron, the first west of the Rocky Mountains, before being shut down by financial difficulties. Several thousand feet of old tunnels still existing under the Lake Oswego Country Club and nearby Iron Mountain are of historic interest. A furnace used for smelting, located near the outlet of Lake Oswego, may still be seen in the city park.

Mercury

In western Oregon cinnabar, an ore of mercury, occurs within a 20-mile-wide belt from Lane, Douglas, and Jackson counties to the California border. Near Cottage Grove, the Black Butte and Bonanza mines were responsible for about one-half of Oregon's quicksilver production, exploiting Eocene marine sediments and volcanics of the Fisher Formation. Black Butte was operated off and on until the early 1970s. Because the waste piles are leaching contaminants into creeks that flow into the municipal Cottage Grove reservoir, the site was placed on the federal superfund cleanup list in 2010.

Cinnabar at the Bonanza Quicksilver Mine in Douglas County was discovered in the 1860s. Eventually ranked as the state's highest producer, the mine peaked between 1940 and 1943, after which output dwindled, and it was closed in 1960.

Oil and Gas

Along with the search for minerals, the hunt for oil has captured the imagination of many since the early 1900s. The western region of the state, in particular, has seen ongoing exploration, both fraudulent and legitimate. Warren D. Smith, among other geologists, wrote letters to newspapers and answered enquiries about possible claims, warning the public against deceptive get-rich-quick schemes. In spite of these efforts, however, investors continued to lose money to fast-talking entrepreneurs.

Oregon laws, which date back to 1923, require that oil and gas operations be regulated by the state. In 1949 and again in 1953 the statutes were modified, authorizing DOGAMI to oversee the drilling, abandonment, and reclamation of sites, as well as the storage of well cuttings and cores. A record of the repository samples and cores is available to the public. Over 500 gas and oil wells have been authorized since the program began, but most were drilled prior to World War II, and few penetrated deeper than a mile. Except for the Mist gas field in Columbia County, exploratory wells have not been successful, although a gas well near Lebanon did produce economic amounts for six months during the mid-1980s.

Surface and Groundwater

The lowland, that extends from the southern Willamette basin into Clark County, Washington, is recognized for its surface and groundwater resources. No part of the basin can be classified as arid, but it experiences rainless periods lasting up to two or more months, during which the supply becomes greatly diminished. Although the state Water Code decrees that Oregon's water belongs to the public, its availability is complicated by that fact that almost all has already been appropriated or allocated.

Surface and groundwater in this province are maintained by the 30 to 60 inches of snowmelt and precipitation that percolates through volcanic rocks of the Cascades to feed the rivers coming off the western slopes. Little is lost during the wet months,

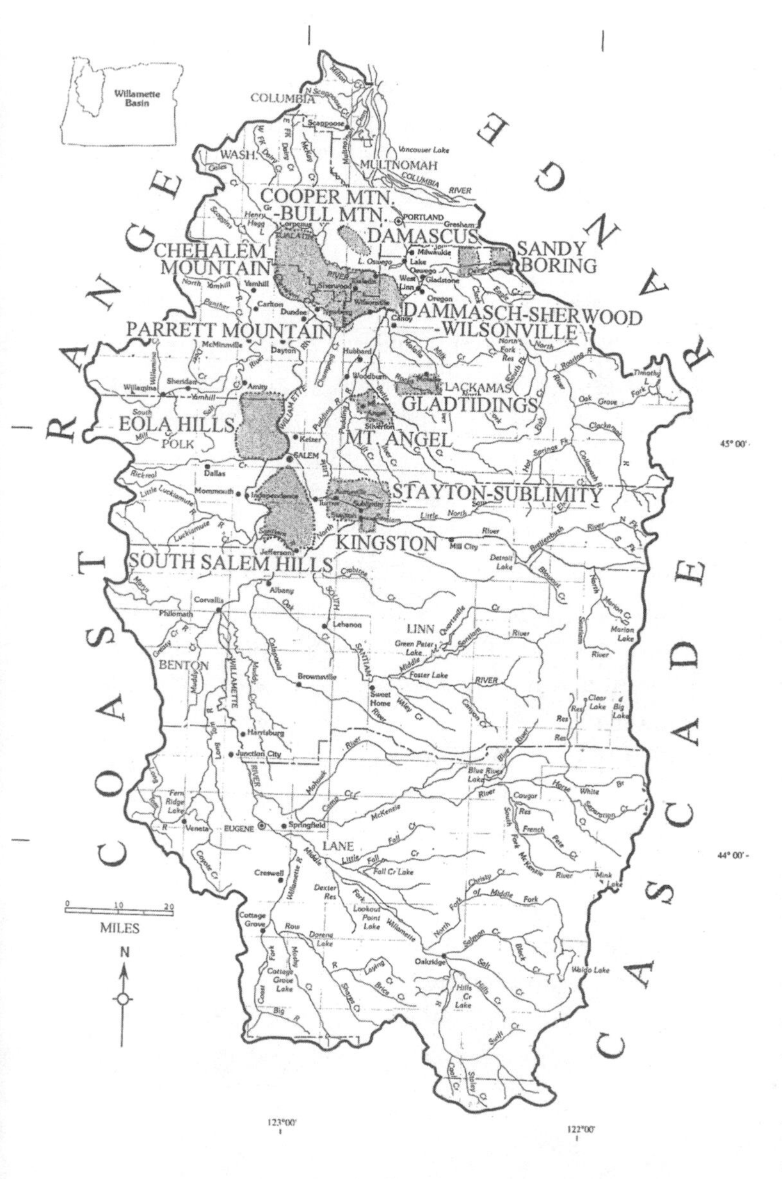

The state Water Resources Department has designated 22 regions throughout Oregon as areas where the amount of groundwater is depleted. Of these, 17 draw from the interbeds of the Columbia River Basalt Group in the northern Willamette basin (shown in pattern), and the remainder are located in eastern Oregon. The department can impose restrictions on future use where drawdown from the aquifer exceeds the estimated natural long-term recharge rate. (After Bastasch, 1997; Burt, et al., 2009; Oregon Water Resources Department, 1984, 1992; Orr and Orr, 2005).

but from August through September discharge occurs by evapo-transpiration through vegetation, with supplying surface streams, and through pumping for agricultural, industrial, or public use. Recharge is slow if the area is small or if the flow is impeded by clay or bedrock.

The groundwater system that underlies 3,700 square miles of the Willamette lowland was divided into separate hydrogeologic units by Marshall Gannett and Rodney Caldwell of the U.S.G.S. The oldest are Eocene to Oligocene marine sedimentary and volcanic rocks, which are occasionally saline and have low permeability. Above these, interbeds between flows of the Miocene Columbia River basalts can produce large quantities of water locally, but they tend to draw down rapidly and recover slowly. A 2009 paper by consultant Walter Burt and coauthors summarizes the hydrology of the basalts in the northern Willamette Valley.

The Willamette aquifer is the principal groundwater repository for the province. Restricted to the coarse-grained Pleistocene alluvial fans and deposits on the valley floor, the sands and gravels of the aquifer range in thickness from 200 to over 400 feet. In the Portland basin, the alluvium was laid down by the ancestral Columbia and Clackamas rivers, and southward it was spread by the Molalla, Santiam, Calapooia, and McKenzie rivers that debouch from the Cascades. The Willamette aquifer is absent in the Tualatin Valley, a region that relies on modest amounts from the Willamette Silt, the Troutdale Formation, or the deeper Columbia River basalts. Comparing the long-term rates of recharge and discharge, Dennis Woodward and others of the U.S.G.S. concluded that on a regional basis the annual level of the Willamette aquifer shows little variation.

The Willamette Silt overlies the aquifer. Deposited by Missoula floods, it is 130 feet thick in the central valley but thins to 10 feet toward the south. The silt provides only a moderate amount of groundwater and is generally used for domestic wells.

Although in recent years some municipalities have begun to tap groundwater wells, historically most have relied on surface water. Exploiting the Willamette River since 2002, the Wilsonville water treatment plant processes the flow and markets it to customers as potable. The state's three largest urban regions, Portland, Salem, and Eugene, are fortunate in that their supplies from Bull Run Lake, from the North Santiam River, and from the McKenzie River are protected by legislation from pollution, contamination, and depletion.

In light of probable global warming and increased water usage, monitoring the western United States snowpack is essential. A 2009 paper by Michael Strobel of the Natural Resources Conservation

Service in Portland stresses the need to compile annual snowpack data as a critical component of the regional water supply for populated areas of the Willamette Valley. In this province at least 50 to 80 percent of the water needs are provided by snowmelt.

Geologic Highlights

Hills, Buttes, and Volcanoes

During the Eocene Epoch, the ocean shoreline along the eastern border of the Willamette Valley was punctuated by volcanic vents that are visible today as topographic features projecting above the floor. Aligned from Salem to Eugene, 14 isolated buttes are composed of 30-to-35-million-year-old basalts. The origin of the buttes varies, but their mineral and chemical composition and age suggest that they are substantially older than the Miocene Columbia River basalts. In the past they have been interpreted as lava flows, but most are now seen as ancient volcanic plugs or sills, where magma invaded the sedimentary layers. Visible from Interstate 5, Hale Butte, Hardscrabble Butte, and Knox Butte are composed of Little Butte volcanics. Once the surrounding ash layers were eroded, the hard basalt was exposed as small rounded hills.

Near Albany, Knox Butte is the most northerly of these, rising 634 feet above the valley floor, while Peterson Butte, with the highest elevation at 1,434 feet, has 12 dikes radiating from the cone. Southward to the Coburg Hills, the buttes average just over 600 feet high. West Point Hill, Rock Hill, and Lenon Hill are spurs of resistant rock from Western Cascade eruptions that extend into the valley.

Silver Falls

Silver Falls was set aside as a state park near Silverton in 1933. It is the largest in Oregon, covering 9,000 acres. Of the 15 cataracts in the park, South Falls drops 177 feet into a beautiful, deep plunge pool, while the North Falls, at 136 feet, has worn away a 300-foot amphitheater in the softer fossiliferous sandstones behind the tumbling water. Chimney-like holes in the overhang are molds where lava surrounded and engulfed standing trees. When the flow cooled, only the shape of the trunks remained.

Thirty million years ago, the region around Silver Falls lay at the edge of a shallow Oligocene ocean. Once the seaway had receded, flows of middle Miocene Columbia River basalts covered the area only to be enveloped by ash and lavas from Cascade volcanoes. Streams began to work their way down through the layers, selectively removing the softer ash beds. Where the large streams easily cut through the basalts, eroding deep canyons,

In Eugene, the face of Skinner Butte, rising 682 feet, displays jointed columnar basalt as does the 602 foot high Gillespie Butte across the river. Spencer Butte dominates south Eugene at 2,065 feet. Both Skinner and Spencer buttes are Tertiary sills that intruded the Eugene Formation. (Photo taken by James G. Houser; courtesy Condon Collection)

North Falls at Silver Falls State Park drops over an edge of Miocene Columbia River basalt into a plunge pool cut into Oligocene marine rocks of the Scotts Mills Formation. (Photo courtesy Oregon State Highway Department)

the smaller creeks with less volume wear away at the rock more slowly. Consequently a lip of basalt remains, which allows the water to spill over.

Table Rock

About 35 miles east of Salem, Table Rock is visible on the skyline of the Western Cascades. Shaped like a cardinal's hat, the distinctive summit and surrounding 5,500 acres are overseen by the Bureau of Land Management, which dedicated the wilderness area in 1984. Access to the crest of the monolith is provided by a moderately strenuous hiking trail through a dense forest that begins on the north face and winds almost completely around the edifice to ascend from the south.

A dissected remnant of late Western Cascades lava flows, Table Rock is capped by isolated exposures of resistant basalt. Columnar fractures in the lava and scattered outcrops on surrounding peaks attest to its volcanic nature.

Urban Geology

There are numerous fieldtrip, hiking, and roadside guidebooks to geology for those traveling far-flung distances by automobile, but few offer walking or biking excursions centered on compact metropolitan areas. One of the earliest proponents of urban geology, Ralph Mason provided an unusual look at the stone facings on Portland's buildings in his several publications. A mining engineer with DOGAMI, Mason's descriptions of structures which have since been altered or torn down make his accounts of particular historic interest. Also with DOGAMI, Ian Madin's 2009 overview and guide to Portland is well-illustrated and thorough.

Local colleges and geologic societies frequently offer tours and talks on urban sites. Formerly at Portland State University, Leonard Palmer's 1973 urban environmental guide to Portland focuses on geohazards.

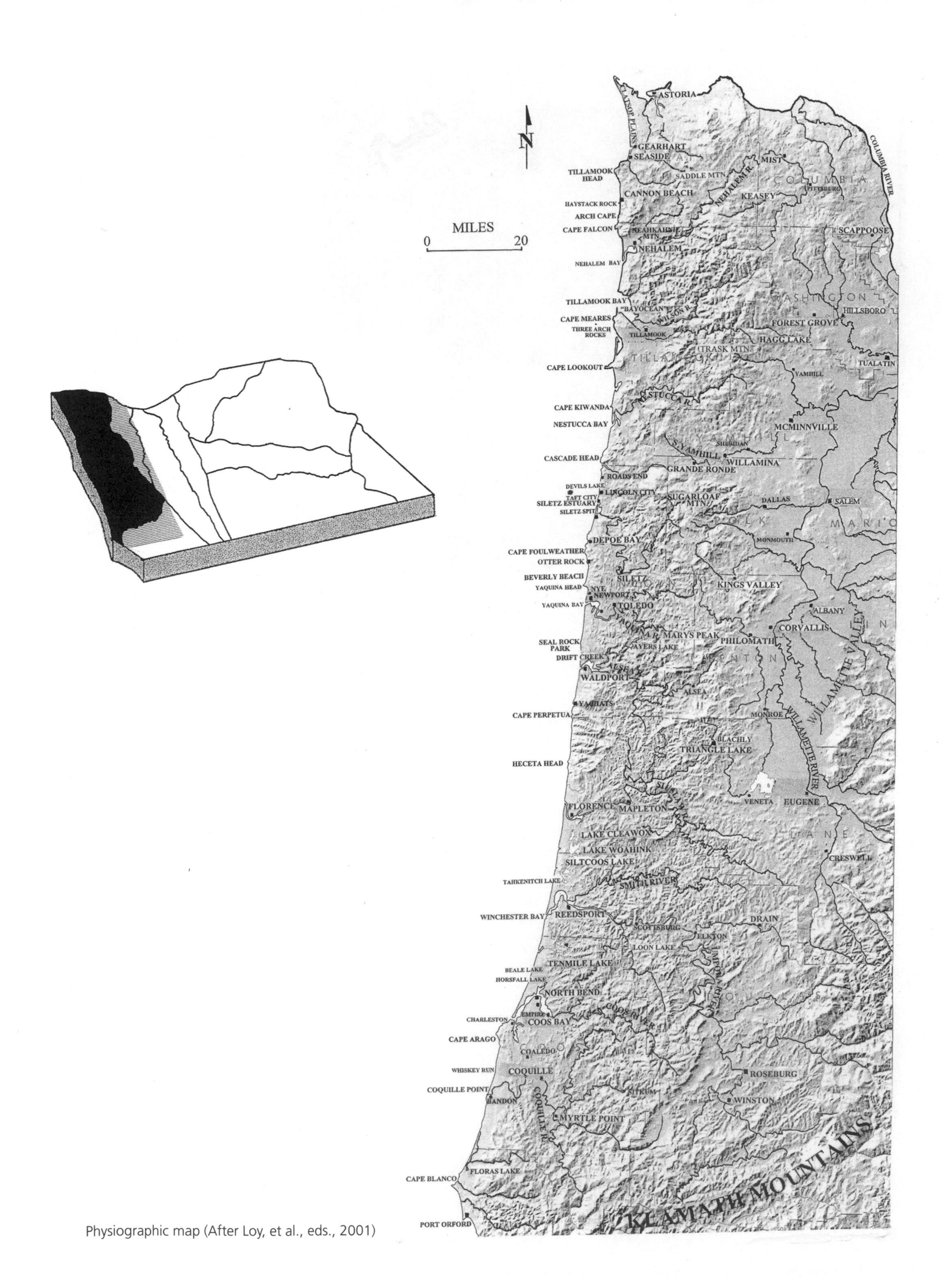

Physiographic map (After Loy, et al., eds., 2001)

Coast Range and Continental Margin

Landscape of the Coast Range and Continental Margin

Moderately high mountains, terraces, rocky headlands, sandy beaches, and an offshore shelf and slope all contribute to the complex topography of the Coast Range and continental margin. From the Olympic Peninsula of Washington, the province continues southward to the Middle Fork of the Coquille River and from the edge of the Willamette Valley west to the base of the continental slope.

The marine influence on western Oregon is responsible for the warmest average winter temperatures, the coolest summers, and the greatest precipitation of any region in the state. Rainfall in excess of 100 inches a year has molded the landscape with intricate stream patterns and dense forests. Because of extensive erosion on the western slope, the crest of the range is offset to the east. The highest elevations are Marys Peak at 4,097 feet, Trask Mountain at 3,423 feet, Sugarloaf Mountain at 3,415 feet, and Saddle Mountain at 3,283 feet.

Most rivers of the Coast Range are moderate in size, and only the Columbia and Umpqua cut entirely across the province. The notable west-flowing Nehalem, Wilson, Siletz, Yaquina, Alsea, Siuslaw, Coos, and Coquille rivers end in broad tidal estuaries. Of those streams flowing into the Willamette Valley, the Long Tom, Marys, Luckiamute, Yamhill, and Tualatin have the greatest discharge and the longest watersheds.

Along the Pacific strand, abrupt headlands are punctuated by bays, estuaries, pocket beaches, sand dunes, and spits. The distinctive terraced promontory at Cape Blanco extends into the Pacific Ocean as Oregon's most westerly point. Offshore, the province continues across the continental shelf and slope.

Past and Present

As with other provinces, the geologic examination of the Coast Range began with the exploration for minerals, the collection of fossils, and mapping. The details of unknown regions had to be filled in before a picture emerged. Joseph Diller's map folios in the late 1800s and early 1900s provided a regional overview as did the stratigraphic study by Ralph Arnold and Harold Hannibal in 1913. Close to the same time oil, gas, and mineral potentials were being assessed by Chester Washburne and Joseph Pardee.

A considerable step forward was initiated with a cooperative mapping and stratigraphic project begun in the 1940s between the U.S. Geological Survey and the Oregon Department of Geology and Mineral Industries (DOGAMI). As part of an investigation for petroleum resources, three Survey geologists W. C. Warren, Hans Norbisrath, and Rex Grivetti worked in the northwest counties. Their efforts were followed by the 1949 publications on the central Coast Range by Harold Vokes and Parke Snavely, also with the Survey. Ewart Baldwin's first revision of Cenozoic stratigraphy appeared about the same time.

First advanced in the 1920s, the theory of plate tectonics, which now dominates West Coast geology, only gained acceptance some 40 years later after research by many geologists and oceanographers confirmed the notion of large moving crustal plates. The concept crossed the continent, where its significance for western North America was ushered in with Robert Dott's "Implications for sea floor spreading" along with Tanya Atwater's papers on tectonics and Cenozoic evolution. William Dickinson's tectonic models of the Paleozoic were published in the late 1960s and early 1970s.

Overview

What is now coastal Oregon had its beginnings in the late Paleocene to early Eocene, with the construction of a volcanic seamount chain along a spreading center between tectonic plates. Transported as a large igneous province, the seamounts of Siletzia collided with and were accreted to the North American plate, forming the basement of the offshore shelf, Coast Range, Willamette Valley, and Western Cascades. Once incorporated around 50 million years ago, Siletzia rotated in a clockwise direction and shifted northward.

As a broad plain and marine shelf west of the emerging Cascade volcanic arc, the subsiding Coast Range slab (Siletzia) accumulated thick sequences of fluvial muds, sands, and volcanic detritus derived from both the Klamath Mountains and western Idaho. Deep-sea fans of layered turbidite sands and muds spread northward in the basin during the early and middle Eocene, but by the late Eocene to early Oligocene the waters had shoaled to deltaic and shoreline environments, when increasing volcanism from the Western Cascades contributed immense volumes of debris.

The edge of the North American plate was folded, elevated, and tilted by the subducting crustal slab, limiting marine waters to Miocene basins at Coos Bay, Newport, and Astoria. At this time, lavas from fissures in eastern Oregon and adjacent Washington and Idaho reached across the entire state, invading the softer coastal sediments.

From the 1940s through the early 2000s, Parke Snavely's geologic examination of the Cenozoic onshore and offshore rocks established the basis for unraveling the tectonic processes of the Pacific continental margin. Born in Yakima, Washington, Snavely completed his graduate work at the University of California, Los Angeles. In a 50-year career, he held positions in both the Pacific Marine Geology and the Western Regional Geology branches of the U.S.G.S. until retirement. Snavely died in November, 2003. (Photo courtesy Condon Collection)

Wendy and Alan Niem were both born in New York and entered the University of Wisconsin, Madison, where Alan completed his PhD and Wendy a Bachelors. Accepting positions in geology at Oregon State University, the Niem Team published numerous regional maps and papers. Alan concentrated on sedimentology and stratigraphy of western Oregon and Washington, exploring the offshore region in a two-man yellow submarine and participating in the Deep Sea Drilling Program aboard the *Glomar Challenger*. Wendy worked on stream systems in the central Coast Range while completing graduate work at Corvallis. The Niems retired in 2002, moving to coastal Lincoln County and continuing to work on U.S.G.S. mapping projects. In the photograph the Niems are holding a 1976 map of the Cape Foulweather and Euchre Mountain quadrangles. (Photo taken in 2010; courtesy A. and W. Niem)

Also from suburban New York, Harvey Kelsey spent much of his early life in the outdoors of New England. Moving to the West Coast in 1972, he finished a PhD at the University of California, Santa Cruz. For his research and teaching at Humboldt State University, Kelsey examines Quaternary surface processes, deposits, and faulting to provide insights to tectonic deformation. In addition to publications on coastal marine terraces, uplift, and faulting, he employs evidence of earthquakes and tsunamis in estuaries to decipher the paleoseismic history of the Cascadia and Sumatran subduction zones. Considering himself a westerner in spirit, Kelsey takes time to indulge his passion for road biking. In the photograph, Kelsey is coring the Yaquina Bay estuary. (Photo courtesy H. Kelsey; 2010)

With roots in the eastern United States and France, Anne Tréhu first came to the Pacific Northwest in high school to build trails in North Cascades National Park. Completing her PhD from MIT, she and her husband John Nabelek accepted positions at Oregon State University in 1987, where her research emphasizes geophysical processes and relationships along plate boundaries in order to understand geologic hazards and the tectonic history along fault systems from California to Alaska. Because of her wide outdoor interests, Tréhu is currently involved with the Oregon Adaptive Skiing program. In the photograph she is preparing an ocean bottom seismometer. (Photo courtesy A. Tréhu)

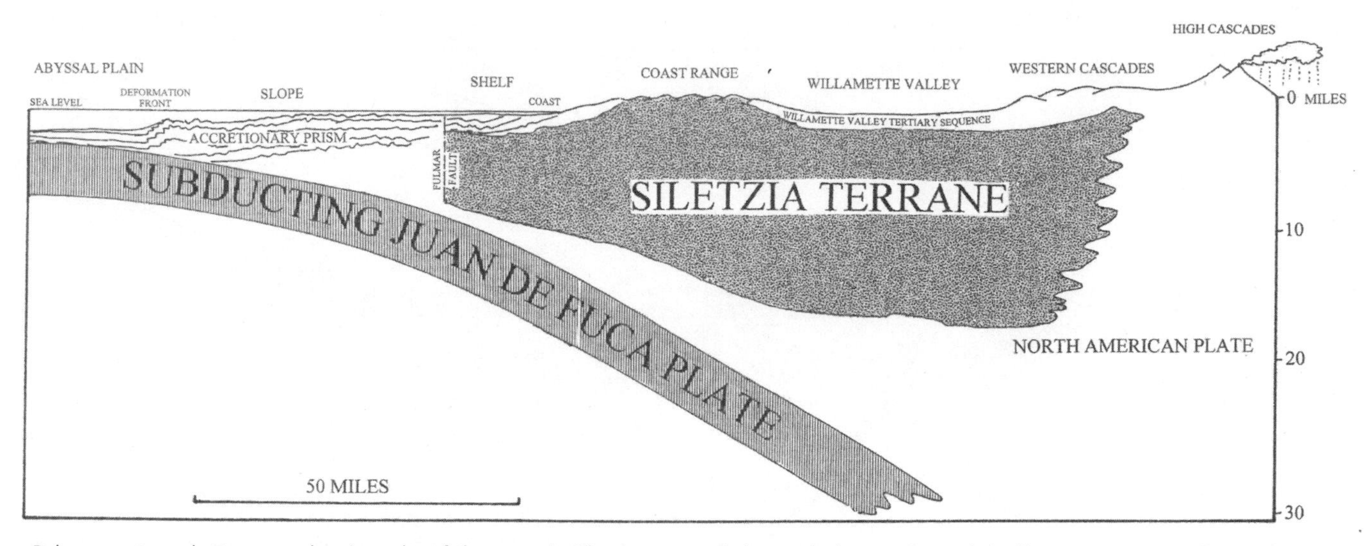

Paleocene to early Eocene volcanic rocks of the oceanic Siletzia terrane lie beneath the continental shelf, Coast Range, Willamette Valley, and Western Cascades. While its origins are still uncertain, Siletzia may have begun as a collection of offshore seamounts that moved eastward as a large igneous province to accrete to the West Coast of North America. (After Duncan, 1982; Fleming and Tréhu, 1999; Schmandt and Humphreys, 2011; Snavely and Wells, 1996; Tréhu, et al., 1994; Wells, Weaver, and Blakely, 1998)

Uplift, in conjunction with subduction and changing sea levels of Pleistocene glacial and interglacial periods, constructed raised terraces along the coast. Deeply eroded and broken into discontinuous sections, the older terraces are highest, while the younger surfaces are at lower elevations.

The offshore portion of the province is distinguished by a shelf, deeper slope, and abyssal plain. The slope is deformed by thrust faults and folds that parallel the coastline. At the base of the slope, the Cascadia subduction zone marks the boundary between the North American and Juan de Fuca plates. Slippage along this interface generates massive earthquakes and tsunamis. At the present time, the earthquake history of the Pacific Northwest is being chronicled with data from coastal marshes and offshore turbidite deposits.

The geomorphology of the coast is continuously altered by uplift, landslides, waves, and erosion, yielding a landscape in a constant state of flux. Sands are seasonally transported on and offshore, while even seemingly durable basalt headlands are slowly being worn down by waves. The province has undergone ill-considered development in the past, and much questionable expansion has been permitted. The present-day recognition of coastal geohazards is guiding many communities to bring building codes in line with an understanding of the risks involved.

Geologic History

Cenozoic

During the Cenozoic Era, the Pacific Northwest experienced major episodes of volcanism, subsidence, and uplift before sedimentation and erosion shaped the final product. The debate as to whether the basalt basement of western Oregon and Washington began as volcanic island archipelagos and ocean crust some distance offshore or from eruptions close to the continental margin is yet to be settled. One model holds that during the Paleocene to early Eocene, between 50 to 60 million years ago, a spreading center along the boundary between the Kula and Farallon plates generated the submarine volcanic plateau of Siletzia. Multiple eruptions built seamount chains that were carried by the Farallon plate to collide with and accrete to North America as the Coast Range block of Siletzia. Following accretion, the old subduction trench was abandoned and the present one established 90 to 110 miles west of the coastline.

An opposing explanation for the origin of the Coast Range block is that the volcanic episodes occurred in-place with rifting and extension along a Mesozoic continental margin. In support of this, geologists cite evidence that the erupted submarine basalts interfinger locally with sediments derived from the North American landmass.

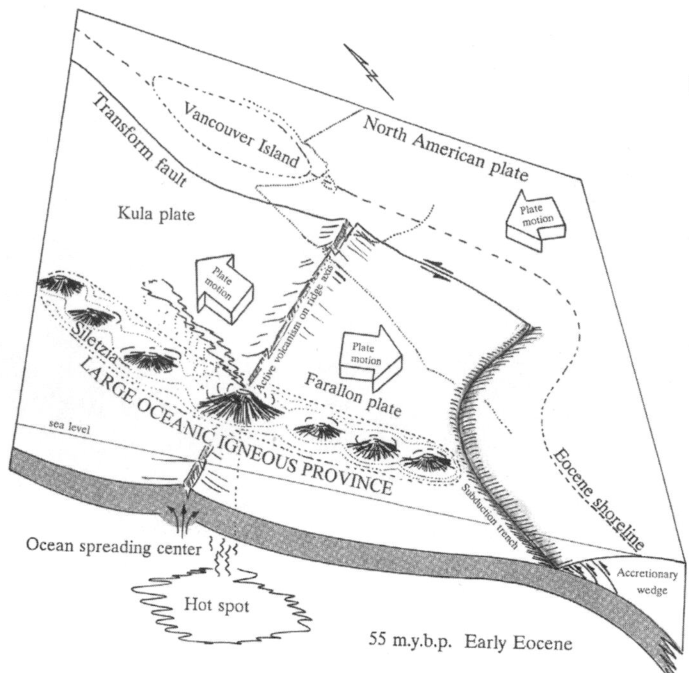

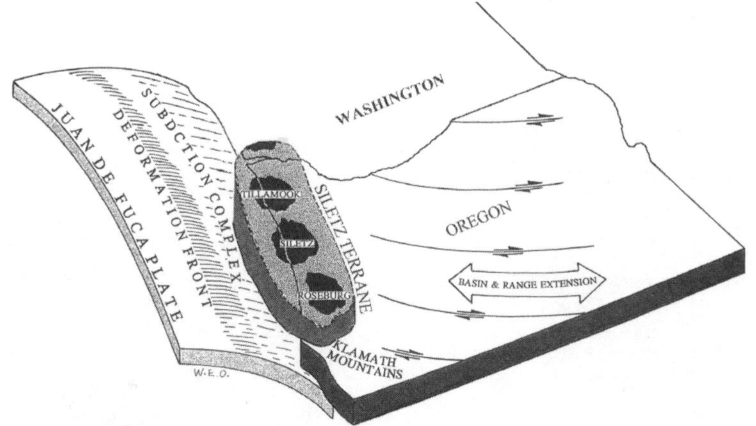

The thickness of the Siletzia terrane varies considerably. Beneath Oregon, it is 15 to 20 miles in depth, while off Vancouver Island the base is less than four miles down. The western boundary of Siletzia is marked by the north-south Fulmar fault, a vertical dextral shear in the crust where the east side moved southward and the west side northward, much like the San Andreas in California. Named by Parke Snavely, the strike-slip fault is estimated to have experienced as much as 120 miles of displacement before motion ceased in the late Eocene, around 37 million years ago. (After Fleming and Tréhu, 1999; Snavely and Wells, 1996; Tréhu, et al., 1994)

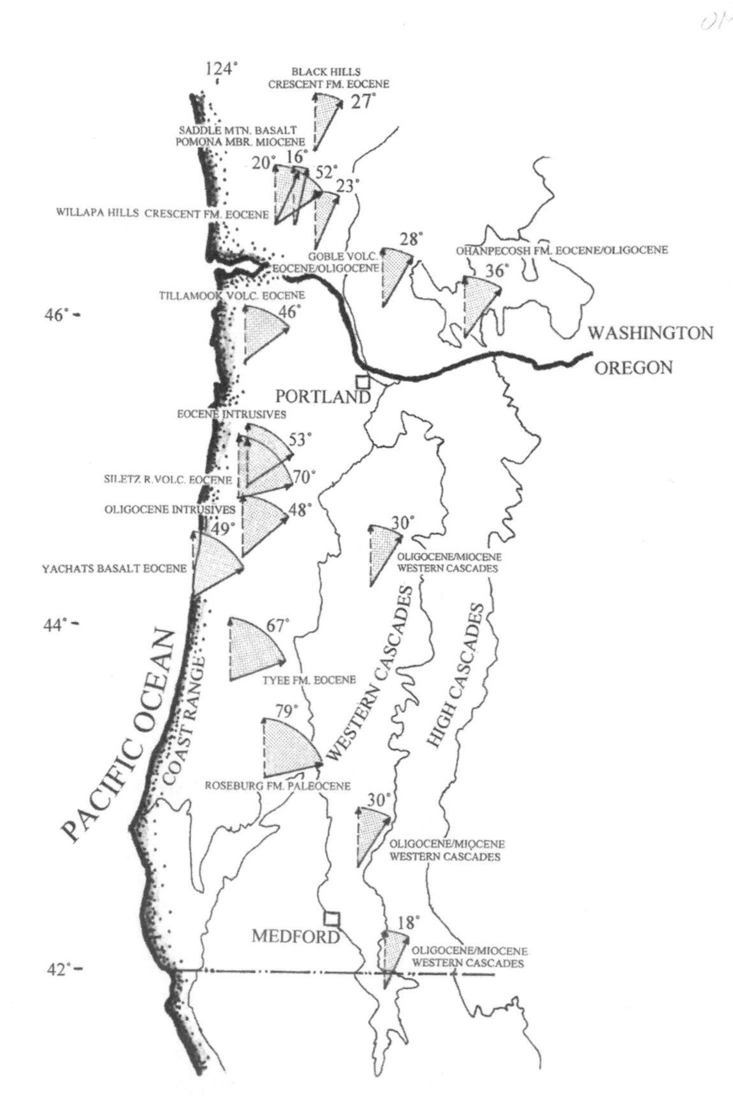

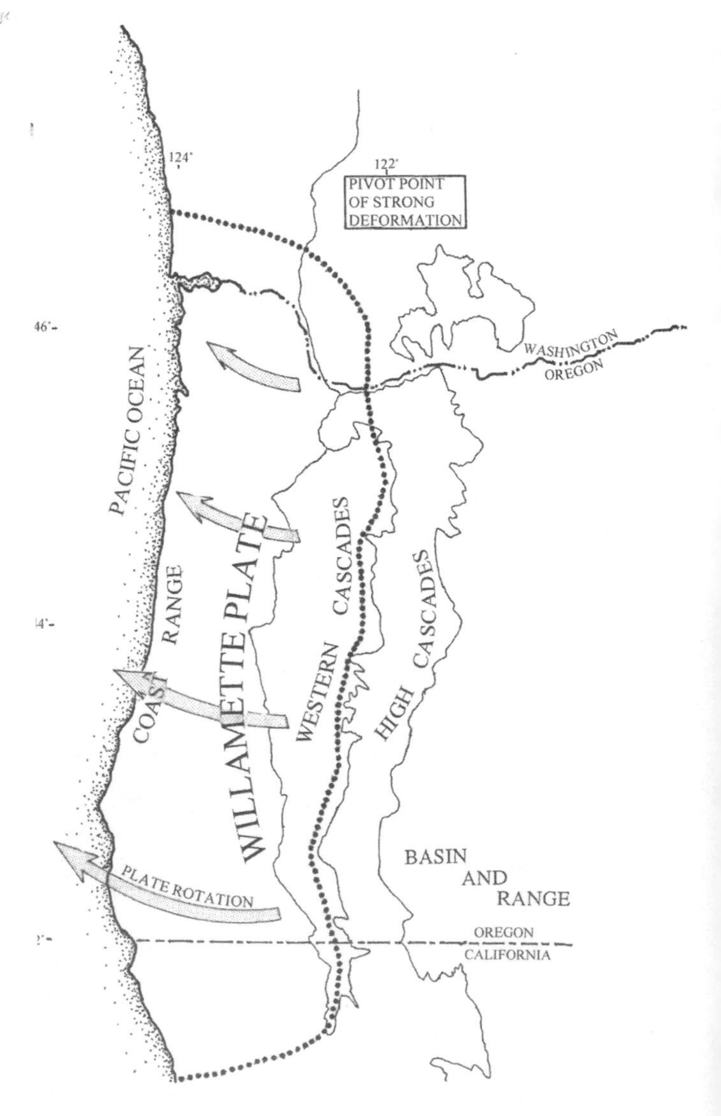

Cenozoic clockwise rotation of the Coast Range block (Willamette plate) was most pronounced in the older rocks and toward the south. (After Guffanti and Weaver, 1988; Magill, and Cox, 1981; Magill, et al., 1982; McCrory, 2002; Sherrod and Smith, 1989; Smith, 1989; Wells, 1990; Wells and Heller, 1988; Wells, Weaver, and Blakely, 1998)

Rotation, faulting, and uplift

The collision of North America with the oceanic Juan de Fuca plate is oblique rather than head-on. Oblique subduction, at a pronounced angle, yielded moderate rotation, faulting, and uplift of the Coast Range between the Eocene and Pleistocene.

With accretion, the rigid Coast Range block rotated in a clockwise direction from a pivotal point in Washington. Ray Wells of the U.S.G.S. has found that rotation was more extensive toward the coast, corresponding with stretching or widening in the Basin and Range and compression of rocks in southwest Washington. It has been calculated that the Coast Range has rotated 51° since late Eocene, up to 44° since the middle Oligocene, and 16° since the middle Miocene, with an average of 1.5° of rotation every 1 million years throughout the 50 million year period.

Clockwise rotation was generated by oblique plate subduction, extension, and dextral shear. Wells concluded that dextral shear was responsible for 40 percent of the rotation and that extension in the Basin and Range was responsible for the remainder. Today the north-northwest movement of the Pacific plate with respect to North America results in right lateral (dextral) shear. With dextral shear, strike-slip faults move sideways past each other, and the block beyond the viewer shifts to the right.

Folding and faulting of the province are an integral part of plate accretion. Structurally the Coast Range is a pair of large, in-line crustal folds with a north-south axis and many smaller wrinkles across the flanks. In Washington, the Willapa Hills is a broad regional fold anchored by volcanics at the core, which continues south into Columbia and Tillamook counties in Oregon, where it domes into

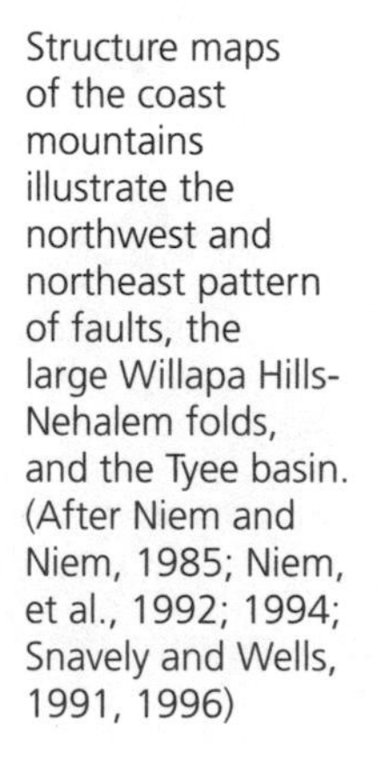

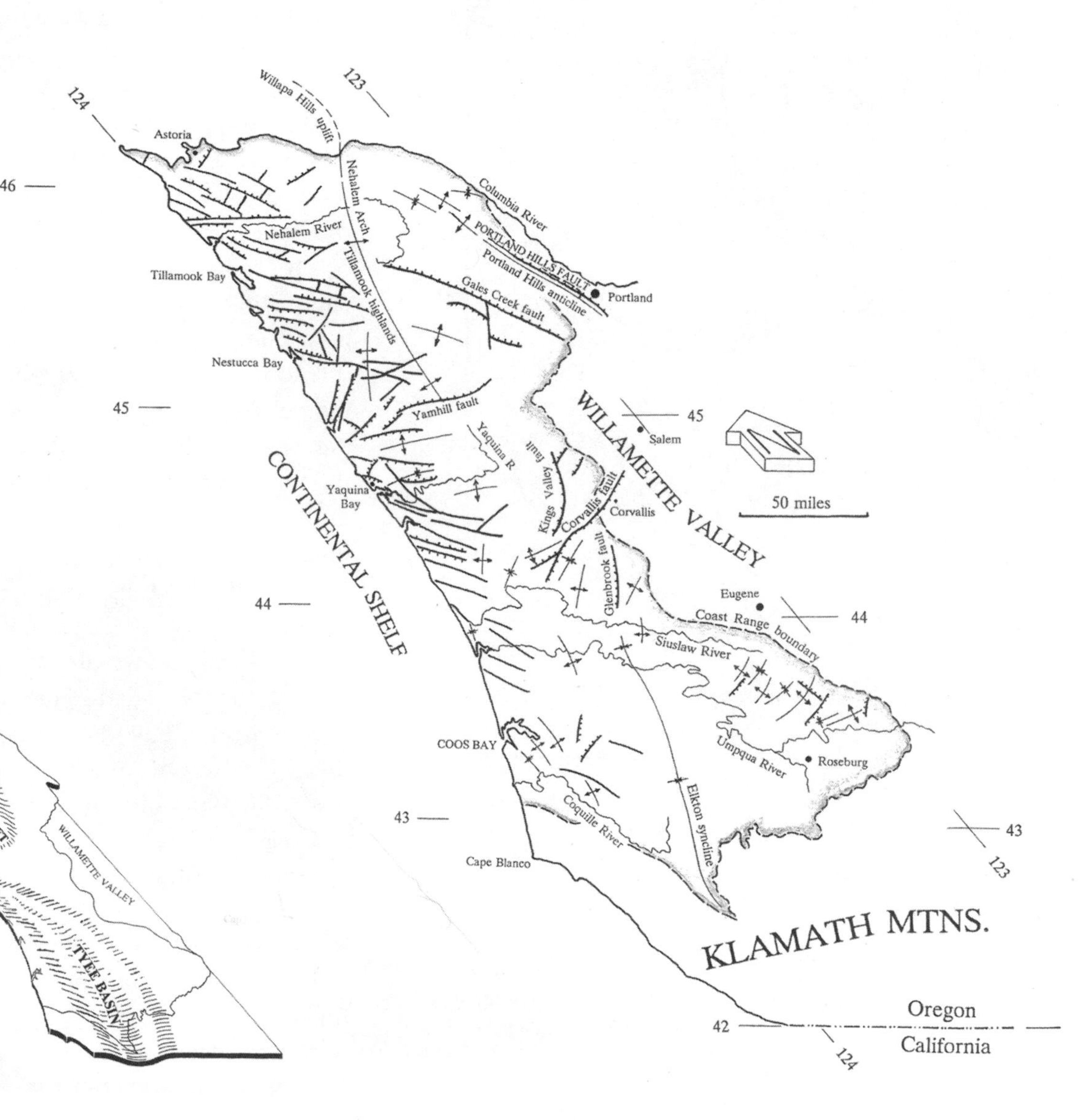

Structure maps of the coast mountains illustrate the northwest and northeast pattern of faults, the large Willapa Hills-Nehalem folds, and the Tyee basin. (After Niem and Niem, 1985; Niem, et al., 1992; 1994; Snavely and Wells, 1991, 1996)

the low Nehalem arch and Tillamook highlands. South of Newport, the upfold inverts to the Elkton syncline or Tyee basin that reaches into the Klamath Mountains.

Uplift of the Coast Range in the Cenozoic brought diminishing marine sediments and increasing terrestrial deposits. In a broad view, the western coastal margin is rising, while the eastern border and the Willamette Valley are either subsiding or are only being minimally elevated.

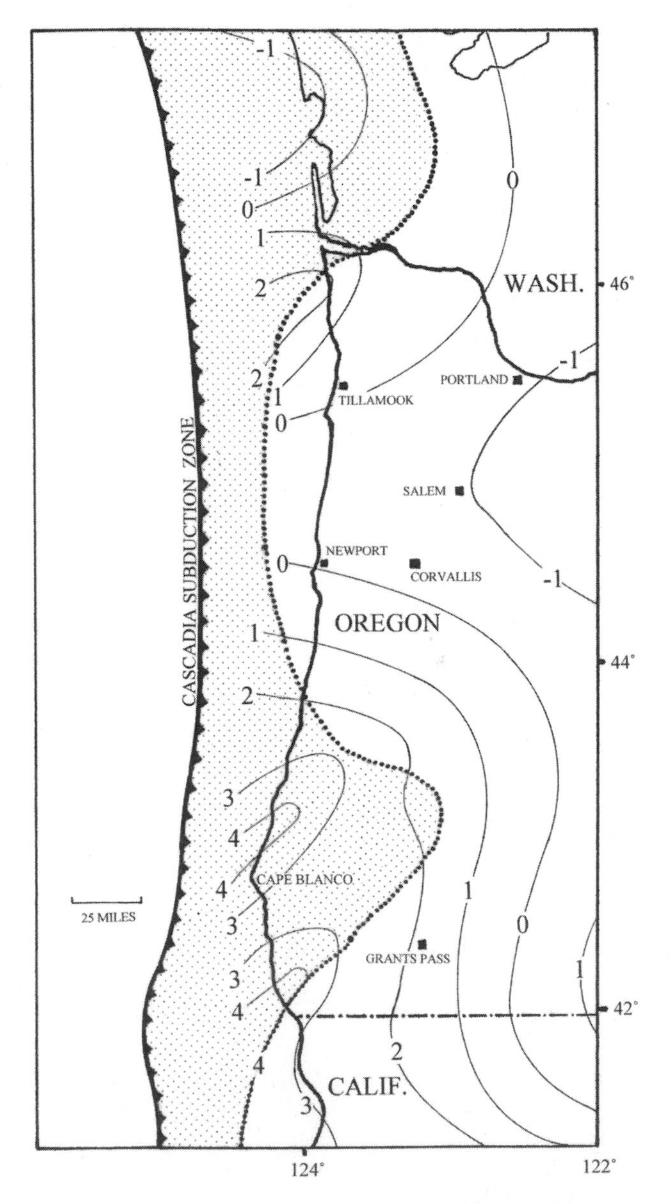

This map shows the apparent rate of uplift along the Pacific coast expressed in millimeters per year. The north-south discrepancies in elevation are a function of interseismic strain accumulating along the subduction boundary between the two plates. In juxtaposition with areas along the Cascadia subduction zone that demonstrate the greatest amounts of accumulated strain, the northern and southern portions of the coast have the highest magnitudes of latitudinal uplift (shaded area). (After Kelsey, et al., 1994; Komar, 1992; Mitchell, et al., 1994; Savage, 1983; Vincent, 1989; Vincent, et al., 1989)

Uplift can be measured by leveling surveys and tide records. Surveys along two axes—latitudinally north-south along the coast and longitudinally east-west from the coast inland to the Willamette Valley—yield data on uplift of the land relative to sea level. When geodetic data from 1987 and 1988 was compared to that taken between 1930 and 1941, it became apparent that a central depression lies between Newport and Tillamook. Higher elevations at the north and south margins are interpreted as evidence of interseismic elastic (temporary) strain causing a bulge between the Juan de Fuca and North American plates. The differences can be seen when assessing rates recorded at Astoria, which is rising only slightly faster than sea level, in comparison to those at Newport which is essentially static. The most profound elevation has been recorded at Cape Blanco.

Paul Komar has pointed out that measuring land submergence or emergence along a coast must take into consideration that the standard itself—sea level—fluctuates. The present-day melting of glaciers and polar ice caps yields a global sea level rise of about one-sixteenth inch a year. By comparing data from tide gauges and rates of tectonic elevation along the Oregon coast, he determined that the northern and southern portions are rising faster than global sea level, while the central region is being submerged, a conclusion that is analogous with the findings of others.

Paleocene-Eocene
Coastal marine basins

The oldest rocks at the core of the Coast Range block (Siletzia terrane) are early Tertiary basalts. Because of their variable age, texture, mineralogy, and distribution, they have been mapped individually as the Metchosin of Vancouver Island, the Crescent in Washington, and the Siletz River and Roseburg in Oregon. Both the Siletz River and the Roseburg formational names have been applied to the Eocene volcanics in southern Oregon, but ongoing discussions between Alan and Wendy Niem and others led to the conclusion that the basalts are continuous, thus giving preference to the Siletz River designation. The Siletz River at the southern extreme and

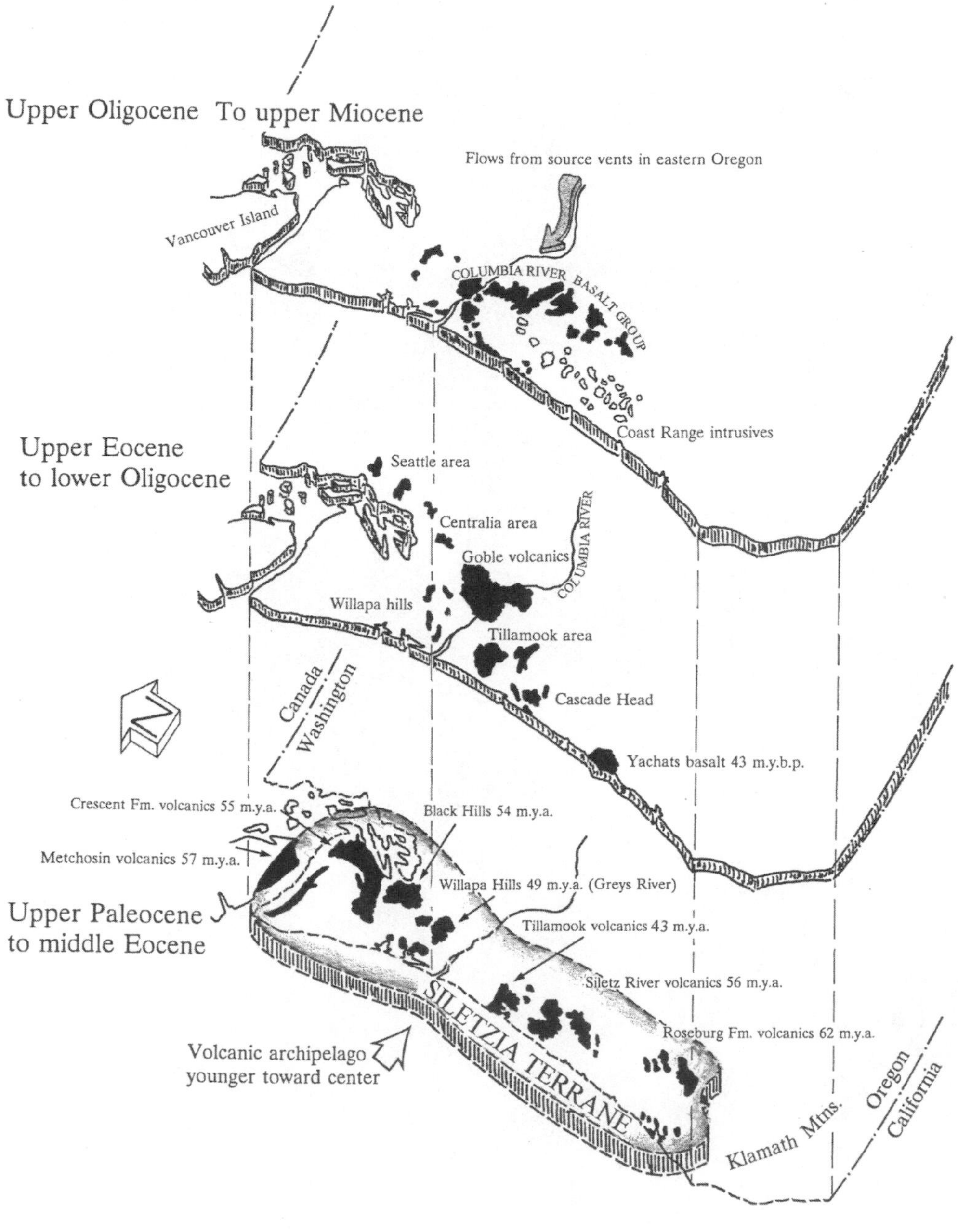

Throughout the Coast Range, Cenozoic volcanism and intrusions emplaced many of the present-day rugged mountains and headlands. Eocene to Oligocene dikes and sills of the Tillamook volcanics, the basalts of Cascade Head, and the Yachats Basalt are older than the Miocene Columbia River lavas from eastern Oregon. (After Armentrout and Suek, 1985; Baldwin, 1976; Christiansen and Yeats, 1992; Niem, et al., 1994; Wells, et al., 2009)

the Metchosin in the north are the oldest, while formations in the middle are up to 10 million years younger. Common to all of these, breccias (angular fragments) and elliptical bodies called pillows (because of their size and shape) formed during submarine eruptions.

By the middle Eocene, the Siletzia platform had subsided into a 400-mile-long forearc basin, which was the repository of massive deep-sea fans and deltaic sediments before uplift brought about a shallowing and eventual closure of the seaway by the late Miocene. Ewart Baldwin's Tertiary stratigraphy and nomenclature of southwest Oregon strata underwent a major revision by In-Chang Ryu and Alan and Wendy Niem in the 1990s, when many new names were proposed, others redefined, and several dropped.

The older Siletz River volcanics were blanketed by fluvial (stream) and marine deposits of the Lookingglass (Tenmile), Flournoy (White Tail Ridge), and Tyee formations. Cherts, metamorphics, and heavy minerals within these sediments indicate sources in the Klamath Mountains and western Idaho. Megafossils (mollusks) and microfossils (ostracods, foraminifera, and coccoliths) typify what were moderate water depths of an inner to

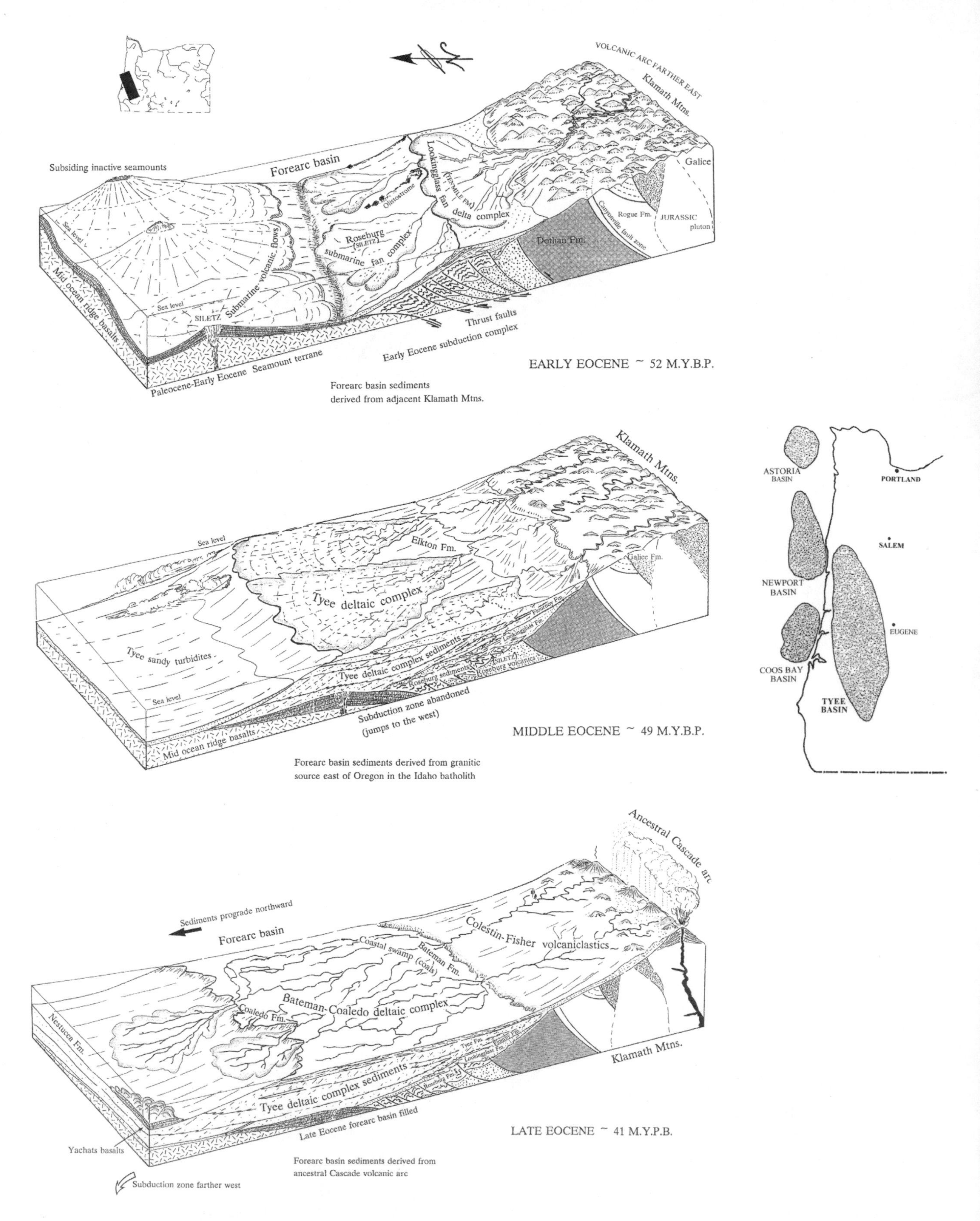

Renewed subsidence 50 million years ago pushed the early Eocene seaway from Cape Blanco northward beyond Newport and brought an influx of characteristic Tyee sandstones and mudstones. But sediments of the middle to late Eocene Elkton, Bateman, and Coaledo Formations, transported by inland streams, were limited to the southwestern forearc region near Coos Bay. Derived from the emerging Western Cascade range and the northern Klamath Mountains, approximately 2,500 feet of dark Elkton mudstones entomb an upper slope fauna of mollusks and microfossils. (After Baldwin, 1974; Brouwers, et al., 1995; Christiansen and Yeats, 1992; Heller and Ryberg, 1983; Orr and Orr, 2009; Ryberg, 1984; Ryu, Niem, and Niem, 1992; Snavely and Wells, 1991)

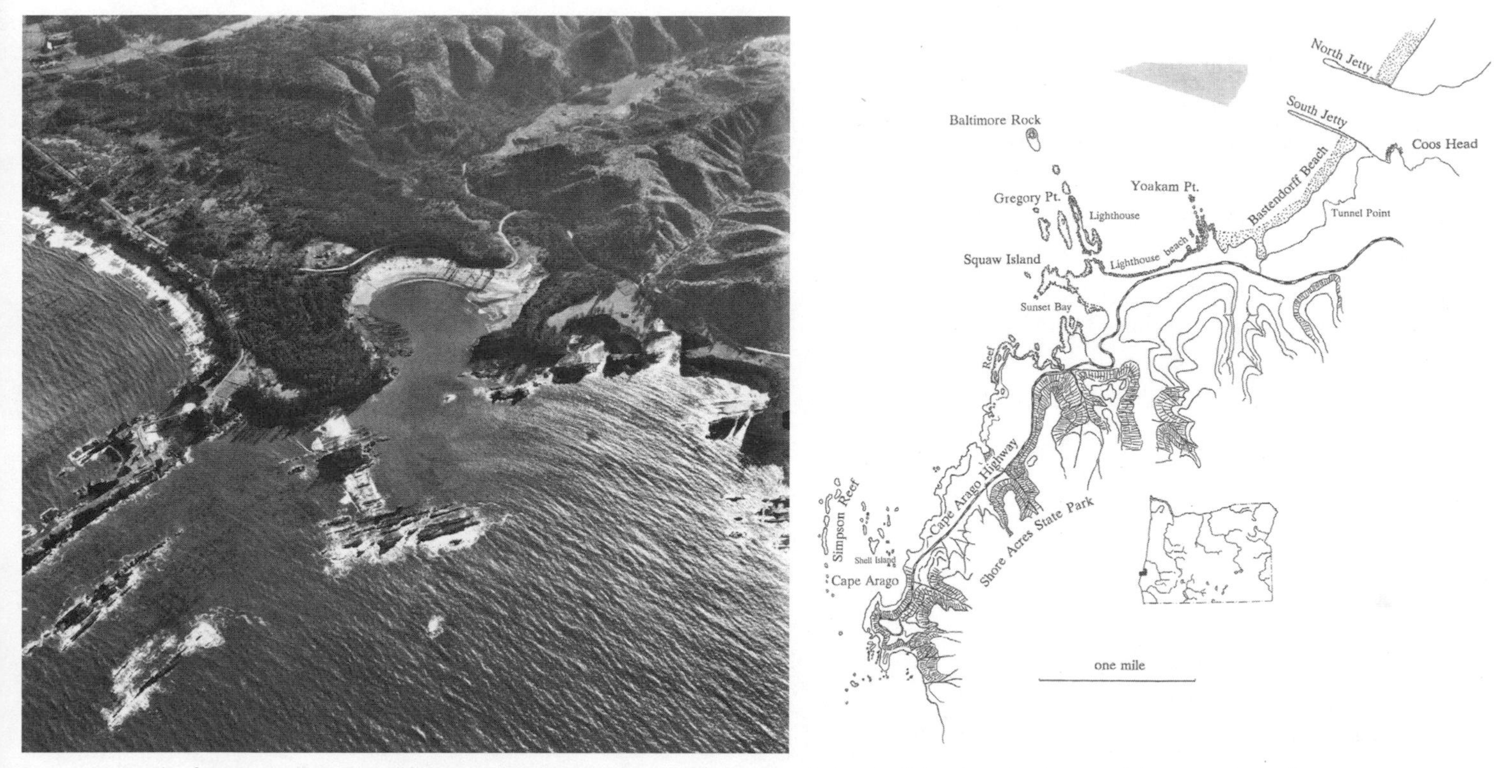

Sunset Bay in the foreground, Gregory Point and Lighthouse Beach in the center, Bastendorff Beach at the top, and Simpson Reef offshore are cut into silts and sandstones of the Coaledo Formation (Photo courtesy W. Robertson).

middle shelf during Siletz River time. Deposited by an advancing sea, deep-water Lookingglass (Tenmile) slope deposits, carried by turbidity currents, grade upward to a shelf setting as the depression filled. During Flournoy time, the retreating seaway was restricted to the area between Coos Bay and Newport. Conglomerates, pebbly sandstones, and siltstones point to stream systems, shallow water, and nonmarine coal beds along the margins of the Flournoy basin.

Distinguished by an abundance of muscovite mica flakes, the 6,000-foot-thick Tyee Formation overlies the Flournoy. Tyee sediments, carried by rivers coming off the Klamaths, built a north-trending delta that merges with deep-water turbidites toward the north. Huge submarine fans of the Tyee lap onto and cover the Siletz River volcanic seamount high. Bathyal microfossils (foraminifera) are present, but overall Tyee faunas are meager because of rapid deposition and the overwhelming volumes of sediment.

The Elkton basin, in turn, is covered by advancing fans of the Bateman with tropical foraminiferal microfossils and the distinctive Eocene shoreline clam *Venericardia*. Shoaling by Coaledo time is reflected in the coarse sands of an immense delta in the Coos Bay basin, layered with the remains of plants that grew on a swampy shore or in estuaries. Where sandstones of the coal-bearing Coaledo Formation have been deeply eroded at Cape Arago, Yoakam Point, and Gregory Point, the strata are honeycombed with sea caves. Offshore from Cape Arago, Simpson reef is a remnant of Coaledo sandstones and silts originally continuous with the headland.

At Coos Bay, swamps, wide coastal plains, and a shallow continental shelf delineate the third and final late Eocene phase of marine sedimentation. Explosive eruptions of the early Western Cascades were the primary source for the ash and pyroclastics of the Tunnel Point Formation overlying the Bastendorff shales. Thin-shelled delicate mollusks and microfossils in Bastendorff deposits are evidence of bathyal depths open to the ocean, well away from the shore, whereas Tunnel Point sandstones were part of a marine embayment. Once a sharp promontory that projected south of the entrance to Coos Bay, Tunnel Point had a natural channel and cave carved through before the cliff partially collapsed.

Following deposition of the Tyee submarine fan, the shoreline retreated progressively to the north and west, reflecting an overall pattern of uplift and the creation of small basins between Newport and Astoria. Along the central coast, Yamhill silts, sands, muds, and volcanic ash were distributed in the deeper offshore, while tropical mollusks, microfossils, and plants, layered within Buell and Rickreall limestone lenses, inhabited the fringes of seamounts and volcanic islands. Sills and sedimentary rocks of the middle Eocene Tillamook Volcanics, which interfinger with the Yamhill Formation, form the basement core of the northern Coast Range anticline.

Above the Yamhill, the bathyal Nestucca Formation lacks planktonic (open ocean) microfossils, an indication that deposition took place in an isolated marine embayment, separated from the ocean by a barrier of offshore islands or by the highlands of Yachats Basalt and Cascade Head. The Nestucca interfingers with nearly horizontal submarine flows of the Yachats Basalt, erupted from volcanoes lining a shallow coastal shelf. Over 2,000 feet thick in

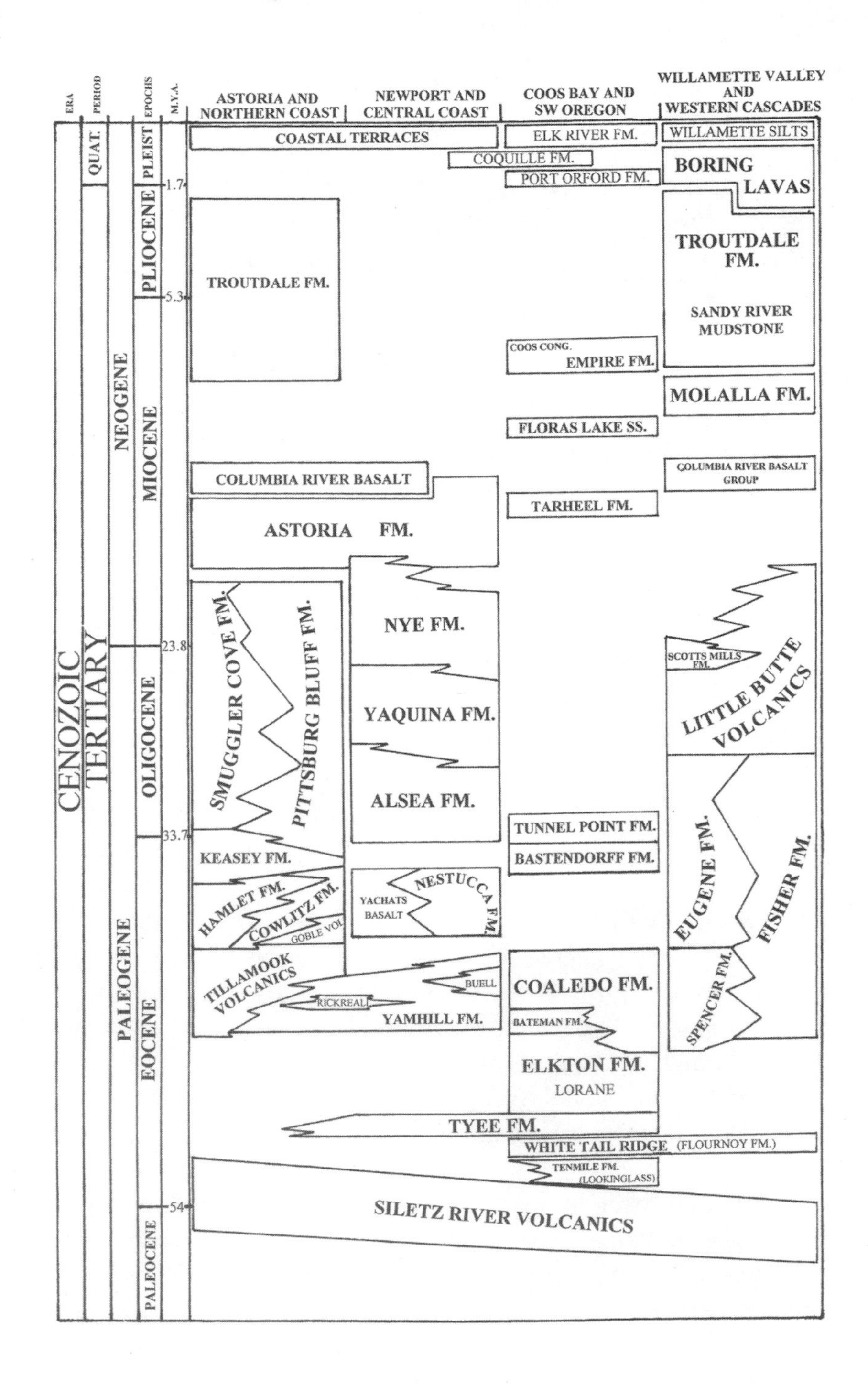

Tertiary stratigraphy of the Oregon Coast Range province. (After Addicott, 1964; Armentrout, ed., 1981; Armentrout, et al., 1983; Baldwin, 1950, 1974; Molenaar, 1985; Niem, Niem, and Snavely, 1992; Niem, et al., 1994; Orr and Orr, 2009; Prothero, ed., 2001, 2003; Ryu, Niem, and Niem, 1992)

places, the Yachats has been interpreted as a small field of seamounts scraped off the subducting Farallon plate. This basalt forms resistant promontories at Sea Lion Caves, Heceta Head, and Cape Perpetua. The Eocene basalts that armor Cascade Head are slightly older. Proposal Rock, now a tree-covered island, was part of Cascade Head before being cut away from the mainland.

In the Nehalem basin, strata of the Hamlet, Cowlitz, and Keasey formations represent swamps, spreading deltas, and deep marine troughs. The high-energy pocket beach conditions of the Hamlet merge into the deep marine Cowlitz delta, which consists of micaceous and feldspar-rich sandstones with abundant upper shelf invertebrates, trace fossils, and microfossils. In sharp contrast, the overlying Keasey Formation, from a continental slope at depths of 1,500 feet or more, preserves unusually delicate fossils such as crinoids, thin-shelled mollusks, corals, sea urchins, and plants.

Cooling, which occurred worldwide throughout the Cenozoic, continues today. Climate changes from the middle Eocene and into the Oligocene triggered two episodes of extinction when warm tropical faunas were replaced by temperate forms. Diversity of invertebrates reached a maximum during the tropical Cowlitz interval, preceding a dramatic regional late Eocene demise. A second extinction occurred when Keasey invertebrates and plants were replaced by the cool-water faunas of the Oligocene to early Miocene Pittsburg Bluff and Eugene formations.

Oligocene
Shoreline Embayments and Deltas

Covering a relatively short time span from 34 to 24 million years before the present, the Oligocene was an epoch when marine sedimentation was limited to embayments along the central and north coast. Strata are typified by siltstones and muds of the Alsea, Yaquina, and Nye formations in the central region, and by the Smuggler Cove and Pittsburg Bluff formations toward the north.

Near Newport, abundant ash, silts, and sands of the Alsea and Yaquina are reflective of an open-ocean, cool-water upper bathyal shelf shoaling to a coastal plain. The Yaquina may have been derived from the underlying Tyee Formation and Eocene basalts, but the persistence of ash and pumice suggests sources from ongoing Western Cascade eruptions.

The Nye Mudstone, spanning the Oligocene to Miocene epochs, overlies the Yaquina in the Newport basin. The Nye thins toward the north, where it may have shoaled against a broad Yaquina delta, however, a cold water bathyal interval reaching 2,000 feet is reflected by the presence of deep-water microfossils. Both formations are famous for their whale and seal remains, fish scales, and shark teeth.

In Clatsop County, the Oligocene Epoch was marked by shallow seas with growing deltas, brackish estuaries, and offshore deeper shelf and slope sands and silts of the Smuggler Cove and Pittsburg Bluff formations. Initially designated as the Oswald West Mudstone, the Smuggler Cove is best known for its trace fossils of tracks, trails, and burrows. The formation reflects cold water, continental slope conditions below 1,000 feet.

Originally regarded as Eocene, the Pittsburg Bluff was assigned to the Oligocene in 1915 before its current placement spanning the Eocene-Oligocene-Miocene boundaries. The most definitive works on these rocks are by Ellen Moore of Oregon State University, who detailed the paleontology and stratigraphy. Even though there are few species overall, mollusks and other invertebrates in the Pittsburg Bluff are abundant, depicting a remarkable range of settings from upper continental shelf, to intertidal, and even terrestrial. Since many of the shells are broken but show little beach wear, Moore surmised that they had been transported some distance by storms before deposition. Fluvial channels and deltas of the Scappoose Formation have been revised and mapped as facies of the upper Pittsburg Bluff by the Niems.

Miocene
Fluctuating Oceans

During the latest Oligocene to early middle Miocene, regional uplift shifted the shoreline close to its present position, and marine waters occupied only narrow inlets at Astoria, Newport, and Coos Bay. Deposition continued uninterrupted as the Western Cascades showered ash directly into the ocean, where currents carried the debris into deeper waters.

Middle Miocene strata on the southwest coast were unknown until 1949, at which time mollusk-bearing sandstones were brought to the surface and piled in a disposal area during U.S. Army Corps dredging operations in the Coos Bay channel. The fossils were inspected by Ewart Baldwin and his student Ellen Moore, who determined they were of Miocene age. In 1966, John Armentrout, also Baldwin's student, discovered similar fossiliferous outcrops near Pigeon Point. Naming them the Tarheel Formation, he surmised that the sandstones were deposited in warm to temperate water at shallow to moderate depths up to 180 feet.

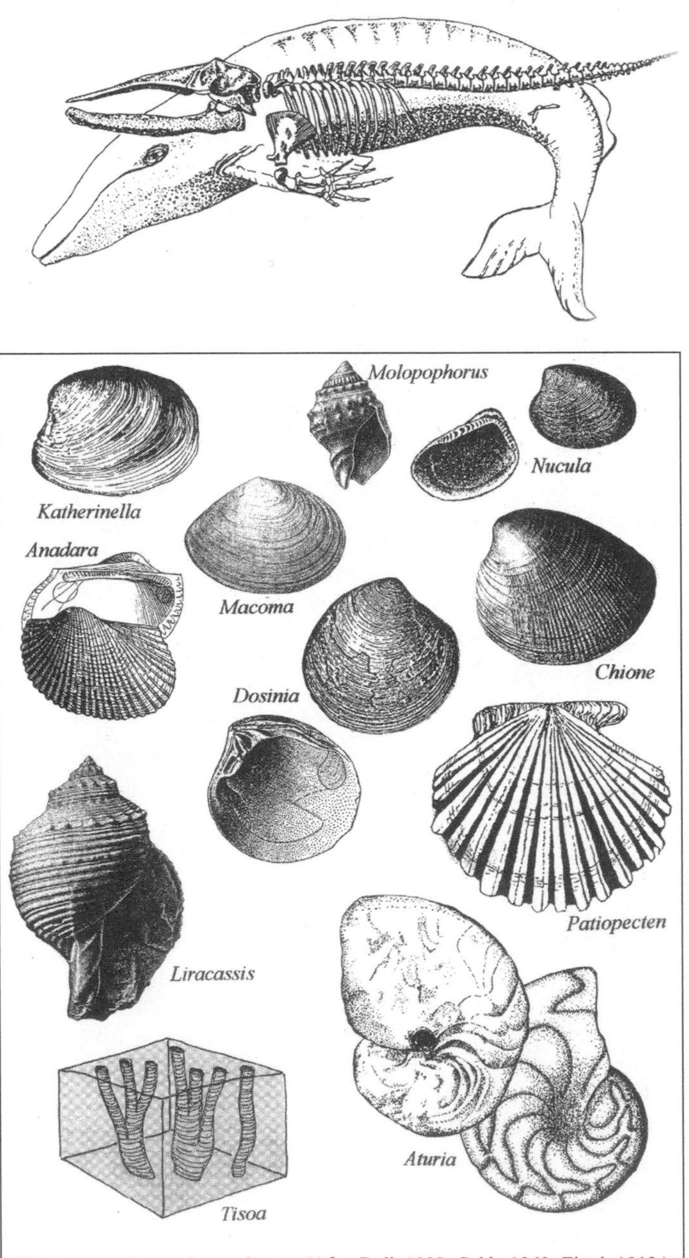

Marine central coast invertebrates. (After Dall, 1909; Gabb, 1869; Zittel, 1913.)

The Oregon coast is famous for its fossils from the Miocene Astoria Formation, which are continuously exposed in eroding sea cliffs. In addition to shells, bone fragments of the primitive baleen whale *Cophocetus* (above) are often encountered. (After Moore, 2000, 2002; Niem and Niem, 1985; Orr and Orr, 2009)

Above the Tarheel, the fossil-rich Empire Formation stretches to Cape Blanco, and at South Slough it is exposed around Coos Head. When he visited Coos Bay, William Dall of the Smithsonian Institution obtained Empire mollusks and marine vertebrate fossils that he described and illustrated in 1902. Armentrout reassessed the invertebrate fauna over 60 years later, concluding that the late Miocene climate and calm estuary were close to the environment of today.

Part of the same formation, the Coos Conglomerate at Fossil Point was described in 1896 by Joseph Diller, who recognized the small patch of less than one acre as the most fossiliferous anywhere along the coast. The lens was a narrow submarine channel cut into upper layers of the Empire Formation in which shallow water mollusks and the remains of whales, seals, walrus, and fish collected.

At Cape Blanco, the Empire was assigned to the middle Miocene until a gap or unconformity was noted within the strata. When Warren Addicott of the U.S.G.S. subsequently divided the formation, he designated the lower section as the Sandstone of Floras Lake, retaining the name Empire for the upper portion. South of Bandon, the state's only known marine diatomite, that projects well out onto the continental shelf, was mapped as a facies of the Empire Formation by Jerry Fowler of Oregon State University and coauthors in 1971.

On the central and northwest coast, fossiliferous sandstones and siltstones of the middle Miocene Astoria Formation are well exposed. The Astoria has not lacked for attention since the first examination of fossils in 1848 by Timothy Conrad of the New York Geological Survey. Astoria shales were formally named by Thomas Condon in 1880 and described by Joseph Diller in 1896. Ellen Moore's research on the Astoria has produced many publications as have Parke Snavely's mapping and descriptions of the strata at Newport, and Wendy and Alan Niem's work on the Astoria, Tillamook, and Nehalem basins.

The diverse warm-temperate fauna of the Astoria Formation includes corals, barnacles, bryozoa, brachiopods, and crabs, along with 97 species and

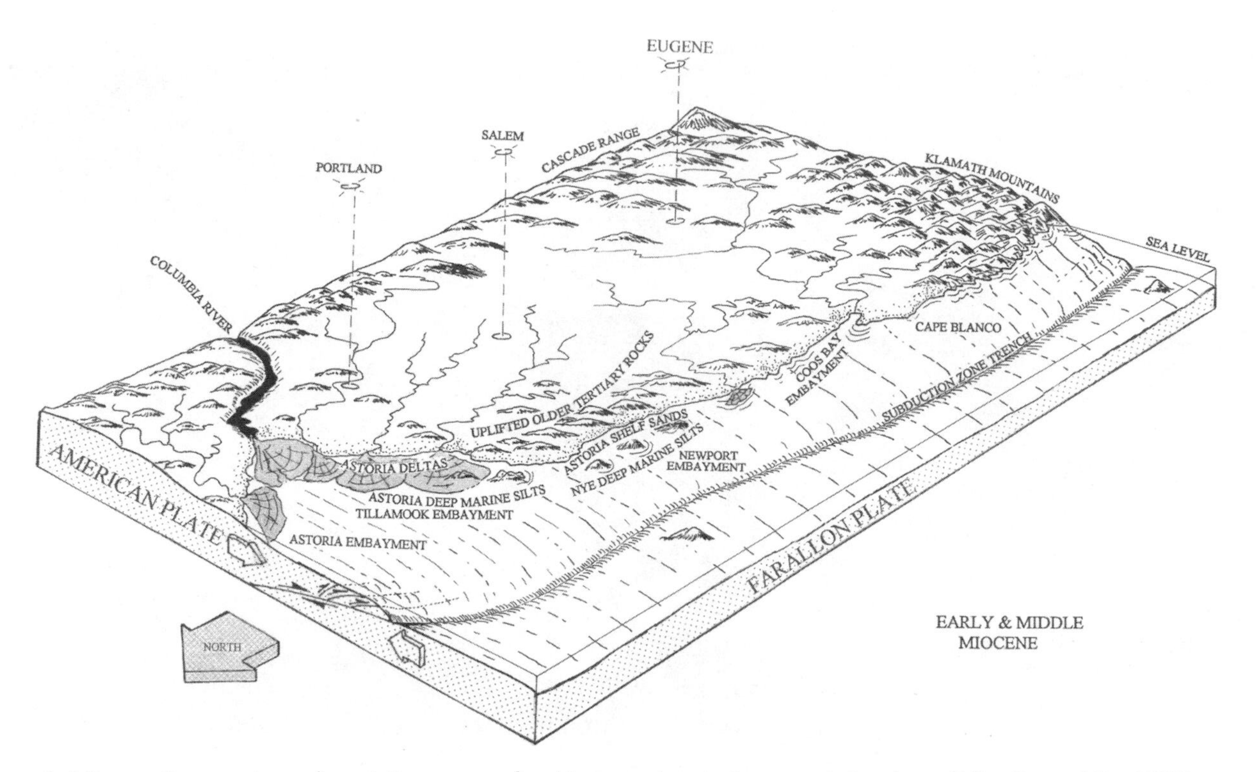

By Miocene time, marine sedimentation was confined to areas close to the present-day shore. (After Orr and Orr, 2009)

73 genera of mollusks, as is characteristic of organisms living on a soft muddy seafloor and continental shelf at depths of 500 feet. Astoria sediments are also the repository of an extraordinary collection of vertebrate bones, shark teeth, and reptilian (turtles) fragments. Even occasional ungulates (hoofed mammals) were washed into the seaway.

Concurrent with deposition of the Astoria Formation, areas of the north and central coast were invaded by Columbia River lavas from eastern Oregon. The flows, sills, and dikes were initially mapped and interpreted by Parke Snavely and others as having erupted from local volcanic vents, but the striking paleomagnetic, chemical, and mineralogical similarities between the basalts making up most promontories between Seal Rock, Oregon, and Grays Harbor, Washington, to those of the Columbia River Group were an enigma until Marvin Beeson demonstrated that both were derived from the same magma source beneath the Grande

Now adjacent to the mainland, Elephant Rock is an elongate invasive sill of Miocene Columbia River basalts at Seal Rock State Park south of Newport. During high interglacial sea levels of the Pleistocene, it would have been an isolated sea stack. (Photo courtesy Oregon Department of Geology and Mineral Industries)

Projecting offshore, Yaquina Head (left) north of Newport and the dramatic promontory of Cape Lookout (right) at Netarts Bay are Miocene Columbia River basalts. Both features are outstanding examples of inverted topography. In such cases, dense fast-moving lavas filled estuaries, but after the softer sediments were eroded away, the resistant basalts remained as elongate headlands, cast in stone. (Photos courtesy Oregon Department of Geology and Mineral Industries and Oregon State Highway Department).

Ronde Valley of Baker County. Voluminous floods of lavas flowed westward along a broad Columbia River paleochannel, through a gap in the Cascade Range, and into the Willamette Valley. Advancing seaward to pond up in marine bays and estuaries, the dense flows penetrated deeply into the soft, soupy sediments. Geophysical transects across the coastal exposures reveal that the basalts are rootless, which further supports a distant origin rather than a local volcanic source.

Outcropping as far as 120 miles south of the entrance to the Columbia River, the Miocene basalts terminated at Seal Rock, close to the same latitude they reached in the Willamette Valley. Yaquina Head, Cape Foulweather, Cape Lookout, Cape Meares, Neahkahnie Mountain, Cape Falcon, and Tillamook Head are all headlands of resistant Columbia River basalts. Most of the offshore stacks, islands, and arches are vestiges of the same flows. These features have been isolated from the mainland by erosion as have Arch Cape and Haystack Rock offshore at Cannon Beach. Pillar Rock, Pyramid Rock, and Three Arch Rocks located near Cape Meares have been cut back as well.

Pliocene-Pleistocene Glaciers and Ocean Waters

There are few confirmed Pliocene sediments in western Oregon as this was a comparatively short time when much of the area was above sea level and erosion predominated. Pliocene gravels of the Troutdale Formation, with Precambrian cobbles derived from well up in the northern Rockies, are encountered intermittently in the Columbia River channel.

Global cooling characterized the Pleistocene or Ice Ages, starting around 2 million years ago. In the Pacific Northwest, continental glaciers from Canada reached into northern Washington, but only ice caps appeared along the crests of the Oregon Cascades. Glaciation, which tied up water as ice, initiated a profound lowering of sea levels worldwide, while in Oregon it brought a widening of the coastal plain and a rapid down-cutting of estuaries and stream valleys. At the termination of the last glacial phase 11,000 years ago, which brought a reversal of Arctic-like conditions, a warming climate melted ice caps that, in turn, elevated ocean levels, submerged the coastline, and created vast new habitats for marine invertebrates.

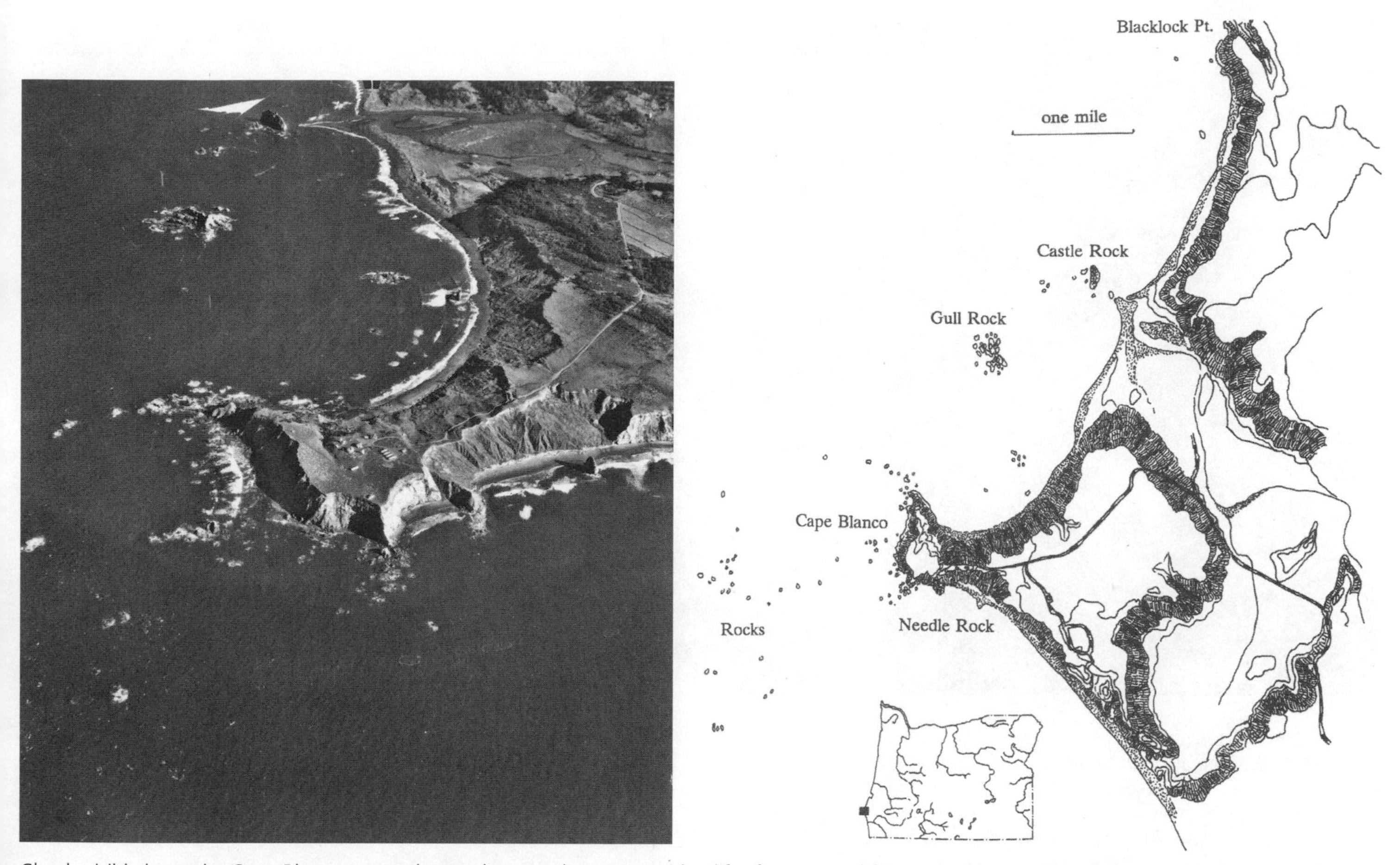

Clearly visible here, the Cape Blanco terrace has undergone the most rapid uplift of any coastal feature in Oregon. This distinct promontory, noted on maps of early explorers, is the state's most westerly point. Named the White Cape in the 16th Century, it is a complex of durable Jurassic Otter Point conglomerates covered by Eocene mudstones and the Miocene Empire Formation. (Photo courtesy U.S. Forest Service).

Although Pleistocene marine fossils are comparatively rare in Oregon, mollusks are prevalent in terrace deposits at Cape Blanco, which takes its name from the white shells that litter the cliffs. Invertebrates in the Port Orford and Elk River formations at Cape Blanco and in the Coquille Formation at Bandon some 20 miles to the northeast were described by Warren Addicott. He characterized the climate as slightly cooler than at present and surmised that the shallow-water conditions differed little from those of the modern offshore. Many of the species inhabiting the region today are similar to those of the Pleistocene.

Elevated Terraces

Along much of the Pacific Northwest coast, a series of stair-step terraces form discontinuous elevated surfaces that chronicle erosion, uplift, and deformation. The oldest terraces are inland, whereas the younger late Pleistocene surfaces are found near sea level. Buried by up to 20 feet of beach gravel and sand, the lower terraces are better preserved than the higher ones. Separated by headlands or bays, the individual surfaces are difficult to correlate because the rates of uplift vary considerably from place to place, and the surfaces have been altered by deformation, faulting, and erosion.

The number of recognized marine platforms has been refined and revised since they were first mapped in 1945 by Allan Griggs of the U.S.G.S. Northward from Coquille Point, the Whisky Run, Pioneer, and Seven Devils terraces were named by Griggs, while the oldest Metcalf was described by John Adams of the Geological Survey of Canada. In his classic 1984 overview paper, Adams calculated the magnitude of uplift and tilting for the Silver Butte, Indian Creek, and Poverty Ridge terraces that lie inland from the Pioneer terrace at Cape Blanco. In 1969 Richard Janda of the U.S.G.S. identified most of the terraces at Cape Blanco, and some 20

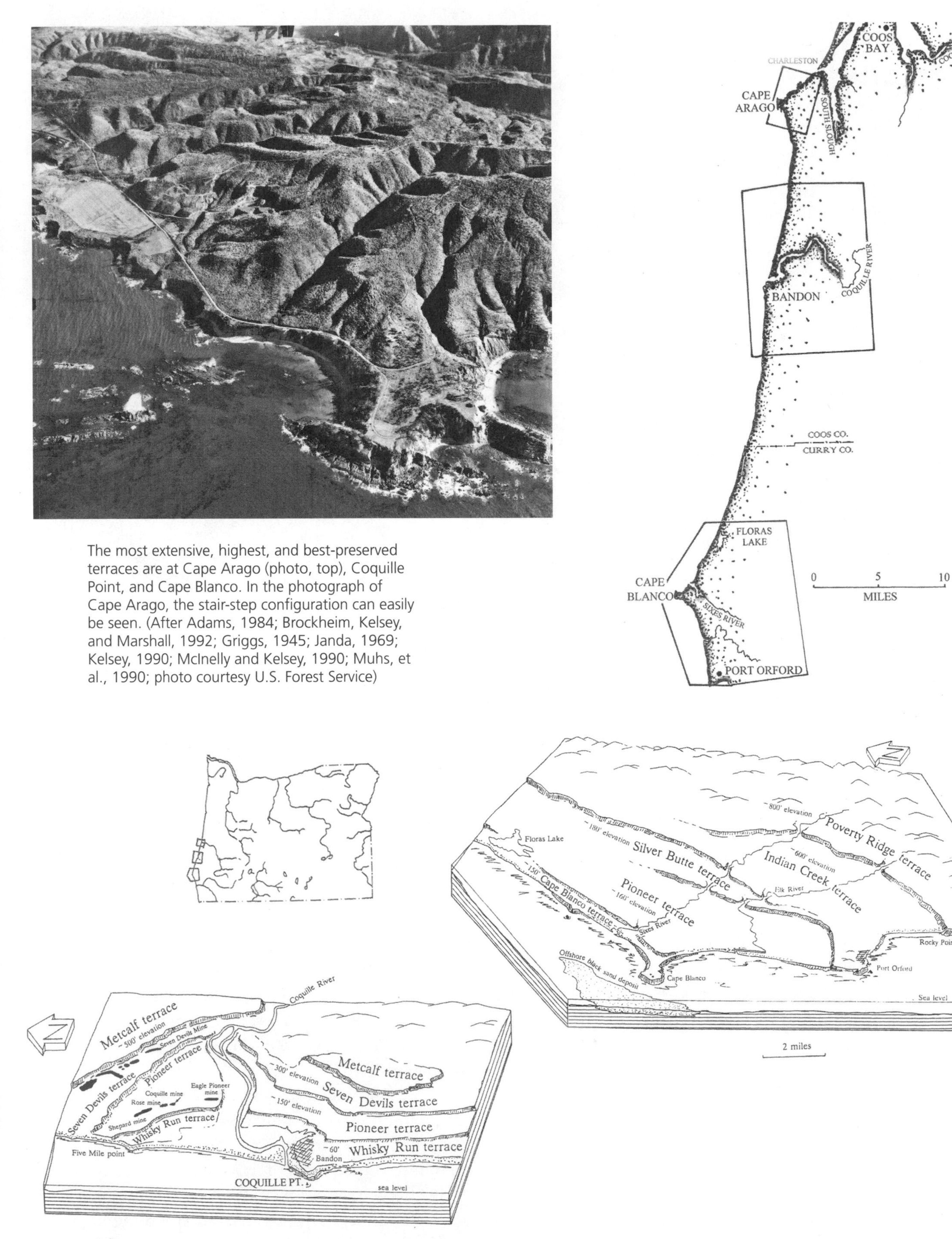

The most extensive, highest, and best-preserved terraces are at Cape Arago (photo, top), Coquille Point, and Cape Blanco. In the photograph of Cape Arago, the stair-step configuration can easily be seen. (After Adams, 1984; Brockheim, Kelsey, and Marshall, 1992; Griggs, 1945; Janda, 1969; Kelsey, 1990; McInelly and Kelsey, 1990; Muhs, et al., 1990; photo courtesy U.S. Forest Service)

years later Harvey Kelsey described the youngest level here. From a height of 150 to 190 feet above sea level in southern Oregon, the lengthy Pioneer decreases to 60 feet at Tillamook.

In 1990, Galan McInelly and Harvey Kelsey recognized and mapped the Arago Peak terrace at Cape Arago, and in 1996 Kelsey named and described six new platforms, the Newport and Wakonda at Yaquina Head, the Yachats and Crestview at Waldport, and the Fern Ridge and Alder Grove at Alsea Bay on the central coast. These range from sea level to almost 800 feet in elevation and extend approximately three miles inland. Based on the depth of the soils, the oldest at Alsea and Yaquina bays were estimated in excess of 200,000 years old, while the youngest terrace at Newport is closer to 80,000 years.

Varying interpretations have been proposed for the observed elevations and tilting of marine terraces. Contrary to earlier descriptions, there are no clear regional trends or orientations to the tilted surfaces. Recent mapping has shown that the three lowest (youngest) surfaces at Cape Banco tilt landward, and the two higher ones seaward. While elevation is seen as due to subduction, tilting and the rate of elevation may be the product of local folding and deformation. Identifying and comparing the distribution of strike-slip faults and folds, Kelsey found that most of the platforms had been uplifted at the comparatively low rate of 4 to 12 inches every 1,000 years throughout the past 125,000 years. Near the folds and faults, however, the elevation was faster, approaching three feet every 1,000 years.

Multiple techniques are used to date and correlate terraces. One method is to measure uranium /thorium radioactive decay from the calcite in fossil shells. Another calculates the chemical changes that amino acids undergo in an animal shell after it dies, commonly employing the thick-shelled clams *Saxidomus* and *Mya*. Using this method, Daniel Muhs inferred dates for the Cape Blanco terrace at 80,000 years, the Whisky Run at 83,000 years, the Pioneer around 100,000 years, the Seven Devils at 124,000 years, and the Metcalf terrace at 230,000 years. A third procedure for distinguishing one terrace from another compares and matches the successions of soils that developed on top of each.

Continental Margin

Topography of the offshore shelf and slope

The Oregon coastal province does not stop at the beach but extends out onto the continental shelf that dips gently seaward to 600 feet and to the lower slope that descends to abyssal seafloor depths of

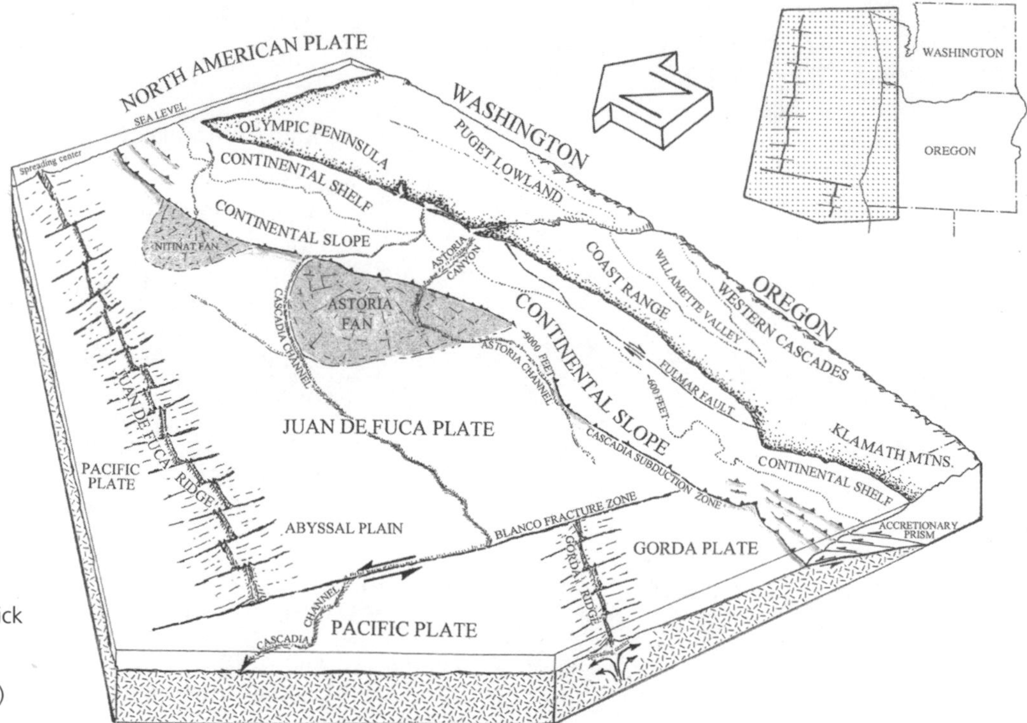

The Oregon submarine landscape of deep canyons, ridges, plains, faults, and a thick sedimentary covering is more intricate than that onshore. (After Kulm and Fowler, 1974)

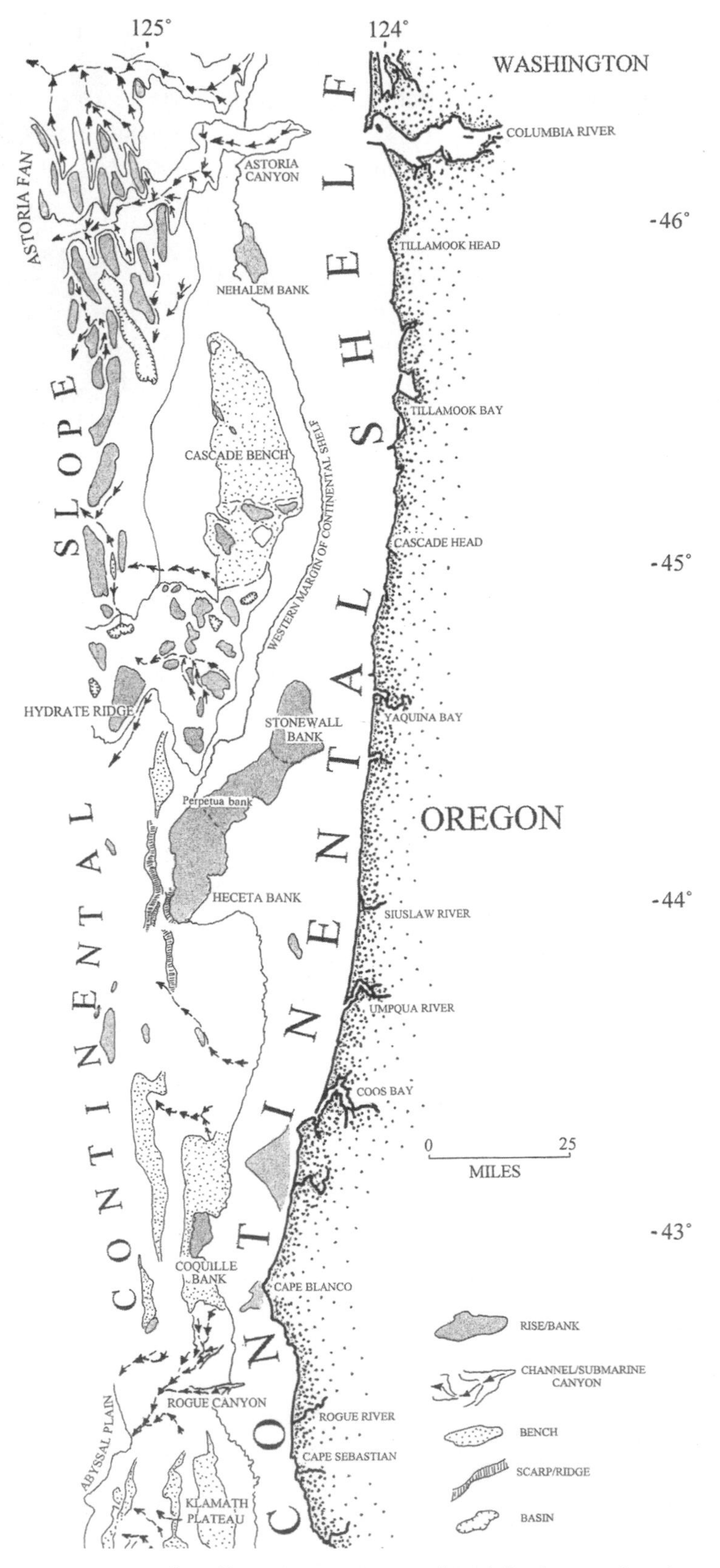

Twenty miles off Yaquina Bay, Stonewall Bank lies in water less than 120 feet deep, and Heceta Bank, west of the mouth of the Siuslaw River, rises to within 360 feet of the surface. Coquille Bank, a shoal approximately three miles wide and eight miles long between Coos Bay and Cape Blanco, exhibits 198 feet of relief. Beneath Heceta Bank a lengthy basalt ridge, buried in the accretionary wedge, marks the western edge of the Eocene Siletzia oceanic terrane. The banks may have resulted when eastward-migrating seamounts "plowed" beneath the continental slope and shelf to bulge up on the seafloor during subduction. The volcanic ridges are presently buried beneath the banks. (After Fleming and Tréhu, 1999; Kulm and Fowler, 1974; von Huene, 2008)

9,000 feet. The offshore topography is mantled by sand and muds, producing flat areas that are folded and broken by banks, ridges, basins, channels, and canyons. The combined width of the shelf and slope varies from approximately 70 miles off Astoria to 40 miles off Cape Blanco. Shelf and slope geology is closely linked with onshore patterns and processes, and the Oregon margin consists mainly of the same Eocene through Pleistocene formations as those exposed in the Coast Range.

With an east-west gradient of only 6 to 10 feet per mile, the nearly horizontal continental shelf is broken by the prominent fault-bounded Nehalem, Stonewall, Perpetua, Heceta, and Coquille banks. These topographic features are capped by Miocene through Pleistocene mudstones, sandstones, and clays.

The upper continental slope drops from 600 to 5,000 feet and is distinguished by benches, low hills, and canyons. The longest Cascade bench—a level platform with steep sides—lies between Cascade Head and Tillamook Bay at depths of 2,000 feet. A similar feature between Cape Sebastian and the California border at 1,500 to 2,000 feet deep is a seaward continuation of the Klamath plateau.

From the edge of the shelf, deep-sea canyons crossing the slope serve as conduits for dispersing sediments to the lower slope and abyssal sea floor. Northwest of the mouth of the Columbia River, the 1,200-mile-long Cascadia channel that eventually cuts the Cape Blanco fracture zone is the longest such feature in the Pacific basin. Off the mouth of the Columbia River, the Astoria Canyon projects westward from the outer shelf across the Astoria fan before reaching the base of the slope at 9,000 feet, at which point it splits into two directions. Adjacent to the mouth of the Rogue River, the Rogue Canyon follows a westerly pathway to abyssal depths.

An analysis of heavy mineral suites in offshore deposits shows that the dominant Pliocene sources were the Klamath Mountains and British Columbia, however, that changed around 2 million years ago with an influx of fine-to-medium-grained sandy turbidites from the Columbia River. Exiting through the Astoria and Willapa canyons, the sediments constructed the Astoria fan that covers more than 3,500 square miles and is somewhat larger than the Nitinat fan off the Washington Olympic Peninsula.

Between the base of the slope and the abyssal plain, the Cascadia subduction trench is the focal point of Oregon's offshore geology. Largely obscured by thicknesses of sediments, the long narrow depression lies at the boundary between the Juan de Fuca and North American plates and runs from Cape Mendocino, California, to Vancouver Island. The Juan de Fuca and Gorda slabs are moving east-northeast into the subduction trench at the rate of one and one-half inches a year.

One by-product of the subduction process is a thick accretionary prism (wedge) of sediment and crust, which was peeled off the top of the Juan de Fuca plate, to become the leading edge of the continent. This mélange has been highly compressed and sheared by thousands of low-angle north-south thrust faults creating long, overlapping anticlinal ridges and intervening synclinal basins that parallel the shoreline. The broad zone of intense faulting reaches from the Cascade subduction zone (deformation front) to the base of the continental shelf. Thrust faults, where the overlying or upper plate has moved up and over the lower block, are the result of extreme crustal shortening or telescoping.

The abyssal plain is a wide nearly horizontal to gently sloping, deep-sea floor, where the older topography is covered by thin layers of marine sediment. Intersecting the abyssal plain, the Juan de Fuca and Gorda plates are bounded by lengthy mid-ocean ridges and fracture zones. At spreading centers along the Juan de Fuca and Gorda ridges, cooling sheets of lava create new oceanic crust. Vents along the Juan de Fuca ridge, releasing hydrogen sulfide, host rich communities of giant clams

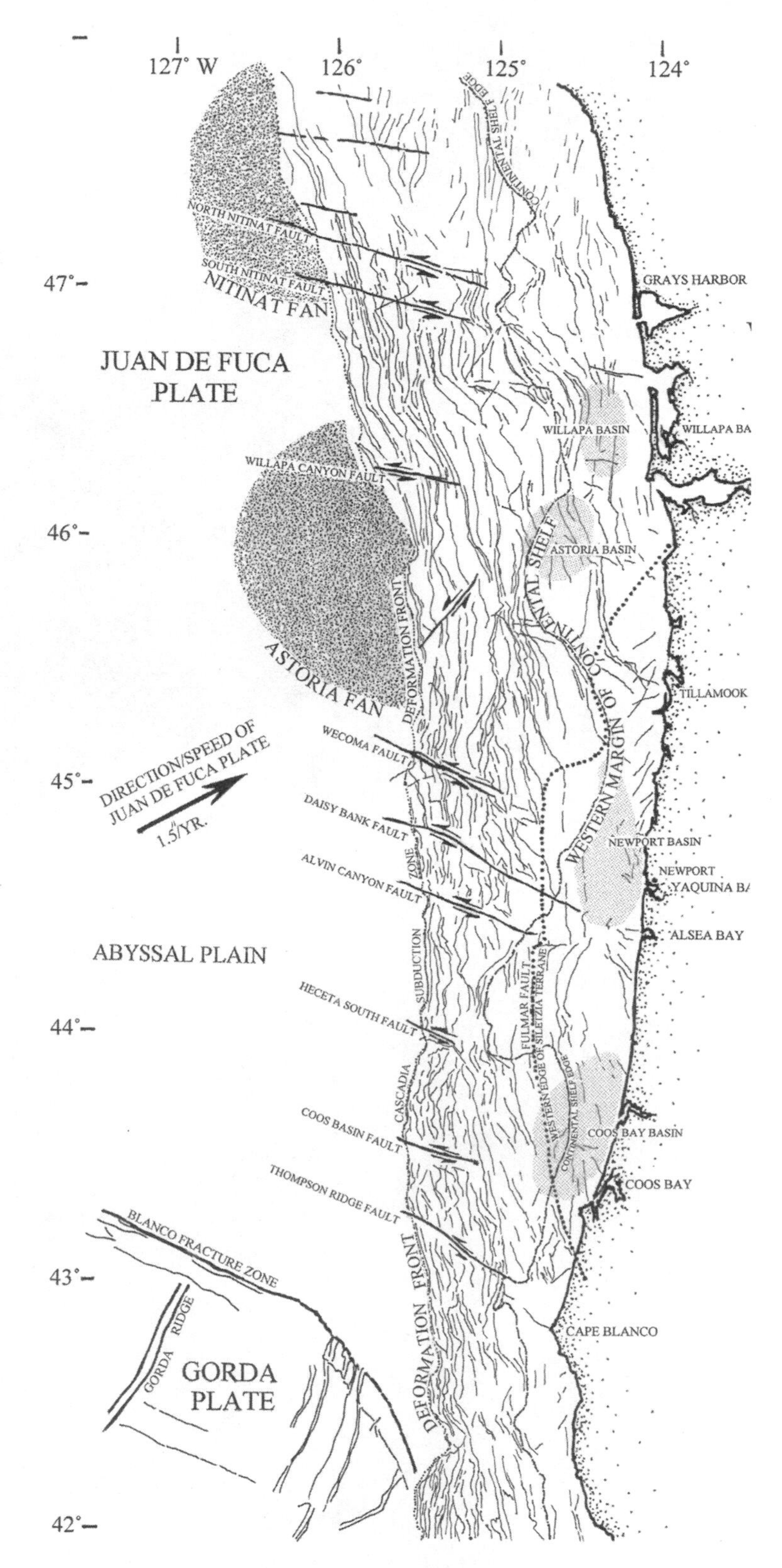

Prominent features of the offshore are west-northwest-trending strike-slip faults (arrows) that run from the abyssal plain across the deformation front and shelf toward the coast. Between 20 to 70 miles in length, at least nine faults from Cape Blanco to the central Washington coast offset both the Juan de Fuca and North American plates and the accretionary prism. Dated at 650,000 to 300,000 years ago, these shear structures define the boundaries of elongate blocks, which are rotating clockwise, driven by the oblique subduction of the Juan de Fuca plate (long dark arrow). The most notable of these are the Wecoma, the Daisy Bank, and the Alvin Canyon faults. At the base of the continental slope, the intersection of the Wecoma fault with the deformation front is marked by a complex of bulges or fault pop-ups with a pronounced indentation in the slope. (After Couch, 1979; Goldfinger, et al., 1996, 1997; Kelsey, et al., 1996; McNeill, et al., 2000; Yeats, et al., 1998)

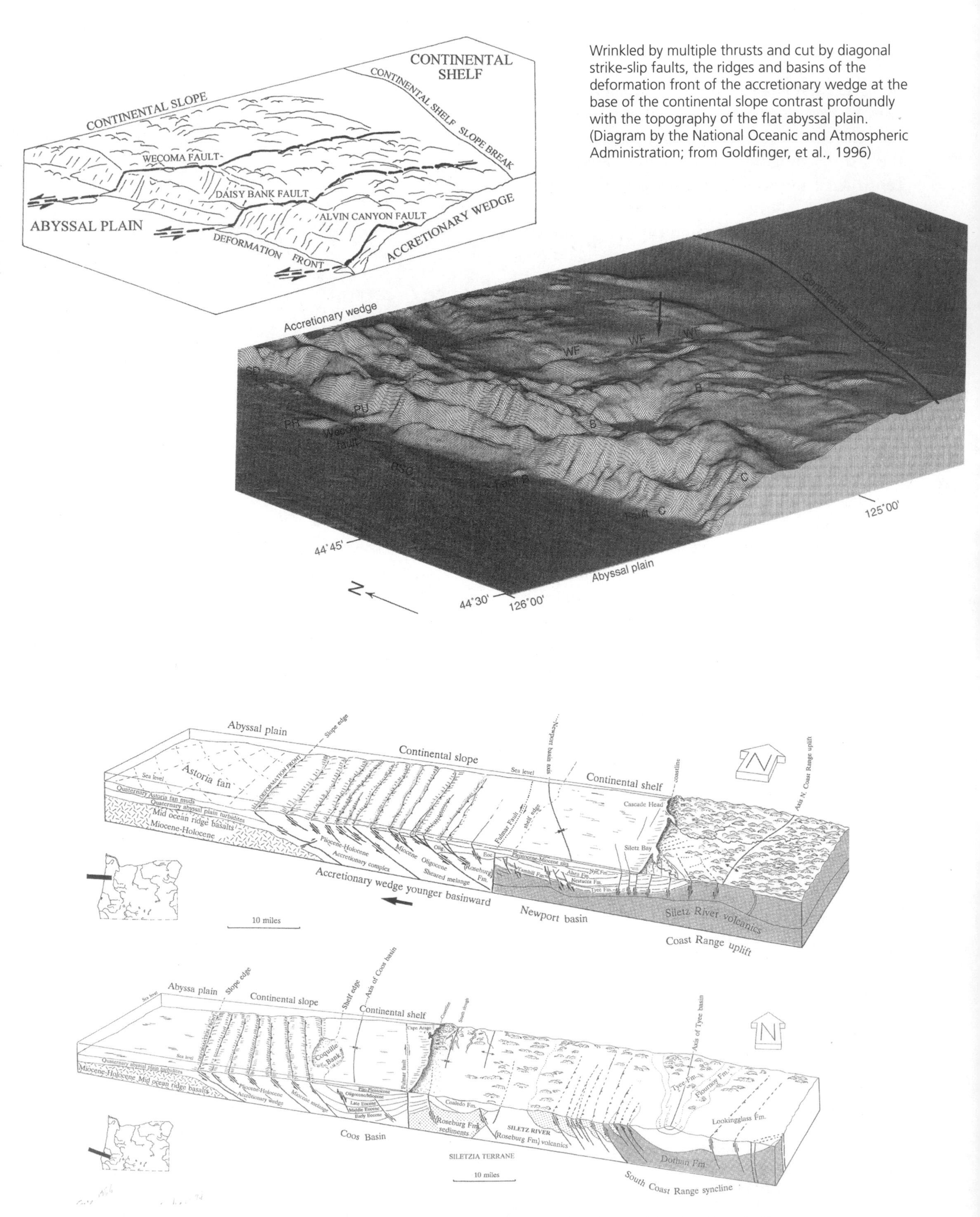

Wrinkled by multiple thrusts and cut by diagonal strike-slip faults, the ridges and basins of the deformation front of the accretionary wedge at the base of the continental slope contrast profoundly with the topography of the flat abyssal plain. (Diagram by the National Oceanic and Atmospheric Administration; from Goldfinger, et al., 1996)

Imbricate sheets of deformed Tertiary sedimentary and basaltic rocks lie at the foundation of the shoreline, shelf, and slope off Newport and Cape Arago. (After Baldwin, 1976; Snavely and Wells, 1996)

and tube worms living in a symbiotic relationship with bacteria.

The Juan de Fuca and Gorda ridges are offset by wide sheared zones of the Blanco and Mendocino transform faults that move laterally past each other and that are generally perpendicular to the ocean ridges. Projecting in a westerly direction from southern Oregon and northern California, the Blanco and Mendocino faults are the sites of frequent earthquakes. The Blanco transform fault separates the Juan de Fuca and Pacific plates.

Geohazards

Typical of an active tectonic margin, the coastal province has been beset by earthquakes, tsunamis, erosion, landslides, and flooding of considerable magnitude throughout its geologic past. Such natural processes are only termed hazardous when human life or property is threatened, but the physical dynamics combined with inadequate land-use planning and population pressures have fostered circumstances of higher-than-normal risks.

Earthquakes

Sources and causes

Episodes of earthquakes are not new to the Pacific Northwest, and most are now being systematically documented through geologic verification of the actual events. Of particular concern is the periodicity of large quakes along the Cascadia subduction zone, coupled with the fact that it has the lowest incidence of documented seismicity of any subduction system worldwide.

Massive earthquakes occur offshore along the Cascadia subduction area at the interface between the Juan de Fuca plate and the North American plates. If the lower subducting slab adheres to the overlying plate, they bind up and lock. As subduction proceeds and stress steadily increases, interseismic strain accumulates, and the distal edge of the North American plate bulges upward. If the bond along the subduction surface fails suddenly, the release of coseismic strain initiates an earthquake and rapid subsidence of the uplifted area.

Faced with a rising coast, which is a sign that strain is building between locked plates, geologists' efforts revolve around understanding the apparent lack of historic megathrust (subduction) quakes. Explanations include aseismic creep, where ongoing slip relieves the pressure, or crustal shortening of the forearc basin by clockwise rotation. Faulting and folding may also absorb large amounts of plate convergence, lessening earthquake intensity. Current research favors a model where interseismic (elastic) strain on the locked plate interface is released at periodic intervals.

In contrast to the locked zone model, other theories maintain that high temperatures may diminish friction and stress along parts of the plate interface. The thick slab of accreted volcanics (Siletzia) along

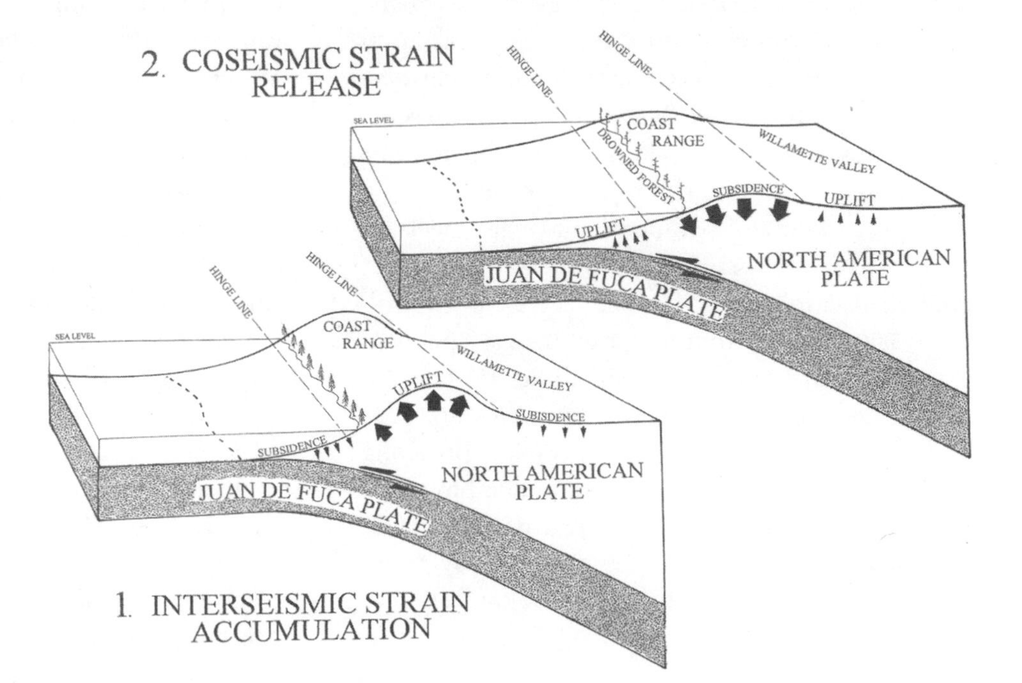

Tied to subduction, the earthquake cycle for Oregon proposes that during an interseismic period of low activity, lasting hundreds of years, strain accumulating at the interface between the Juan de Fuca and North American plates causes coastal uplift. When the strain is released, an earthquake, rapid subsidence of the coastal region, and a tsunami drowning the area take place simultaneously (coseismic). (After Kelsey et al., 1996; Mitchell, et al., 1994; McNeill, et al., 1998)

the central Oregon margin might act like a thermal blanket to confine heat to the interface between the two plates and thus reduce friction.

While gaps of seismic activity on the order of 300 to 1,000 years may occur between major quakes, interseismic strain alone might not account for these extended lapses. In 1996, John Adams correlated and dated 13 turbidite sequences from deep-sea piston cores in the Cascadia channel using volcanic ash from the 7,700-year-old Mt. Mazama eruption as a marker horizon. Following the eruption, turbidity currents, with ash-laden sediments, were dispersed from the mouth of the Columbia River into the channel. An examination of the successive thin clay layers atop the turbidites led him to speculate, that since they were deposited on average every 600 years, the swiftly-moving currents were triggered by strong Cascadia subduction zone earthquakes, which show the same periodicity.

Chris Goldfinger, a researcher at Oregon State University, offers evidence that the gaps between subduction quakes occur because the different segments of the Cascadia zone may act independently of each other. In his chronology, 19 quakes of magnitude 9.0 have ruptured along the entire zone, while those of magnitude 8.0 took place on shorter portions. Also examining turbidite frequencies, he assessed the 10,000-year history of subduction events to estimate that southern Oregon and northern California have experienced megaquakes every 250 years, central Oregon about every 450 years, and northern Oregon and Washington every 500 years. Because the quakes may occur in clusters, the question to be asked is whether the present-day hiatus is part of a grouping within a short time frame, or whether the state is experiencing a 1,000-year gap. The last major quake around 1700 A.D. was in the vicinity of northern Oregon and southern Washington.

In 2008, Anne Tréhu reported that in 2004 and 2007 two earthquakes of magnitude 4.9 and 4.8 and at depths between five to ten miles occurred offshore from central Oregon. Such seismicity along the Juan de Fuca-North American plate boundary may represent slippage on a weak portion of the Cascadia subduction zone. Further work by Tréhu relates seismic activity to deeply subducted seamounts (volcanoes).

Seismologists are presently struggling to determine the probability of when the next quake will strike, where it will take place, and what the magnitude will be. A comparison to subduction systems elsewhere shows striking similarities between the Cascadia zone and that of south-central Chile, which suffered a 9.5 magnitude event in 1960. Buried under tsunami debris, Chilean coastal areas were subjected to as much as six feet of coseismic subsidence.

Historic earthquakes

Oregon's historic record of earthquakes is minimal, going back only 175 years, but evidence of catastrophic Holocene seismic activity has been found in lowland sediments. In 1987 Brian Atwater of the U.S.G.S. was the first to use bog stratigraphy to demonstrate that earthquakes, tsunamis, and subsidence occurred simultaneously (coseismically) and repeatedly along the Pacific Northwest coast. Using shallow cores from tidal areas and bays, he identified marsh and soil layers covered by sandy tsunami deposits as signs of a sudden down-dropping. Preserved in an upright position, drowned trees also point to the rapidity of the subsidence.

In light of Atwater's discovery, dozens of coastal subsidence sites from British Columbia to northern California have now been identified. Thin tsunami sand layers over paleomarsh settings have been documented from 12 estuaries between the Columbia and the Sixes River. Each of the bays may have experienced multiple events, but the timing and number of the megaquakes, or whether subsidence at some of the sites was the product of local faulting in conjunction with subduction zone earthquakes, is unclear.

A quake in the Siletz River valley 5,500 years ago is the oldest dated in Oregon. Changes in microfossil assemblages prior to the quake and after submergence substantiate extensive paleoenvironmental alterations at South Slough in Coos County. Four separate marsh-sand transitional phases at Netarts Bay that took place over the past 3,000 years show similar changes. Cycles of episodic coastal submergence in the Alsea Bay estuary have been chronicled by Curt Peterson and Mark Darienzo at Portland State University, who concluded that coseismic subsidence is the only explanation for peat-burial sequences.

Mapping by the Oregon Department of Geology and Mineral Industries has predicted likely scenarios for inland reaches (run-up heights) of tsunami waves. Its conclusions show that many coastal communities are much more vulnerable than previously realized. In this photo, Cannon Beach, at sea level, lies completely unprotected from waves. Haystack Rock is offshore, and the Ecola State Park landslide area is in the center back just in front of Tillamook Head. DOGAMI hopes to complete its mapping south to Bandon by 2018. (Photo taken in 1965; courtesy Oregon State Archives)

Tsunamis

The sudden displacement of the ocean floor can trigger ocean waves called tsunamis that are catastrophic in scale. Not to be confused with wind-generated waves or lunar tides, tsunamis can travel up to 600 miles per hour, appearing as low, broad waves in the open ocean and piling up as high as 100 feet when the wave encounters shallower water. A succession of destructive waves from a single earthquake may last for several hours.

Since a seismometer was installed in 1971 at the Hatfield Marine Science Center in Newport, only three minor tsunamis have been recorded. In one case, the groups of waves caused by an earthquake in northern Japan lasted up to two days. The largest tsunamis to reach the Oregon coast were in March, 1964, and May, 1968. Generated by the 1964 earthquake in Alaska, waves tossed logs and debris onto beaches, across highways, and into nearby buildings. At one point, the water level dropped extensively, exposing vast areas of the upper shelf before cascading back. Although property damage in Oregon was light, four children sleeping at Beverly Beach State Park were drowned. The town of Seaside was badly damaged, and the bridge at Cannon Beach was displaced 1,000 feet inland.

Submarine landslides could also generate tsunamis. Brian McAdoo of Vassar College and Phil Watts, a private consultant, reported in 2004 that cohesive sediments, which built high escarpments offshore from Siletz Bay, pose a significant potential for initiating strong tsunamis in the event of slope failures.

Coastal Processes

The first explorers to the Oregon coast encountered hidden shoals and offshore rocks, shifting bars, fog, storms, and treacherous currents, and as late as 1951 the U.S. Coast Pilot warned of "dangers too numerous for description." Today these elements still play a dominant role in shaping what Paul

Komar characterizes as one of the highest wave-energy climates in the world.

Inventories of ocean processes and hazards compiled in the 1970s by Jim Stembridge, a resource specialist, have been updated and augmented by Paul Komar's ongoing research. Jonathan Allan and Robert Witter at DOGAMI and Peter Ruggiero at Oregon State University are currently examining the impact of climate changes on Pacific Northwest shoreline erosion and flooding.

Paul Komar's first introduction to coastal processes came during visits to Lake Michigan, where he saw homes, that had been undercut by erosion, slide into the surf. At Scripps Institute of Oceanography, he completed a PhD on the longshore transport of sand by currents and waves. After joining the Oceanography Department at Oregon State University in 1970, Komar concentrated on western coastal geomorphology and evolution, although his interests expanded to other countries. His 1997 book *The Northwest Coast* provides an invaluable and readable account of coastal Oregon and Washington. Maintaining an office at OSU since retiring in 1998, Komar is currently working on global climates and sea level changes. (Photo courtesy P. Komar)

Reducing the beaches, terraces, and headlands

Waves, prevailing winds, tides, climate, currents, and sea levels are all working to modify the coastal morphology. While the sandy areas are being rapidly reduced, the level terraces and protruding bluffs are also being steadily diminished. Erosive processes are tied to cycles of a turbulent wintertime ocean that contrasts with moderate summers conditions.

A forceful November surf, on reaching the beach, strips away sand and distributes it offshore in submerged bars, restoring it during the summer months. Overall, the amount of sand moved north

During the 1982 to 1983 El Niño, Alsea Spit was reduced dramatically when the longshore movement of sand shifted the channel leading into the bay northward (top, left). This action continued for three years, bringing losses to houses on top of the spit (bottom, left). On-going erosion has similarly threatened housing on Siletz Spit (right), cutting away at its width. Rip-rap will protect the houses only until the next onset of high storm waves. (Photos courtesy Oregon Department of Geology and Mineral Industries)

Tillamook Spit was the site of the city of Bayocean, advertised as "The Queen of Oregon Resorts." Houses, streets, a natatorium, and the beginnings of a hotel, grocery, and bowling alley were erected in 1912, but erosion opened three gaps in the peninsula bringing an end to hopes for a stable settlement. The last of the residents left in 1952. Several years later the Army Corps of Engineers diked the breach to protect eroding farmland and constructed the south jetty in 1974. In the photograph, Bayocean peninsula and the jetty are in the center, and Cape Meares is in the lower left corner. (Photo courtesy Oregon Department of Geology and Mineral Industries)

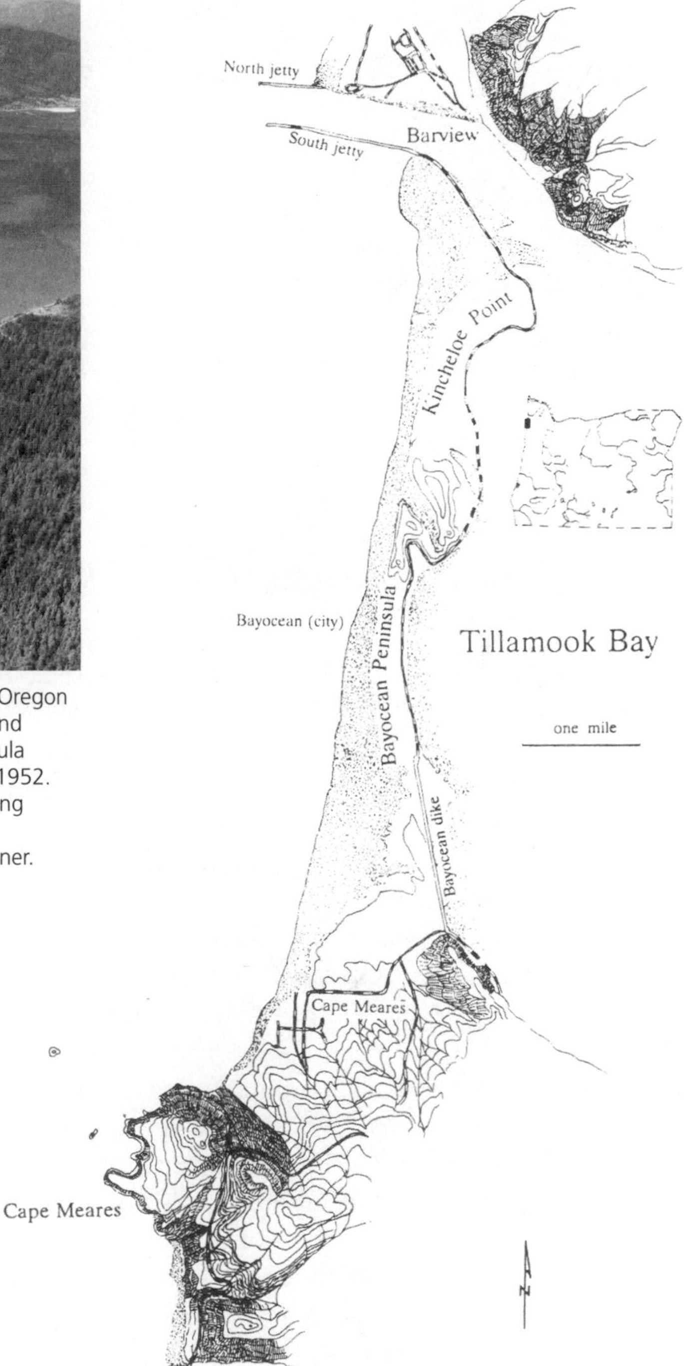

is about the same as that carried south, and the zero net drift is due to projecting volcanic headlands that trap and prevent it from being removed. Isolated between the rock barriers, each of the pocket beaches contains its own stock of sand within what Komar terms beach cells.

Except where sand is confined by headlands, the waves and currents shape it into elongate north-south peninsulas called spits, which parallel the general orientation of the beach. Characterized as "Here today, gone tomorrow," the transitory nature of sand spits makes them particularly vulnerable to erosion, while their scenic location makes them attractive to builders. Situated between the bay and ocean, the fragile environment of sand spits is easily unbalanced when the natural pattern of waves and currents is altered by jetty construction, by the placement of rip-rap, or by roadways.

The location of a mile-long jetty on the north side of Tillamook Bay by the Army Corps of Engineers in 1917 is one of the most notorious examples of human activity initiating sand spit erosion. Constricting the natural channel, the jetty caused sand to accumulate and gaps to appear through the south end of Tillamook Spit. The community of Bayocean, placed at a midway point, was doomed and even the village at Cape Meares was threatened. With completion of the jetty, the beach disappeared at the rate of six feet a year from the 1930s to 1940s.

Cliffs seem comparatively durable, but, in reality, they are subject to degradation by rock falls,

Headlands such as the 1,000-foot-high basalt monolith at Cape Meares (above left) appear deceptively durable, but, as with other coastal promontories, it is being steadily diminished by rockfalls. Boulders line the toe of a landslide that slumped off the face. A spectacular slump on Cascade Head (above right) carried over 20 acres of rock and terrace debris down to the ocean, as waves dissect the promontory itself. (Photos courtesy Oregon State Highway Department and Oregon Department of Geology and Mineral Industries)

landslides, and undercutting. Waves, attacking and removing the lower strata of bluffs and terraces, expose them to over-steepening and slumping. The composition and structure of the rocks, the groundwater, rainfall, and waves all play roles in the differing rates of erosion.

Because of their level surface, marine terraces are especially favorable spots for the placement of highways, parks, communities, and homes, but a variety of factors such as wave run-up, a small beach, or high sea levels lead to erosion and retreat. Beverly Beach in Lincoln County is a particularly visible case in which the cliff is open to the surf. Above the terrace, Highway 101 is not protected by the narrow strip of sand at the base, and the road had to be moved inland several times when the coastline retrograded. Between 1940 and 1980, as much as 50 feet of the cliff was lost, and since then another 30 feet has been cut away.

Paul Komar relates the measures taken when a 39-home condominium complex was to be situated at Jump-Off Joe. Approved by the Newport Planning Department in 1964 and substantiated by a geotechnical report from the firm Shannon and Wilson, stabilization was to be accomplished by the installation of a drainfield and seawall and by decreasing the angle of the slope. Even after additional problems came to light, construction began in 1982. Slumping and sliding occurred almost immediately, many of the units were never completed or sold, and the builders filed for bankruptcy.

Based on DOGAMI landslide maps and reports, which show much of the coast as hazard-prone, Newport officials proposed code changes to assure that buildings can be moved and that deeds disclose specific risks. Opposition to the new code came from builders and property owners fearing loss of sales, reduced property values, lawsuits, and increased construction costs.

Landslides

Earthflows are common in sedimentary rocks of the coastal mountains, while debris flows are more frequent along steep stream gradients. Large earthflows are deep-seated and occur when blocks, covering several acres, move downslope. By contrast, debris flows are water-saturated, rapid, shallow, and generally activated by heavy rains to move as a muddy slurry.

North of Florence, the soft clay-rich Astoria, Nye, and Nestucca formations are responsible for many historic and present-day earthflows. The Astoria Formation is particularly susceptible, and, after heavy March rains in 2007, deep fissures appeared,

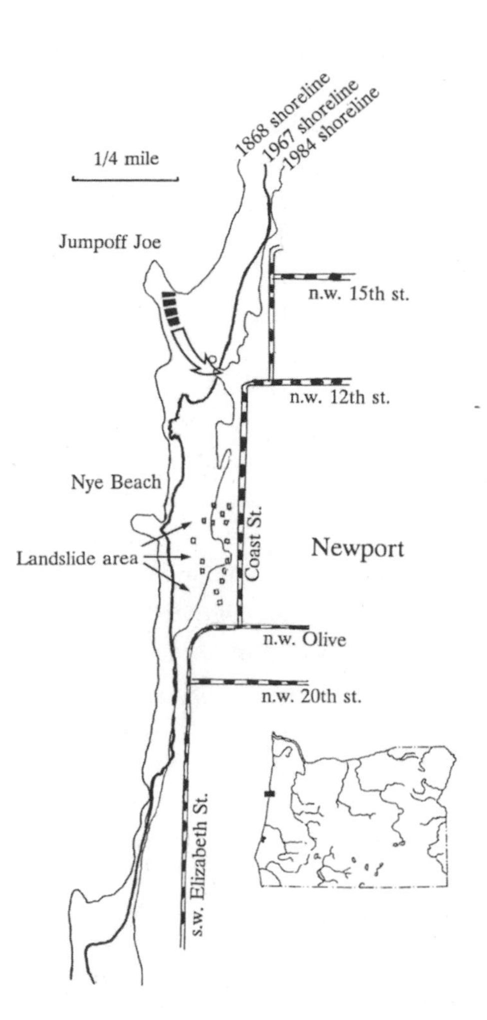

Newport has seen a remarkable combination of slumping, storm wave erosion, and landsliding that demolished the peninsula at Jump-Off Joe (above right) and moved the strand at Nye Beach inland as much as seven feet a year. The promontory and adjacent cliff are composed of poorly consolidated, seaward-dipping Astoria Formation and Nye Mudstones, topped by a terrace. Adjacent to Jump-Off Joe on 14th Street (left), an active slump block undercut housing in 1982. (The photos of Jump-Off Joe were taken in 1880, 1915, and 1926; courtesy Oregon Department of Geology and Mineral Industries)

A 1982 evaluation by John Gentile, then at Oregon State University, pinpointed 153 prehistoric and present-day slides within a one-and-one-half-mile-wide strip between Roads End and Yachats in Lincoln County. The most susceptible unit was the clay-rich Nye Mudstone underlying marine terraces. Erosion of the Nestucca, Astoria, and Nye formations is particularly evident in this county, which is additionally subsiding relative to sea level. (Photo at Roads End; courtesy J. Orr)

spreading mud and rock across Astoria city streets. Just a few years earlier a slide of spectacular proportions threatened the The Capes housing development near Netarts Bay. In this case, a Holocene paleovalley fill of the Astoria Formation was reactivated by El Nino storms. Ancient and recent slides at Ecola State Park and Silver Point near Cannon Beach sent a mass of Astoria debris downslope and into the sea at both localities.

Clatsop, Columbia, and Tillamook counties experience winter landslides on a regular basis, but no one spot is more notorious than the Wilson River Highway. Here rainfall averages 100 inches a year, and catastrophic slope failures in steep exposures of Eocene Tillamook basalts interlayered with sedimentary rocks repeatedly cover the highway and even dam the river. In 1991, after 500,000 cubic yards of rock and soil cascaded downslope, the road was closed for nearly two months. This was one of the largest earthflows in recent history.

Debris flows in rugged mountainous areas involve a similar set of characteristics. Where soft, loosely consolidated strata becomes water-saturated, the hillside moves more or less continuously in waves as a viscous mass. Hummocky surfaces, tilted trees, small ponds, and swampy depressions

Extensive clear-cuts and roadbuilding on Neahkahnie Mountain adjacent to Oswald West State Park (summit trees) expose the steep slopes to landslides (Photo courtesy Oregon Department of Transportation)

are all signs of such activity. Major storms in 1996 activated over 350 debris flows in tributaries along the Umpqua River in Douglas County, and the Clatskanie region in Columbia County experienced a similar devastation in December, 2008. Long identified on DOGAMI maps and known to forestry officials as posing extremely hazardous conditions, the highway near Clatskanie was overwhelmed with rubble, mud, and debris that moved downhill from Oregon State University clear-cuts after heavy rains. The water was dammed into a lake that broke loose to bury homes and roadways. Although there was over $12 million in damage, no one was injured. A similar slide here in 1933 killed four people.

While precipitous topography and a high rate of precipitation were the main contributors to Pleistocene slope movements in the southern Coast Range, University of Oregon professor Joshua Roering concluded that other factors may have played a role. By mapping the geometric curvature, gradient, and distribution of large landslides active between 18,000 to 40,000 years ago, he found that they were deep-seated (thick), pervasive, and long-lived. Most occurred in the failure-prone soft siltstone and sandstone layers of the Tyee Formation, and he estimates that there are hundreds if not thousands of large paleo-landslides and lakes with similar origins in this region.

Loosened Tyee debris was responsible for the landslide-created Triangle Lake near Blachly, for Loon Lake near Scottsburg, and for the lake basin at Sitkum in Lane County. Dammed during the Pleistocene, most are only remnants of what were formerly much larger bodies of water, which have since filled in with alluvium. In December, 1975, Drift Creek in the Alsea River watershed was impounded by a thick mass of Tyee material that covered 40 acres. This was one of the largest movements in the state, involving reactivation of an ancient slide and creating a dam that that impounded Ayers Lake. Road building and logging of the slope preceded the failure.

Flooding

In a province bordered by the ocean, with widespread wetlands, estuaries, tributary streams, and heavy precipitation, flooding is an annual phenomenon, most frequent during the winter months. Many factors such as topography, logging, and obstruction of the channel can cause rivers to rise. During wet seasons runoff is rapid, and peak flows are short. Records since the mid-1800s, listed in the invaluable Oregon weather book by George Taylor of Oregon State University and Raymond Hatton from Central Oregon Community College, show flooding and peak discharges of coastal rivers almost always follow rain on snowpack.

Ocean flooding is brought on by storm conditions that result in tidal surges and fluctuating currents. When ocean water piles up against the beach and backs into estuaries, low-lying areas are inundated. If heavy runoff and a high surf collide, the results may be calamitous.

Natural Resources

Coal

Although coal seams are widely scattered, there have been only two productive regions, one in the Newport-Beaver Hill basin and the other at Eden Ridge, both in Coos County. Mining in the Coaledo Formation at Newport-Beaver Hill near Coos Bay began in 1855, whereas Tyee sediments at Eden Ridge south of Powers were not examined extensively until the 1950s. Coal from this field in the rugged southeastern corner of the county has a greater heat value but extraction was more difficult. Most of Oregon's coal is high in ash and ranked as sub-bituminous, making commercial exploitation unlikely. At Coos Bay, production reached its peak year in 1904, and in 1985 the potentially dangerous shafts and portals were sealed and covered as a safety measure, closing the Oregon coal mining era.

Oil and Gas

The great thickness of Cenozoic sediments in the Coast Range have been viewed as potential sources of hydrocarbons since the first holes were drilled, but the results were disappointing. Gas shows, but no oil, were encountered in prospects near Bandon as early as 1913, and exploration in the offshore over the next 40 years was equally unsatisfactory.

Discovery of the state's first commercial gas field in 1979 near Mist in Columbia County is credited to Wes Bruer, consulting geologist for Reichold Energy Corporation. A 1950 graduate of Oregon State

Oregon's coal mining era was relatively brief, and all production took place in Coos County. The upper photo shows the remaining wooden pillars at Empire, where the rail line ran coal out from the Libby Mine in the Newport-Beaver Hill district to be loaded on ships. Few traces are left of the mine entrance in this district, seen in the 1905 photograph (right). (Photos courtesy Oregon Department of Geology and Mineral Industries)

University, Bruer spent many years investigating Oregon's oil and gas potential. To date, 18 wells in the Mist field have produced 65 billion cubic feet of gas. Most of the reservoirs are fault traps within a larger anticline, where hydrocarbons are present in sandstones of the upper Eocene Cowlitz Formation. Today the depleted Mist field is being used to store imported gas.

Likely porous hydrocarbon reservoir sands are those deposited in high energy upper upper shelf environments. Fine-grained sedimentary rocks of the Cowlitz, Yamhill, Coaledo, Spencer, Eugene, Yaquina, and Astoria formations meet these conditions to some extent, but the presence of clays and zeolite minerals plugs the sandstones. In February, 2010, the Oregon legislature passed a bill banning oil and gas drilling within the three miles off the coast.

About 60 miles offshore from Newport, seismic profiles taken by Ann Tréhu at Oregon State University in 1989 show evidence of widespread gas hydrates along the margin of the accretionary wedge. Gas hydrates occur when cold methane gas is locked into a lattice of water in an ice-like mixture, and Hydrate Ridge is especially rich in the hydrocarbons. As with the hydrothermal vents along the Juan de Fuca ridge, there are similar colonies of worms and clams living off the hydrogen sulfide in a symbiosis with bacteria.

Gold

An assortment of economic minerals occur in black sands from Cape Blanco to Cape Arago. Concentrated at the mouths to rivers and on elevated terraces, the sands were conveyed from high in the Coast Range by Pleistocene streams. Sorted by the winnowing action of the waves, the heavier gold, platinum, chromite, magnetite, garnet, and zircon grains are deposited along the beach, whereas the lighter quartz, feldspar, and mica are carried out onto the continental shelf. As indicated by the black color of the sands, the Jurassic Galice Formation, intruded by small grantitic dikes, was the host rock.

The discovery of gold in 1852 north of the Coquille River sparked the settlement of Randolph, a boom town of tents, stores, and saloons. The placers at Whiskey Run failed to pay, and mining was abandoned after a storm destroyed most of the operations two years later. Miners attempted to tunnel into black sand lenses of the older terraces, but the 50-to-60-foot covering of loose material above the layers and the extremely fine-grain of the minerals made recovery difficult. North of there, deposits of gold, platinum, and chromite were accessed by tunnels into the Pioneer terrace, but the Seven Devils or Last Chance Mine saw only brief activity before closing after World War II.

Production amounts from beach placers in Coos and Curry counties were never systematically reported, but $60,000 in gold and platinum—around $2 million in today's dollars—is estimated to have been mined between 1903 and 1929.

In 1988, federal and state agencies examined the Oregon shelf for economic deposits of metalliferous black sands, but only limited amounts were found.

Surface and Groundwater

Surface and groundwater resources in the Coast Range are marginal, although the province experiences greater precipitation and stream runoff than elsewhere in Oregon and even more than in most areas of the United States. The streams that drain the western slopes have modest watersheds and flows diminish greatly during the late summer when demand is highest. Surface amounts are augmented by Pacific storms and groundwater springs originating from Tertiary lavas.

Groundwater percolates through gravel and sand along river floodplains, in marine terraces, and in dune aquifers, but the quantity is low and generally only sufficient for domestic needs. The underlying Tertiary sediments have equally low yields, and wells drilled into these layers often produce saline water that has a bad taste or odor. Elevated marine terraces provide a better supply, but, as with dunes or where Tertiary rocks are encountered, the quality may be poor, high in salt and mineral content, or polluted by septic systems.

Historically, the diversion of water for industry, municipalities, and agriculture has come through pumping from streams or lakes, but increasing demands and a decrease in levels have forced users to try other means. Consequently, dunal aquifers have been targeted. Rumored to contain unlimited quantities of water, dune reservoirs are recharged directly by precipitation and discharged through springs or into small lakes. In actuality, over-pumping can pull deep brackish water up into the aquifer. Wells have only a brief life span of five-to-seven-years and frequently must be abandoned. Resorting to dunal aquifers in 2000, Coos County completed a 20-well system that draws water from beneath the Oregon Dunes National Recreation Area. Other coastal communities suffer similar problems.

Geologic Highlights

Long famous for its scenery, the Oregon coast provides a combination of pleasing beaches, dramatic headlands, sea stacks, and high terraces interspersed with quiet coves and immense sand fields. For travelers along the coast, vistas appear around each curve to rival those in the Cascades. Flat marine terraces dominate from Cape Blanco to Coos Bay.

The harbor and spouting horn at Depoe Bay are among the most visited attractions on the coast. Spherical basalt pillows are common in the Columbia River lavas, which compose the rocky landscape, including the wall that protects the picturesque inner harbor. (Photos courtesy Oregon Department of Geology and Mineral Industries and Oregon State Highway Department)

Ever-changing sand dunes, secluded lakes, and dense forests come together at Cleawox Lake (foreground), Woahink Lake (upper center), and Siltcoos Lake (upper right). (Photo courtesy Oregon State Highway Department)

Towering dunes stretch from Coos Bay to Florence, and north of there volcanic cliffs are home to marine birds and sea mammals. From Tillamook Bay to the Columbia River, the rough coastline is interspersed with arcuate coves and sandspits, which end at Clatsop Plains.

As James Stovall frequently remarked to his geology classes at the University of Oregon, the Devil nominally owns much of the coastal real estate. Devils Elbow, Devils Punchbowl, and Devils Lake are among some of the more scenic spots. Many of the imaginative names were given because of unusual rock configurations or other physical features.

Caves and Trenches

Rocks, cut by fractures and faults and exposed to pounding surf, have been eroded into caves, long straight trenches, or tunnels. If the roof collapses, as happened at Depoe Bay and the Devils Punchbowl, a spout, churn, or punchbowl may result. During high tide and on windy days, waves force their way through an opening cut into fractured and faulted Columbia River basalts at Depoe Bay to erupt upward, sending a spray of water high into the air. At Devils Punchbowl State Park, two adjacent sea caves, eroded into the Astoria Formation, collapsed to form a pit. During high tides or stormy conditions, sea water foams and froths as in a boiling pot.

Along the south central coast, Devils Churn, Cooks Chasm, Devils Elbow, and Sea Lion Caves are deep erosional cuts into Yachats Basalt where the roof has fallen in along a trench. The tunnel at Devils Elbow is 600 feet long and home to a noisy and smelly population of marine mammals. At Sea Lion Caves, the openings at the intersection of a system of fractures are visible in the ceiling. The southern part of the tunnel is below sea level, while the public viewing area is in the northern section.

Sand dunes and freshwater lakes

Impressive lengthy sand dunes cover almost half of the Oregon shoreline and about one-third of Washington's. Of these, the 31,566-acre Oregon Dunes National Recreation Area from Coos Bay to Florence is the longest at 50 miles. The field averages two miles in width, and many of the dunes reach 180 feet in height.

Within the recreational area, a chain of freshwater dunal lakes are the youngest and perhaps most ephemeral of the features. These bodies of water were impounded when streams were blocked by sand advancing inland or by sediments choking the valleys. At 3,164 acres, Siltcoos Lake is similar in size to nearby Woahink, Tahkenitch, and Tenmile, making them among Oregon's largest within the dunes. Devils Lake, at 678 acres, is the only freshwater lake of any size on the central coast. Even though they encompass less than 10 acres, Clear Lake near Roseburg and the similarly named Clear Lake north of Florence are the two deepest reaching 119 feet and 80 feet respectively. Dunal lakes vary considerably in elevation. The surface of Siltcoos Lake is five feet above and that of Woahink 38 feet below sea level. The lakes are fed by small streams, rainfall, and groundwater, and the interface between fresh and salt water in the subsurface gives them an exceedingly fragile environment. Particularly susceptible to pollution, they have been deleteriously impacted by all-terrain vehicles, logging, septic fields, housing, and pumping for municipal and industrial water supplies.

From Tillamook Head to the Columbia River, Clatsop Plains is a strip of dunes about a mile wide and 17 miles long, dropping from 100 feet in elevation to sea level. The Plains accumulated during the Holocene, when sand built up along the marine shoreline, which lay east of where it is now. As in-filling continued, the sand extended the coast westward. Dunal ridges run parallel to the shore for miles, marking the position of the beachfront from the transgressing sea. Within the elongate depressions, lakes are supplied by groundwater, and most are independent of local streams.

Coastal Mountains

A familiar shape on the skyline east of Seaside, the high twin peaks of Saddle Mountain were first noted and named by John Wilkes of the U.S. Exploring Expedition in 1841. Composed of Columbia River

Probably the most extreme example of inverted topography in Oregon, Saddle Mountain exhibits vertical dikes hundreds of feet long, pillows from underwater volcanic eruptions, and breccias. (Photo courtesy K. Sayce; flown by pilot Frank Wolfe)

The 4,097-foot-high Marys Peak is the highest in the Coast Range. Capped by a resistant 1,000-foot-thick sill of medium-grained gabbro that intruded Eocene sandstones of the Tyee Formation, the mountain is cut by the Kings Valley fault. The magnificent view from Marys Peak encompasses the Willamette Valley and Cascade volcanoes to the east and the intruded sills of Flat Mountain and Green Peak to the south. Westward, the Coast Range and Pacific Ocean stretch to the horizon. (Photo courtesy U.S. Forest Service)

basalts, the mountain and surrounding forests are now part of Saddle Mountain State Park. A climb to the top provides scenes of the Cascades, Mount St. Helens, the Columbia River, and the city of Astoria on the Pacific Ocean.

References

Aalto, K.R., 1988. Sandstone petrology and tectonostratigraphic terranes of the northwest California and southwest Oregon coast ranges. Journal Sedimentary Petrology, v.59, no.4, pp.561-571.

———, 2006. The Klamath peneplain: a review of J.S. Diller's classic erosion surface. *In*: Snoke, A.W., and Barnes, C.G., eds., Geological studies in the Klamath Mountains province, California and Oregon. Geological Society America, Special Paper 410, pp.451-463.

Adams, J., 1984. Active deformation of the Pacific Northwest continental margin. Tectonics, v.3, no.4, pp.449-472.

———, 1990. Paleoseismicity of the Cascadia subduction zone; evidence from turbidites off the Oregon-Washington margin. Tectonics, v.9, no.4, pp.569-583.

———, 1996. Great earthquakes recorded by turbidites off the Oregon-Washington coast. U.S. Geological Survey, Professional Paper 1560, pp.147-158. Also: Tectonics, 1990, v.9, no.4, pp.569-583.

Allen, C.M., and Barnes, C.G., 2006. Ages and some cryptic sources of Mesozoic plutonic rocks in the Klamath Mountains, California and Oregon. *In*: Snoke, A.W., and Barnes, C.G., eds., Geological studies in the Klamath Mountains province, California and Oregon. Geological Society America, Special Paper 410, pp.223-245.

Allen, J.C., et al., 2009. Coastal geomorphology, hazards, and management issues along the Pacific Northwest coast of Oregon and Washington. *In*: O'Connor, J.E., Dorsey, R.J., and Madin, I.P., eds., Volcanoes to vineyards; field trips through the . . . Pacific Northwest. Geological Society America, Field Guide 15, pp.495-519.

Allen, J. E., 1932. Contributions to the structure, stratigraphy, and petrography of the lower Columbia River Gorge. Masters, University of Oregon, 96p.

———, 1946. Reconnaissance geology of limestone deposits in the Willamette Valley, Oregon. Oregon Department Geology and Mineral Industries, GMI Short Paper 15, 15p.

———, 1947. Bibliography of the geology and mineral resources of Oregon (supplement), July 1, 1936 to December 31, 1945. Oregon Department Geology and Mineral Industries, Bull.33, 108p.

———, 1966. The Cascade Range volcano-tectonic depression of Oregon. Oregon Department of Geology and Mineral Industries. Lunar Geological Field Conference, Trans., pp.21-23.

———, 1975. Volcanoes of the Portland area, Oregon. Ore Bin, v.37, no.9, pp.145-157.

———, 1979. The magnificent gateway; a layman's guide to the geology of the Columbia River gorge. Portland, Timber Press, 144p.

———, and Baldwin, E.M., 2944. Geology and coal resources of the Coos Bay Quadrangle, Oregon. Oregon Department Geology and Mineral Industries, Bull.27, 153p.

———, and Beaulieu, J.D., 1976. Plate tectonic structures in Oregon. Ore Bin, v.38, no.6, pp.87-99.

———, Burns, M., and Burns, S., 2009. Cataclysms on the Columbia. Rev. second ed., Portland, Oregon, Ooligan Press, 204p.

Allen, R., et al., 2009. Structure and seismicity along Cascadia. Geological Society America, Abstracts with Program., v.41, no.7, p.64.

Allison, I. S., 1935. Glacial erratics in Willamette Valley. Geological Society America, Bull., v.46, pp.615-632.

———, 1953. Geology of the Albany Quadrangle, Oregon. Oregon Department of Geology and Mineral Industries, Bull.37, 18p.

———, 1966. Pumice beds at Summer Lake, Oregon –A correlation. Geological Society America, Bull., v.77, pp.239-330.

———, 1966a. Fossil Lake, Oregon, its geology and fossil faunas. Oregon State University Monographs, Studies in Geology, 9, 48p.

———, 1978. Late Pleistocene sediments and floods in the Willamette Valley. Ore Bin, v.40, no.11, pp.177-191; pt.2, p.193-202.

———, 1979. Pluvial Fort Rock Lake, Lake County, Oregon. Oregon Department Geology and Mineral Industries, Special Paper 7, 72p.

———, 1982. Geology of pluvial Lake Chewaucan, Lake County, Oregon. Oregon State University, Studies in Geology, 11, 78p.

Anderson, D.A., DeRoo, T.G., and Hedeen, C.D., 2006. Response to 1998 debris flow in the upper White River valley, Oregon. Oregon Geology, v.67, no.1, pp.7-10.

Anderson, D.L., and Schramm, K.A., 2005. Global hotspot maps. *In*: Foulger, et al., eds., Plates, plumes, and paradigms. Geological Society America, Special Paper 388, pp.19-29.

Anderson, J.L., ed., 1990. The nature and origin of Cordilleran magmatism. Geological Society America, Memoir 174, 414p.

———, and Vogt, B.F., 1987. Intracanyon flows of the Columbia River Basalt Group and adjacent Cascade Range, Oregon and Washington. Washington Division of Geology and Earth Resources, Bull.77, pp.249-267.

———, et al., 1987. Distribution maps of stratigraphic units of the Columbia River basalt group. Washington Division of Geology and Earth Resources, Bull.77, pp.183-195.

Ando, M., and Balazs, F.I., 1979. Geodetic evidence for aseismic subduction of the Juan de Fuca plate. Journal Geophysical Research, v.84, pp.3023-3028.

Armentrout, J.M., 1967. The Tarheel and Empire formations . . . Coos Bay Oregon. Masters, University of Oregon, Eugene, 155p.

———, and Suek, D.H., 1985. Hydrocarbon exploration in western Oregon and Washington. American Association Petroleum Geology, Bull.69, no.4, pp.627-643.

———, Cole, M. R., and TerBest, H., eds., 1979. Cenozoic paleogeography of the Western United States. Society Economic Paleontologists and Mineralogists, Pacific Coast Paleography Symposium 3, 335p.

———, ed., 1981. Pacific Northwest Cenozoic biostratigraphy. Geological Society America, Special Paper 184, pp.137-148.

———, et al., 1983. Correlation of Cenozoic stratigraphic units of western Oregon and Washington. Oregon Department Geology and Mineral Industries, Oil and Gas Investigation 7, 90p.

Arnold, R., and Hannibal, H., 1913. The marine Tertiary stratigraphy of the north coast of America. American Philosophical Society, Proceedings, v.52, no.212, pp.559-605.

Ashley, R.P., 1995. Petrology and deformation history of the Burnt River Schist . . . *In*: Vallier, T.L., and Brooks, H.C., eds., 1995. Geology of the Blue Mountains region of Oregon, Idaho, and Washington. U.S. Geological Survey, Professional Paper 1438, pp.457-495.

Ashwill, M. S., 1979. An exposure of limestone at Gray Butte, Jefferson County, Oregon. Oregon Geology, v.41, no.7, pp.107-109.

———, 1987. Paleontology in Oregon; workers of the past. Oregon Geology, v.49, no.12, pp.147-153.

Atwater, B. F., 1987. Evidence for great Holocene earthquakes along the outer coast of Washington state. Science, v.236, pp.942-944.

———, et al., 1995. Summary of coastal geologic evidence for past great earthquakes at the Cascadia subduction zone. Earthquake Spectra, v.11, no.1, pp.1-18.

Atwater, T., 1970. Implications of plate tectonics for the Cenozoic tectonic evolution of Western North America. Geological Society America, Bull., v.81, pp.3518-3536.

———, 1989. Plate tectonic history of the northeast Pacific and western North America. Geological Society America, The Geology of North America, v.N, The eastern Pacific and Hawaii, pp.21-72.

Ave Lallemant, H. G., 1976. Structure of the Canyon Mountain (Oregon) ophiolite complex and its implication for sea-floor spreading. Geological Society America, Memoir 173, pp.1-49.

———, 1995. Pre-Cretaceous tectonic evolution of the Blue Mountains Province, northeastern Oregon. *In*: Vallier, T.L., and Brooks, H.C., eds., 1995. Geology of the Blue Mountains region of Oregon, Idaho, and Washington. U.S. Geological Survey, Professional Paper 1438, pp. 271-304.

———, Schmidt, W.J., and Kraft, J.L., 1985. Major late-Triassic strike-slip displacement in the Seven Devils terrane, Oregon and Idaho: a result of left-oblique plate convergence. Tectonophysics 119, pp.299-328.

Avolio, G.W., 1975. Investigation of the Six Soldier slide. *In*: Engineering geology and soils engineering symposium, Proceedings, 13th Annual Symposium, pp.1-12.

Avramenko, W., 1981. Volcanism and structure in the vicinity of Echo Mountain, central Oregon Cascade Range. Masters, University of Oregon, Eugene, 156p.

Bacon, C.R., and Nathenson, M., 1996. Geothermal resources in the Crater Lake area, Oregon. U.S. Geological Survey, Open-File Report 96-663, 34p.

———, and Lanphere, M.A., 2006. Eruptive history and geochronology of Mount Mazama and the Crater Lake region. Oregon. Geological Society America, Bull.118, no.11/12, pp.1331-1359.

———, et al., 2002. Morphology, volcanism, and mass wasting in Crater Lake, Oregon. Geological Society America, Bull., v.114, no.6, pp.675-692.

Badger, T.C., and Watters, R.J., 2009. Landslides along the Winter Rim fault, Summer Lake, Oregon. *In*: O'Connor, J.E., Dorsey, R.J., and Madin, I.P., eds., Volcanoes to vineyards; field trips through the . . . Pacific Northwest. Geological Society America, Field Guide 15, pp.203-220.

Bailey, D.G., 1989. Calc-alkalic volcanism associated with crustal extension in northeastern Oregon. New Mexico Bureau of Mines and Mineral Resources, Bulletin 131, p.12.

Bailey, M.M., 1989. Evidence for magma recharge and assimilation in the Picture Gorge Basalt subgroup, Columbia River Basalt group. *In*: Reidel, S.P., and Hooper, P.R., eds., 1989. Volcanism and tectonism in the Columbia flood- basalt province. Geological Society America, Special Paper 239, pp.332-355.

———, 1989a. Revisions to stratigraphic nomenclature of the Picture Gorge Basalt subgroup, Columbia River basalt group. *In:* Reidel, S.P., and Hooper, P.R., eds., Volcanism and tectonism in the Columbia flood-basalt province. Geological Society America, Special Paper 239, pp.67-84.

Baker, V.R., and Bunker, R., 1985. Cataclysmic late Pleistocene flooding from Glacial Lake Missoula: a review. Quaternary Science Reviews, v.4, pp.1-41.

Baksi, A.K., 1989. Reevaluation of the timing and duration of extrusion of the Imnaha, Picture Gorge, and Grande Ronde basalts, Columbia River Basalt group. *In*: Reidel, S.P., and Hooper, P.R., eds., Volcanism and tectonism in the Columbia River flood-basalt province. Geological Society America, Special Paper 239, pp.105-110.

———, 2010. Distribution and geochronology of the Oregon plateau (U.S.A.) flood basalt volcanism; the Steens revisited. Journal Volcanology and Geothermal Research, v.196, no.1-2, pp.134-138.

Baldwin, E.M., 1945. Some revisions of the late Cenozoic stratigraphy of the southern Oregon Coast. Journal Geology, v.52, pp.35-46.

———, 1957. Drainage changes of the Willamette River at Oregon City and Oswego, Oregon. Northwest Science, v.31, no.3, pp.109-117.

———, 1964. Geology of the Dallas and Valsetz quadrangles, Oregon. Oregon Department Geology and Mineral Industries, Bull.35, 52p.

———, 1969. Thrust faulting along the lower Rogue River, Klamath Mountains, Oregon. Geological Society America, Bull., v.80, pp.2047-2052.

———, 1974. Eocene stratigraphy of southwestern Oregon. Oregon Department Geology and Mineral Industries, Bull.83, 40p.

———, 1976. Geology of Oregon. Third ed., Dubuque, Kendall/Hunt Publ., 170p.

———, 1980. The gold dredge at Whisky Run north of the mouth of the Coquille River, Oregon. Oregon Geology, v.42, no.8, pp.145-146.

———, and Howell, P.W., 1949. The Long Tom, a former tributary of the Siuslaw River. Northwest Science, v.23, p.112-124.

———, et al., 1973. Geology and mineral resources of Coos County, Oregon. Oregon Department Geology and Mineral Industries, Bull.80, 82p.

Balster, C.A., and Parsons, R.B., 1969. Late Pleistocene stratigraphy, southern Willamette Valley, Oregon. Northwest Science, v.43, no.3, pp.116-129.

Barnes, C.G., et al., 1995. The Grayback pluton; magmatism in a Jurassic back-arc environment, Klamath Mountains,Oregon. Journal Petrology, v.36, pp.397-415.

———, et al., 2006. Petrology and geochemistry of the middle Jurassic Ironside Mountain batholith. *In*: Snoke, A.W., and Barnes, C.G., eds., Geological Studies in the Klamath Mountains province, California and Oregon. Geological Society American, Special Paper 410, pp.199-221.

Barrash, W., and Venkatakrishnan, R., 1982. Timing of late Cenozoic volcanic and tectonic events along the western margin of the North American plate. Geological Society America, Bull., v.93, pp.977-989.

———, et al., 1980. Geology of the La Grande area, Oregon. Oregon Department Geology and Mineral Industries, Special Paper 6, 47p.

———, et al, 1983. Structural evolution of the Columbia Plateau in Washington and Oregon. American Journal Science, v.283, pp.897-935.

Beaulieu, J.D., 1971. Geologic formations of western Oregon. Oregon Department Geology and Mineral Industries, Bull.70, 72p.

———, 1972. Geologic formations of the southern half of the Huntington quadrangle, Oregon Department of Geology and Mineral Industries, Bull.73, 80p.

———, 1973. Geologic field trips in northern Oregon and southern Washington. Oregon Department of Geology and Mineral Industries, Bull.77, 206p.

———, 1974. Geologic hazards inventory of the Oregon coastal zone. Oregon Department of Geology and Mineral Industries, Miscellaneous Paper 17, 94p.

———, 1976. Geologic hazards in Oregon. Ore Bin, v.38, no.5, pp.67-83.

———, 1977. Geologic hazards of parts of northern Hood River, Wasco, and Sherman counties, Oregon. Oregon Department of Geology and Mineral Industries, Bull. 91, 95p.

——— and Hughes, P.W., 1975. Environmental geology of western Coos and Douglas counties, Oregon. Oregon Department of Geology and Mineral Industries, Bull.87, 148p.

——— and Hughes, P.W., 1976. Land use geology of western Curry County, Oregon. Oregon Department Geology and Mineral Industries, Bull.90, 148p.

——— and Hughes, P. W., 1977. Land use geology of central Jackson County, Oregon. Oregon Department Geology and Mineral Industries, Bull.94, 83p.

——— and Olmstead, D.L., 1999. Mitigating geologic hazards in Oregon. Oregon Department Geology and Mineral Industries, Special Paper 31, 60p.

———, Hughes, P.W., and Mathiot, R.K.,1974. Environmental geology of western Linn County, Oregon. Oregon Department of Geology and Mineral Industries, Bull.84, 117p.

Beebee, R.A., O'Connor, J.E., and Grant, G.E., 2002. Geology and geomorphology of the lower Deschutes River Canyon. *In*: Moore, G., ed., Field guide to geologic processes in Cascadia. Oregon

Department of Geology and Mineral Industries, Special Paper 36, pp.91-109.

Beeson, M.H., and Moran, M.R., 1979. Columbia River Basalt group stratigraphy in western Oregon. Oregon Geology, v.41, no.1, pp.11-14.

——— and Tolan, T.L., 1985. Regional correlations within the Frenchman Springs Member of the Columbia River Basalt group: New insights into the middle Miocene tectonics of north-western Oregon. Oregon Geology, v.47, no.8, pp.87-96.

——— and Tolan, T.L., 1987. Columbia River gorge: The geologic evolution of the Columbia River in northwestern Oregon and southwestern Washington. *In*: Hill, M.L., ed., Decade of North American geology; Centennial field guide. Geological Society America, Cordilleran Section, pp.321-326.

——— and Tolan, T, L., 1990. The Columbia River Basalt Group in the Cascade Range: a middle Miocene reference datum for structural analysis. Journal Geophysical Research, v.95, no.B12, pp.19,547-19,559.

———, Perttu, R., and Perttu, J., 1979. The origin of the Miocene basalts of coastal Oregon and Washington: an alternative hypothesis. Oregon Geology, v.41 no.10, pp.159-166.

———, Tolan, T.L., and Anderson, J.L., 1989. The Columbia River Basalt Group in western Oregon; geologic structures and other factors that controlled flow emplacement patterns. *In*: Reidel, S.P., and Hooper, P.R., eds., Volcanism and tectonism in the Columbia River flood-basalt province. Geological Society America, Special Paper 239, pp.223-246.

———, Tolan, T. L., and Maden, I. P., 1989. Geologic map of the Lake Oswego Quadrangle, Clackamas, Multnomah, and Washington counties, Oregon. Oregon Department of Geology and Mineral Industries, Geologic Map Series, GMS-59.

———, Tolan, T.L., and Maden, I.P., 1991. Geologic map of the Portland Quadrangle, Multnomah and Washington counties, Oregon, and Clark County, Washington. Oregon Department of Geology and Mineral Industries, Geologic Map Series, GMS-75.

Bela, J.L., 1979. Geologic hazards of eastern Benton County, Oregon. Oregon Department Geology and Mineral Industries, Bull.98, 122p.

Benito, G., and O'Connor, J.E., 2003. Number and size of last-glacial Missoula floods in the Columbia River valley . . . Geological Society America, Bull., v.115, pp.624-638.

Berg, J.W., and Baker, C.D., 1963. Oregon earthquakes, 1841 through 1958. Seismological Society America, Bull., v.53, no.1, pp.95-108.

Berggren, W.A., and Van Couvering, J.A., 1974. The late Neogene biostratigraphy, geochronology, and paleoclimatology of the last 15 m.y. in marine and continental sequences. Paleogeography, Paleoclimatology, and Paleoecology, v.16, no.1 and 2.

Bergquist, J.R., et al., 1990. Mineral resources of the Badlands Wilderness study area . . . U.S. Geological Survey, Bull., pp.B1-B14.

Bestland, E. A., 1987. Volcanic stratigraphy of the Oligocene Colestin Formation in the Siskiyou Pass area of southern Oregon. Oregon Geology, v.49, no.7, pp.79-86.

Bezore, S.P., 1969. The Mount Saint Helena ultramafic-mafic complex of northern California Coast Ranges. Geological Society America, Abstracts with Programs, pt.3, p.5-6.

Bishop, E.M., 1989. Smith Rock and the Gray Butte complex. Oregon Geology, v.51, no.4, pp.75-80.

———, 1995. Mafic and ultramafic rocks of the Baker terrane, eastern Oregon, and their implications for terrane origin. *In*: Vallier, T.L., and Brooks, H.C., eds., 1995. Geology of the Blue Mountains region of Oregon, Idaho, and Washington. U.S. Geological Survey, Professional Paper 1438, pp.221-245.

——— and Smith, G.A.,1990. A field guide to the geology of Cove Palisades State Park and the Deschutes basin in central Oregon. Oregon Geology, v.52, no.1, pp.3-12, 16.

Black, G.L., 1996. Earthquake intensity maps for the March 25, 1993, Scotts Mills, Oregon, earthquake. Oregon Geology, v.58, no.2, pp.35-41.

———, et al., 2000. Relative earthquake hazard map of the Eugene-Springfield metropolitan area, Lane County, Oregon. Oregon Department of Geology and Mineral Industries, Interpretive Map Series, IMS-14.

Blackwell, D. D., et al., 1978. Heat flow of Oregon. Oregon Department Geology and Mineral Industries, Special Paper 4, 42p.

———, et al., 1982. Heat flow, arc volcanism, and subduction in northern Oregon. Journal Geophysical Research, v.87, no.B10, pp.8735-8754.

———, et al., 1990. Heat flow in the Oregon Cascade range and its correlation with regional gravity, Curie Point depths, and geology. Journal Geophysical Research, v.95, no. B12, pp.19,475-19,493.

Blake, M.C., et al., 1985. Tectonostratigraphic terranes in south-west Oregon. *In*: Howell, David G., ed., Tectonostratigraphic terranes of the circum-Pacific region. Earth Science Series no.1, Circum-Pacific Council for Energy and Mineral Resources, pp.147-157.

Blakely, R.J. 1994. Extent of partial melting beneath the Cascade Range, Oregon. Journal Geophysical Research, v.99, no.B2, pp.2757-2773.

——— and Jachens, R.C., 1990. Volcanism, isostatic residual gravity, and regional tectonic setting of the Cascade volcanic province. Journal Geophysical Research, v.95, no.B12, pp.19,439-19,451.

———, et al., 1985. Tectonic setting of the southern Cascade Range as interpreted from its magnetic and gravity fields. Geological Society America, Bull., v.96, pp.43-48.

———, et al., 1995. Tectonic setting of the Portland-Vancouver area, Oregon and Washington. Geological Society America, Bull., v.107, no.9, pp.1051-1062.

———, et al., 2000. New aeromagnetic data reveal large strike-slip (?) faults in the northern Willamette Valley, Oregon. Geological Society America, Bull., v.112, no.7, pp.1225-1233.

Blodgett, R.B., and Stanley, G.D., eds., 2008. The terrane puzzle: new perspectives on paleontology and stratigraphy from the North American Cordillera. Geological Society America, Special Paper 442, 326p.

Blome, C.D., and Irwin, W.P., 1983. Tectonic significance of late Paleozoic to Jurassic radiolarians from the North Fork terrane, Klamath Mountains, California. *In*: Stevens, C., ed., Prejurassic rocks in western North American suspect terranes. Society Economic Paleontology and Mineralogy, Pacific Section, pp.77-89.

———, and Nestell, M. K., 1991. Evolution of a Permo-Triassic sedimentary melange, Grindstone terrane, east-central Oregon. Geological Society America, Bull., v.103, pp.1280-1296.

———, and Nestell, M.K., 1992. Field guide to the geology and paleontology of pre-Tertiary volcanic arc and mélange rocks, Grindstone, Izee, and Baker terranes, east-central Oregon. Oregon Geology, v.54, no.6, pp.123-141.

———, et al., 1986. Geologic implications of radiolarian-bearing Paleozoic and Mesozoic rocks from the Blue Mountains province, eastern Oregon. *In*: Vallier, T.L., and Brooks, H.C., eds., Geology of the Blue Mountains region of Oregon, Idaho, and Washington. U.S. Geological Survey, Professional Paper 1435, pp.79-93.

Bloomer, S.H., et al., 1995. Early arc volcanism and the ophiolite problem. *In*: Taylor, B., and Natland, J., eds., Active margins and marginal basins of the western Pacific. American Geophysical Union, Geophysical Monograph 88, pp.1-30.

Bockheim, J.G., Kelsey, H.M., and Marshall, J.G., 1992. Soil development, relative dating, and correlation of late Quaternary marine terraces in southwestern Oregon. Quaternary Research, v.37, pp.60-74.

Boetius, A., and Suess, E., 2004. Hydrate Ridge; a natural laboratory . . . Chemical Geology, v.205, no.3-4, pp.291-310.

Bogen, N.L., 1986. Paleomagnetism of the Upper Jurassic Galice Formation, southwestern Oregon: evidence for differential rotation of the eastern and western Klamath Mountains. Geology, v.14, no.3, pp.335-338.

Bondre, N.R., and Hart, W.K., 2006. Morphological and textural diversity of Steens basalt lava flows, southeastern Oregon, USA. Eos, Trans., American Geophysical Union, 87, p.52.

Bott, J.D.J., and Wong, I.G., 1993. Historical earthquakes in and around Portland, Oregon. Oregon Geology, v.55, no.5, pp.116-122.

Bourgeois, J., 1980. A transgressive shelf sequence exhibiting hummocky stratification: the Cape Sebastian sandstone (upper Cretaceous), southwestern Oregon. Journal Sedimentary Petrology, v.50, no.3, pp.0681-0702.

———, and Leithold, E.L., 1983. Sedimentation, tectonics and sea-level change as reflected in four wave-dominated shelf sequences in Oregon and California. *In*: Larue, D.K., et al., eds., Cenozoic marine sedimentation, Pacific margin, U.S.A., Society Economic Paleontologists and Mineralogists, Pacific Section, Los Angeles, pp.1-16.

Bowen, R. G., 1978. Low- to intermediate-temperature thermal springs and wells in Oregon. Oregon Department Geology and Mineral Industries, Geologic Map Series, GMS-10.

———, Blackwell, D.D., and Hull, D.A., 1977. Geothermal exploration studies in Oregon. Oregon Department Geology and Mineral Industries, Miscellaneous Paper 19, 50p.

Brand, B.D., and Heiken, G., 2009. Tuff cones, tuff rings, and maars of the Fort Rock-Christmas Valley, basin, Oregon. *In*: O'Connor, J.E., Dorsey, R.J., and Madin, I.P., eds., Volcanoes to vineyards; field trips through the . . . Pacific Northwest. Geological Society America, Field Guide 15, pp.495-538.

Brandon, A.D., and Goles, G.G., 1988. A Miocene subcontinental plume in the Pacific Northwest: geochemical evidence. Earth and Planetary Science Letters, 88, pp.273-283.

Brittain, R.C., 1986. Eagle-Pitcher diatomite mine and processing plant, eastern Oregon. Oregon Geology, v.48, no.9, pp.108-109.

Brogan, P.F., 1964. East of the Cascades. Bindfords and Mort, Portland, Oregon, 304p.

Brooks, H.C., 1957. Oregon's opalite mining district active again. Ore Bin, v.19, no.10, pp.83-88.

———, 1963. Quicksilver in Oregon. Oregon Department Geology and Mineral Industries, Bull.55, 223p.

———, 1979. Plate tectonics and the geologic history of the Blue Mountains. Oregon Geology, v.41, no.5, pp.71-80.

———, 1990. Limestone deposits in Oregon. Oregon Department Geology and Mineral Industries, Special Paper 23, pp.8-10.

———, 1990a. Mining and exploration in Oregon in 1989. Oregon Geology, v.52, no.2, pp.37-42.

———, 2007. A pictorial history of gold mining in the Blue Mountains of eastern Oregon. Baker City, Baker County Historical Society, 200p.

———, and Ramp, L., 1968. Gold and silver in Oregon. Oregon Department Geology and Mineral Industries, Bull.61, 337p.

———, and Vallier, T.L., 1967. Progress report on the geology of part of the Snake River Canyon, Oregon and Idaho. Oregon Department Geology and Mineral Industries, v.29, no.12, pp.233-266.

———, and Vallier, T.L., 1978. Mesozoic rocks and tectonic evolution of eastern Oregon and western Idaho. *In*: Howell, D., and McDougall, K., eds., Mesozoic Paleogeography of the western United States. Pacific Coast Paleogeography Symposium 2, Society Economic Paleontologists and Mineralogists, Pacific Sect., pp.133-145.

Brouwers, E., et al., 1995. Paleogeography, paleoecology, and biostratigraphy of upper Paleocene to middle Eocene units of the Tyee basin, southwest Oregon. *In*: Fritsche, A.E., ed., Cenozoic paleogeography of the western United States – II: Society Economic Paleontologists and Mineralogists, Pacific Section, Book 75, pp.246-256.

Brown, D.E., and Petros, J.R., 1985. Geochemistry, geochronology, and magnetostratigraphy of a measured section of the Owyhee basalt, Malheur County, Oregon. Oregon Geology, v.47, no.2, pp.15-20.

———, Peterson, N.V., and McLean, G.D., 1980. Preliminary geology and geothermal resource potential of the Lakeview area, Oregon. Oregon Department Geology and Mineral Industries, Open-file Report 0-80-9, 108p.

———, et al., 1980. Preliminary geology and geothermal resource potential of the Willamette Pass area, Oregon. Oregon Department Geology and Mineral Industries, Open-file Rept. 0-80-3, 65p.

Bruer, W.G., 1980. Mist gas field, Columbia County, Oregon. American Association of Petroleum Geologists, Pacific Section. 55th Annual Meeting, Bakersfield, Calif., 10p.

Brueseke, M.E., and Hart, W.K., 2004. The physical and petrologic evolution of a multi-vent volcanic field associated with Yellowstone-Newberry volcanism. Eos Trans., American Geophysical Union, 85, p.47.

———, et al., 2007. Distribution and geochronology of Oregon Plateau (U.S.A.) flood basalt volcanism: the Steens basalt revisited. Journal of Volcanology and Geothermal Research, v.161, pp.187-214.

Bryan, K., 1929. Geology of reservoir and dam sites with a report on Owyhee irrigation project, Oregon. U.S. Geological Survey, Water Supply Paper 597-A, 89p.

Burchfiel, B.C., Cowan, D.S., and Davis, G.A., 1992. Tectonic overview of the Cordilleran orogen in the western United States. *In*: Burchfiel, B.C., Lipman, P.W., and Zoback, M.L., eds., The Cordilleran orogen: conterminous U.S. Geological Society America, The Geology of North America, v.G-3, pp.407-479.

———, Lipman, P.W., and Zoback, M.L., eds., The Cordilleran orogen: conterminous U.S. Geological Society America, The Geology of North America, v.G-3, 724p.

Burns, S., 1995. Environmental site analysis of Newell Creek Canyon, Oregon City, Oregon: influence of local decision making. Geological Society America, Cordilleran Section, Abstracts with Programs, v.27, no.5, pp.8.

———, 1996. Landslides in the West Hills of Portland: a preliminary look. Oregon Geology, v.58, no.2, pp.42-43.

———, 1998. Landslide hazards in Oregon. *In*: Burns, S., ed., Environmental, groundwater, and engineering geology; applications from Oregon. Belmont, Calif., Star Publ., pp.303-315.

———, 1998a. Landslides in the Portland area resulting from the storm of February, 1996. *In*: Burns, S., ed., Environmental, groundwater, and engineering geology; applications from Oregon. Belmont, Calif., Star Publ., pp.353-365.

———, Kuper, H.T., and Lawes, J.L., 2002. Landslides at Kelso, Washington, and Portland, Oregon. *In:* Moore, G., ed., Field guide to geologic processes in Cascadia. Oregon Department of Geology and Mineral Industries, Special Paper 36, pp.257-272.

———, ed., 1998. Environmental, groundwater, and engineering geology; applications from Oregon. Belmont, Calif., Star Publ., 689p.

———, et al., 1997. Map showing faults, bedrock geology, and sediment thickness of the western half of the Oregon City Quadrangle. Oregon Department of Geology and Mineral Industries, Interpretive Map Series, IMS-4.

———, et al., 1998. Landslides in the Portland, Oregon metropolitan area resulting from the storm of February, 1996. Portland State University, Dept. Geology, Metro Contract 905828, 37p.

Burns, W.J., and Mickelson, K..A., 2010. Landslide inventory maps of the Astoria Quadrangle, Clatsop County, Oregon. Oregon Department of Geology and Mineral Industries, IMS-31.

———, and Wang, Y., comp., 2007. Landslide symposium; proceedings and field trip guide. Oregon Department Geology and Mineral Industries, Open-File Report 0-07-06, 140p.

Burt, W., et al., 2009. Hydrogeology of the Columbia River Basalt Group in the northern Willamette Valley, Oregon. *In*: O'Connor, J.E., Dorsey, R.J., and Madin, I.P., eds., Volcanoes to vineyards; field trips through the . . . Pacific Northwest. Geological Society America, Field Guide 15, pp.pp.697-736.

Busby, E., 1998. The Canyonville landslide of January 16, 1974, Douglas County. *In*: Burns, S., ed., Environmental, groundwater, and engineering geology; applications from Oregon. Belmont, Calif., Star Publ., pp.391-398.

Byerly, P., 1952. Pacific coast earthquakes. Oregon State System of Higher Education, Eugene, Condon Lectures, 38p.

Byrne, J. V., 1962. Geomorphology of the continental terrace off the central coast of Oregon. Ore Bin, v.24, no.5, pp.65-74.

———, 1963. Geomorphology of the continental terrace off the northern coast of Oregon. Ore Bin, v.25, no.12, pp.201-209.

———, 1963a. Geomorphology of the Oregon continental terrace south of Coos Bay. Ore Bin, v.25, no.9, pp.149-155.

Callaghan, E., and Buddington, A.F., 1938. Metalliferous mineral deposits of the Cascade Range in Oregon. U.S. Geological Survey, Bull.893, 141p.

Cameron, K. A., 1991. Prehistoric buried forests of Mount Hood. Oregon Geology, v.53, no.2, pp.34-43.

———, and Pringle, P.T., 1987. A detailed chronology of the most recent major eruptive period at Mount Hood, Oregon. Geological Society America, Bull., v.99, pp.845-851.

Camp, V.E., 1981. Geologic studies of the Columbia Plateau: Part II. Upper Miocene basalt distribution, reflecting source locations, tectonism, and drainage history in the Clearwater embayment, Idaho. Geological Society America, Bull., Part I, v.92, pp.669-678.

———, 1995. Mid-Miocene propagation of the Yellowstone mantle plume head beneath the Columbia River basalt source region. Geology, v.23, pp.435-438.

———, and Hanan, B.B., 2008. A plume-triggered delamination origin for the Columbia River Basalt Group. Geosphere, v.4, no.3, pp.480-495.

———, and Hooper, P.R., 1981. Geologic studies of the Columbia Plateau: Part I. Late Cenozoic evolution of the southeast part of the Columbia River basalt province. Geological Society America, Bull., Part I, v.92, pp.659-688.

———, and Ross, M.E., 2004. Mantle dynamics and genesis of mafic magmatism in the intermontane Pacific Northwest. Journal Geophysical Research, v.109, Issue B8, pp.B08204.

———, Ross, M.E., and Hanson, W.E., 2003. Genesis of flood basalts and Basin and Range volcanic rocks from Steens Mountain to the Malheur River Gorge, Oregon. Geological Society America, Bull.115, pp.105-128.

———, et al., in press. The Steens basalt; earliest lavas of the Columbia River Basalt Group. Geological Society America, Special Paper.

Carlson, P.R., and Nelson, C.H., 1987. Marine geology and resource potential of Cascadia Basin. *In*: Scholl, D.W., Grantz, A., and Vedder, J.G., eds., Geology and resource potential of the continental margin of western North America and adjacent ocean basins-Beaufort Sea to Baja California. Circum-Pacific Council for Energy and Mineral Research, Earth Science Series, v.6, pp.523-535.

Carlson, R.W., 1981. Late Cenozoic rotations of the Juan de Fuca ridge and the Gorda rise: a case study. Tectonophysics, v.77, pp.171-188.

———, 1984. Isotopic constraints on Columbia River flood basalt genesis and the nature of the sub-continental mantle. Geochemica et Cosmochimica, Acta, v.48, pp.2357-2372.

———, and Hart, W.K., 1983. Geochemical study of the Steens Mountain flood basalt. Carnegie Institute, Washington, D.C., Yearbook 82, pp.475-480.

———, and Hart, W.K., 1987. Crustal genesis on the Oregon plateau. Journal Geophysical Research, v.92, no.B7, pp.6191-6206.

———, and Hart, W.K., 1988. Flood basalt volcanism in the northwestern United States. *In*: Mcdougall, J.D., ed., Continental flood basalts. Boston, Kluwer Academic Publ., pp.35-61.

Carson, R.J., 2001. Where the Rockies meet the Columbia Plateau . . . Oregon Geology, v.63, no.1, pp.13-33.

———, ed., 2008. Where the Great River Bends. Keokee Books, Sandpoint, Idaho, 220p.

Carter, D.T., et al., 2006. Late Pleistocene outburst flooding from pluvial Lake Alvord into the Owyhee River, Oregon. Geomorphology, v.75, no.3-4, pp.346-367.

Cashman, K.V., et al., 2009. Fire and water; volcanology, geomorphology, and hydrogeology of the Cascade Range, central Oregon. *In*: O'Connor, J.E., Dorsey, R.J., and Madin, I.P., eds., Volcanoes to vineyards; field trips through the . . . Pacific Northwest. Geological Society America, Field Guide 15, pp.539-582.

Castor, S. B., and Berry, M. R., 1981. Geology of the Lakeview uranium district, Oregon. *In*: Goodell, P., and Waters, A., Uranium in volcanic and volcaniclastic rocks. America Association Petroleum Geologists, Studies in Geology, v.13, pp.55-62.

———, and Henry, C.D., 2000. Geology, geochemistry, and origin of volcanic rock-hosted uranium deposits in northwestern Nevada and southeastern Oregon, USA. Ore Geology Reviews, no.16, pp.1-40.

Catchings, R.D., and Mooney, W.D., 1988. Crustal structure of the Columbia Plateau: evidence for continental rifting. Journal Geophysical Research, v.93, no.B1, pp.459-474.

Chan, M.A., 1985. Correlations of diagenesis with sedimentary facies in Eocene sandstones, western Oregon. Journal Sedimentary Petrology, v.55, no.3, pp.0322-0333.

———, and Dott, R.H., 1983. Shelf and deep-sea sedimentation in Eocene forearc basin, western Oregon - fan or non-fan? American Association Petroleum Geologists, Bull., v.67, pp.2100-2116.

———, and Dott, R.H., 1986. Depositional facies and progradational sequences in Eocene wave-dominated deltaic complexes, southwestern Oregon. American Association Petroleum Geologists, Bull., v.70, pp.415-429.

———, and Dott, R.H., 1986a. Wave-dominated deltaic complexes, southwestern Oregon. American Association Petroleum Geologists, Bull.,vol.70, p.460-472.

Chitwood, L.A., 1994. Inflated basaltic lava – examples of processes and landforms from central and southeast Oregon. Oregon Geology, v.56, no.1, pp.11-20.

———, and McKee, E. H., 198l. Newberry Volcano, Oregon. U.S. Geological Survey, Circular 838, pp.85-91.

———, Jensen, R.A., and Groh, E.A., 1977. The age of Lava Butte. Ore Bin, v.39, no.10, pp.157-164.

Choiniere, S. R., and Swanson, D. A., 1979. Magnetostratigraphy and correlation of Miocene basalts of the northern Oregon coast and Columbia Plateau, southeast Washington. American Journal Science, v.279, pp.755-777.

Christiansen, R. L., and McKee, E.H., 1978. Late Cenozoic volcanic and tectonic evolution of the Great Basin and Columbia intermontane regions. *In*: Smith, R., and Eaton, G., eds., Cenozoic tectonic and regional geophysics of the western Cordillera. Geological Society America, Memoir 152, pp.283-312.

———, and Yeats, R.S., 1992. Post-Laramide geology of the U.S. Cordilleran region. *In*: Burchfiel, B.C., Lipman, P.C., and Zoback, M.L., eds., The Cordilleran orogen: geology of North America, conterminous U.S. Geological Society America, v.G-3, pp.261-406.

———, Foulger, G.R., and Evans, J.R., 2002. Upper-mantle origin of the Yellowstone hotspot. Geological Society American, Bull.114, pp.1245-1256.

Church, S.E., et al., 1986. Lead-isotopic data from sulfide minerals from the Cascade Range, Oregon and Washington. Geochimica et Cosmochimica Acta, v.50, pp.317-328.

Ciesiel, R.F., and Wagner, N.S., 1969. Lava-tube caves in the Saddle Butte area of Malheur County, Oregon. Ore Bin, v.31, no.8, pp.153-171.

Clague, J.J., 1997. Evidence for large earthquakes at the Cascadia subduction zone. Reviews of Geophysics, v.35, pp.439-460.

———, et al., 2000. Penrose Conference; great Cascadia earthquake tricentennial, Seaside, Oregon, June 4-8. Oregon Department Geology and Mineral Industries, Special Paper 33, 156p.

Clark, J.L., and Beck, N., 2000. Through the years of Oregon Geology. Oregon Geology, v.62, no.1, pp.3-10.

Clark, R.D., 1989. The odyssey of Thomas Condon. Portland, Oregon, Oregon Historical Society Press, 569p.

Clifton, H.E., and Luepke, G., 1987. Heavy-mineral placer deposits of the continental margin of Alaska and the Pacific coast states. *In*: Scholl, D.W., Grantz, A., and Vedder, J.G., eds., Geology and resource potential of the continental margin of western North America and adjacent ocean basins-Beaufort Sea to Baja California. Circum-Pacific Council for Energy and Mineral Research., Earth Science Series, v.6, pp.691-738.

Cloos, M., et al., eds., 2007. Convergent marine terranes and associated regions: a tribute to W.G. Ernst. Geological Society America, Special Paper 419, 273p.

Cohen, A.S., et al., 2000. A paleoclimate record for the past 250,000 years from Summer Lake, Oregon, USA: II. Sedimentology, paleontology, and geochemistry. Journal Paleolimnology, v.24, pp.151-182.

Cohn, L., 1991. The big one. Old Oregon Magazine, v.71, no.1, pp.18-22.

Colbath, G. K., and Steele, M.J., 1982. The geology of economically significant lower Pliocene diatomites in the Fort Rock basin near Christmas Valley, Lake County, Oregon. Oregon Geology, v.44, no.10, pp.111-118.

Cole, M.R., and Armentrout, J.M., 1979. Neogene paleogeography of the western United States. *In*: Armentrout, J.M., Cole, M. R., and TerBest, H.J., eds., Cenozoic paleogeography of the western United States. Pacific Coast Paleogeography Symposium 3, Society Economic Paleontologists and Mineralogists, pp.297-323.

Coleman, R.G., 1972. The Colebrooke Schist of southwestern Oregon. U.S. Geological Survey, Bull.1339, 61p.

———, and Irwin, W.P., eds., 1977. North American ophiolites. Oregon Department of Geology and Mineral Industries, Bull.95, pp.74-105.

Colgan, J.G., Shuster, D.L., and Reiners, P.W., 2008. Two-phase Neogene extension in the northwestern Basin and Range recorded in a single thermochronology sample. Geology, v.36, no.8, pp.631-634.

———, et al., 2006. Cenozoic tectonic evolution of the Basin and Range province in northwestern Nevada. American Journal Science, v.306, pp.616-654.

Condon, T., 1902. The two islands. J.K. Gill, Portland, Oregon, 211p.

Coney, P.J., Jones, D.L., and Monger, J.W.H., 1980. Cordilleran suspect terranes. Nature, v.288, pp.329-333.

Conrey, R. M., 1991. Geology and petrology of the Mt. Jefferson area, High Cascade Range, Oregon. PhD., Washington State University, 357p.

———, Grunder, A., and Schmidt, M., 2004. State of the Cascade arc. Oregon Department of Geology and Mineral Industries, Open-File Report OFR 0-04-04. SOTA Field Trip Guide, 39p.

———, et al., 1996. Potassium-argon ages from Mount Hood area of Cascade Range, northern Oregon. Isochron/West, no.63, pp.10-20.

———, et al., 1996a. Potassium-argon ages of Boring lava, northwest Oregon and southwest Washington. Isochron/West, no.63, pp.3-9.

———, et al., 1997. Diverse primitive magmas in the Cascade arc, northern Oregon and southern Washington. Canadian Mineralogist, v.35, pp.367-396.

———, et al., 2001. Trace element and isotopic evidence for two types of crustal melting beneath a High Cascade volcanic center, Mt. Jefferson. Contributions to Mineral Petrology, v.141, pp.710-732.

———, et al., 2002. North-central Oregon Cascades: exploring petrologic and tectonic intimacy in a propagating intra-arc rift. *In*: Moore, G., ed., Field guide to geologic processes in Cascadia. Oregon Department of Geology and Mineral Industries, Special Paper 36, pp.47-90.

———, et al., 2003. The Boring volcanic field of the Portland Basin. Eos Transactions, American Geophysical Union, v.84, no.46, Supplement, Abstracts, no. V31E-0981.

Corcoran, R.E., 1965. Geology of Lake Owyhee State Park and vicinity, Malheur County, Oregon. Ore Bin, v.27, no.5, pp.81-98.

———, 1969. Geology of the Owyhee Upland province. Oregon Department Geology and Mineral Industries, Bull.64, pp.80-88.

———, and Libbey, F.W., 1955. Investigation of Salem hills bauxite deposits. Ore Bin, v.17, no.4, pp.23-27.

Couch, R.W., and Braman, D. E., 1980. Geology of the continental margin near Florence, Oregon. Oregon Geology, v.41, on.11, pp.171-179.

———, and Foote, R., 1985. The Shukash and Lapine basins: Pleistocene depressions in the Cascade Range of central Oregon. Eos Transactions, v.66, no.3, p.24.

———, and Johnson, S., 1968. The Warner Valley earthquake sequence: May and June, 1968. Ore Bin, v.30, no.10, pp.191-204.

———, and Pitts, S.G., 1980. The structure of the continental margin near Coos Bay, Oregon. *In*: Newton, V.C., et al., Prospects for oil and gas in the Coos, Douglas, and Lane counties, Oregon. Oregon Department Geology and Mineral Industries, Oil and Gas Inv.6, pp.23-28.

———, and Riddihough, R.P., 1989. The crustal structure of the western continental margin of North America. *In:* Pakiser, L.C., and Mooney, W.D., Geophysical framework of the continental United States. Geological Society America, Memoir. 172, pp.103-128.

———, and Whitsett, R., 1969. The North Powder earthquake of August 14, 1969. Ore Bin, v.31, no.12, pp.239-244.

———, Thrasher, G., and Keeling, K., 1976. The Deschutes valley earthquake of April 12, 1976. Ore Bin, v.38, no.10, pp.151-154.

Crandell, D. R., 1965. The glacial history of western Washington. *In*: Wright, H.E., and Frey, D.G., eds., The Quaternary of the United States. Princeton University Press, Princeton, N.J., pp.341-354.

———, 1967. Glaciation at Wallowa Lake, Oregon. U.S. Geological Survey, Professional Paper 575-C, pp.145-153.

———, 1980. Recent eruptive history of Mount Hood, Oregon, and potential hazards from future eruptions. U.S. Geological Survey, Bull.1492, 81p.

———, and Mullineaux, D.R., 1975. Appraising volcanic hazards of the Cascade Range. Ore Bin, v.37, no.11, pp.173-183.

Crenna, P.A., Yeats, R.S., and Levi, S., 1994. Late Cenozoic tectonics and paleogeography of the Salem metropolitan area, central Willamette Valley, Oregon. Oregon Geology, v.56, no.6, pp.129-136.

Crowell, John C., 1985. The recognition of transform terrane dispersion within mobile belts. *In:* Howell, D. G., ed., Tectonostratigraphic terranes of the circum-Pacific region. Pacific Council for Energy and Mineral Resources, Earth Science Series, no.1, pp.51-61.

Cummings, M.L., 1991. Geology of the Deer Butte Formation, Malheur County, Oregon; faulting, sedimentation, and volcanism in a post-caldera setting. Sedimentary Geology, v.74, pp.345-362.

———, and Conaway, J.S., 2009. Landscape and hydrologic response in the Williamson River basin following the Holocene eruption of Mount Mazama. *In*: O'Connor, J.E., Dorsey, R.J., and Madin, I.P., eds., Volcanoes to vineyards; field trips through the . . . Pacific Northwest. Geological Society America, Field Guide 15, pp.271-294.

———, and Growney, L.P., 1988. Basalt hydrovolcanic deposits in the Dry Creek arm area of the Owyhee Reservoir, Malheur County, Oregon: stratigraphic relations. Oregon Geology, v.50, no.7/8, pp.75-82.

———, Cady, S.L., and Perkins, R.B., 2009. A tale of two basins; comparisons and contrasts in the high desert hydrology of Steens Mountain . . . *In*: O'Connor, J.E., Dorsey, R.J., and Madin, I.P., eds., Volcanoes to vineyards; field trips through the . . . Pacific Northwest. Geological Society America, Field Guide 15, pp.295-308.

———, Evans, J.G., and Ferns, M.L., 1994. Stratigraphic and structural evolution of the middle Miocene Oregon-Idaho graben, Malheur County, Oregon. *In*: Swanson, D.A., and Haugerud, R.S., eds., Geologic field trips in the Pacific Northwest. Geological Society America, Annual Meeting, 1994, Seattle Washington, v.1, pp.1G-1G-20.

———, Johnson, A.G., and Cruikshank, K.M., 1996. Intragraben fault zones and hot springs deposits in the Oregon-Idaho graben. *In*: Conyer, A.R., and Fahey, P.I., eds., Geology and ore deposits of the

American Cordillera. Geological Society Nevada, Symposium, Proceedings, 1995, pp.1047-1062.

———, et al., 1990. Stratigraphic development and hydrothermal activity in the central Western Cascade range, Oregon. Journal Geophysical Research, v.95, no.B12, pp.19,601-19,610.

———, et al., 2000. Stratigraphic and structural evolution of the middle Miocene syn-volcanic Oregon-Idaho graben. Geological Society America, Bull.112, pp.668-682.

Curray, J.R., 1965. Late Quaternary history, continental shelves of the United States. *In*: Wright, H.E., and Frey, D.G., eds., The Quaternary of the United States. Princeton University Press, Princeton, N.J., 723-735.

Dake, H.C., 1938. The gem minerals of Oregon. Oregon Department Geology and Mineral Industries, Bull.7, 16p.

Danner, W.R., 1977. Recent studies of the Devonian of the Pacific Northwest. *In*: Murphy, M.A., Berry, W.B.N., and Sandberg, C.A., eds. Western North American: Devonian. University of California, Riverside. Campus Museum, Contribution 4, pp.220-225.

Darienzo, M.E., and Peterson, C. D., 1990. Episodic tectonic subsidence of late Holocene salt marshes, northern Oregon central Cascadia margin. Tectonics, v.9, no.1, pp.1-22.

———, and Peterson, C.D., 1995. Magnitude and frequency of subduction-zone earthquakes along the northern Oregon coast in the past 3,000 years. Oregon Geology, v.57, no.1, pp.3-11.

———, Peterson, C.D., and Clough, C., 1994. Stratigraphic evidence for great subduction-zone earthquakes at four estuaries in northern Oregon, U.S.A. Journal Coastal Research, no.10, no.4, pp.850-876.

Davidson, G., 1896. Coast pilot of California, Oregon, and Washington territory. U.S. Coast Survey, Pacific Coast, 262p.

Davis, G. A., 1979. Problems of intraplate extensional tectonics, western United States, with special emphasis on the Great Basin. *In*: Newman, G.W., and Goode, H.D., eds., Basin and Range symposium and Great Basin field conference, pp.41-54.

———, Monger, J.W.H., Burchfiel, B.C., 1978. Mesozoic construction of the Cordilleran "collage", central British Columbia to central California. *In*: Howell, D.G., and McDougall, K., eds., Mesozoic paleogeography of the western United States. Society Economic Paleontologists and Mineralogists, Pacific Coast Paleogeography Symposium 2, p.1-32.

Davis, L.G., et al., 2009. Geoarchaeological themes in a dynamic coastal environment, Lincoln and Lane counties, Oregon. *In*: O'Connor, J.E., Dorsey, R.J., and Madin, I.P., eds., Volcanoes to vineyards; field trips through the . . . Pacific Northwest. Geological Society America, Field Guide 15, pp.319-336.

Derkey, R. E., 1980. The Silver Peak volcanogenic massive sulfide, northern Klamath Mountains, Oregon. Geological Society America, Abstracts with Program, v.12, no.3, p.104.

DeRoo, T.G., 2006. Recent geologic history of the upper White River valley, Oregon. Oregon Geology, v.67, no.1, pp.3-6.

———, Smith, D.A., and Anderson, D.A., 1998. Factors affecting landslide incidence after large storm events during the winter of 1995-1996 in the upper Clackamas River . . . *In*: Burns, S., ed., Environmental, groundwater, and engineering geology; applications from Oregon. Belmont, California, Star Publ., pp.379-390.

Dicken, S.N., 1973. Oregon geography. Eugene, University of Oregon Bookstore, 147p.

———, 1980. Pluvial lake Modoc, Klamath County, Oregon, and Modoc and Siskiyou counties, California. Oregon Geology, v.42, no.11, pp.179-187.

———, and Dicken, E.F., 1985. The legacy of ancient Lake Modoc: a historical geography of the Klamath Lakes basin Oregon and California. Eugene, University of Oregon Bookstore, various pagings.

Dickinson, W.R., 1977. Paleozoic plate tectonics and the evolution of the Cordilleran continental margin. *In*: Stewart, J.H., et al., eds., 1977. Paleozoic paleogeography of the western United States. Society Economic Paleontologists and Mineralogists, Pacific Coast Paleogeography Symposium 1, pp.137-155.

———, 1979. Cenozoic plate tectonic settings of the Cordilleran region in the United States. *In*: Armentrout, J.M., Cole, M.R., and TerBest, H.J., eds, Cenozoic paleogeography of the western United States. Society Economic Paleontologists and Mineralogists, Pacific Coast Paleogeography Symposium 3, pp.1-13.

———, 1979a. Mesozoic forearc basin in central Oregon. Geology (Boulder), v.7, no.4, pp.166-170.

———, 1995. Forearc basins. *In*: Busby, C.J., and Ingersoll, R.V., eds. Tectonics of sedimentary basins. Cambridge, Mass., Blackwell Science Publ., pp.221-262.

———, 2004. Evolution of the North American Cordillera. Annual Review Earth and Planetary Science Letters, v.32, pp.13-45.

———, and Thayer, T.P., 1978. Paleogeographic and paleotectonic implications of Mesozoic stratigraphy and structure in the John Day inlier of central Oregon. *In*: Howell, D.G., and McDougall, K.A., eds., Mesozoic paleogeography of the western United States. Society Economic Paleontologists and Mineralogists, Pacific Section, Pacific Coast Paleography Symposium 2, pp.147-161.

———, and Vigrass, L.W., 1965. Geology of the Suplee-Izee area Crook, Grant, and Harney counties, Oregon. Oregon, Department Geology and Mineral Industries, Bull.58, 109p.

———, Hopson, C.A., and Saleeby, J.B., 1996. Alternate origins of the coast Range ophiolite (California). GSA Today, v.6, no.2, pp.2-10.

Dignes, T.W., and Woltz, D., 1981. West coast. American Association Petroleum Geologists, Bull.65, no.10, pp.1781-1791.

Diller, J.S., 1896. Geological reconnaissance in northwestern Oregon. U.S. Geological Survey, 17th Annual Report, p.1-80.

———, 1898. Description of the Roseburg quadrangle. U.S. Geological Survey, Atlas Folio, no.49.

———, 1899. The Coos Bay coalfield, Oregon. U.S. Geological Survey, 19th Annual Report, Pt.3, pp.309-370.

———, 1902. Topographic development of the Klamath Mountains. U.S. Geological Survey, Bull.196, pp.9-63.

———, 1903. Description of the Port Orford quadrangle. U.S. Geological Survey, Atlas Folio, no.89.

———, and Kay, G.F., 1924. Description of the Riddle quadrangle, Oregon. U.S. Geological Survey, Folio Atlas, no.218.

Dillhoff, R.M., et al., 2009. Cenozoic paleobotany of the John Day Basin, central Oregon. *In*: O'Connor, J.E., Dorsey, R.J., and Madin, I.P., eds., Volcanoes to vineyards; field trips through the . . . Pacific Northwest. Geological Society America, Field Guide 15, pp.135-164.

Dokka, R.K., and Travis, C.J., 1990. Role of the eastern California shear zone in accommodating Pacific-North American plate motion. Geophysical Research Letters, v.17, no.9, pp.1323-1326.

———, 1903. Description of the Port Orford quadrangle. U.S. Geological Survey, Bull. 196, 69p.

Donato, M.M., 1991. Geologic map showing part of the May Creek Schist and related rocks, Jackson County, Oregon. U.S. Geological Survey, Miscellaneous Field Studies Map MF- 2171.

———, Barnes, C.G., and Tomlinson, S.L., 1996. The enigmatic Applegate Group of southwestern Oregon: age, correlation, and tectonic affinity. Oregon Geology, v.58, no.4, pp.79-91.

———, Coleman, R.G., and Kays, M.A., 1980. Geology of the Condrey Mountain Schist, northern Klamath Mountains, California and Oregon. Oregon Geology, v.42, no.7, pp.125-129.

Donnelly-Nolan, J.M., and Jensen, R.A., 2009. Ice and water on Newberry Volcano, central Oregon. *In*: O'Connor, J.E., Dorsey, R.J., and Madin, I.P., eds., Volcanoes to vineyards; field trips through the . . . Pacific Northwest. Geological Society America, Field Guide 15, pp.81-90.

Dorsey, R.J., and LaMaskin, T.A., 2007. Stratigraphic record of Triassic-Jurassic collisional tectonics in the Blue Mountains province, northeastern Oregon. American Journal Science, v.307, pp.1167-1193.

———, and Lenegan, R.J., 2007. Structural controls on middle Cretaceous sedimentation in the Toney Butte area of the Mitchell inlier, Ochoco basin, central Oregon. *In*: Cloos, M., et al., eds., Convergent margin terranes and associated regions: a tribute to W.G. Ernst. Geological Society America, Special Paper 419, pp.97-115.

Dott, R.H., 1962. Geology of the Cape Blanco area, southwest Oregon. Ore Bin, v.24, no.8, pp.121-133.

———, 1966. Eocene deltaic sedimentation at Coos Bay, Oregon. Journal Geology, v.74, pp.373-420.

———, 1971. Geology of the southwestern Oregon coast west of the 124th meridian. Oregon Department Geology and Mineral Industries, Bull.69, 63p.

———, and Bird, K. J., 1979. Sand transport through channels across an Eocene shelf and slope in southwestern Oregon, U.S.A. Society Economic Paleontologists and Mineralogists, Special Publication 27, pp.327-342.

Draper, D.S., 1991. Late Cenozoic bimodal magnetism in the northern Great Basin and Range Province in southeastern Oregon. Journal Volcanology and Geothermal Research, v.47, pp.299-328.

Driedger, C. L., and Kennard, P. M., 1984. Ice volumes on Cascade volcanoes: Mount Rainier, Mount Hood, Three Sisters, and Mount Shasta. U.S. Geological Survey, Open- File Report 84-581, 42p.

Duncan, R.A., 1982. A captured island chain in the Coast Range of Oregon and Washington. Journal Geophysical Research, v.87, no.B13, pp.10,827-10,837.

———, and Kulm, L.D., 1989. Plate tectonic evolution of the Cascades arc-subduction complex. *In*: Winterer, E.L., Hussong, D.M., and Decker, R.W., eds., The Eastern Pacific Ocean and Hawaii. Geological Society America, The Geology of North America, v.N, pp.413-438.

———, Fowler, G., and Kulm, L.D., 1970. Planktonic foraminiferan-radiolarian ratios and Holocene-Late Pleistocene deep-sea stratigraphy off Oregon. Geological Society America, Bull., v.81, pp.561-566.

Durbin, K., 1998. Logging and landslides. Oregon Quarterly, Winter, pp.20-26.

Dymond, J., Collier, R., and Watwood, M.E., 1989. Bacterial mats from Crater Lake, Oregon, and their relationship to possible deep-lake hydrothermal venting. Nature, v.341, no.6250, pp.673-675.

———, et al., 1987. Hydrothermal activity in Crater Lake: evidence from sediments. Eos Transactions, v.68, no.50, p.1771.

Dziak, R.P., et al., 2000. Recent tectonics of the Blanco Ridge, eastern Blanco transform fault zone. Marine Geophysical Researches, v.21, pp.423-450.

Easterbrook, D. J., 1986. Stratigraphy and chronology of Quaternary deposits of the Puget lowland and OlympicMountains of Washington and the Cascade Mountains of Washington and Oregon. Quaternary glaciation in the northern hemisphere. Quaternary Science Reviews, v.5, pp.145-159.

———, ed., 2003. Quaternary geology of the United States. INQUA Field Guide Volume. Reno, Desert Research Institute, 438p.

Eaton, G. P., 1979. Regional geophysics, Cenozoic tectonics and geologic resources of the Basin and Range province and adjoining regions. *In:* Newman, G.W., et al., eds., Basin and Range Symposium and Great Basin field conference, pp.11-39.

———, 1982. The Basin and Range province: origin and tectonic significance. Annual Review Earth and Planetary Science Letters, v.10, pp.409-440.

———, 1984. The Miocene Great Basin of western North America as an extending back-arc region. Tectonophysics, 102, pp.275-295.

———, et al., 1978. Regional gravity and tectonic patterns. *In*: Smith, R.B., and Eaton, G.P., Cenozoic tectonics and regional geophysical of the western Cordillera. Geological Society America, Memoir 152, pp.51-91.

Ehlen, J., 1967. Geology of state parks near Cape Arago, Coos County, Oregon. Ore Bin, v.29, no.4, pp.61-82.

Ekren, E.B., et al., 1981. Geologic map of Owyhee County, Idaho, west of longitude 116° W. U.S. Geological Survey, Miscellaneous Investigations Series, I-1256.

Ellis, W.L., et al., 2007. Hydrogeologic investigations of the Johnson Creek landslide. *In*: Burns, W.J., and Wang, Y., comp., Landslide symposium; proceedings and field trip guide. Oregon Department Geology and Mineral Industries, Open-File Report 0-07-06, pp.34-35.

Engebretson, D. C., Cox, A., and Gordon, R.G., 1985. Relative motions between oceanic and continental plates in the Pacific Basin. Geological Society America, Special Paper 206, 59p.

Enlows, H. E., 1976. Petrography of the Rattlesnake Formation the type area, central Oregon. Oregon, Department Geology and Mineral Industries, Short Paper 25, 34p.

Evans, J.G., 1984. Structure of part of the Josephine peridotite, northwestern California and southwestern Oregon. U.S. Geological Survey, Bull.1546 A-D, pp.7-37.

———, 1996. Geologic map of the Monument Peak Quadrangle, Malheur County, Oregon. U.S. Geological Survey, Miscellaneous Field Studies Map MF 2317.

———, and Geisler, T.M., 2001. Geologic field-trip guide to Steens Mountain Loop Road, Harney County, Oregon. U.S. Geological Survey, Bull.2183, 15p.

———, et al., 1987. Mineral resources of the Owyhee Canyon Wilderness study area, Malheur County, Oregon. U.S. Geological Survey, Bull.1719-E, 18p.

Evarts, R.C., et al., 2009. The Boring volcanic field of the Portland-Vancouver area, Oregon and Washington. *In*: O'Connor, J.E., Dorsey, R.J., and Madin, I.P., eds., Volcanoes to vineyards; field trips through the . . . Pacific Northwest. Geological Society America, Field Guide 15, pp.253-270.

Farooqui, S. M., et al., 1981. Dalles Group: Neogene formations overlying the Columbia River Basalt group in north-central Oregon. Oregon Geology, v.43, no.10, pp.131-140.

Farr, L.C., 1989. Stratigraphy, diagenesis, and depositional environment of the Cowlitz Formation (Eocene) northwest, Oregon. Masters, Portland State University, 168p.

Faul, H., and Faul, C., 1983. It began with a stone: a history of geology from the stone age to the age of plate tectonics. New York, Wiley, 270p.

Faulds, J.E., and Stewart, J.H., eds., 1998. Accommodation zones and transfer zones; the regional segmentation of the Basin and Range province. Geological Society America, Special Paper 323, 256p.

———, Henry, C.D., and Hinz, N.H., 2005. Kinematics of the northern Walker Lane. Geology, v.33, pp.505-508.

Faustini, J.M., and Jones, J.A., 2003. Influence of large woody debris on channel morphology and dynamics in steep, boulder-rich mountain streams, western Cascades, Oregon. Geomorphology, v.51, no.1-3, pp.187-205.

Fecht, K.R., Reidel, S.P., and Tallman, A.M., 1987. Paleodrainage of the Columbia River system on the Columbia Plateau of Washington state - a summary. *In*: Schuster, J.E., ed., Selected papers on the geology of Washington. Washington Division of Geology and Earth Resources, Bull.77, pp.238-245.

Ferns, M. L., 1989. Geology and mineral resources map of the Owyhee Dam quadrangle, Malheur County, Oregon. Oregon Department of Geology and Mineral Industries, Geologic Map Series GMS-55.

———, and Brooks, H.C., 1995. The Bourne and Greenhorn subterranes of the Baker terrane, eastern Oregon. *In*: Vallier, T.L., and Brooks, H.C., eds., 1995. Geology of the Blue Mountains region of Oregon, Idaho, and Washington. U.S. Geological Survey, Professional Paper 1438, pp.331-358.

———, and McClaughry, J.D., 2006. Preliminary geologic map of the Powell Buttes 7 ½ quadrangle, Crook County, Oregon. Oregon Department of Geology and Mineral Industries, Open-File Report, 0-06-24.

———, Evans, J.G., and Cummings, M.L., 1993. Geologic map of the Mahogany Mountain 30x60 minute quadrangle, Malheur County, Oregon, and

Owyhee County, Idaho. Oregon Department of Geology and Mineral Industries, GMS-77.

———, Madin, I.P., and Taubeneck, W.H., 2001. Reconnaissance geologic map of the LaGrande 30″x60′ quadrangle, Baker, Grant, Umatilla, and Union counties, Oregon. Oregon Department of Geology and Mineral Industries, Reconnaissance Map Series, RMS-1, 52p.

———, et al., 1993. Geologic map of the Vale 30x60 minute quadrangle, Malheur County, Oregon, and Owyhee County, Idaho. Oregon Department Geology and Mineral Industries, GMS-77.

Fisher, R., and Rensberger, J.M., 1972. Physical stratigraphy of the John Day Formation, central Oregon. University California, Publications Geological Sciences, v.101, pp.1-45.

Fitterman, D.V., 1988. Overview of the structure and geothermal potential of Newberry Volcano, Oregon. Journal Geophysical Research, v.93, no.B9, pp.10,059-10,066.

Fleming, S.W., and Trehu, A.M., 1999. Crustal structure beneath the central Oregon convergent margin . . . evidence for a buried basement ridge . . . Journal Geophysical Research, v.104, no.B9, pp.20431-20447.

Follo, M.F., 1992. Conglomerates as clues to the sedimentary and tectonic evolution of a suspect terrane, Wallowa Mountains, Oregon. Geological Society America, Bull., v.104, pp.1561-1576.

———, 1994. Sedimentology and stratigraphy of the Martin Bridge Limestone and Hurwal Formation . . . from the Wallowa Terrane, Oregon. *In*: Vallier, T.L., and Brooks, H.C., eds., Stratigraphy physiography, and mineral resources of the Blue Mountains region. U.S. Geological Survey, Professional Paper 1439, pp.1-27.

Foulger, G.R., 2007. The "plate" model for the genesis of melting anomalies. Geological Society America, Special Paper 430, pp. 1-28.

———, et al., eds., 2005. Plates, plumes, and paradigms. Geological Society America, Special Paper 388, 881p.

Fowler, G.A., Orr, W.N., and Kulm, L.D., 1971. An upper Miocene diatomaceous rock unit on the Oregon continental shelf. Journal Geology, v.79, pp.603-608.

Frank, F.J., 1973. Ground water in the Eugene-Springfield area, southern Willamette Valley, Oregon. U.S. Geological Survey, Water Supply Paper 2032, 48p.

———, 1974. Groundwater in the Corvallis-Albany area, central Willamette Valley, Oregon. U.S. Geological Survey, Water Supply Paper 2032, 48p.

Freed, M., 1979. Silver Falls State Park. Oregon Geology, v.41, no.1, pp.3-10.

Frost, C.D., Barnes, C.G., and Snoke, A.W., 2006. Nd and Sr isotopic data from argillaceous rocks of the Galice Formation and Rattlesnake Creek terrane, Klamath Mountains. *In*: Snoke, A.W., and Barnes, C.G., eds., Geological studies in the Klamath Mountains province, California and Oregon. Geological Society America, Special Paper 410, pp.103-120.

Fryberger, J.S., 1959. The geology of Steens Mountain, Oregon. Masters, University of Oregon, 65p.

Gallino, G.L., and Pierson, T.C., 1985. Polallie Creek debris flow and subsequent dam-break flood of 1980, East Fork Hood River basin, Oregon. U.S. Geological Survey, Water Supply Paper 2273, 22p.

Gandera, W.E., 1977. Stratigraphy of the middle to late Eocene formations of southwestern Willamette Valley. Masters, University of Oregon, 71p.

Gannett, M.W., 1987. Groundwater availability in the Powell Buttes area, central Oregon. Oregon Water Resources Department, 31p.

———, and Caldwell, R.C., 1998. Geologic framework of the Willamette lowland aquifer system, Oregon and Washington. U.S. Geological Survey, Professional Paper 1424-A, 32p.

———, et al., 2001. Ground-water hydrology of the upper Deschutes basin, Oregon. U.S. Geological Survey, Water Resources Investigations Report 00-4162, 77p.

———, et al., 2007. Ground-water hydrology of the upper Klamath Basin, Oregon and California. U.S. Geological Survey, Scientific Investigations Report 2007-5050, 84p.

Gaona, M.T., 1984. Stratigraphy and sedimentology of the Osburger Gulch sandstone member of the upper Cretaceous Hornbrook Formation, northern California and southern Oregon. *In*: Nilsen, T. H., ed., 1984. Geology of the upper Cretaceous Hornbrook Formation, Oregon and California. Society Economic Paleontologists and Mineralogists, Pacific Sect., v.42, pp.141-148.

Garcia, M.O., 1979. Petrology of the Rogue and Galice formations, Klamath Mountains, Oregon. Journal Geology, v.87, pp.29-41.

———, 1982. Petrology of the Rogue River island-arc complex, southwest Oregon. American Journal Science, v.282, pp.783-807.

Geist, D., and Richards, M., 1993. Origin of the Columbia Plateau and Snake River Plain: deflection of the Yellowstone plume. Geology, v.21, pp.789-792.

Geitgey, R.P., 1987. Oregon sunstones. Oregon Geology, v.49, no.2, pp.23-24.

———, 1989. Industrial minerals in Oregon. Oregon Geology, v.51, no.6, pp.123-129.

———, 1995. Mineral industry in Oregon, 1993-1994. Oregon Geology, v.57, no.3, pp.61-65.

———, and Vogt, B.F., 1990. Industrial rocks and minerals of the Pacific northwest. Oregon Department of Geology and Mineral Industries, Special Paper 23, 110p.

Gentile, J.R., 1982. The relationship of morphology and material to landslide occurrence along the coastline in Lincoln County, Oregon. Oregon Geology, v.44, no.9, pp.99-102.

George, A.J., 1985. The Santiam mining district of the Oregon Cascades; a cultural property inventory and historical survey. Lyons, Oregon, Shiny Rock Mining Corp., 336p.

Gerlach, D.C., Ave Lallemant, H.G., and Leeman, W. P., 1981. An island arc origin for the Canyon Mountain ophiolite complex, eastern Oregon, U.S.A. Earth and Planetary Science Letters, 53, pp.255-265.

Giaramita, M.J., and Harper, G.D., 2006. Geochemistry of ophiolitic rocks associated with the western part of the Elk outlier of the western Klamath terrane, southwestern Oregon. *In*: Snoke, A.W., and Barnes, C.G., eds., Geological studies in the Klamath Mountains province, California and Oregon. Geological Society America, Special Paper 410, pp.153-176.

Gibson, C.N., Vallier, T.L., and O'Connor, J.E., 1990. The Rush Creek landslide in Hells Canyon of the Snake River, Oregon and Idaho. Geological Society America, Abstracts with Programs, v.22, no.6, p.12.

Gilluly, J., 1933. Copper deposits near Keating, Oregon. U.S. Geological Survey Bull.830, pp.1-32.

———, 1937. Geology and mineral resources of the Baker Quadrangle, Oregon. U.S. Geological Survey, Bull.879, 119p.

Glen, J.M.G., and Ponce, D.A., 2002. Large-scale fractures related to the inception of the Yellowstone hotspot. Geology, v.30, pp.647-650.

Glenn, J.L., 1965. Late Quaternary sedimentation and geologic history of the north Willamette Valley, Oregon. PhD, Oregon State University, 231p.

Goldfinger, C., 1994. Active deformation of the Cascadia forearc . . . PhD, Oregon State University.

———, 2000. Super-scale failure of the southern Oregon Cascadia margin. Pure and Applied Geophysics, v.157, no.6-8, pp.1189-1226.

———, Kulm, L.D., and Yeats, R.S., 1992. Neotectonic map of the Oregon continental margin and adjacent abyssal plain. Oregon Department of Geology and Mineral Industries, Open-File Report 0-92-4, 17p.

———, Kulm, L.D., and Yeats, R.S., 1992a. Transverse structural trends along the Oregon convergent margin: implications for Cascadia earthquake potential and crustal rotations. Geology, v.20, pp.141-144.

———, et al., 1986. Oregon subduction zone: venting, fauna, and carbonates. Science, Reprint Series, v.231, pp.561-566.

———, et al., 1992. Neotectonic map of the Oregon continental margin and adjacent abyssal plain. Oregon Department of Geology and Mineral Industries, Open-File Report 0-92-4, 17p.

———, et al., 1996. Active strike-slip faulting and folding of the Cascadia plate boundary and forearc in central and northern Oregon. U.S. Geological Survey, Professional Paper 1560, v.1, pp.223-256.

———, et al., 1996a. Oblique strike-slip faulting of the Cascadia submarine forearc: the Daisy Bank fault zone off central Oregon. *In*: Bebout, G.E., et al., eds., Subduction: top to bottom. Geophysical Monograph Series, v.96, American Geophysical Union, pp.65-74.

———, et al., 1997. Oblique strike-slip faulting of the central Cascadia submarine forearc. Journal Geophysical Research, v.102, no.B4, pp.8217-8243.

———, et al., 2008. Late Holocene rupture of the northern San Andreas fault and possible stress linkage to the Cascadia subduction zone. Seismological Society America, Bull., 98, no.2, pp.861-889.

Goldstrand, P.M., 1994. The Mesozoic geologic evolution of the northern Wallowa terrane, northeastern Oregon and western Idaho. *In*: Vallier, T.L., and Brooks, H.C., eds., Stratigraphy physiography, and mineral resources of the Blue Mountains region. U.S. Geological Survey, Professional Paper 1439, pp.29-53.

Goles, G. G., Brandon, A.D., and Lambert, R.St J., 1989. Miocene basalts of the Blue Mountains province in Oregon; Part 2, Sr isotopic ratios and trace element features of little-known Miocene basalts of central and eastern Oregon. *In*: Reidel, S.P., and Hooper, P.R., eds. Volcanism and tectonism in the Columbia River flood-basalt province. Geological Society America, Special Paper 239, pp.357-365p.

Good, J.W., and Ridlington, S.S., 1992. Coastal natural hazards. Corvallis, Oregon Sea Grant, Oregon State University, 162p.

Goodge, J. W., 1989. Evolving early Mesozoic convergent margin deformation, central Klamath Mountains, northern California. Tectonics, v.8, no.4, pp.845-864.

Graven, E.P., 1990. Structure and tectonics of the southern Willamette Valley, Oregon. Masters, Oregon State University, 119p.

Gray, G.G., 1977. A geological field trip guide from Sweet Home, Oregon, to the Quartzville mining district. Ore Bin, v.39, no.6, pp.93-108.

———, 1978. Overview of the Bohemia mining district. Ore Bin, v.40, no.5, pp.77-91.

Gray, (Gary) G., 1986. Native terranes of the central Klamath Mountains, California. Tectonics, v.5, no.7, pp.1043-1054.

———, 2006. Structural and tectonic evolution of the western Jurassic belt along the Klamath River corridor . . . California. *In*: Snoke, A.W., and Barnes, C.G., eds., Geological studies in the Klamath Mountains province, California and Oregon. Geological Society America, Special Paper 410, pp.141-152.

Gray, K.D., and Oldow, J.S., 2005. Contrasting structural histories of the Salmon River belt and

Wallowa terrane. Geological Society America, Bull., v.117, no.5/6, pp.687-706.

Gray, L.B., Sherrod, D.R., and Conrey, R.M., 1996. Potassium-argon ages from the northern Oregon Cascade Range. Isochron/west, no.63, pp.21-28.

Greeley, R., 1971. Geology of selected lava tubes in the Bend area, Oregon. Oregon Department Geology and Mineral Industries, Bull.71, 47p.

Gresens, R.L., and Stewart, R.J., 1981. What lies beneath the Columbia Plateau? Oil and Gas Journal, August, pp.157-164.

Griffiths, R.W., and Campbell, I.H., 1991. Interaction of mantle plume heads with the earth's surface and onset of small-scale convection. Journal Geophysical Research, v.96, pp.18295-18310.

Griggs, A.B., 1945. Chromite-bearing sands of the southern part of the coast of Oregon. U.S. Geological Survey, Bull.945-E., pp.113-150.

Griggs, G.B., and Kulm, L.D., 1970. Sedimentation in Cascadia deep-sea channel. Geological Society America, Bull.81, pp.1361-1384.

———, et al., 1970. Deep-sea gravel from Cascadia Channel. Journal Geology, v.78, no.5, pp.611-619.

———, et al., 1970. Holocene faunal stratigraphy and paleoclimatic implications of deep-sea sediments in Cascadia Basin. Paleogeography, Paleoclimatology, and Paleoecology, v.7, pp.5-12.

Guffanti, M., and Weaver, G.S., 1988. Distribution of late Cenozoic volcanic vents in the Cascade Range. Journal of Geophysical Research, v.93, p.6513-6529.

Gutmanis, J.C., 1989. Wrench faults, pull-apart basins, and volcanism in central Oregon: a new tectonic model based on image interpretation. Journal Geology, v.24, pp.183-192.

Haeussler, P.J., et al., 2003. Life and death of the Resurrection plate. Geological Society America, Bull., v.115, pp.867-880.

Hamilton, W.B., 1978. Mesozoic tectonics of the western United States. *In*: Howell, D.G., and McDougall, K.A., eds., Mesozoic paleogeography of the western United States, Society Economic Paleontologists and Mineralogists, Pacific Coast Symposium 2, p.33-70.

Hammond, C.M., and Vessely, D.A., 1998. Engineering geology of the ancient Highlands landslide, Portland, Oregon. *In*: Burns, S., ed., Environmental, groundwater, and engineering geology; applications from Oregon. Belmont, Calif., Star Publ., pp.3-14.

———, Meier, D., and Backstrand, R., 2009. Paleo-landslides in the Tyee Formation and highway construction, central Oregon Coast Range. *In*: O'Connor, J.E., Dorsey, R.J., and Madin, I.P., eds., Volcanoes to vineyards; field trips through the . . . Pacific Northwest. Geological Society America, Field Guide 15, pp.pp.481-494.

Hammond, P. E., 1979. A tectonic model for evolution of the Cascade Range. *In*: Armentrout, J.M., Cole, M.R., and TerBest, H., eds., Cenozoic paleogeography of the western United States. Society Economic Paleontologists and Mineralogists, Pacific Coast Paleogeography Symposium 3, pp.219-237.

Hampton, E.R., 1972. Geology and ground water of the Molalla-Salem slope area, northern Willamette Valley, Oregon. U.S. Geological Survey, Water Supply Paper 1997, 83p.

Hannan, R.W., and Devine, H., 1998. Ruckel slide, Columbia River Gorge, Cascade Locks, Oregon. *In*: Burns, S., ed., Environmental, groundwater, and engineering geology; applications from Oregon. Belmont, Calif., Star Publ., pp.373-377.

———, and Margolin, L.H., 1998. Needle Rock slide, Lost Creek Lake . . . *In*: Burns, S., ed., Environmental, groundwater, and engineering geology; applications from Oregon. Belmont, Calif., Star Publ., pp.295-300.

Hansen, H.P., 1946. Postglacial forest succession and climate in the Oregon Cascades. American Journal Science, v.244, pp.710-734.

———, 1947. Postglacial vegetation of the northern Great Basin. American Journal Botany, v.31, pp.164-171.

Hanson, L.G., 1986. Scenes from ancient Portland; sketches of major events in the city's geologic history. Oregon Geology, v.48, no.11, pp.130-131.

Harper, G.D., 1983. A depositional contact between the Galice Formation and a late Jurassic ophiolite in northwestern California and southwestern Oregon. Oregon Geology, v.45, no.1, pp.3-7,9.

———, 1984. The Josephine ophiolite, northwestern California. Geological Society America, Bull., v.95, pp. 1009-1026.

———, 2006. Structure of syn-Nevadan dikes and their relationship to deformation of the Galice Formation, western Klamath terrane, northwestern California. *In*: Snoke, A.W., and Barnes, C.G., eds., Geological studies in the Klamath Mountains province, California and Oregon. Geological Society America, Special Paper 410, pp.121-140.

———, and Wright, J.E., 1984. Middle to late Jurassic tectonic evolution of the Klamath Mountains, California-Oregon. Tectonics, v.3, no.7, pp.759-772.

———, Giaramita, M.J., and Kosanke, S.B., 2002. Josephine and Coast Range ophiolites, Oregon and California. *In*: Moore, G., ed., Field guide to geologic processes in Cascadia. Oregon Department of Geology and Mineral Industries, Special Paper 36, pp.1-22.

———, Grady, K., and Wakabayashi, J., 1990. A structural study of a metamorphic sole beneath the Josephine ophiolite, western Klamath terrane, California-Oregon. *In*: Harwood, D.S, and Miller, M.M, eds., Paleozoic and early Mesozoic paleographic relations; Sierra Nevada, Klamath

Mountains, and related terranes. Geological Society America, Special Paper 255, pp.379-396.

———, Saleeby, J. B., and Norman, E.A.S., 1985. Geometry and tectonic setting of sea-floor spreading for the Josephine ophiolite, and implications for Jurassic accretionary events along the California margin. *In*: Howell, D.G., ed., Tectonostratigraphic terranes of the circum-Pacific region. Circum Pacific Council for Energy and Mineral Resources, Earth Science Series no.l, pp.239-257.

Harris, S.L., 2005. Fire mountains of the west; the Cascade and Mono Lake volcanoes. Missoula, Mountain Press, 454p.

Hart, D.H., and Newcomb, R.C., 1965. Geology and groundwater of the Tualatin Valley, Oregon. U.S. Geological Survey, Water Supply Paper 1697, 172p.

Hart, W. K., and Carlson, R.W., 1985. Distribution and geochronology of Steens Mountain-type basalts from the northwestern Great Basin. Isochron/West 43, pp.5-10.

———, and Mertzman, S.A., 1983. Late Cenozoic volcanic stratigraphy of the Jordan Valley area, south-eastern Oregon. Oregon Geology, v.45, no.2, pp.15-19.

Harvey, A.F., and Peterson, G.L., 1998. Water-induced landslide hazards, western portion of the Salem Hills, Marion County, Oregon. Oregon Department of Geology and Mineral Industries, Interpretive Map Series IMS-6.

Harwood, D.S, and Miller, M.M., eds., 1990. Paleozoic and early Mesozoic paleographic relations; Sierra Nevada, Klamath Mountains, and related terranes. Geological Society America, Special Paper 255, 422p.

Haskell, D.C., 1968. The United States Exploring Expedition, 1833-1842 and its publications 1844-1874. New York, Greenwood, 188p.

Hatton, R. R., 1988. Oregon's big country. Bend, Maverick Publ., 137p.

Hawkes, A.D., et al., 2005. Evidence for possible precursor events of megathrust earthquakes on the west coast of North America. Geological Society America, Bull., v.117, no.7/8, pp.996-1008.

Hawkesworth, C., et al., 1995. Calc-alkaline magmatism, lithospheric thinning and extension in the Basin and Range. Journal Geophysical Research, v.100, pp.10271-10286.

Heaton, T. H., and Hartzell, S. H., 1987. Earthquake hazards on the Cascadia subduction zone. Science, v.236, p.162-168.

Heiken, G.H., Fisher, R.V., and Peterson, N.V., 1981. A field trip to the maar volcanoes of the Fort Rock – Christmas Lake Valley basin, Oregon. U.S. Geological Survey, Circular 838, p.119-140.

Heller, P. L., and Ryberg, P. T., 1983. Sedimentary record of subduction to forearc transition in the rotated Eocene basin of western Oregon. Geology, v.11, pp.380-383.

———, Tabor, R.W., and Suczek, C. A., 1987. Paleogeographic evolution of the United States PacificNorthwest during Paleogene time. Canadian Journal Earth Science, v.24, pp.1652-1667.

Hemphill-Hailey, E., 1996. Diatoms as an aid to identifying late-Holocene tsunami deposits. The Holocene, v.6, pp.439-448.

———, et al., 1989. Holocene activity of the Alvord fault, Steens Mountain, southeastern Oregon. U.S. Geological Survey, Report, Grant no. 14-08-0001,45p.

Hering, C.W., 1981. Geology and petrology of the Yamsay Mountain complex, south-central Oregon: a study of bimodal volcanism. PhD, University of Oregon, 194p.

Hickman, C.J.S., 1969. The Oligocene marine molluscan fauna of the Eugene Formation in Oregon. University of Oregon Museum Natural History, Bull.16, 112p.

———, 2003. Evidence for abrupt Eocene-Oligocene molluscan faunal change in the Pacific Northwest. *In*: Prothero, D., Ivany, L., and Nesbitt, E., eds., From greenhouse to icehouse; the marine Eocene-Oligocene transition. New York, Columbia University Press, pp.71-87.

Hildreth, W., 2007. Quaternary magmatism in the Cascades; geologic perspectives. U.S. Geological Survey, Professional Paper 1744, 136p.

Hill, B. E., 1992. Geology and geothermal resources of the Santiam Pass area of the Oregon Cascade Range, Deschutes, Jefferson, and Linn counties, Oregon. Oregon Department of Geology and Mineral Industries, Open-File Report 0-92-3, pp.5-18.

———, and Priest, G.R., 1991. Initial results from the 1990 geothermal drilling program at Santiam Pass, Cascade Range, Oregon. Oregon Geology, v.53, no.5, pp.101-103.

———, and Scott, W.E., 1990. Field trip guide to the central Oregon High Cascades. Part 2 (conclusion): Ash-flow tuffs in the Bend area. Oregon Geology, v.52, pp.123-124.

———, and Taylor, E.M., 1990. Paper connected with field trip Oregon High Cascade pyroclastic units in the vicinity of Bend, Oregon. Oregon Geology, v.52, no.6, pp.125-126, 139.

Hill, M., ed., 1987. Decade of North American Geology, Project Series; Centennial field guides. Geological Society America, Cordilleran Section, 490p.

Hill, R.I., et al., 1992. Mantle plumes and continental tectonics. Science, v.256, pp.186-193.

Hill, R.L., 2004. Volcanoes of the Cascades. Helena, Montana, Falcon Press, 90p.

Hines, G., 1887. Wild life in Oregon . . . New York, Worthington Co., 437p.

Hines, P.R., 1969. Notes on the history of the sand and gravel industry in Oregon. Ore Bin, v.31, no.12, pp.225-237.

Hoblitt, R.P., Miller, C.D., and Scott, W.D., 1987. Volcanic hazards with regard to siting nuclear-power plants in the Pacific Northwest. U.S. Geological Survey, Open-File Report 87-297, 196p.

Hodge, E.T., 1925. Mount Multnomah; ancient ancestor of the Three Sisters. University of Oregon Publication, v.2, no.10, 158p.

———, 1938. Geology of the lower Columbia River. Geological Society America, Bull., v.49, pp.831-930.

———, 1942. Geology of north central Oregon. Oregon State College, Studies in Geol., no.3, 76p.

Hofmeister, R.J., 1999. Earthquake-induced slope stability: a relative hazard map for the vicinity of the Salem Hills, Oregon. Oregon Geology, v.61, no.3, pp.55-63.

———, 2000. Slope failures in Oregon; GIS inventory for three 1996/1997 storm events. Oregon Department of Geology and Mineral Industries, Special Paper 34, 20p.

———, Wang, Y., and Keefer, D.K., 2000. Earthquake-induced slope instability; methodology of relative hazard mapping, western portion of the Salem Hills. Oregon Department of Geology and Mineral Industries, Special Paper 30, 73p.

———, et al., 2002. Geographic Information System (GIS) overview map of potential rapidly moving landslides in western Oregon. Oregon Department of Geology and Mineral Industries, IMS-22, various pagings.

Holmes, K. L., 1968. New horizons in historical research: the geologists' frontier. Ore Bin, v.30, no.8, pp.151-159.

Hooper, P.R., 1997. The Columbia River flood basalt province: current statue. *In*: Mahoney, J.J., and Coffin, M.F., eds., 1997. Large igneous provinces: continental, oceanic, and planetary flood volcanism. Washington, D.C., American Geophysical Union, Geophysical Monograph 100, pp.1-27.

———, and Camp, V.E., 1981. Deformation of the southeast part of the Columbia Plateau. Geology, v.7, pp.322-328.

———, and Conrey, R.M., 1989. A model for the tectonic setting of the Columbia River basalt eruptions. *In*: Reidel, S.P., and Hooper, P.R., eds., Volcanism and tectonism in the Columbia River flood-basalt province. Geological Society America, Special Paper 239, pp.293-305.

———, and Hawkesworth, C.J., 1993. Isotopic and geochemical constraints on the origin and evolution of the Columbia River basalt. Journal Petrology, v.34, pp.1203-1246.

———, and Swanson, D.A., 1987. Evolution of the eastern part of the Columbia Plateau. Washington Division Geology and Earth Resources, Bull.77, pp.197-214.

———, and Swanson, D.A., 1990. The Columbia River Basalt Group and associated volcanic rocks of the Blue Mountains Province. *In*: Walker, G. ed., Geology of the Blue Mountains region of Oregon, Idaho, and Washington: Cenozoic geology of the Blue Mountains. U.S. Geological Survey, Professional Paper 1437, pp.63-99.

———, Bailey, D.G., and McCarley-Holder, G.A., 1995. Tertiary calc-alkaline magmatism associated with lithospheric extension in the Pacific Northwest. Journal Geophysical Research, v.100, no.B7, pp.10,303-10,319.

———, Binger, G.B., and Lees, K.R., 2002. Age of the Steens and Columbia River flood basalts and their relationship to extension-related calc-alkalic volcanism in eastern Oregon. Geological Society America, Bulletin 114, pp.43-50.

———, Knowles, C.R., and Watkins, N.D., 1979. Magnetostratigraphy of the Imnaha and Grande Ronde basalts in the southeast part of the Columbia Plateau. American Journal Science, v.279, pp.737-754.

———, et al., 1993. The Prineville basalt, north-central Oregon. Oregon Geology, v.5, no.1, pp.3-12.

———, et al., 1995. Geology of the northern part of the Ironside Mountain inlier, northeastern Oregon. *In*: Vallier, T.L., and Brooks, H.C., eds., Geology of the Blue Mountains region of Oregon, Idaho, and Washington. U.S. Geological Survey, Professional Paper 1438, pp.415-455.

———, et al., 2007. Origin of the Columbia River flood basalts: plume vs. nonplume models. *In*: Foulger, G.R., et al., eds., Plates, plumes, and paradigms. Geological Society America, Special Paper 388, pp.635-668.

Hotz, P.E., 1971. Geology of lode gold districts in the Klamath Mountains, California and Oregon. U.S. Geological Survey, Bull.1290, 91p.

———, 1971a. Plutonic rocks of the Klamath Mountains, California and Oregon. U.S. Geological Survey, Professional Paper 684-B, pp.B1-B20.

———, 1979. Regional metamorphism in the Condrey Mountain Quadrangle, north-central Klamath Mountains, California. U.S. Geological Survey, Professional Paper 1086, 25p.

Housen, R.A., and Dorsey, R.J., 2005. Paleomagnetism and tectonic significance of Albian and Cenomanian turbidites, Ochoco basin, Mitchell inlier, central Oregon. Journal Geophysical Research, v.110, B-07102, pp.1-22.

Howard, J.K., and Dott, R.H., 1961. Geology of Cape Sebastian State Park and its regional relationships. Ore Bin, v.23, no.8, pp.75-81.

Howell, D.G., ed., 1985. Tectonostratigraphic terranes of the circum-Pacific region. Circum-Pacific Council for Energy and Mineral Resources, Earth Science Series, No.1, 581p.

———, and McDougall, K.A., ed., 1978. Mesozoic paleogeography of the Western United States. Society Economic Paleontologists and

Mineralogists, Pacific Section, Pacific Coast, Paleogeography Symposium 2, 573p.

Hughes, J.M., Stoiber, R.E., and Carr, M.J., 1980. Segmentation of the Cascade volcanic chain. Geology, v.8, pp.15-17.

Hughes, S.S., 1990. Mafic magmatism and associated tectonism of the central High Cascade Range, Oregon. Journal Geophysical Research, v.95, no.B12, pp.19,623-19,638.

———, and Taylor, E.M., 1986. Geochemistry, petrogenesis, and tectonic implications of central High Cascade mafic platform lavas. Geological Society America, Bull., v.97, pp.1024-1036.

Humphreys, E.D., and Dueker, K.G., 1994. Physical state of the western U.S. upper mantle. Journal Geophysical Research, v.99, pp.9635-9650.

———, and Coblentz, D.D., 2007. North American dynamics and western U.S. tectonics. Reviews of Geophysics, v.45, 30p.

———, et al., 2000. Beneath Yellowstone: evaluating plume and nonplume models using teleseismic images of the upper mantle. GSA Today, no.10, pp.1-6.

Hunter, R.E., 1980. Depositional environments of some Pleistocene coastal terrace deposits, southwestern Oregon-case history of a progradational beach and dune sequence. Sedimentary Geology, v.27, pp.241-262.

———, and Clifton, H..E, 1982. Cyclic deposits and hummocky cross-stratification of probable storm origin in upper Cretaceous rocks of the Cape Sebastian area, southwestern Oregon. Journal Sedimentary Petrology, v.27, no.4, pp.241-262.

———, Clifton, H.E., and Phillips, R.L., 1970. Geology of the stacks and reefs off the southern Oregon coast. Ore Bin, v.32, no.10, pp.185-201.

Hyndman, R.D., and Wang, K., 1993. Thermal constraints on the zone of major thrust earthquake failure, the Cascadia subduction zone. Journal Geophysical Research, v.98, pp.2039-2060.

———, 1995. The rupture zone of Cascadia great earthquakes from current deformation and the thermal regime. Journal Geophysical Research, v.100, no.B11, pp.22,133-22,154.

Imlay, R.W., 1986. Jurassic ammonites and biostratigraphy of eastern Oregon and western Idaho. *In*: Vallier, T.L., and Brooks, H.C., eds., Geology of the Blue Mountains region of Oregon, Idaho, and Washington. U.S. Geological Survey, Professional Paper 1435, pp.53-57.

Ingebritsen, E.D., Sherrod, D.R., and Mariner, R.H., 1989. Heat flow and hydrothermal circulation in the Cascade Range, north-central Oregon. U.S. Geological Survey, Open-File Report 89-178, pp.122-141. (also: Science, no.243, pp.1458-1462).

———, Mariner, R.H., and Sherrod, D.R., 1994. Hydrothermal systems of the Cascade Range, north-central Oregon. U.S. Geological Survey, Professional Paper 1044-L, 86p.

———, Sherrod, D.R., and Mariner, R.H., 1992. Rates and patterns of groundwater flow in the Cascade Range volcanic arc . . . Journal Geophysical Research, v.97, no.B4, pp.4599-4627.

Ingersoll, R.V., 1982. Triple-junction instability as cause for late Cenozoic extension and fragmentation of the western United States. Geology, v.10, no.12, pp.621-624.

Irwin, W. P., 1960. Geologic reconnaissance of the northern Coast Ranges and Klamath Mountains, California, with a summary of the mineral resources. California Division Mines, Bull.179, 80p.

———, 1964. Late Mesozoic orogenies in the ultramafic belts of northwestern California and south-western Oregon. U.S. Geological Survey, Professional Paper 501-C, pp.C1-C9.

———, 1966. Geology of the Klamath Mountains Province. *In*: Bailey, E.N., ed., Geology of northern California. California Division Mines, Bull.190, pp.19-38.

———, 1972. Terranes of the western Paleozoic and Triassic belt in the southern Klamath Mountains, California. U.S. Geological Survey, Professional Paper 800-C, pp.103-111.

———, 1977. Review of Paleozoic rocks of the Klamath Mountains. *In*: Stuart, J.H., Stevens, C.H., and Frische, A.E., eds., Paleozoic paleogeography of the western United States. Society Economic Paleontologists and Mineralogists, Pacific Section Paleogeography Symposium 1, pp.441-454.

———, 1985. Age and tectonics of plutonic belts in accreted terranes of the Klamath Mountains, California and Oregon. *In*: Howell, D.G., ed.,Tectonostratigraphic terranes of the circum-Pacific region. Circum-Pacific Council for Energy and Mineral Research, Earth Science Series, No.1, pp.187-198.

———, 1997. Preliminary map of selected post-Nevadan geologic features of the Klamath Mountains and adjacent areas, California and Oregon. U.S. Geological Survey, Open-File Report 97-465, 29p.

———, Jones, D.L., and Kaplan, T.A., 1978. Radiolarians from pre-Nevadan rocks of the Klamath Mountains, California and Oregon. *In*: Howell, D. G., and McDougall, K.A., eds., Mesozoic paleogeography of the western United States. Society Economic Paleontologists and Mineralogists, Pacific Coast Paleogeography Symposium 2, pp.303-310.

———, Wardlaw, B.R., and Kaplan, T.A., 1983. Conodonts of the western Paleozoic and Triassic belt, Klamath Mountains, California and Oregon. Journal Paleontology, v.57, pp.1030-1039.

Ivany, L.C., Newbitt, E.A., and Prothero, D.R., 2003. The marine Eocene-Oligocene transition: a synthesis. *In*: Prothero, D.R., Ivany, L.C., and Nesbitt, E.A., eds., From greenhouse to icehouse. New York, Columbia University Press, pp.522-534.

Jacobson, R.S., 1985. The 1984 landslide and earthquake activity on the Baker-Homestead highway near Halfway, Oregon. Oregon Geology, v.74, no.5, pp.51-58.

———, et al., 1986. Map of Oregon seismicity, 1841-1986. Oregon Department Geology and Mineral Industries, Geological Map Series, GMS-49.

Jacoby, G.C., Bunker, D.E., and Benson, B.E., 1997. Tree-ring evidence for an A.D. 1700 Cascadia earthquake in Washington and northern Oregon. Geology, v.25, pp.999-1002.

Janda, R.J., 1969. Age and correlation of marine terraces near Cape Blanco, Oregon. Geological Society America, Cordilleran Section, Abstracts with Programs, v.1, pt.3, p.19-20.

Jefferson, A., Grant, G., and Rose, T., 2006. Influence of volcanic history on groundwater patterns on the west slope of the Oregon High Cascades. Water Resources Research, v.42, no.W12411, 15p.

Jennings, A.E., and Nelson, A.R., 1992. Foraminiferal assemblage zones in Oregon tidal marshes. Journal Foraminiferal Research, v.22, pp.13-29.

Jensen, R.A., 2006. Roadside guide to the geology and history of Newberry Volcano. 4th ed., Bend, CenOreGeoPub, 182p.

———, Donnelly-Nolan, J.M., and Mckay, D., 2009. A field guide to Newberry Volcano, Oregon. *In*: O'Connor, J.E., Dorsey, R.J., and Madin, I.P., eds., Volcanoes to vineyards; field trips through the . . . Pacific Northwest. Geological Society America, Field Guide 15, pp.53-79.

Johnson, D.M., et al., 1985. Atlas of Oregon Lakes. Corvallis, Oregon State University Press, 317p.

Johnson, H. P., and Holmes, M.L., 1989. Evolution in plate tectonics; the Juan de Fuca Ridge. *In*: Winterer,E.L., Hussong, D.M., and Decker, R.W., eds., The eastern Pacific Ocean and Hawaii. Decade of North Americana Geology, Vol.N., pp.73-90.

Johnson, J.A., 1998. Geologic map of the Frederick Butte volcanic center, Deschutes and Lake counties, south-central Oregon. U.S. Geological Survey, Open File Report 98-208.

———, and Grunder, A.L., 2000. The making of intermediate composition magma in a bimodal suite: Duck Butte eruptive center, Oregon, USA. Journal of Volcanology and Geothermal Research, v.95, pp. 175-195.

Johnson, K., and Barnes, C.G., 2006. Magma mixing and mingling in the Grayback pluton, Klamath Mountains, Oregon. *In*: Snoke, A.W., and Barnes, C.G., eds., Geological studies in the Klamath Mountains province, California and Oregon. Geological Society America, Special Paper 410, pp.247-267.

Johnson, K. E., and Ciancanelli, E. V., 1984. Geothermal exploration at Glass Buttes, Oregon. Oregon Geology, v.46, no.2, pp.15-18.

Johnston, D.A., and Donnelly-Nolan, J.M., eds., 1981. Guides to some volcanic terranes in Washington, Idaho, Oregon, and northern California. U.S. Geological Survey, Circular 838, 189p.

Jones, D.L., Silberling, N.J., and Hillhouse, J., 1977. Wrangellia - A displaced terrane in northwestern North America. Canadian Journal Earth Sciences, v.14, pp.2565-2577.

———, et al., 1982. The growth of western North America. Scientific American, v.247, no.5, pp.70-85.

Jordan, B.T., 2002. Basaltic volcanism and tectonics of the High Lava Plains, southeastern Oregon. Oregon State University, Corvallis, PhD, 218p.

———, 2005. Age-progressive volcanism of the Oregon High Lava Plains: overview and evaluation of tectonic models. *In*: Foulger, G.R., et al., eds., Plates, plumes, and paradigms. Geological Society America, Special Paper 388, pp.505-515.

———, Streck, M.J., and Grunder, A.L., 2002. Bimodal volcanism and tectonism of the High Lava Plains, Oregon. *In*: Moore, G., ed., Field guide to geologic processes in Cascadia. Oregon Department of Geology and Mineral Industries, Special Paper 36, pp.23-46.

———, et al., 2004. Geochronology of age-progressive volcanism of the Oregon High Lava Plains: implications for the plume interpretation of Yellowstone. Journal of Geophysical Research, v.109, pp.B10, 202.

Kadri, M.K., 1982. Structure and influence of the Tillamook uplift on the stratigraphy of the Mist area, Oregon. Masters, Portland State University, 105p.

———, et al., 1983. Geochemical evidence for changing provenance of Tertiary formations in northwestern Oregon. Oregon Geology, v.45, no.2, pp.20-22.

Karl, H.A., Hampton, M.A., and Kenyon, N.H., 1989. Lateral migration of Cascadia Channel in response to accretionary tectonics. Geology, v.17, pp.144-147.

Karson, J.A., Tivey, M.A., and Delaney, J.R., 2002. Internal structure of uppermost oceanic crust along the western Blanco transform scarp. Journal Geophysical Research, v.107, no.B9, pp.EPM-1.

Katsura, K. T., 1988. The geology and epithermal vein mineralization at the Champion mine, Bohemia mining district, Oregon. Masters, University of Oregon, 254p.

Kays, A.M., 1970. Western Cascades volcanic series, south Umpqua Falls region, Oregon. Ore Bin, v.32, no.5, pp.81-94.

———, and Ferns, M.L., 1980. Geologic field trip guide through the north-central Klamath Mountains. Oregon Geology, v.42, no.2, pp.23-35.

———, Stimac, J.P., and Goebel, P.M., 2006. Permian-Jurassic growth and amalgamation of the Wallowa composite terrane, northeastern Oregon. *In*: Snoke, A.W., and Barnes, C.G., eds., Geological studies in the Klamath Mountains province, California and

Oregon. Geological Society America, Special Paper 410, pp.465-494.

Keech, C.E., and Sanford, B.A., 1998. Tooth Rock landslide, Columbia Gorge, Oregon. *In*: Burns, S., ed., Environmental, groundwater, and engineering geology; applications from Oregon. Belmont, Calif., Star Publ., pp.367-372.

Keefer, D.K., 1984. Landslides caused by earthquakes. Geological Society America, Bull., v.95, no.4, pp.406-421.

———, and Wang, Y., 1997. A method for predicting slope instability for earthquake hazard maps. *In*: Wang, Y., and Neuendorf, K.K.E., eds., Earthquakes – Converging at Cascadia. Oregon Department Geology and Mineral Industries, Special Paper 28, pp.39-52.

Kelsey, H.M., 1990. Late Quaternary deformation of marine terraces on the Cascadia subduction zone near Cape Blanco, Oregon. Tectonics, v.9, pp.983-1014.

———, and Bockheim, J.G., 1994. Coastal landscape evolution as a function of eustasy and surface uplift rate, Cascadia margin, southern Oregon. Geological Society America, Bull., v.106, pp.840-854.

———, Witter, R.C., and Hemphill-Haley, E., 1998. Response of a small Oregon estuary to coseismic subsidence and postseismic uplift in the past 300 years. Geology, v.26, pp.231-234.

———, Witter, R.C., and Hemphill-Haley, E., 2002. Plate-boundary earthquakes and tsunamis of the past 5500 yr, Sixes River estuary, southern Oregon. Geological Society America, Bull., v.114, no.3, pp.298-314.

———, et al., 1994. Topographic form of the Coast Ranges of the Cascadia margin in relation to coastal uplift rates and plate subduction. Journal Geophysical Research, v.99, no.B6, pp.12,245-12,255.

———, et al., 1996. Quaternary upper plate deformation in coastal Oregon. Geological Society America, Bull., v.108, no.7, pp.843-860.

———, et al., 2005. Tsunami history of an Oregon coastal lake reveals a 4600 yr record of great earthquakes on the Cascadia subduction zone. Geological Society America, Bull., v.117, no.7/8, pp.1009-1032.

Kienle, C.F., Nelson, C.A., and Lawrence, R.D., 1981. Faults and lineaments of the southern Cascades, Oregon. Oregon Department of Geology and Mineral Industries, Special Paper 13, 23p.

Kittleman, L. R., 1973. Guide to the geology of the Owyhee region of Oregon. Museum Natural History, University of Oregon, Bull.21, 61p.

———, et al., 1965. Cenozoic stratigraphy of the Owyhee region, southeastern Oregon. Museum Natural History, University of Oregon, Bull.1, 45p.

———, et al., 1967. Geologic map of the Owyhee region, Malheur County, Oregon. Museum of Natural History, University of Oregon, Bull.8.

Kleck, W.D., 1972. The geology of some zeolite deposits in the southern Willamette Valley, Oregon. Ore Bin, v.34, no.9, pp.145-151.

Kleinhans, L. C., Balcells-Baldwin, E.A., and Jones, R.A., 1984. A paleogeographic reinterpretation of some middle Cretaceous units, north-central Oregon: evidence for a submarine turbidite system. *In*: Nilsen, T.H., ed., Geology of the upper Cretaceous Hornbrook Formation, Oregon and California. Society Economic Paleontologists and Mineralogists, Pacific Section, v.42, pp.239-257.

Koch, J.G., 1966. Late Mesozoic stratigraphy and tectonic history, Port Orford-Gold Beach area, southwestern Oregon coast. American Association Petroleum Geology, v.50, no.1, pp.25-71.

Komar, P.D., 1976. Beach processes and sedimentation. Prentice-Hall, New Jersey, 429p.

———, 1978. Wave conditions on the Oregon coast during the winter of 1977-1978 . . . Shore and Beach, v.46, pp.3-8.

———, 1983. The erosion of Siletz Spit, Oregon. *In*: Komar, P.D., ed., CRC Handbook of coastal processes and erosion. CRC Press, Boca Raton, Florida, pp.65-76.

———, 1992. Ocean processes and hazards along the Oregon coast. Oregon Geology, v.54, no.1, pp.3-19.

———, 1997. The Pacific Northwest coast. Durham, Duke University Press, 195p.

———, 2004. Oregon's coastal cliffs: processes and erosion impacts. U.S. Geological Survey, Professional Paper 1693, pp.65-80.

———, and Rea, C.C., 1976. Beach erosion on Siletz Spit, Oregon. Ore Bin, v.38, no.8, pp.119-134.

———, and Schlicker, H., 1980. Beach processes and erosion problems on the Oregon Coast. Oregon Department Geology and Mineral Industries, Bull.101, pp.169-173.

Koski, R.A. and Derkey, R.E., 1981. Massive sulfide deposits in oceanic-crust and island-arc terranes of south-western Oregon. Oregon Geology, v.43, no.9, pp.119-125.

Kulm, L.D., 1980. Sedimentary rocks of the central Oregon continental shelf. *In*: Newton, V.C., et al., Prospects for oil and gas in the Coos basin, western Coos, Douglas,and Lane counties, Oregon. Oregon, Department Geology and Mineral Industries, Oil and Gas Investigations 6, pp.29-34.

———, and Fowler, G.A., 1974. Oregon continental margin structure and stratigraphy: a test of the imbricate thrust model. *In*: Burk, C.A., and Drake, M.J., eds., The geology of continental margins. New York, Springer-Verlag, p.261-283.

———, and Scheidegger, K.F., 1979. Quaternary sedimentation on the tectonically active Oregon continental slope. Society Economic Paleontologists and Mineralogists, Special Publication, 27, pp.247-264.

———, et al., 1968. Evidence for possible placer accumulations on the southern Oregon continental shelf. Ore Bin, v.30, no.5, pp.81-104.

———, et al., 1975. Oregon continental shelf sedimentation interrelationships of facies distribution and sedimentary processes. Journal Geology, v.83, pp.145-175.

———, et al., 1986. Oregon subduction zone: venting, fauna, and carbonates. Science, v.231, pp.561-566.

LaMaskin, T.A., 2008. Late Triassic (Carnian-Norian) mixed carbonate-volcaniclastic facies of the Olds Ferry terrane, eastern Oregon and western Idaho. *In*: Blodgett, R.B., and Stanley, G.D., eds., The terrane puzzle: new perspectives on paleontology and stratigraphy from the North American Cordillera. Geologic Society America, Special Paper 442, pp.251-267.

———, Dorsey, R.J., and Vervoort, J.D., 2008. Tectonic controls on mudrock geochemistry, Mesozoic rocks of eastern Oregon and western Idaho, U.S.A. Journal Sedimentary Research, v.78, pp.787-805.

———, Dorsey, R.J., and Vervoort, J.D., 2009. Initiation of the Cretaceous Andean-type margin of the western U.S. Cordillera. Geological Society America, Abstracts with Programs, v.41, no.7, pp.183.

———, et al., 2009. Masozoic sedimentation, magmatism, and tectonics in the Blue Mountains province, northeastern Oregon. *In*: O'Connor, J.E., Dorsey, R.J., and Madin, I.P., eds., Volcanoes to vineyards; field trips through the . . . Pacific Northwest. Geological Society America, Field Guide 15, pp.187-202.

———, et al., (in press) Early Mesozoic paleogeography and tectonic evolution of the western United States. Geological Society America, Bull.

Lange, E.F., 1985. The Willamette meteorite, 1902-1962. West Linn, Oregon, West Linn Fair Board, 24p.

Lawrence, D.C., 1988. Geologic field trip guide to the northern Succor Creek area, Malheur County, Oregon. Oregon Geology, v.50, no.2, pp.15-21p.

Lawrence, R.D., 1976. Strike-slip faulting terminates the Basin and Range province in Oregon. Geological Society America, Bull. 87, pp.846-850.

———, 1980. Marys Peak field trip: structure of the eastern flank of the central Coast Range, Oregon. Oregon Department Geology and Mineral Industries, Bull.101, pp.121-167.

Leeman, W.P., et al., 1995. Petrology of the Canyon Mountain complex, eastern Oregon. *In*: Vallier, T.L., and Brooks, H.C., eds., Geology of the Blue Mountains region of Oregon, Idaho, and Washington. U.S. Geological Survey, Professional Paper 1438, pp.1-43.

Leithold, E.L., 1984. Characteristics of coarse-grained sequences deposited in nearshore, wave-dominated environments-examples from the Miocene of south-central Oregon. Sedimentology, v.31, pp.749-775.

———, and Bourgeois, J., 1983. Sedimentology of the sandstone of Floras Lake (Miocene) transgressive, high-energy shelf deposition, SW Oregon. *In*: Larue, D.K., et al., eds., Cenozoic marine sedimentation, Pacific margin, U.S.A. Society Economic Paleontologists and Mineralogists, Pacific Section, pp.17-28.

Leppert, D., 1990. Developments in applications for southeast Oregon bentonites and natural zeolites. Oregon Department Geology and Mineral Industries, Special Paper 23, pp.19-24.

Libbey, F.W., 1962. The Oregon King mine, Jefferson County, Oregon. Ore Bin, v.24, no.7, pp.101-115.

———, 1967. The Almeda Mine Josephine County, Oregon. Oregon Department Geology and Mineral Industries, Short Paper 24, 53p.

———, 1976. Lest we forget. Ore Bin, v.38, no.12, pp.179-195.

Liberty, L.M., et al., 1999. Integration of high-resolution seismic and aeromagnetic data for earthquake hazards evaluations . . . Willamette Valley, Oregon. Seismological Society America, Bull., v.89, no.6, pp.1473-1483.

Lillie, R.J., and Couch, R.W., 1979. Geophysical evidence of fault termination of the Basin and Range province in the vicinity of the Vale, Oregon, geothermal area. *In*: Newman, G.W., et al., eds., Basin and Range symposium and Great Basin field conference, pp.175-184.

Lindsey, K., et al., 2009. Hydrogeology of the Columbia River Basalt Group in the Columbia Plateau. *In*: O'Connor, J.E., Dorsey, R.J., and Madin, I.P., eds., Volcanoes to vineyards; field trips through the . . . Pacific Northwest. Geological Society America, Field Guide 15, pp.673-696.

Lindsley-Griffin, N. and Kramer, J.C., eds., 1977. Geology of the Klamath Mountains, northern California. Geological Society America, Cordilleran Section, 73rd Annual Mtg. [fieldtrip guide], 156p.

Loy, W.G., et al., eds., 2001. Atlas of Oregon. 2nd edition. Eugene, University of Oregon Press, 301p.

Luedke, R.G., and Smith, R.L., 1982. Map showing distribution, composition, and age of late Cenozoic volcanic centers in Oregon and Washington. U.S. Geological Survey, Miscellaneous Investigations Series, I-1091-D.

———, Smith, R. L., and Russell-Robinson, S. L., 1983. Map showing distribution, composition, and age of late Cenozoic volcanoes and volcanic rocks of the Cascade range and vicinity, northwestern United States. U.S. Geological Survey, Miscellaneous Invertigations Series, I-1507.

Lund, E.H., 1971. Coastal landforms between Florence and Yachats, Oregon. Ore Bin, v.33, no.2, pp.21-44.

———, 1972. Coastal landforms between Tillamook Bay and the Columbia River, Oregon. Ore Bin, v.34, no.11, pp.173-194.

———, 1972a. Coastal landforms between Yachats and Newport, Oregon. Ore Bin, v.34, no.5, pp.73-91.

———, 1973. Landforms along the coast of southern Coos County, Oregon. Ore Bin, v.35, no.12, pp.189-210.

———, 1973a. Oregon coastal dunes between Coos Bay and Sea Lion Point. Ore Bin, v.35, no.5, pp.73-92.

———, 1974. Coastal landforms between Roads End and Tillamook Bay, Oregon. Ore Bin, v.36, no.11, pp.173-194.

———, 1974a. Rock units and coastal landforms between Newport and Lincoln City, Oregon. Ore Bin, v.36, no.5, pp.69-90.

———, 1977. Geology and hydrology of the Lost Creek glacial trough. Ore Bin, v.39, no.9, pp.141-156.

———, and Bentley, E., 1976. Steens Mountain, Oregon. Ore Bin, v.38, no.4, pp.51-66.

Lund, J.W., et al., 2009. Geothermal geology and utilization in Oregon. *In*: O'Connor, J.E., Dorsey, R.J., and Madin, I.P., eds., Volcanoes to vineyards; field trips through the . . . Pacific Northwest. Geological Society America, Field Guide 15, pp.583-598.

Lux, D. R., 1983. K-Ar and 40 Ar-39 Ar ages of mid-Tertiary volcanic rocks from the Western Cascade Range, Oregon. Isochron/West, no.33, p.27-32.

Mabey, M.A., et al., 1993. Earthquake hazard maps of the Portland quadrangle . . . Oregon Department of Geology and Minerals Industries, GMS-79.

MacLeod, N.S., and Sherrod, D.R., 1988. Geological evidence for a magma chamber beneath Newberry Volcano, central Oregon. Journal Geophysical Research, v.93, pp.10059-10066.

———, Walker, G.W., and McKee, E.H., 1975. Geothermal significance of eastward increase in age of upper Cenozoic rhyolitic domes in southeastern Oregon. U.S. Geological Survey, Open-File Report 75-348, 21p.

———, et al., 1995. Geologic map of Newberry volcano, Deschutes, Klamath, and Lake Counties, Oregon. U.S. Geological Survey, Miscellaneous Investigations Map, Map I-2455.

Madin, I. P., 1989. Evaluating earthquake hazards in the Portland, Oregon, metropolitan area: mapping potentially hazardous soils. Oregon Geology, v.51, no.5, pp.106-110, 118.

———, 1990. Earthquake-hazard geology maps of the Portland metropolitan area, Oregon. Oregon Department of Geology and Mineral Industries, Open-File Report 0-90-2, 21p.

———, 1992. Seismic hazards on the Oregon coast. Corvallis, Oregon Sea Grant, Oregon State University, pp.3-27.

———, 1994. Geologic map of the Damascus Quadrangle, Clackamas and Multnomah Counties, Oregon. Oregon Department of Geology and Mineral Industries, Geological Map Series, GMS-60.

———, 2007. LIDAR and landslides—New technology supports a new landslide hazard mapping program. *In*: Burns, W.J., and Wang, Y., comp., Landslide symposium proceedings and field trip guide. Oregon Department of Geology and Mineral Industries, Open-File Report 0-07-06.

———, 2009. Portland, Oregon, geology by tram, train, and foot. Oregon Geology, v.69, no.1, pp.1-19.

———, and Burns, W.J., 2007. A new landslide database for Oregon. *In*: Burns, W.J., and Wang, Y., comp. Landslide symposium proceedings and field trip guide. Oregon Department of Geology and Mineral Industries, Open-File Report 0-07-06.

———, and Murray, R.B., 2007. Giant landslides in the Coburg Hills; implications to urban rural development. *In*: Burns, W.J., and Wang, Y., comp., 2007 landslide symposium proceedings and field trip guide. Oregon Department of Geology and Mineral Industries, Open-File Report 0-07-06, p.99.

———, and Mabey, M.A., eds., 1996. Earthquake hazards maps for Oregon. Oregon Department of Geology and Mineral Industries. GMS-100.

———, et al., 1993. March 25, 1993, Scotts Mills earthquake – western Oregon's wake-up call. Oregon Geology, v.55, no.3, pp.51-57.

Magill, J. and Cox, A., 1980. Tectonic rotation of the Oregon Western Cascades. Oregon Department Geology and Mineral Industries, Special Paper 10, 67p.

———, and Cox, A., 1981. Post-Oligocene tectonic rotation of the Oregon Western Cascade Range and the Klamath Mountains. Geology, v.9, pp.127-131.

Mahoney, J.J., and Coffin, M.F., eds., 1997. Large igneous provinces: continental, oceanic, and planetary flood volcanism. American Geophysical Union, Geophysical Monograph 100, pp.1-27.

Mahoney, K.A., and Steere, M.L., 1983. Index to the Ore Bin (1939-1978) and Oregon Geology (1979-1982). Oregon Department Geology and Mineral Industries, Special Paper 16, 46p.

Mamay, S.H., and Read, C.B., 1956. Additions to the flora of the Spotted Ridge Formation in central Oregon. U.S. Geological Survey, Professional Paper 274-I, pp.211-226.

Mangan, M.T., et al., 1986. Regional correlation of Grande Ronde basalt flows, Columbia River Basalt group, Washington, Oregon, and Idaho. Geological Society America, Bull. 97, pp.1300-1318.

Mangum, D., 1967. Geology of Cape Lookout State Park, near Tillamook, Oregon. Ore Bin, v.29, no.5, pp.85-109.

Mankinen, E. A., and Irwin, W.P., 1990. Review of paleomagnetic data from the Klamath Mountains, Blue Mountains, and Sierra Nevada. *In*: Harwood, D.S, and M.M. Miller, eds., Paleozoic and early Mesozoic paleographic relations; Sierra Nevada, Klamath Mountains, and related terranes. Geological Society America, Special Paper 255, pp. 397-409.

———, Irwin, W.P., and Sherman, G., 1984. Implications of paleomagnetism for the tectonic history of the eastern Klamath and related terranes in California and Oregon. *In*: Nilsen, T., ed., Geology of the upper Cretaceous Hornbrook Formation, Oregon and California. Society Economic Paleontologists and Mineralogists, Pacific Section, v.42, pp.221-229.

———, et al., 1987. The Steens Mountain (Oregon) geomagnetic polarity transition; 3, Its regional significance. Journal Geophysical Research, v.92, no.B8, pp.8057-8076.

Mann, G.M., and Meyer, C.E., 1993. Late Cenozoic structure and correlations to seismicity along the Olympic-Wallowa lineament, northwest United States. Geological Society America, Bull., v.105, pp.853-871.

Mariner, R.H., et al., 1990. Discharge rates of fluid and heat by thermal springs of the Cascade Range, Washington, Oregon, and northern California. Journal Geophysical Research, v.95, no.B12, pp.19,517-19,531.

Marsh, S.P., Kropschot, S.J., and Dickinson, R.G., 1984. Wilderness mineral potential; assessment of mineral-resource potential in U.S. Forest Service lands studied, 1964-1984. U.S. Geological Survey, Professional Paper 1300, v.2, 1183p.

Mason, R.S., 1965. The walls of Portland. Ore Bin, v.27, no.4, pp.65-74.

———, 1985. Walls worth walking by: a tour of the South Park Blocks area of downtown Portland. Oregon Geology, v.47, no.11, pp.127-134.

Maxime, M., and Boudier, F., 1985. Structures in the Canyon Mountain ophiolite indicate an island-arc intrusion. Tectonophysics, v.120, pp.191-209.

Maynard, L.C., 1974. Geology of Mt. McLoughlin, Oregon. Masters, University of Oregon, 139p.

McAdoo, B.G., and Watts, P., 2004. Tsunami hazard from submarine landslides on the Oregon continental slope. Marine Geology, v.203, pp.235-245.

McArthur, L.A., 2003. Oregon Geographic Names. 7th ed. rev. and enl. Portland, Oregon Historical Society, 1073p.

McBirney, A. R., 1968. Compositional variations of the climactic eruption of Mount Mazama. *In*: Dole, Hollis, ed., Andesite conference guidebook - international mantle project; science report 16-S. Oregon Department Geology and Mineral Industries, Bull.62, pp.53-56.

———, 1978. Volcanic evolution of the Cascade Range. Annual Review Earth and Planetary Science Letters, v.6, pp.437-456.

———, et al., 1974. Episodic volcanism in the central Oregon Cascade Range. Geology, v.2, pp.585-590.

McBride, J.N., 1976. Science at a land grant college: the science controversy in Oregon, 1931-1942. Masters, Oregon State University.

McCaffrey, R., and Goldfinger, C., 1995. Forearc deformation and great subduction earthquakes. Science, no.267, pp.856-859.

———, et al., 2000. Rotation and plate locking at the southern Cascadia subduction zone. Geophysical Research Letters, v.27, no.19, pp.3117-3120.

McClaughry, J.D., and Ferns, M.L., 2006. Field trip guide to the geology of the lower Crooked River basin, Redmond and Prineville areas, Oregon. Oregon Geology, v.67, no.1, pp.15-23.

———, and Ferns, M.L., 2006a. Preliminary geologic map of the Prineville 7 ½ quadrangle, Crook County, Oregon. Oregon Department of Geology and Mineral Industries, Open-File Report 0-06-22.

———, and Ferns, M.L., 2006b. Preliminary geologic map of the Ochoco Reservoir 7 ½ quadrangle, Crook County, Oregon. Oregon Department of Geology and Mineral Industries, Open-File Report 0-06-23.

———, Gordon, C.L., and Ferns, M.L., 2009. Field trip guide to the middle Eocene Wildcat Mountain Caldera, Ochoco National Forest, Crook County, Oregon. Oregon Geology, v.69, no.1, pp.5-24.

———, et al., 2009. Field trip guide to the Oligocene Crooked River caldera. Oregon Geology, v.69, no.1, pp.25-44.

———, et al., 2009a. Paleogene calderas of central and eastern Oregon; eruptive sources of widespread tuffs in the John Day and Clarno formations. *In*: O'Connor, J.E., Dorsey, R.J., and Madin, I.P., eds., Volcanoes to vineyards; field trips through the . . . Pacific Northwest. Geological Society America, Field Guide 15, pp. 91-110.

McCornack, E.C., 1928. Thomas Condon, pioneer geologist of Oregon. Eugene, University of Oregon Press, 255p.

McCrory, P.A., et al., 2002. Crustal deformation at the leading edge of the Oregon Coast Range block, offshore Washington. U.S. Geological Survey, Professional Paper 1661-A, 47p.

McDonald, J.H., Harper, G.D., and Zhu, B., 2006. Petrology, geochemistry, and provenance of the Galice Formation, Klamath Mountains, Oregon and California. *In*: Snoke, A.W., and Barnes, C.G., eds., Geological studies in the Klamath Mountains province, California and Oregon. Geological Society America, Special Paper 410, pp. 77-101.

McDougall, K.A., 1980. Paleoecological evaluation of late Eocene biostratigraphic zonations of the Pacific Coast of North America. Journal Paleontology, v.54, no.4, Supplement, 75p.

McDowell, P.F., 1991. Quaternary stratigraphy and geomorphic surfaces of the Willamette Valley, Oregon. *In*: Morrison, R.B., ed., Quaternary nonglacial geology – Conterminous United States. Geological Society America, Decade of North American Geology, v.K-2, pp.156-164.

———, 1992. An overview of Harney Basin geomorphic history, climate, and hydrology. *In*: Raven, C., and Elston, R.G., eds., Land and life at Malheur Lake. U.S. Dept. of Fish and Wildlife Service, Cultural Resources Series, no.8, pp.13-24.

———, and Dugas, D.P., 1993. Holocene lake-level variations and eolian activity in the Oregon Great Basin. *In*: Benson, L.V., ed., Ongoing paleoclimatic studies in the northern Great Basin, Proceedings, Reno, Nevada, May, 1993. U.S. Geological Survey, Circular 1119, pp.53-54.

———, and Roberts, M.C., 1987. Field guidebook to the Quaternary stratigraphy, geomorphology, and soils of the Willamette Valley. Association American Geographers, Annual Meeting., Portland, Oregon, 75p.

McElderry, S., 1998. Vanport conspiracy rumors and social relations in Portland, 1940-1950. Oregon Historical Quarterly, v.99, no.2, pp.134-163.

McFarland, W.D., and Morgan, D.S., 1996. Description of the ground-water flow system in the Portland Basin, Oregon and Washington. U.S. Geological Survey, Water-Supply Paper 2470-A, 58p.

McHugh, M.H., 1986. Landslide occurrence in the Elk and Sixes River basins, southwest Oregon. Masters, Oregon State University, 106p.

McInelly, G.W., and Kelsey, H.M., 1990. Age estimates and uplift rates for late Pleistocene marine terraces, southern Oregon portion of the Cascadia forearc. Journal Geophysical Research, no.95, pp.6685-6698.

———, and Kelsey, H.M., 1990a. Late Quaternary tectonic deformation in the Cape Arago-Bandon region of coastal Oregon as deduced from wave-cut platforms. Journal Geophysical Research, v.95, no.B5, pp.6699-6713.

Mckay, D., et al., 2009. The post-Mazama northwest rift zone eruption at Newberry Volcano, Oregon. *In*: O'Connor, J.E., Dorsey, R.J., and Madin, I.P., eds., Volcanoes to vineyards; field trips through the . . . Pacific Northwest. Geological Society America, Field Guide 15, pp. 91-110.

McKee, B., 1972. Cascadia; the geologic evolution of the Pacific Northwest. McGraw-Hill, New York, 394p.

McKee, E.H., Duffield, W.A., and Stern, R.J., 1983. Late Miocene and early Pliocene basaltic rocks and their implications for crustal structure, northeastern California and southcentral Oregon. Geological Society America, Bull.94, pp.292-304.

McKeel, D.R., 1983. Subsurface biostratigraphy of the east Nehalem Basin, Columbia County, Oregon. Oregon Department of Geology and Mineral Industries, Oil and Gas Investigations 9, 33p.

———, 1984. Biostratigraphy of exploratory wells, northern Willamette Basin, Oregon. Oregon Department of Geology and Mineral Industries, Oil and Gas Investigations 12, 19p.

———, 1985. Biostratigraphy of exploratory wells, southern Willamette Basin, Oregon. Oregon Department Geology and Mineral Industries, Oil and Gas Investigations 13, 17p.

McKnight, B.K., 1984. Stratigraphy and sedimentation of the Payne Cliffs Formation, southwestern Oregon. *In*: Nilsen, T.H., ed., Geology of the upper Cretaceous Hornbrook Formation, Oregon and California. Society of Economic Paleontologists and Mineralogists, Pacific Section, v.42, pp.187-194.

McNeill., L.C., 2000. Tectonics of the Neogene Cascadia forearc basin. Geological Society America, Bull., v.112, no.8, pp.1209-1224.

———, et al., 1997. Listric normal faulting on the Cascadia continental margin. Journal Geophysical Research, v.102, no.B6, pp.12,123-12,138.

———, et al., 1998. The effects of upper plate deformation on records of prehistoric Cascadia subduction zone earthquakes. *In*: Steward, I.S., and Vita-Finzi, C., eds., Coastal tectonics. London, Geological Society, Special Publication 146, pp.321-342.

Meigs, A., et al., 2009. Geological and geophysical perspectives on the magmatic and tectonic development, High Lava Plains and northwest Basin and Range. *In*: O'Connor, J.E., Dorsey, R.J., and Madin, I.P., eds., Volcanoes to vineyards; field trips through the . . . Pacific Northwest. Geological Society America, Field Guide 15, pp.435-470.

Merrill, G.P., 1924. First hundred years of American geology. New Haven, Yale University Press, 773p.

Metcalf, R.V., and Shervais, J.W., 2008. Suprasubduction-zone ophiolites: is there really an ophiolite conundrum? *In*: Snoke, A.W., and Barnes, C.G., eds., Geological studies in the Klamath Mountains province, California and Oregon. Geological Society America, Special Paper 410, pp.191-222.

Miller, D.M., Nilsen, T.H., and Bilodeau, W.L., 1992. Late Cretaceous to early Eocene geologic correlation of the U.S. Cordillera. *In*: Burchfiel, B.C., Lipman, P.W., and Zoback, M.L., eds., The Cordilleran orogen: conterminous U.S. Geological Society America, Decade of North American Geology, v.G-3, pp.205-260.

Miller, M.M., 1987. Dispersed remnants of a northeast Pacific fringing arc: upper Paleozoic terranes of Permian McCloud faunal affinity, western U.S. Tectonics, v.6, no.6, pp.807-930.

Miller, P. R., 1984. Mid-Tertiary stratigraphy, petrology, and paleogeography of Oregon's central Western Cascade Range. Masters, University of Oregon, 187p.

———, and Orr, W.N., 1986. The Scotts Mills Formation: mid-Tertiary geologic history and paleogeography of the central Western Cascade range, Oregon. Oregon Geology, v.48, no.12, pp.139-151p.

———, and Orr, W. N., 1988. Mid-Tertiary transgressive rocky coast sedimentation: central Western Cascade range, Oregon. Journal Sedimentary Petrology, v.58, no.6, pp.959-968.

Minerals Yearbook; Oregon, various years. U.S. Bureau of Mines.

Mitchell, C.E., Vincent, P., and Weldon, R.J., 1994. Present-day vertical deformation of the Cascadia margin, Pacific Northwest, United States. Journal Geophysical Research, v.99, no.B6, pp.12,257-12,277.

Molenaar, C.M., 1985. Depositional relations of Umpqua and Tyee formations (Eocene), southwestern Oregon. American Association of Petroleum Geologists, Bull., v.69, pp.1217-1229.

Mooney, W. D., and Weaver, C. S., 1989. Regional crustal structure and tectonics of the Pacific coastal states; California, Oregon, and Washington. *In:* Pakiser, L.C., and Mooney, W.D. Geophysical framework of the continental United States. Geological Society America, Memoir 172, pp.129-161.

Moore, B.N., 1937. Nonmetallic mineral resources of eastern Oregon. U.S. Geological Survey, Bull.875, 180p.

Moore, G.W., ed., 1994, Geologic catastrophes in the Pacific Northwest. Oregon Geology, v.56, no.1, pp.3-6.

———, ed., 2002. Field guide to geologic processes in Cascadia. Oregon Department Geology and Mineral Industries, Special Paper 36, 324p.

Moore, R.L., 2002. Mercury contamination associated with abandoned mines. Oregon Geology, v.64, no.1, pp.29-31.

Moores, E.M., Kellogg, L.H., and Dilek, Y., 2000. Tethyan ophiolites, mantle convection, and tectonic "historical contingency"; a resolution of the "ophiolite conundrum." *In*: Dilek, Y., et al., eds., Ophiolites and oceanic crust. Geological Society America, Special Paper 349, pp.3-12.

Morgan, D.S., and McFarland, W.D., 1996. Simulation analysis of the ground-water flow system in the Portland Basin, Oregon and Washington. U.S. Geological Survey, Water Supply Paper 2470-B, 83p.

Mortimer, N., 1985. Structural and metamorphic aspects of middle Jurassic terrane juxtaposition, northeastern Klamath Mountains, California. *In*: Howell, D.G., ed., Tectonostratigraphic terranes of the circum-Pacific region, Circum-Pacific Council for Energy and Mineral Resources, Earth Science, Series, no.l, pp.201-214.

———, and Coleman, R.G., 1984. A Neogene structural dome in Klamath Mountains, California and Oregon. *In*: Nilsen, T., ed., Geology of the upper Cretaceous Hornbrook Formation, Oregon and California. Society Economic Paleontologists and Mineralogists, Pacific Section, v. 42, pp.179-186.

Muhs, D.R., et al., 1990. Age estimates and uplift rates for late Pleistocene marine terraces: southern Oregon portion of the Cascadia forearc. Journal Geophysical Research, v.95, no.B5, pp.6685-6698.

Mullen, E. D., 1983. Paleozoic and Triassic terranes of the Blue Mountains, northeast Oregon: Discussion and field trip guide. Part I. A new consideration of old problems. Oregon Geology, v.45, no.6, pp.65-68.

———, 1983a. Paleozoic and Triassic terranes of the Blue Mountains, northeast Oregon: discussion and field trip guide. Part II: Road log and commentary. Oregon Geology, v.45, no.7/8, pp.75-82.

Munts, S.R., 1973. Platinum in Oregon. Ore Bin, v.35, no.9, pp.141-152.

———, 1981. Geology and mineral deposits of the Quartzville mining district, Linn County, Oregon. Oregon Geology, v.43, no.2, p.18.

Murphy, J.B., Oppliger, G.L., and Brimhall, G.H., 1998. Plume-modified orogeny; an example from the western United States. Geology, v.26, no.8, pp.731-734.

Musicar, J., 2007. Plans given for slipping slope. The World Newspaper, December 31, p.1.

Negrini, R.M., 2002. Pluvial lake sizes in the northwestern Great Basin throughout the Quaternary Period. *In*: Hershler, R., Madsen, D.B., and Curry, D.R., eds., Great Basin aquatic systems history. Smithsonian Institution, Washington, D.C., pp. 11-52.

———, et al., 2000. A paleoclimate record for the past 250,000 years from Summer Lake, Oregon, USA. I: Chronology and magnetic proxies for lake level. Journal of Paleolimnology, v.24, pp.125-149.

Nelson, A.R., 1992. Holocene tidal-marsh stratigraphy in south-central Oregon . . . *In*: Fletcher, C.H., and Wehmiller, J.F., eds., Quaternary coasts of the United States: marine and lacustrine systems. Society for Sedimentary Geology (SEPM), Special Publication no.48, pp.287-301.

———, and Personius, S.F., 1996. Great-earthquake potential in Oregon and Washington – an overview . . . U.S. Geological Survey, Professional Paper 1560, v.1, pp.91-114.

———, Jennings, A.E., and Kashima, K., 1996. An earthquake history derived from stratigraphic and microfossil evidence of relative sea-level change at Coos Bay, southern coastal Oregon. Geological Society America, Bull., v.108, no.2, pp.141-154.

———, Shennan, I., and Long, A.J., 1996. Identifying coseismic subsidence in tidal-wetland stratigraphic sequences at the Cascadia subduction zone of western North America. Journal Geophysical Research, no.101, pp.6115-6135.

———, et al., 1995. Radiocarbon evidence for extensive plate-boundary rupture about 300 years ago at the Cascadia subduction zone. Nature, v.378, pp.371-374.

Nelson, C. H., 1984. The Astoria fan: an elongate type fan. Geo-Marine Letters, v.3, pp.65-70.

———, et al., 1970. Development of the Astoria Canyon-fan physiography and comparison with similar systems. Marine Geology, v.8, pp.259-291.

———, et al., 1994. The volcanic, sedimentologic, and paleolimnologic history of the Crater Lake caldera floor, Oregon. Geological Society America, Bull., v.106, pp.648-704.

Neuendorf, K.E., ed., 1978. Bibliography of the geology and mineral resources of Oregon, January 1, 1971 to December 31, 1975. Oregon Department Geology and Mineral Industries, Bull.97, 74p.

———, 1981. Bibliography of the geology and mineral resources of Oregon. January 1, 1976 to December 31, 1979. Oregon Department Geology and Mineral Industries, Bull.102, 68p.

———, 1987. Bibliography of the geology and mineral resources of Oregon, January 1, 1980 to December 31, 1984. Oregon Department Geology and Mineral Industries, Bull.103, 176p.

Newcomb, R.C., 1966. Lithology and eastward extension of The Dalles Formation, Oregon and Washington. U.S. Geological Survey, Professional Paper 550-D, pp.D59-D63.

———, 1967. The Dalles-Umatilla syncline, Oregon and Washington. U.S. Geological Survey, Professional Paper 575 B, pp.B88-B93.

Newton, C. R., 1990. Significance of "Tethyan" fossils in the American Cordillera. Science, v.249, pp.385-390.

Newton, V.C., 1969. Subsurface geology of the lower Columbia and Willamette basins, Oregon. Oregon Department Geology and Mineral Industries, Oil and Gas Investigations 2, 121p.

———, 1979. Oregon's first gas wells completed. Oregon Geology, v.41, no.6, pp.89-90.

———, and Van Atta, R.O., 1976. Prospects for natural gas production and underground storage of pipe-line gas in the upper Nehalem River basin Columbia-Clatsop counties, Oregon. Oregon Department Geology and Mineral Industries, Oil and Gas Investigations 5, 56p.

———, et al., 1980. Prospects for oil and gas in the Coos basin, western Coos, Douglas, and Lane counties, Oregon. Oregon Department Geology and Mineral Industries, Oil and Gas Investigations 6, 74p.

Niem, A.R., 1974. Wright's Point, Harney County, Oregon. An example of inverted topography. Ore Bin, v.36, no.3, pp.33-49.

———, 1975. Geology of Hug Point state park northern Oregon coast. Ore Bin, v.37, no.2, pp.17-36.

———, and Niem, W. A., 1985. Oil and gas investigations of the Astoria basin, Clatsop and northernmost Tillamook counties, northwest Oregon. Oregon Department Geology and Mineral Industries, Oil and Gas Investigations 14.

———, and Niem, W.A., 1990. Geology and oil, gas, and coal resources, southern Tyee basin, southern Coast Range, Oregon. Oregon Department Geology and Mineral Industries, Open-file report O-89-3, various pagings.

———, and Van Atta, R.O., 1973. Cenozoic stratigraphy of northwestern Oregon and adjacent southwestern Washington. Oregon Department Geology and Mineral Industries, Bull.77, pp.75-92.

———, Snavely, P.D., and Niem, W.A., 1990. Onshore-offshore geologic cross section from the Mist gas field, northern Oregon Coast Range, to the northwest Oregon continental shelf. Oregon Department Geology and Mineral Industries, Oil and Gas Investigations 17, 46p.

———, et al., 1992. Onshore-offshore geologic cross section, northern Oregon Coast Range to continental slope. Oregon Department of Geology and Mineral Industries, Special Paper 26, 10p.

———, et al., 1994. Sedimentary, volcanic, and tectonic framework of forearc basins and the Mist gas field, northwest Oregon. *In*: Swanson, D.A., and Haugerud, R.A., eds., Geologic field trips in the Pacific Northwest. Geological Society America, Annual Meeting, pp.IF-1-IF-41.

Niem, W.A., Niem, A.R., and Snavely, P.D., 1992. Sedimentary embayments of the Washington-Oregon coast. *In*: Burchfiel, B.C., Lipman, P.W., and Zoback, M.L., eds., The Cordilleran orogen: conterminous U.S. Geological Society America, Decade of North American Geology, v.G-3, pp.314-319.

Niewendorp, C.A., and Neuhaus, M.E., 2003. Map of selected earthquakes for Oregon, 1841 through 2002. Oregon Department Geology and Mineral Industries, Open-File Report 03-02.

Nilsen, T.H., 1984. Stratigraphy, sedimentology, and tectonic frame-work of the upper Cretaceous Hornbrook Formation, Oregon and California. *In*: Nilsen, T.H., ed., Geology of the upper Cretaceous Hornbrook Formation, Oregon and California. Society Economic Paleontologists and Mineralogists, Pacific Sect., v.42, pp.51-88.

———, ed., 1984. Geology of the upper Cretaceous Hornbrook Formation, Oregon and California. Society Economic Paleontologists and Mineralogists, Pacific Section, v.42, pp.239-257.

Noblett, Jeffrey B., 1981. Subduction-related origin of the volcanic rocks of the Eocene Clarno Formation near Cherry Creek, Oregon. Oregon Geology, v.43, no.7, pp.91-99.

Nolf, B., 1966. Broken Top breaks: flood released by erosion of glacial moraine. Ore Bin, v.28, no.10, pp.182-188.

North, W.B., and Byrne, J.V., 1965. Coastal landslides of northern Oregon. Ore Bin, v.27, no.11, pp.217-241.

Norton, M.A., and Bartholomew, W.S., 1984. Update of ground water conditions and declining water levels in the Butter Creek area, Morrow and Umatilla Counties, Oregon. Oregon Water Resources Department, Groundwater Report 30, 203p.

Noson, L.L., Quamar, A., and Thorsen, G.W., 1988. Washington State earthquake hazards. Washington Division Geology and Earth Resources, Information Circular 85, 77p.

Nur, A., 1983. Accreted terranes. Reviews of Geophysical and Space Physics, v.21, no.8, pp.1779-1785.

Obermeier, S.F., 1995. Preliminary estimates of the strength of prehistoric shaking in the Columbia River valley . . . with emphasis for a Cascadia subduction zone earthquake about 300 years ago. U.S. Geological Survey, Open-File Report 94-589, 46p.

———, and Dickenson, S.E., 2000. Liquefaction evidence for the strength of ground motions resulting from late Holocene Cascadia subduction earthquakes . . . Seismological Society America, v.90, no.4, pp.876-896.

Obermiller, W. A., 1987. Geologic, structural and geochemical features of basaltic and rhyolitic volcanic rocks of the Smith Rock/Gray Butte area, central Oregon. Masters, University of Oregon, 189p.

Obrebski, M., et al., 2011. Lithosphere-aesthenosphere interaction beneath the western United States from the joint inversion of body-wave travel times and surface-wave phase velocities. Geophysical Journal International, v.185, no.2, pp.1003-1021.

O'Connor, J.E., 1993. Hydrology, hydraulics, and geomorphology of the Bonneville flood. Geological Society America, Special Paper 274, 83p.

———, and Baker, V.R., 1992. Magnitudes and implications of peak discharges from glacial Lake Missoula. Geological Society America, Bull., v.104, pp.267-279.

———, and Burns, S.F., 2009. Cataclysms and controversy-aspects of the geomorphology of the Columbia River Gorge. *In*: O'Connor, J.E., Dorsey, R.J., and Madin, I.P., eds., Volcanoes to vineyards; field trips through the . . . Pacific Northwest. Geological Society America, Field Guide 15, pp.237-251.

———, and Waitt, R.B., 1995. Beyond the channeled scabland. Parts 1-3. Oregon Geology, v.57, no.3, pp.4, 5.

———, Dorsey, R.J., and Madin, I.P., eds., 2009. Volcanoes to vineyards: geologic field trips through the dynamic landscape of the Pacific Northwest. Geological Society America, Field Guide 15, 874p.

———, Grant, G.E., and Haluska, T.L., 2003. Overview of geology, hydrology, geomorphology, and sediment budget of the Deschutes River basin, Oregon. *In*: O'Connor, J.E., and Grant, G.E., eds., A peculiar river. American Geophysical Union, Washington, D.C., Water Science and Application 7, pp.7-29.

———, et al., 1995. Beyond the channeled scabland. Part I, Part II, Part III. Oregon Geology, v.57, no.3, pp.51-60; v.57, no.4, pp.75-86; v.57, no.5, pp.99-115.

———, et al., 2001. Origin, extent, and thickness of Quaternary geologic units in the Willamette Valley, Oregon. U.S. Geological Survey, Professional Paper 1620, 52p.

Oles, K.F., and Enlows, H. E., 1971. Bedrock geology of the Mitchell Quadrangle, Oregon. Oregon Department Geology and Mineral Industries, Bull.72, 62p.

———, et al., 1980. Geologic field trips in western Oregon and southwestern Washington. Oregon Department Geology and Mineral Industries, Bull.101, 232p.

Olmstead, D.L., 2003. Development in Oregon's tsunami inundation zone. Oregon Department of Geology and Mineral Industries, Open-File Report 0-03-05, 17p.

———, 1988. Hydrocarbon exploration and occurrences in Oregon. Oregon Department Geology and Mineral Industries, Oil and Gas Investigations 15, 78p.

Oregon Department of Environmental Quality, 2009. Site details environmental cleanup; site information (ECSI) database (The Opalite Mine). Internet, February 20, 5p.

Oregon Water Resources Board, Salem, 1961. Deschutes River basin, Salem, Oregon, 188p.

Oregon Water Resources Department, 1985. Rogue River basin. Salem, Oregon 293p.

———, 1986. John Day River basin. Salem, Oregon, 264p.

———, 1988. Umatilla basin report. Salem, Oregon, 246p.

———, 1989. Goose and Summer lakes basin report. Salem, Oregon, 112p.

———, 1992. Willamette basin report. Salem, Oregon, 350p.

Oregonian Newspaper, December 14, 1994. Hill, R.L., Hazards of Mount Hood, p.D-1.

———, February 24, 1996. Bernton, H., Survey ties some slides to roads, clear-cuts, p.A-1.

———, January 4, 2008. Brettman, A., and Learn, S., Residents predicted landslide, p.C-1.

———, February 25, 2008. Preusch, M., Trains won't get back on Cascades track for weeks; landslide repairs, p.B1.

———, March 9, 2008. Milstein, M., Not so high and dry after all, p.A-1.

———, October 29, 2008. Rainey, S., The Secret's out, p.A1.

———, January 3, 2009. Mayes, S., Rain brings new destruction, p.B-1.

———, January 7, 2009. Tomlinson, S., and Wilson, K.A.C., Deluge, damage loom for region, p.A-1.

———, April 27, 2009. Zaitz, L., Bold grab for water, p.A1.

———, June 11, 2010. McCullen, K., Water levels dropping in Columbia Plateau aquifer, p.E3.

———, October 1, 2011. Cockle, R., Columbia water to pump up aquifers, p.A-1.

Orr, E.L., and Orr, W.N., 1984. Bibliography of Oregon paleontology, 1792-1983. Oregon Department Geology and Mineral Industries, Special Paper 17, 82p.

———, and Orr, W.N., 1999. The other face of Oregon: geologic processes that shape our state. Oregon Geology, v.61, no.6, pp.131-150.

———, and Orr, W.N., 2005. Oregon water, an environmental history. Portland, Inkwater Press, 279p.

———, and Orr, W.N., 2009. Oregon Fossils. 2nd edition. Corvallis, Oregon State University Press, 300p.

Orr, W.N., 1986. A Norian (late Triassic) ichthyosaur from the Martin Bridge Limestone, Wallowa Mountains, Oregon. U.S. Geological Survey, Professional Paper 1435, pp.41-47.

———, and Orr, E.L., 1981. Handbook of Oregon plant and animal fossils, Eugene, Ore., 285p.

———, Ehlen, J., and Zaitzeff, J.B., 1971. A late Tertiary diatom flora from Oregon. California Academy Sciences, Proceedings, 4th Ser., v.37, no.16, pp.489-500.

Otto, B.R., and Hutchinson, D.A., 1977. The geology of Jordan Craters, Malheur County, Oregon. Ore Bin, v.39, no.8, pp.125-140.

Parker, G.G., Shown, L.M., and Ratzlaff, K.W., 1964. Officer's Cave, a pseudokarst feature in altered tuff and volcanic ash of the John Day Formation in eastern Oregon. Geological Society America, Bull. v.75, pp.393-402.

Palmer, L., 1998. Engineering geology of the Oregon coast. *In*: Burns, S., ed., Environmental, groundwater and engineering geology; applications from Oregon. Belmont, Calif., Star Publication, Co., pp.439-450.

———, and Redfern, R., 1973. Urban environmental geology and planning, Portland, Oregon. *In*: Beaulieu, J.D., ed., Geologic field trips in northern Oregon and southern Washington. Oregon Department of Geology and Mineral Industries, Bull., 77, pp.163-170.

Parsons, R.B., 1969. Geomorphology of the Lake Oswego area, Oregon. Ore Bin, v.31, no.9, pp.186-192.

Parsons, T., Thompson, G.S., and Sleep, N.H., 1994. Mantle plume influence on the Neogene uplift and extension of the U.S. western Cordillera. Geology, v.22, pp.83-86.

Peck, D.L., et al., 1964. Geology of the central and northern parts of the Western Cascade Range in Oregon. U.S. Geological Survey, Professional Paper 449, 56p.

———, Imlay, R.W., and Popenoe, W.P., 1956. Upper Cretaceous rocks of parts of south-western Oregon and northern California. American Association Petroleum Geologists, Bull., v.40, pp.1968-1984.

Pegasus Gold Corporation, 1990. Heap leach technology workshop, Portland, Oregon, July 9, July 16, 1990. Presented by Dirk Van Zyl and William M. Schafer. Various pagings.

Perkins, M.E., and Nash, B.P., 2002. Explosive silicic volcanism of the Yellowstone hotspot: the ash fall tuff record. Geological Society America, Bull.114, pp.367-381.

———, et al., 1998. Sequence, age, and source of silicic fallout tuffs in middle to late Miocene basins of the northern Basin and Range province. Geological Society America, Bull.110, no.3, pp.344-360.

Personius, S.F., 1995. Late Quaternary stream incision and uplift in the forearc of the Cascadia subduction zone, western Oregon. Journal Geophysical Research, v.100, pp.20,193-20,210.

———, Kelsey, H.M., and Grabau, P.C., 1993. Evidence for regional stream aggradation in the central Oregon Coast Range during the Pleistocene-Holocene transition. Quaternary Research, v.40, pp.297-308.

Perttu, R.K., and Benson, G.T., 1980. Deposition and deformation of the Eocene Umpqua group, Sutherlin area, southwestern Oregon. Oregon Geology, v.42, no.8, pp.135-140,146.

Pessagno, E. A., 1990. Implications of new Jurassic stratigraphic, geochronometric, and paleo-latitudinal data from the western Klamath terrane (Smith River and Rogue Valley subterranes). Geology, v.18, pp.665-668.

———, 2006. Faunal evidence for the tectonic transport of Jurassic terranes in Oregon, California, and Mexico. *In*: Snoke, A.W., and Barnes, C.G., eds., Geological studies in the Klamath Mountains province, California and Oregon. Geological Society America, Special Paper 410, pp. 31-52.

———, and Blome, C. D., 1986. Faunal affinities and tectonogenesis of Mesozoic rocks in the Blue Mountains province of eastern Oregon and western Idaho. U.S. Geological Survey, Professional Paper 1435, pp.65-78.

Peterson, C.D., and Darienzo, M.E., 1997. Discrimination of flood, storm and tectonic subsidence events in coastal marsh records of Alsea Bay, central Cascadia margin, USA. *In*: Rogers, A.M., et al., eds., Assessing and reducing earthquake hazards in the Pacific Northwest. U.S. Geological Survey, Professional Paper 1560, pp.115-146.

———, and Madin, I.P., 1997. Coseismic liquefaction evidence in the central Cascadia margin, USA. Oregon Geology, v.59, no.3, pp.51-74.

———, and Priest, G.R., 1995. Preliminary reconnaissance survey of Cascadia paleotsunami deposits at Yaquina Bay, Oregon. Oregon Geology, v.57, no.2, pp.33-40.

———, Darienzo, M.E., and Parker, M., 1988. Coastal neotectonic field trip guide for Netarts Bay, Oregon. Oregon Geology, v.50, no.9/10, pp.99-106,117.

———, Gleeson, G. W., and Wetzel, N., 1987. Stratigraphic development, mineral sources and preservation of marine placers from Pleistocene terraces in southern Oregon, U.S.A. Sedimentary Geology, v.53, pp.203-229.

———, Kulm, L. D., and Gray, J.J., 1986. Geologic map of the ocean floor off Oregon and the adjacent continental margin. Oregon Department Geology and Mineral Industries, Geological Map Series, GMS-42.

———, et al., 2002. Pleistocene and Holocene dunal landscapes of the central Oregon Coast: Newport to Florence. *In*: Moore, G., ed., Field guide to geologic processes in Cascadia. Oregon Department of Geology and Mineral Industries, Special Paper 36, pp.201-222.

———, et al., 2008. Minimum runup heights of paleotsunami from evidence of sand ridge overtopping at Cannon Beach, Oregon, central Cascadia margin, U.S.A. Journal Sedimentary Research, v.78, pp.390-409.

Peterson, G.L., et al., 1998. Engineering geology and drainage of the Arizona Inn Landslide . . . *In*: Burns, S., ed., Environmental, groundwater, and engineering geology; applications from Oregon. Belmont, Calif., Star Publ., pp.231-247.

Peterson, N.V., 1959. Lake County's new continuous geyser. Ore Bin, v.21, no.9, pp.83-88.

———, 1962. Geology of Collier State Park area, Klamath County, Oregon. Ore Bin, v. 24, no.6, pp.88-97.

———, and Groh, E.A., 1961. Hole-in-the-Ground. Ore Bin, v.23, no.10, pp.95-100.

———, and Groh, E.A., 1963. Maars of south-central Oregon. Ore Bin, v.25, no.5, pp.73-88.

———, and Groh, E.A., 1963a. Recent volcanic landforms in central Oregon. Ore Bin, v.25, no.3, pp.33-34.

———, and Groh, E.A., 1964. Crack-in-the-Ground, Lake County, Oregon. Ore Bin, v.26, no.9, pp.158-166.

———, and Groh, E.A., 1964a. Diamond Craters, Oregon. Ore Bin, v.26, no.2, pp.17-34.

———, and Groh, E.A., 1965. Lunar geological field conference guidebook. Oregon Department of Geology and Mineral Industries, Bull.57, 51p.

———, and Groh, E. A., 1970. Geologic tour of Cove Palisades State Park near Madras, Oregon. Ore Bin, v.32, no.8, pp.141-168.

———, and Groh, E.A., 1972. Geology and origin of the Metolius Springs Jefferson County, Oregon. Ore Bin, v.34, no.3, pp.41-51.

———, and Ramp, L., 1978. Soapstone industry in southwest Oregon. Ore Bin, v.40, no.9, pp.149-157.

———, et al., 1976. Geology and mineral resources of Deschutes County Oregon. Oregon Department Geology and Mineral Industries, Bull.89, 66p.

Pezzopane, S.K., 1993. Active faults and earthquake ground motions in Oregon. PhD, University of Oregon, 208p.

———, and Weldon, R.J., 1993. Tectonic role of active faulting in central Oregon. Tectonics, v.12, no.5, pp.1140-1169.

Pfaff, V.J., and Beeson, M.H., 1989. Miocene basalt near Astoria, Oregon; geophysical evidence for Columbia Plateau origin. *In*: Reidel, S.P., and Hooper, P.R., eds., 1989. Volcanism and tectonism in the Columbia River flood-basalt province. Geological Society America, Special Paper 239, pp.143-156.

Pfeifer, T., D'Agnese, S., and Pfeifer, A., 1998. Wilson River rockslide . . . *In*: Burns, S., ed., Environmental, groundwater, and engineering geology; applications from Oregon. Belmont, Calif., Star Publ., pp.249-266.

Phelps, D.W., 1978. Petrology, geochemistry, and structural geology of Mesozoic rocks in the Sparta Quadrangle and Oxbow and Brownlee Reservoir areas, eastern Oregon and western Idaho. PhD, Rice University, 229p.

———, and Ave Lallemant, H.G., 1980. The Sparta ophiolite complex, northeast Oregon: a plutonic equivalent to low K20 island-arc volcanism. American Journal Science, v.280-A, pp.345-358.

Pierce, K.I., and Morgan, L.A., 1992. The track of the Yellowstone hotspot: volcanism, faulting, and uplift. Geological Society America, Memoir 179, pp.1-53.

———, and Morgan, L.A., 2009. Is the track of the Yellowstone hotspot driven by a deep mantle plume? – Review of volcanism, faulting, and uplift in light of new data. Journal of Volcanology and Geothermal Research, v.188, no.1-3, pp.1-25.

Pierson, T.C., et al., 2009. Eruption-related lahars and sedimentation response downstream of Mount Hood. *In*: O'Connor, J.E., Dorsey, R.J., and Madin, I.P., eds. Volcanoes to vineyards: geologic field trips through the dynamic landscape of the Pacific Northwest. Geological Society America, Field Guide 15, pp.221-236.

Piper, A.M., 1942. Ground-water resources of the Willamette Valley, Oregon. U.S. Geological Survey, Water Supply Paper 890, 194p.

Pollock, J. M., and Cummings, M.L., 1985. North Santiam mining area, Western Cascades - relations between alteration and volcanic stratigraphy: discussion and field trip guide. Oregon Geology, v.47, no.12, pp.139-145.

Ponce, D.A., and Glen, J.M.G., 2000. Relationship of epithermal gold deposits to large-scale fractures in northern Nevada. Economic Geology, v.97, pp.1-7.

Popenoe, W.P., Imlay, R.W., and Murphy, M.A., 1960. Correlation of the Cretaceous formations of the Pacific Coast (United States and northwestern Mexico). Geological Society America, Bull.71, pp.1491-1540.

Priest, G.R., 1990. Geothermal exploration in Oregon, 1990. Oregon Geology, v.52, no.3, pp.81-86.

———, 1990a. Volcanic and tectonic evolution of the Cascade volcanic arc, central Oregon. Journal of Geophysical Research, v.95, no.B12, pp.19,583-19,599.

———, 1998. The Capes landslide, Tillamook County, Oregon. Oregon Department of Geology and Mineral Industries, Open-File Report 0-98-02, 10p.

———, 2003. Memorandum: Cape Cove landslide . . . Oregon Department of Geology and Mineral Industries, Open-File Report 0-00-03, 14p.

———, and Allan, J.C., 2007. Research investigation at the Johnson Creek landslide. *In*: Burns, W.J., and Wang, Y., comp., 2007 landslide symposium proceedings and field trip guide. Oregon Department of Geology and Mineral Industries, Open-File Report 0-07-06, pp.25-26.

———, and Vogt, B.F., 1983. Geology and geothermal resources of the central Oregon Cascade Range. Oregon Department Geology and Mineral Industries, Special Paper 15, 123p.

———, Hladky, F.R., and Murray. R.B., 2008. Geologic map of the Klamath Falls area, Klamath County, Oregon. Oregon Department of Geology and Mineral Industries, Geologic Map Series, GMS-118.

———, et al., 1983. Overview of the geology of the central Oregon Cascade range. Oregon Department Geology and Mineral Industries, Special Paper 15, pp.3-28.

———, et al., 1997. Cascadia subduction zone tsunamis: hazard mapping at Yaquina Bay, Oregon. Oregon Department of Geology and Mineral Industries, Open-File Report 0-97-34, 144p.

Prothero, D.R., 2003. Pacific coast Eocene-Oligocene marine chronostratigraphy: a review and an update. *In*: Prothero, D.R., Ivany, L.C., and Nesbitt, E.A., eds., 2003. From greenhouse to icehouse. New York, Columbia University Press, pp.1-13.

———, and Donohoo, L.L., 2001. Magnetic stratigraphy and tectonic rotation of the middle Eocene Coaledo Formation, southwestern Oregon. Geophysical Journal International, v.145, pp.223-232.

———, and Haskins, K., 2000. Magnetic stratigraphy and tectonic rotation of the Eocene-Oligocene Keasey Formation, northwest Oregon. Journal Geophysical Research, v.105, no.B7, pp.16,473-16,480.

———, Ivany, L.C., and Nesbitt, E.A., eds., 2003. From greenhouse to icehouse. New York, Columbia University Press, 541p.

———, ed., 2001. Magnetic stratigraphy of the Pacific coast Cenozoic. Society for Sedimentary Geology (SEPM), Pacific Section, Book 91, 394p.

Purcell, D., 1978. Guide to the lava tube caves of central Oregon. Corvallis, High Mountain Press, 55p.

Purdom, W.B., 1977. Guide to the geology and lore of the wild reach of the Rogue River, Oregon. University of Oregon, Museum of Natural History, Bull.22, 67p.

Pyle, D., et al., 2009. Siletzia, an oceanic large igneous province in the Pacific Northwest (abstract). Geological Society America, Annual Mtg., v.41, p.369.

Pyles, M.R., and Skaugset, A.E., 1998. Landslides and forest practice regulation in Oregon. *In*: Burns, S., ed., Environmental, groundwater, and engineering geology; applications from Oregon. Belmont, Calif., Star Publ., pp.481-492.

Ramos, F.C., and Conrey, R.M., 2003. The mafic Holocene San Mountain-Nash Crater chain, Oregon Cascade Range. Eos Transactions, American Geophysical Union, 84, no.46.

Ramp, L., 1953. Structural data from the Chrome Ridge area Josephine County, Oregon. Ore Bin, v.18, no.3, pp.19-25.

———, 1961. Chromite in southwestern Oregon. Oregon Department Geology and Mineral Industries, Bull.52, 169p.

———, 1962. Jones marble deposit, Josephine County, Oregon. Ore Bin, v.24, no.10, pp.153-158.

———, 1975. Geology and mineral resources of Douglas County, Oregon. Oregon Department Geology and Mineral Industries, Bull.75, 106p.

———, 1975a. Geology and mineral resources of the upper Chetco drainage area, Oregon. Oregon Department Geology and Mineral Industries, Bull.88, 195p.

———, 1978. Investigations of nickel in Oregon. Oregon Department Geology and Mineral Industries, Miscellaneous Paper 20, 68p.

———, 1980. Sheeted dikes of the Wild Rogue Wilderness, Oregon. Oregon Geology, v.42, no.7, pp.119-124.

———, and Peterson, N.V., 1979. Geology and mineral resources of Josephine County, Oregon. Oregon Department Geology and Mineral Industries, Bull.100, 45p.

———, Schlicker, H.G., and Gray, J. J., 1977. Geology, mineral resources, and rock material of Curry County, Oregon. Oregon Department Geology and Mineral Industries, Bull. 93, 79p.

Raven, C., and Elston, R.G., eds., 1992. Land and life at Malheur Lake. U.S. Fish and Wildlife Service, Cultural Resource Series, no.8, 151p.

Reckendorf, F., 1998. Geologic hazards of development on sand dunes along the Oregon coast. *In*: Burns, S., ed., Environmental, groundwater, and engineering geology, Star Publ., Belmont, Calif., pp.429-438.

———, 1998a. Use of geomorphic surfaces in floodplain mapping as modified by land-use changes . . . *In*: Burns, S., ed., Environmental, groundwater, and engineering geology, Star Publ., Belmont, Calif., pp.411-424.

Register-Guard Newspaper, August 26, 2008. Palmer, Susan, Dam plan confronts formidable obstacles, p.A-1.

———, January 3, 2009. Ross, W., Storms send slopes crashing down, p.A-1.

Reid, H.F., 1905. Glaciers of Mt. Hood, Oregon and Mt. Adams, Washington. Mazama, v. 2, pp.195-200.

Reidel, S.P., and Hooper, P. R., 1989. Volcanism and tectonism in the Columbia River flood-basalt province. *In*: Reidel, S.P., and Hooper, P.R., eds. Volcanism and tectonism in the Columbia River flood-basalt province. Geological Society America, Special Paper 239, 386p.

———, Kauffman, J.D., and Garwood, D., 2007. The Grande Ronde basalt after 30 years of study. Geological Society America, Abstracts with Programs, p.72.

———, et al., 1989. The Grande Ronde Basalt, Columbia River Basalt group; stratigraphic descriptions and correlations in Washington, Oregon, and Idaho. *In*: Reidel, S.P., and Hooper, P.R., eds. Volcanism and tectonism in the Columbia River flood-basalt province. Geological Society America, Special Paper 239, pp.21-52.

Reilinger, R., and Adams, J., 1982. Geodetic evidence for active landward tilting of the Oregon and Washington coastal ranges. Geophysical Research Letters, v.9, no.4, pp.402-403.

Retallack, G. J., 1991. A field guide to mid-Tertiary paleosols and paleoclimatic changes in the high desert of central Oregon. Oregon Geology, Part 1, v.53, no.3, pp.51-59; Part 2, v.53, no.4, pp.75-80.

———, Bestland, E.A., and Fremd, T.J., eds., 2000. Eocene and Oligocene paleosols of central Oregon. Geological Society America, Special Paper 344, 192p.

———, et al., 2004. Eocene-Oligocene extinction and paleoclimatic change near Eugene, Oregon. Geological Society America, Bull., v.116, no.7/8, pp.817-839.

Ribe, N.M., and Christensen, U.R., 1994. Three-dimensional modeling of plume-lithosphere interaction. Journal Geophysical Research, v.99, pp.669-682.

Riddihough, R.P., Finn, C., and Couch, R., 1986. Klamath-Blue Mountain lineament, Oregon. Geology, v.14, pp.528-531.

Roberts, M.C., 1984. The late Cenozoic history of an alluvial fill: the southern Willamette Valley, Oregon. *In*: Mahaney, W.C., ed., Correlation of Quaternary chronologies. Geo Books, Norwich, England, pp.491-503.

———, and Whitehead, D.R., 1984. The palynology of a non-marine Neogene deposit in the Willamette Valley, Oregon. Reviews of Palaeobotany and Palynology, v.41, pp.1-12.

Roberts, M.S., comp., 1970. Bibliography of the geology and mineral resources of Oregon, January 1, 1956 to December 31, 1960. Oregon Department Geology and Mineral Industries, Bull.78, 198p.

———, Steere, M.L., and Brookhyser, C.S., comp., 1973. Bibliography of the geology and mineral resources of Oregon, January 1, 1961 to December 31, 1970. Oregon Department Geology and Mineral Industries, Bull.78, 198p.

Robertson, R. D., 1982. Subsurface stratigraphic correlations of the Eocene Coaledo Formation, Coos Bay basin, Oregon. Oregon Geology, v.44, no.7, pp.75-78,82.

Robinson, P. T., 1987. John Day Fossil Beds National Monument, Oregon: Painted Hills unit. Geological Society America, Cordilleran Section Centennial Field Guide, pp.317-310.

———, Brem, G.F., and McKee, E.H., 1984. John Day Formation of Oregon: A distal record of early Cascade volcanism. Geology, p.229-232.

———, Walker, G.W., and McKee, E. H., 1990. Eocene (?), Oligocene, and lower Miocene rocks of the Blue Mountains region. U.S. Geological Survey, Professional Paper 1437, pp.29-61.

Robyn, T.L., and Hoover, J.D., 1982. Late Cenozoic deformation and volcanism in the Blue Mountains of central Oregon: microplate interactions? Geology (Boulder), v.10, no.11, pp.572-576.

Rodgers, D.W., Hackett, W.R., and Ore, T.H., 1990. Extension of the Yellowstone plateau, eastern Snake River plain, and Owyhee plateau. Geology, v.18, pp.1138-1141.

Roering, J.J., Kirchner, J.W., and Dietrich, W.E., 2005. Characterizing structural and lithologic controls on deep-seated landsliding. Geological Society America, Bull., v.117, no.5/6, pp.654-668.

Rogers, A.M., et al., eds., 1996. Assessing earthquake hazards and reducing risk in the Pacific Northwest. U.S. Geological Survey, Professional Paper 1560, v.1-2, 515p.

Rogers, J.J.W., and Novitsky-Evans, J. M., 1977. The Clarno Formation of central Oregon, U.S.A. - volcanism on a thin continental margin. Earth and Planetary Science Letters 34, pp.56-66.

———, and Ragland, P.C., 1980. Trace elements in continental-margin magmatism: Part I. Trace elements in the Clarno Formation of central Oregon and the nature of the continental margin on which eruption occurred: summary. Geological Society America, Bull., Part I, v.91, pp.196-198.

Rogers, S.M., 1992. A discussion of "Shore protection and engineering . . . " Corvallis, Oregon Sea Grant, Oregon State University, pp.101-103.

Rosenfeld, C.L., 1992. Natural hazards of the Pacific Northwest, past, present, and future. Oregon Geology, v.54, no.4, pp.75-86.

Ross, C.P., 1938. The geology of part of the Wallowa Mountains. Oregon Department Geology and Mineral Industries, Bull.3, 74p.

Ross, M.E., 1980. Tectonic controls of topographic development within Columbia River basalts in a portion of the Grande Ronde River-Blue Mountains region, Oregon and Washington. Oregon Geology, v.42, no.10, pp.167-174.

Roure, F., and Blanchet, R., 1983. A geological transect between the Klamath Mountains and the Pacific Ocean (southwestern Oregon): a model for paleosubductions. Tectonophysics, v.91, pp.53-72.

Ruddiman, W.F., and Wright, H.E., 1987. North America and adjacent oceans during the last deglaciation. Geological Society America, The Geology of North America, v.K-3, 501p.

Russell, I.C., 1883. A geological reconnaissance in southern Oregon. U.S. Geological Survey, 4th Annual Report, pp.431-464.

———, 1893. A geological reconnaissance in central Washington. U.S. Geological Survey, Bull.108, pp.21-24.

———, 1905. Preliminary report on the geology and water-resources of central Oregon. U.S. Geological Survey Bull.252, 138p.

Ryberg, P.T., 1978. Lithofacies and depositional environments of the Coaledo Formation, Coos County, Oregon. Masters, University of Oregon, 159p.

———, 1984. Sedimentation, structure and tectonics of the Umpqua Group (Paleocene to early Eocene) southwestern Oregon. PhD, University of Arizona, 180p.

Rytuba, J. J., 1989. Volcanism, extensional tectonics, and epithermal mineralization in the northern Basin and Range province, California, Nevada, Oregon, and Idaho. U.S. Geological Survey, Circular 1035, pp.59-61.

———, and McKee, E.H., 1984. Perkaline ash flow tuffs and calderas of the McDermitt volcanic field, southeast Oregon and north central Nevada. Journal Geophysical Research, v.89, no.B10, pp.8616-8628.

———, Glanzman, R.K., and Conrad, W.K., 1979. Uranium, thorium, and mercury distribution through the evolution of the McDermitt caldera complex. *In*: Newman, G.W., and Goode, H.D., eds., Basin and Range symposium and Great Basin field conference, pp.405-412.

———, Vander Meulen, D.B., and Barlock, V.E., 1991. Tectonic and stratigraphic controls on epithermal precious metal mineralization in the northern part of the Basin and Range, Oregon, Idaho, and Nevada. *In*: Buffa, R.H., and Coyner, A.F., eds., Geology and ore deposits of the Great Basin field trip guidebook compendium. Geological Society Nevada, v.2, pp.633-644.

———, 1990. et al., Field guide to hot-spring gold deposits in the Lake Owyhee volcanic field, eastern Oregon. Geological Society Nevada and U.S. Geological Survey, Spring field trip no.10, 15p.

Saleeby, J. B., 1992. Petrotectonic and paleogeographic settings of U.S. Cordilleran ophiolites. *In*: Burchfiel, B.C., Lipman, P.W., and Zoback, M.L., eds., The Cordilleran orogen:conterminous U.S. Geological Society America, The Geology of North America, v.G-3, pp.653-682.

———, and Harper, G.D., 1993. Tectonic relations between the Galice Formation and the Condrey Mountain Schist, Klamath Mountains, northern California. *In*: Dunn, G.C., and McDougall, K.A., eds., Mesozoic paleogeography of the western United States – II. Society of Economic Paleontologists and Mineralogists, Pacific Section, v.71, pp.61-80.

———, et al., 1982. Time relations and structural-stratigraphic patterns in ophiolite accretion, west central Klamath Mountains, California. Journal Geophysical Research, v.87, no. B5, pp.3831-3848.

Sammel, E.A., and Craig, R.W., 1981. The geothermal hydrology of Warner Valley, Oregon. U.S. Geological Survey, Professional Paper 1044- I, 46p.

Sanborn, E.I., 1937. The Comstock flora of west central Oregon; Eocene flora of western America. Carnegie Institute, Washington, D.C., Publication 465, pp.1-28.

Sarewitz, D., 1983. Seven Devils terrane: is it really a piece of Wrangellia? Geology, v.11, pp.634-637.

Scarberry, K.C., 2007. Extension and volcanism; tectonic development of the northwestern margin of the Basin and Range province in southern Oregon. PhD, Oregon State University, 168p.

———, Meigs, A.J., and Grunder, A.L., 2010. Faulting in a propagating continental rift; insight from the late Miocene structural development of the Abert Rim fault, southern Oregon, USA. Tectonophysics, v.488, no.1-4, pp.71-86.

Scharf, D.W., 1935. A Miocene mammalian fauna from Sucker Creek, southeastern Oregon. Carnegie Institute, Washington, D.C., Publication 435, pp.97-118.

Schermer, E.R., Howell, D.G., and Jones, D.L.,1984. The origin of allochthonous terranes: perspectives on the growth and shaping of continents. Annual Review, Earth Planetary Science Letters, v.12, pp.107-131.

Schlicker, H.G., 1956. Landslides. Ore Bin, v.18, no.5, pp.39-43.

———, 1961. Geology of the Ecola State Park landslide area, Oregon. Ore Bin, v.23, no.9, pp.85-88.

———, 1977. Geologic restraints to development in selected areas of Marion County, Oregon. Oregon Department Geology and Mineral Industries, Open-File Report 0-77-4, 59p.

———, and Deacon, R.J., 1967. Engineering geology of the Tualatin Valley region, Oregon. Oregon Department Geology and Mineral Industries, Bull.60, 103p.

———, and Finlayson, C.T., 1979. Geology and geologic hazards of northwestern Clackamas County, Oregon. Oregon Department Geology and Mineral Industries, Bull.99, 79p.

———, Deacon, R.J., and Twelker, N.H., 1964. Earthquake geology of the Portland area, Oregon. Ore Bin, v.29, no.12, pp.206-230.

———, et al., 1973. Environmental geology of Lincoln County, Oregon. Oregon Department Geology and Mineral Industries, Bull.81, 171p.

———, et al., 1974. Environmental geology of coastal Lane County Oregon. Oregon Department Geology and Mineral Industries, Bull.85, 116p.

Schmandt, B., and Humphreys, E., 2011. Seismically imaged relic slab from the 55 Ma Siletzia accretion to the northwest United States. Geology, v.39, no.2, pp.175-178.

Schmidt, M.E., and Grunder, A.L., 2009. The evolution of North Sister. Geological Society America, Bull., v.121, no.5/6, pp.643-662.

Schneer, C.J., ed., 1979. Two hundred years of geology in America. Hanover, New Hampshire, University of New Hampshire, 385p.

Scholl, D. W., Grantz, A., and Vedder, J.G., eds., 1987. Geology and resource potential of the continental margin of western North America and adjacent ocean basins-Beaufort Sea to Baja California. Circum-Pacific Council for Energy and Mineral Research, Earth Science Series, v.6, 799p.

Schoonmaker, A., Harper, G.D., and Heizler, M.T., 2006. Tectonic emplacement of the Snow Camp remnant of the Coast Range ophiolite near Game Lake, southwestern Oregon. *In*: Snoke, A.W., and Barnes, C.G., eds., Geological studies in the Klamath Mountains province, California and Oregon. Geological Society America, Special Paper 410, pp.177-197.

Schroeder, N.A.M., Kulm, L.D., and Muehlberg, G. E., 1987. Carbonate chimneys on the outer continental shelf: evidence for fluid venting on the Oregon margin. Oregon Geology, v.49, no.8, pp.91-98.

Schuster, J. E., ed., 1987. Selected papers on the geology of Washington. Washington Division Geology and Earth Resources, Bull.77, 395p.

Schwartz, J.J., et al., 2010. Analysis of the Wallowa-Baker terrane boundary . . . Geological Society America, Bull., v.122, no.3/4, pp.517-536.

Scott, W. E., 1990. Geologic map of the Mount Bachelor volcanic chain and surrounding area, Cascade Range, Oregon. U.S. Geological Survey, Miscellaneous Investigations Series, Map I-1967.

———, 1990a. Temporal relations between eruptions of the Mount Bachelor volcanic chain and fluctuations of late Quaternary glaciers. Oregon Geology, v.52, no.5, pp.114-117.

———, and Gardner, C. A., 1990. Field trip guide to the central Oregon High Cascades. Part l: Mount Bachelor-South Sister area. Oregon Geology, v.52, no.5, pp.99-114.

———, and Mullineaux, D.R., 1981. Late Holocene eruptions, South Sister volcano, Oregon. U.S. Geological Survey, Geological Research, p.217.

———, et al., 1997. Geologic history of Mount Hood volcano, Oregon – a field-trip guidebook. U.S. Geological Survey, Open-File Report, 38p.

———, et al., 1997a. Volcanic hazards in the Mount Hood region, Oregon. U.S. Geological Survey, Open-File Report 97-89, 14p.

———, et al., 1999. Volcano hazards in the Three Sisters region, Oregon. U.S. Geological Survey, Open-File Report 99-437, 13p.

Seabridge Gold of Toronto, Inc., Internet, May 1, 2007.

Seiders, V.M., and Blome, C.D., 1987. Stratigraphy and sedimentology of Upper Cretaceous rocks in coastal south-west Oregon: evidence for wrench-fault tectonics in a postulated accretionary terrane: alternative interpretation and reply. Geological Society America, Bull., 98, pp.739-744.

Self, S., Thordarson, T., and Keszthelyi, L., 1997. Emplacement of continental flood basalt lava flows. *In:* Mahoney, J.J., and Coffin, M.F., eds. Large igneous provinces: continental, oceanic, and planetary flood volcanism. Washington, D.C., American Geophysical Union, Geophysical Monograph 100, pp.381-410.

Seward, J.H., 1998. Forest landslide analysis with limited data. *In*: Burns, S., ed., Environmental, groundwater, and engineering geology, Star Publ., Belmont, Calif., pp.517-520.

Shennan, I., et al., 1998. Tidal marsh stratigraphy, sea-level change and large earthquakes—II: submergence events during the last 3500 years at Netarts Bay, Oregon, USA. Quaternary Science Reviews, v.17, pp.365-393.

Sherrod, D. R., and Pickthorn, L.B.G., 1992. Geologic map of the west half of the Klamath Falls 1° by 2° Quadrangle, south-central Oregon. U.S. Geological Survey, Miscellaneous Investigations Series, I-2152.

———, and Scott, W.E., 1995. Preliminary geologic map of the Mount Hood 30-by-60-minute quadrangle, Cascade Range, north-central Oregon. U.S. Geological Survey, Open-File Report 95-219.

———, and Smith, J.G., 1990. Quaternary extrusion rates of the Cascade Range, northwestern United States and southern British Columbia. Journal Geophysical Research, v.95, no.B12, pp.19,465-19,474.

———, and Smith, J.G., 2001. Geologic map of upper Eocene to Holocene volcanic and related rocks of the Cascade Range, Oregon. U.S. Geological Survey, Geologic Investigations Series, I-2569.

———, Gannett, M.W., and Lite, K.E., 2002. Hydrogeology of the upper Deschutes Basin, central Oregon. *In*: Moore, George, ed., Field guide to geologic processes in Cascadia. Oregon Department of Geology and Mineral Industries, Special Paper 36, pp.109-144.

———, ed., 1988. Geology and geothermal resources of the Breitenbush-Austin Hot Springs area, Clackamas and Marion Counties, Oregon. Oregon Department of Geology and Mineral Industries, Open-File Report O-88-5, pp.1-14.

———, et al., 1996. Water, rocks, and woods – a field excursion to examine the geology, hydrology, and geothermal resources in the Clackamas, North Santiam, and McKenzie River drainages, Cascade Range, Oregon. Oregon Geology, v.58, no.5, pp.103-124.

———, et al., 1997. Volcanic hazards at Newberry volcano, Oregon. U.S. Geological Survey, Open-File Report 97-513, 14p.

———, et al., 2004. Geologic map of the Bend 30-60-minute quadrangle, central Oregon. U.S. Geological Survey, Geologic Investigations Series I-2683.

Shervais, J.W., 2001. Birth, death, and resurrection; the life cycle of suprasubduction zone ophiolites. Geochemistry, Geophysics, Geosystems, v.2.

Sibrava, W., et al., eds., 1986. Quaternary glaciations in the northern hemisphere. Quaternary Science Reviews 5, 510p.

Silberling, N.J., et al., 1987. Lithotectonic terrane map of the western conterminous United States. U.S. Geological Survey, Miscellaneous Field Studies, MF 1874-C.

Silver, E.A., 1978. Geophysical studies and tectonic development of the continental margin off the western United States, lat. 34 to 48 N. *In*: Smith, R.B., and Eaton, G.P., eds. Cenozoic tectonic and regional geophysics of the western Cordillera. Geological Society America, Memoir 152, pp.251-262.

Simpson, R. W., and Cox, A., 1977. Paleomagnetic evidence for tectonic rotation of the Oregon Coast range. Geology, v.5, pp.585-589.

Sliter, W.V., Jones, D.L., and Throckmorton, C.K., 1984. Age and correlation of the Cretaceous Hornbrook Formation, California and Oregon. *In*: Nilsen, T.H., ed., Geology of the upper Cretaceous Hornbrook Formation, Oregon and California. Society Economic Paleontologists and Mineralogists, Pacific Section, v.42, pp.89-98.

Smith, G. A., 1986. Simtustus Formation: paleogeographic and stratigraphic significance of a newly defined Miocene unit in the Deschutes basin, central Oregon. Oregon Geology, v.48, no.6, pp.63-72.

———, 1986a. Stratigraphy, sedimentology, and petrology of Neogene rocks in the Deschutes basin, central Oregon: a record of continental-margin volcanism and its influence on fluvial sedimentation in an arc-adjacent basin. PhD, Oregon State University, 467p.

———, 1987. The influence of explosive volcanism on fluvial sedimentation: the Deschutes Formation (Neogene) in central Oregon. Journal Sedimentary Petrology, v.57, no.4, pp.613-629.

———, 1991. A field guide to depositional processes and facies geometry of Neogene continental volcaniclastic rocks, Deschutes basin, central Oregon. Oregon Geology, v.53, no.1, pp.3-20.

———, 1998. Geology along U.S. Highways 197 and 97 between The Dalles and Sunriver, Oregon. Oregon Geology, v.60, no.1, pp.3-17.

———, Bjornstad, B.N., and Fecht, K.R., 1989. Neogene terrestrial sedimentation on and adjacent to the Columbia Plateau; Washington, Oregon, and Idaho. *In*: Reidel, S.P., and Hooper, P.R., eds., Volcanism and tectonism in the Columbia River flood-basalt province. Geological Society America, Special Paper 239, pp.187-197.

———, Snee, L.W., and Taylor, E.M., 1987. Stratigraphic, sedimentologic, and petrologic record of late Miocene subsidence of the central Oregon High Cascades. Geology, v.15, pp.389-392.

Smith, R.B., and Braile, I.W., 1994. The Yellowstone hotspot. Journal Volcanology and Geothermal Research, v.61, pp.121-188.

———, and Eaton, G.P., 1978. Cenozoic tectonic and regional geophysics of the western Cordillera. Geological Society America, Memoir 152, 388p.

Smith, W.D.P., 1941. Geology and physiography of the northern Wallowa Mountains, Oregon. Oregon Department Geology and Mineral Industries, Bull.12, 64p.

———, and Ruff, L.L., 1938. The Geology and mineral resources of Lane County, Oregon. Oregon Department Geology and Mineral Industries, Bull.11, 65p.

Snavely, P.D., 1987. Depoe Bay, Oregon. *In:* Hill, Mason L., ed., Decade of North American geology; Centennial Field Guide. Cordilleran Section of the Geological Society of America, pp.307-310.

———, 1987a. Tertiary geologic framework, neotectonics, and petroleum potential of the Oregon-Washington continental margin. *In:* Scholl, D.W., Grantz, Arthur, and Vedder, John G., eds. Geology and resource potential of the continental margin of western North America and adjacent ocean basins-Beaufort Sea to Baja California. Circum-Pacific Council for Energy and Mineral Resources, Earth Science Series, v.6, pp.305-335.

———, and Baldwin, E.M., 1948. Siletz River Volcanic series, northwestern Oregon. American Association Petroleum Geologists, v.32, no.5, pp.805-812.

———, and MacLeod, N.S., 1971. Visitor's guide to the geology of the coastal area near Beverly Beach State Park, Oregon. Ore Bin, v.33, no.5, pp.85-106.

———, and Wagner, H.C., 1963. Tertiary geologic history of western Oregon and Washington. Washington, Division of Mines and Geology, Report of Investigations 22, 25p.

———, and Wells, R.E., 1991. Cenozoic evolution of the continental margin of Oregon and Washington. U.S. Geological Survey, Open-File Report 91-441-B, 34p.

———, MacLeod, N.S., and Wagner, H.C., 1973. Miocene tholeiitic basalts of coastal Oregon and Washington and their relations to coeval basalts of the Columbia Plateau. Geological Society America, Bull., v.84, pp.387-424.

———, Pearl, J.E., and Lander, D.L., 1977. Interim report on petroleum resources potential and geologic hazards in the outer continental shelf - Oregon and Washington Tertiary province. U.S. Geological Survey, Open-file Report. 77-282, 57p.

———, Rau, W.W., and Wagner, H.C., 1964. Miocene stratigraphy of the Yaquina Bay area, Newport, Oregon. Ore Bin, v.26, no.8, pp.133-151.

———, Wagner, H.C., and MacLeod, N.S., 1965. Preliminary data on compositional variations of Tertiary volcanic rocks in the central part of the Oregon Coast range. Ore Bin, v.27, no.6, pp.101-117.

———, et al., 1975. Alsea Formation-an Oligocene marine sedimentary sequence in the Oregon Coast Range. U.S. Geological Survey, Bull. 1395-F, pp.F1-F21.

———, et al., 1980. Geology of the west-central part of the Oregon Coast Range. Geologic field trips in western Oregon and southwestern Washington. Oregon Department Geology and Mineral Industries., Bull.101, pp.39-76.

Snoke, A.W., and Barnes, C.G., 2006. The development of tectonic concepts for the Klamath Mountains province, California and Oregon. *In*: Snoke, A.W., and Barnes, C.G., eds., Geological studies in the Klamath Mountains province, California and Oregon. Geological Society America, Special Paper 410, pp.1-29.

———, and Barnes, C.G., eds., 2006. Geological studies in the Klamath Mountains province, California and Oregon. Geological Society America, Special Paper 410, 505p.

Socolow, A.A., ed., 1988. The state geological surveys; a history. [n.p.] American Association of State Geologists, 499p.

Sonder, L.J., and Jones, C.H., 1999. Western United States extension: how the west was widened. Annual Reviews Earth and Planetary Science Letters, v.27, pp.417-462.

Stanley, G.D., 1986. Late Triassic coelenterate faunas of western Idaho and northeastern Oregon: implications for biostratigraphy and paleogeography. U.S. Geological Survey, Professional Paper 1435, pp.23-36.

———, 1987. Travels of an ancient reef. Natural History, v.11, pp.35-42.

———, McRoberts, C.A., and Whalen, M.T., 2008. Stratigraphy of the Triassic Martin Bridge Formation, Wallowa terrane: stratigraphy and depositional setting. *In*: Blodgett, R.B., and Stanley, G.D., eds., The terrane puzzle. Geological Society America, Special Paper 442, pp.227-250.

Stanley, W.D., Mooney, W.D., and Fuis, G.S., 1990. Deep crustal structure of the Cascade Range and surrounding regions from seismic refraction and magnetotelluric data. Journal Geophysical Research, v.95, no.B12, pp.19,419-19,438.

Steere, M.L., comp., 1953. Bibliography of the geology and mineral resources of Oregon, January 1, 1946 to December 31, 1950. Oregon Department Geology and Mineral Industries, Bull.44, 61p.

———, and Owen, L.F., comp., 1962. Bibliography of the geology and mineral resources of Oregon, January 1, 1951 to December 31, 1955. Oregon Department Geology and Mineral Industries, Bull.53, 97p.

Stembridge, J.E., 1975. Shoreline changes and physiographic hazards on the Oregon coast. PhD, University of Oregon, 202p.

———, 1978. Inventory: Oregon coastal shoreline erosion. Salem, Oregon Soil and Water Conservation Commission, 109p.

Stern, R.J., and Bloomer, S.H., 1992. Subduction zone infancy . . . Geological Society America, Bull., v.104, no.12, pp.1621-1636.

Stevens, C.H., ed., 1983. Pre-Jurassic rocks in western North American suspect terranes. Society Economic Paleontologists and Mineralogists, Pacific Section, 141p.

Stevens, I.I., 1855. Report of explorations for a route for the Pacific Railroad near the forty-seventh and forty-ninth parallels of north latitude from St. Paul to Puget Sound. *In*: Henry, J., and Baird, S., Reports of Explorations and surveys to ascertain the most practicable and economical route from the Mississippi River to the Pacific Ocean. Washington, D.C., U.S. War Office, v. l, 651p.

Stewart, J. H., 1978. Basin-range structure in western North America: A review. Geological Society America, Memoir 152, pp.1-28.

———, 1988. Tectonics of the Walker Lane belt, western Great Basin. *In*: Ernst, W.G., ed., The geotectonic development of California. Prentice-Hall, pp.683-713.

———, 1998. Regional characteristics, tilt domains, and extensional history of the late Cenozoic Basin and Range province, western North America. *In*: Faulds, J.E., and Stewart, J.H., eds., 1998. Accommodation zones and transfer zones the regional segmentation of the Basin and Range province. Geological Society America, Special Paper 323, pp.47-74.

———, Stevens, C.H., and Fritsche A. E., eds., 1977. Paleozoic paleogeography of the western United States. Society Economic Paleontologists and Mineralogists, Pacific Section, Paleography Symposium 1, 502p.

Stickney, M.C., and Bartholomew, M.J., 1987. Seismicity and late Quaternary faulting of the northern Basin and Range province, Montana and Idaho. Seismological Society America, Bulletin, v.77, p.1602-1625.

Stokes, W.L., 1979. Paleostratigraphy of the Great Basin region. *In*: Newman, G.W., and Goode, H.D., eds., Basin and Range symposium and Great Basin field conference, pp.196-217.

Streck, M.J., 2002. Partial melting to produce high-silica rhyolites of a young bimodal suite: compositional constraints . . . Harney Basin, Oregon. International Journal of Earth Sciences, v.91, pp.583-593.

———, and Grunder, A.L., 1997. Compositional gradients and gaps in high-silica rhyolites of the Rattlesnake tuff, Oregon. Journal Petrology, v.38, no.1, pp.133-163.

———, Johnson, J.A., and Grunder, A.L., 1999. Field guide to the Rattlesnake tuff and High Lava Plains near Burns, Oregon. Oregon Geology, v.61, no.3, pp.64-76.

Streiff, R.E., The geology and mineralization of the northern portion of the Blue River mining district, Lane and Linn Counties, Oregon. Oregon Geology, v.62, no.3, pp.69-83.

Strickler, M.D., 1986. Geologic setting of the Turner-Albright massive sulfide deposit, Josephine County, Oregon. Oregon Geology, v.48, no.10, pp.115-122.

Strobel, M.L., et al., 2009. Snowpack data collection in the Mount Hood area . . . *In*: O'Connor, J.E., Dorsey, R.J., and Madin, I.P., eds., Volcanoes to vineyards; field trips through the . . . Pacific Northwest. Geological Society America, Field Guide 15, pp.471-480.

Stroud, P.L., 1998. Hydrology baseline study, Newberry geothermal project, Newberry Volcano, central Oregon. *In*: Burns, S., ed., Environmental, groundwater and engineering geology; applications from Oregon. Belmont, Calif., Star Publ., pp.593-609.

Sutton, K.G., 1974. Geology of Mt. Jefferson. Masters, University of Oregon, 120p.

Swanson, D.A., et al., 1979. Revisions in stratigraphic nomenclature of the Columbia River Basalt Group. U.S. Geological Society, Bull., 1457-G, 59p.

Swanson, F.J., 1976. Geology, geomorphology, and forest land use in the Elk and Sixes River basins, S.W. Oregon. Oregon Academy Sciences, Proceedings for 1975, v.XI, p.85.

———, and Dyrness, C.T., 1975. Impact of clear-cutting and road construction on soil erosion by landslides in the western Cascade Range, Oregon. Geology, v.3, no.1, pp.393-396.

———, and Jones, J.A., 2002. Geomorphology and hydrology of the H.J. Andrews Experimental Forest, Blue River, Oregon. *In*: Moore, G., ed., Field guide to geologic processes in Cascadia. Oregon Department of Geology and Mineral Industries, Special Paper 36, p.289-314.

Swanston, D.N., et al., 1987. Effects of timber harvesting on progressive hillslope deformation in southwest Oregon. *In:* Beschta, R.L., et al., eds., Erosion and Sedimentation in the Pacific Rim, IAHS Publication no.165, pp. 141-144.

Taubeneck, W.H., 1955. Age of the Bald Mountain batholith, northeastern Oregon. Northwest Science, v.29, pp.93-96.

———, 1987. The Wallowa Mountains, northeast Oregon. *In*: Hill, Mason L., ed., Cordilleran Section, Geological Society America, Centennial field guide, pp.327-332.

Taylor, D.,1988. Middle Jurassic (late Aalenian and early Bajocian) ammonite biochronology of the Snowshoe Formation, Oregon. Oregon Geology, v.50, no.11/12, pp.123-138.

Taylor, E. M., 1965. Recent volcanism between Three Fingered Jack and north Sister Oregon Cascade Range. Ore Bin, v.27, no.7, pp.121-147.

———, 1968. Roadside geology Santiam and McKenzie Pass highways, Oregon. Oregon Department Geology and Mineral Industries, Bull. 62, pp.3-33.

———, 1978. Field geology of S.W. Broken Top Quadrangle, Oregon. Oregon Department Geology and Mineral Industries, Special Paper 2, 50p.

———, 1981. Central High Cascade roadside geology. *In*: Johnston, D.A., and Donnelly-Nolan, J.M., eds., Guides to some volcanic terranes in Washington, Idaho, Oregon, and northern California. U.S. Geological Survey, Circular 838, pp.55-83.

———, 1987. Field geology of the northwest quarter of the Broken Top 15' Quadrangle, Deschutes County, Oregon. Oregon Department Geology and Mineral Industries, Special Paper 21, 20p.

———, 1987a. Late High Cascade volcanism from summit of McKenzie Pass, Oregon: Pleistocene composite cones on platform of shield volcanoes: Holocene eruptive centers and lava fields. *In:* Hill, M.L., ed., Cordilleran Section, Geological Society America, Centennial Field Guide, v.1, pp.313-315.

———, 1990. Volcanic history and tectonic development of the central High Cascade Range, Oregon. Journal Geophysical Research, v.95, no.B12, pp.19,611-19,622.

———, and Ferns, M.L., 1994. Geology and mineral resource map of the Tumalo Dam Quadrangle, Deschutes County, Oregon. Oregon Department Geology and Mineral Industries, Geological Map Series GMS-81.

———, and Ferns, M.L., 1995. Geologic map of the Three Creek Butte Quadrangle, Deschutes County, Oregon. Oregon Department of Geology and Mineral Industries, Geological Map Series, GMS-87.

———, et al., 1987. Geologic map of the Three Sisters Wilderness, Deschutes, Lane, and Linn counties, Oregon. U.S. Geological Survey, Miscellaneous Field Studies Map, MF-1952.

Taylor, S.B., 2007. Watershed assessment, river restoration, and the geoscience profession in Oregon. Oregon Geology, v.68, no.1, pp.27-32.

———, Dutton, B.E., and Poston, P.E., 2002. Luckiamute River watershed, Upper Willamette basin. *In*: Moore, G., ed., 2002. Field guide to geologic processes in Cascadia. Oregon Department of Geology and Mineral Industries, Special Paper 36, pp. 167-186.

TenBrink, U., 1991. Volcano spacing and plate rigidity. Geology, v.19, pp.397-400.

Testa, S.M., 2011. Wildfires and debris flows: federal mud. Earth, March, pp.78-79.

Thayer, T. P., 1939. Geology of the Salem Hills and the North Santiam River basin, Oregon. Oregon Department Geology and Mineral Industries, Bull.15, 40p.

———, 1977. The Canyon Mountain complex, Oregon, and some problems of ophiolites. Oregon Department Geology and Mineral Industries, Bull.95, pp.93-106.

Thompson, G.G., Yett, J.R., and Green, K.E., 1984. Subsurface stratigraphy of the Ochoco Basin, Oregon. Oregon Department Geology and Mineral Industries, Oil and Gas Investigations 8, 22p.

Throop, A. H., 1986. Fishing and placer mining: are they compatible? Oregon Geology, v.48, no.3, pp.27-28,34.

———, 1989. Cyanide in mining. Oregon Geology, v.51, no.1, pp.9-11,20.

———, 1991. Should the pits be filled? Geotimes, v.36, no.11, pp.20-22. *Also*: Oregon Geology, v.52, no.4, pp.82-84.

Tolan, T.L., and Beeson, M.H., 1984. Intracanyon flows of the Columbia River Basalt Group in the lower Columbia River Gorge and their relationship to the Troutdale Formation. Geological Society America, Bull., v.95, p.463-477.

———, Beeson, M.H., and Vogt, B.F., 1984. Exploring the Neogene history of the Columbia River: discussion and geologic field trip guide to the Columbia River gorge. Part I: discussion. Oregon Geology, v.46, no.8, pp.87-97.

———, Beeson, M. H., and Vogt, B.F., 1984a. Exploring the Neogene history of the Columbia River: discussion and geologic field trip guide to the Columbia River gorge. Part II: Road log and comments. Oregon Geology, v.46, no.8, pp.103-112.

———, et al., 1989. Revisions to the estimates of the areal extent and volume of the Columbia River Basalt group. *In*: Reidel, S.P., and Hooper, P.R., eds. Volcanism and tectonism in the Columbia River flood-basalt province. Geological Society America, Special Paper 239, pp.1-20.

———, et al., 2009. An introduction to the stratigraphy, structural geology and hydrogeology of the Columbia River flood-basalt province. *In*: O'Connor, J.E., Dorsey, R.J., and Madin, I.P., eds., Volcanoes to vineyards; field trips through the . . . Pacific Northwest. Geological Society America, Field Guide 15, pp.599-643.

———, et al., 2009a. Stratigraphy and tectonics of the central and eastern portions of the Columbia River flood-basalt province. *In*: O'Connor, J.E., Dorsey, R.J., and Madin, I.P., eds., Volcanoes to vineyards; field trips through the . . . Pacific Northwest. Geological Society America, Field Guide 15, pp.645-672.

Townley, S.D., and Allen, M.W., 1939. Descriptive catalog of earthquakes of the Pacific Coast of the United States, 1769 to 1928. Seismological Society America, Bull.29, no.1, 297p.

Tréhu, A.M., Blakely, R.J., and Williams, M.C., in press. Subducted seamounts and recent earthquakes beneath the central Cascadia forearc. Geology.

———, Braunmiller, J., and Nabelek, J.L., 2008. Probable low-angle thrust earthquakes on the Juan de Fuca-North American plate boundary. Geology, v.36, no.2, pp.127-130.

———, et al., 1994. Crustal architecture of the Cascadia forearc. Science, v.266, pp.237-243.

———, et al., 1995. A seismic subduction zone offshore central Oregon. Journal Geophysical Research, v.100, no.B8, pp.15101-15116.

———, et al., 1999. Temporal and spatial evolution of a gas-hydrate-bearing accretionary ridge on the Oregon continental margin. Geology, v.27, pp.939-942.

Trimble, D.E., 1963. Geology of Portland, Oregon and adjacent areas. U.S. Geological Survey, Bulletin 1119, 119p.

Tyler, D.B., 1968. The Wilkes Expedition; the first United States Exploring Expedition (1838-1842). Philadelphia, The American Philosophical Society, 435p.

U.S. Coast Pilot, 1951. Pacific Coast, California, Oregon, and Washington. U.S. Coast and Geodetic Survey. 7th ed., no.750.

Vallier, T.L., 1977. The Permian and Triassic Seven Devils Group, western Idaho and northeastern Oregon. U.S. Geological Survey, Bull.1437, 58p.

———, 1995. Petrology of pre-Tertiary igneous rocks in the Blue Mountains region of Oregon, Idaho, and Washington. *In*: Vallier, T.L., and Brooks, H.C., eds., Geology of the Blue Mountains region of Oregon, Idaho, and Washington. U.S. Geological Survey, Professional Paper 1438, pp.125-209.

———, 1998. Islands & rapids; a geologic story of Hells Canyon. Lewiston, Idaho, Confluence Press, 150p.

———, and Brooks, H.C., 1970. Geology and copper deposits of the Homestead area, Oregon and Idaho. Ore Bin, v.32, no.3, pp.37-57.

———, and Brooks, H.C., eds., 1986. Geology of the Blue Mountains region of Oregon, Idaho, and Washington. U.S. Geological Survey, Professional Paper 1435, 93p.

———, and Brooks, H.C., eds., 1995. Geology of the Blue Mountains region of Oregon, Idaho, and Washington. U.S. Geological Survey, Professional Paper 1438, 540p.

———, Brooks, H.C., and Thayer, T.P., 1977. Paleozoic rocks of eastern Oregon and western Idaho. *In*: Stewart, J.H., et al., eds. Paleozoic paleogeography of the western United States. Pacific Coast paleogeography symposium 1, Society Economic Paleontologists and Mineralogists, pp.455-466.

VanderMeulen, D.B., et al., 1990. Mineral resources of the Blue Canyon and Owyhee breaks wilderness study areas, Malheur County, Oregon. U.S. Geological Survey, Professional Paper 1741-G, 28p.

Venkatakrishnan, R., Bond, J.G., and Kauffman, J.D., 1980. Geological linears of the northern part of the Cascade Range, Oregon. Oregon Department Geology and Mineral Industries, Special Paper 12, 25p.

Verplanck, E.P., and Duncan, R.A., 1987. Temporal variations in plate convergence and eruption rates in the Western Cascades, Oregon. Tectonics, v.6, no.2, pp.197-209.

Vessely, D.A., Riemer, M., and Aarango, I., 1996. Liquefaction susceptibility of soft alluvial silts in the Willamette Valley. Oregon Geology, v.58, no.6, pp.142-149.

Vincent, P., 1989. Geodetic deformation of the Oregon Cascadia margin. Masters, University of Oregon, 86p.

———, Richards, M.A., and Weldon, R.J., 1989. Vertical deformation of the Oregon Cascadia margin. Eos Transactions, American Geophysical Union, no.70, p.1332.

Vokes, H.E., Snavely, P.D., and Myers, D.A., 1951. Geology of the southern and southwestern border areas, Willamette Valley, Oregon. U.S. Geological Survey Oil and Gas Investigations Map, OM-110.

von Huene, R., 2008. When seamounts subduct. Science, v.321, pp.1165-1166.

Wagner, N.S., 1952. Catlow Valley crevice. Ore Bin, v.14, no.6, pp.37-41.

———, 1956. Historical notes on the Standard Mine, Grant County, Oregon. Ore Bin, v.18, no.9, pp.75-77.

———, 1958. Limestone occurrences in eastern Oregon. Ore Bin, v.20, no.5, pp.43-47.

———, 1959. Mining in Baker County, 1861-1959. Ore Bin, v.21, no.3, pp.21-27.

Waitt, R.B., 1985. Case for periodic colossal jokulhlaups from Pleistocene glacial Lake Missoula. Geological Society America, Bull., v.96, no.10, pp.1271-1286.

———, 1987. Evidence for dozens of stupendous floods from glacial Lake Missoula in eastern Washington, Idaho, and Montana. *In*: Hill., M.L., ed., Decade of North American geology; Centennial field guides, Geological Society America, Cordilleran Section, pp.345-350.

———, Denlinger, R.P., and O'Connor, J.E., 2009. Many monstrous Missoula floods down the channeled scabland and Columbia Valley. *In*: O'Connor, J.E., Dorsey, R.J., and Madin, I.P., eds., Volcanoes to vineyards; field trips through the . . . Pacific Northwest. Geological Society America, Field Guide 15, pp.775-844.

Walder, J.S., et al., 1999. Volcanic hazards in the Mount Jefferson region, Oregon. U.S. Geological Survey, Open-File Report 99-24, 14p.

Walker, G. W., 1974. Some implications of late Cenozoic volcanism to geothermal potential in the High Lava Plains of south-central Oregon. Ore Bin, v.36, no.7, pp.109-119.

———, 1977. Geologic map of Oregon east of the 121st Meridian. U.S. Geological Survey, Miscellaneous Investigation Series, Map I-902.

———, 1979. Revisions to the Cenozoic stratigraphy of Harney Basin, southeastern Oregon. U.S. Geological Survey, Bull. 1475, 34p.

———, 1981. High Lava Plains, Brothers fault zone to Harney Basin, Oregon. *In*: Johnston, D., et al., eds., Guides to some volcanic terranes in Washington, Idaho, and Northern California, U .S. Geological Survey, Circular 838, pp.105-111.

———, 1990. Geology of the Blue Mountains region of Oregon, Idaho, and Washington: Cenozoic geology of the Blue Mountains. U.S. Geological Survey, Professional Paper 1437, 135p.

———, 1990a. Miocene and younger rocks of the Blue Mountains region, exclusive of the Columbia River basalt group and associated mafic lava flows. U.S. Geological Survey, Professional Paper 1437, pp.101-118.

———, and Duncan, R.A., 1989. Geologic map of the Salem 1 by 2 Quadrangle, western Oregon. U.S. Geological Survey, Miscellaneous Investigation Series, Map I-1893.

———, and MacLeod, N.S., 1991. Geologic map of Oregon. Reston, U.S. Geological Survey, 2 sheets.

———, and Robinson, P.T., 1990. Cenozoic tectonism and volcanism of the Blue Mountains region.

U.S. Geological Survey, Professional Paper 1437, pp.119-134.

———, and Robinson, P.T., 1990a. Paleocene (?), Eocene, and Oligocene (?) rocks of the Blue Mountains region. U.S. Geological Survey, Professional Paper 1437, pp.13-27.

———, and Swanson, D.A., 1963. Summary report on the geology and mineral resources of the Poker Jim Ridge . . . Hart Mountain . . . Lake County, Oregon. U.S. Geological Survey, Bull.1260-L, 17p.

———, and Swanson, D.A., 1968. Summary report on the geology and mineral resources of the Harney Lake and Malheur Lake areas . . . Oregon. U.S. Geological Survey, Bull.1260-L,M, 17p.

———, Greene, R.C., and Pattee, E.C., 1966. Mineral resources of the Mount Jefferson Primitive area, Oregon. U.S. Geological Survey, Bull. 1230-D, pp.D1-D32.

Walker, J.R., and Naslund, H. R., 1986. Tectonic significance of mildly alkaline Pliocene lavas in Klamath River gorge, Cascade Range, Oregon. Geological Society America, Bull., v.97, pp.206-212.

Wallace, R.E., 1984. Patterns and timing of late Quaternary faulting in the Great Basin province and relation to some regional tectonic features. Journal Geophysical Research, v.89, no.B7, pp.5763-5769.

Wang, K., et al., 1995. Case for very low coupling stress on the Cascadia subduction fault. Journal Geophysical Research, v.100, no.B7, pp.12,907-12,918.

Wang, Y., and Clark, J.L., 1999. Earthquake damage in Oregon: preliminary estimates of future earthquake losses. Oregon Department Geology and Mineral Industries, Special Paper 29, 59p.

———, Keefer, D.K., and Wang, Z., 1998. Seismic hazard mapping in Eugene-Springfield, Oregon. Oregon Geology, v.60, no.2, pp.31-41.

———, et al., 2002. Columbia River gorge landslides. *In*: Moore, G., ed., Field guide to geologic processes in Cascadia. Oregon Department Geology and Mineral Industries, Special Paper 36, pp.273-288.

Wardlaw, B. R., Nestell, M.K., and Dutro, J.T., 1982. Biostratigraphy and structural setting of the Permian Coyote Butte Formation of central Oregon. Geology, v.10, pp.13-16.

Waring, G.A., 1908. Geology and water resources of a portion of south-central Oregon. U.S. Geological Survey, Water Supply Paper 220, 83p.

———, 1965. Thermal springs of the United States and other countries of the world - a summary. U.S. Geological Survey, Professional Paper 492, 383p.

Warren, W.C., and Norbisrath, H., 1946. Stratigraphy of upper Nehalem River basin, northwestern Oregon. American Association Petroleum Geologists, v.30, no.2, pp.213-237.

Waters, A.C., 1961. Stratigraphic and lithologic variations in the Columbia River Basalt. American Journal Science, v.259, pp.583-611.

———, 1966. Stein's Pillar area, central Oregon. Ore Bin, v.28, no.8, pp.137-144.

———, 1973. The Columbia River gorge: basalt stratigraphy, ancient lava dams, and landslide dams. Oregon Department Geology and Mineral Industries., Bull.77, pp.133-162.

Watkins, N.D., and Baksi, A.K., 1974. Magnetostratigraphy and oroclinal folding of the Columbia River, Steens, and the Owyhee basalts in Oregon, Washington, and Idaho. American Journal Science, v.274, pp.148-189.

Weaver, C.E., 1937. Tertiary stratigraphy of western Washington and northwestern Oregon. University of Washington, Publications in Geology, v.4, 266p.

———, 1942. Paleontology of the marine Tertiary formations of Oregon and Washington. University of Washington Publications in Geology, v.5, pts.I, II, and III, 790p.

Webb, T., Bartlein, P. J., and Kutzbach, J.E., 1987. Climatic change in eastern North America during the past 18,000 years; comparisons of pollen data with model results. *In*: Ruddiman, W.F., and Wright, H.E., eds., North America and adjacent oceans during the last deglaciation. Geological Society America, The Geology of North America, v.K-3, pp.447-462p.

Wegener, A.L., 1966. The origin of continents and oceans. New York, Dover (Translation of 4th ed., 1929).

Weins, D.A., and Smith, G.P., 2003. Seismological constraints on structure and flow patterns within the mantle wedge. *In*: Eiler, J., ed., Inside the subduction factory . . . American Geophysical Union, Geophysical Monograph 138, pp.59-82.

Weissenborn, A.E., ed., 1969. Mineral and water resources of Oregon. Oregon Department Geology and Mineral Industries, Bull.64, 462p.

———, and Snavely, P.D., 1968. Oregon Islands Wildlife Refuge, Oregon. U.S. Geological Survey, Bull., 1260-H, pp.G1-G4.

Wells, F.G., 1956. Geology of the Medford quadrangle, Oregon-California. U.S. Geological Survey, Geologic Quadrangle Map GQ-89.

———, and Peck, D.L., 1961. Geologic map of Oregon west of the 121st Meridian. U.S. Geological Survey, Miscellaneous Geologic Investigations Map I-325.

———, and Walker, G.W., 1953. Geology of the Galice quadrangle, Oregon. U.S. Geological Survey, Geologic Quadrangle Map GQ-25.

———, Hotz, P.E., and Cater, F.W., 1949. Preliminary description of the geology of the Kerby quadrangle, Oregon. Oregon Department Geology and Mineral Industries, Bull.40, 23p.

Wells, R.E., 1984. Paleomagnetic constraints on the interpretation of early Cenozoic Pacific Northwest paleogeography. *In*: Nilsen, T., ed., Geology of the upper Cretaceous Hornbrook Formation, Oregon and California. Society Economic Paleontologists and Mineralogists, Pacific Section, v.42, pp.231-237.

———, and Heller, P.L., 1988. The relative contribution of accretion, shear, and extension to Cenozoic tectonic rotation in the Pacific Northwest. Geological Society America, Bull., v.100, pp.325-338.

———, and Simpson, R.W., 2001. Northward migration of the Cascadia forearc in the northwestern U.S. and implications for subduction deformation. Earth, Planets, Space, v.53, pp.275-283.

———, Weaver, C.S., and Blakely, R.J., 1998. Forearc migration in Cascadia and its neotectonic significance. Geology, v.26, pp.759-762.

———, et al., 1984. Cenozoic plate motions of the volcano-tectonic evolution of western Oregon and Washington. Tectonics, v.3, no.2, pp.275-294.

———, et al., 1989. Correlation of Miocene flows of the Columbia River Basalt group from the central Columbia River plateau to the coast of Oregon and Washington. *In*: Reidel, S.P., and Hooper, P.R., eds., Volcanism and tectonism in the Columbia River flood-basalt province. Geological Society America, Special Paper 239, pp.113-128.

———, et al., 2009. The Columbia River Basalt Group-from the gorge to the sea. *In*: O'Connor, J.E., Dorsey, R.J., and Madin, I.P., eds., Volcanoes to vineyards; field trips through the . . . Pacific Northwest. Geological Society America, Field Guide 15, pp.737-774..

Werner, K.S., 1990. Direction of maximum horizontal compression in western Oregon. Masters, Oregon State University, 159p.

———, et al., 1992. The Mount Angel fault; implications of seismic-reflection data and the Woodburn, Oregon, earthquake sequence of August 1990. Oregon Geology, v.54, pp.112-117.

West, D.O., and McCrumb, D.R., 1988. Coastline uplift in Oregon and Washington and the nature of Cascadia subduction-zone tectonics. Geology, v.16, pp.169-172.

Westhusing, J.K., 1973. Reconnaissance surveys of near-event seismic activity in the volcanoes of the Cascade Range, Oregon. Bulletin Volcanology, ser.2, pp.258-285.

Whalen, M.T., 1988. Depositional history of an upper Triassic drowned carbonate platform sequence: Wallowa terrane, Oregon and Idaho. Geological Society America, Bull. v.100, pp.1097-1110.

Wheeler, G., 1981. A major Cretaceous discontinuity in north-central Oregon. Oregon Geology, v.43, no.2, pp.15-17.

———, 1982. Problems in the regional stratigraphy of the Strawberry Volcanics. Oregon Geology, v.44, no.1, pp.3-7.

White, C.M., 1980. Geology and geochemistry of Mt. Hood volcano. Oregon Department Geology and Mineral Industries, Special Paper 8, 26p.

———, 1980. Geology of the Breitenbush hot springs quadrangle, Oregon. Oregon Department Geology and Mineral Industries, Special Paper 9, 26p.

White, J.D.L., 1994. Intra-arc basin deposits within the Wallowa terrane, Pittsburg Landing area, Oregon and Idaho. *In*: Vallier, T.L., and Brooks, H.C., eds. Stratigraphy, physiography, and mineral resources of the Blue Mountains Region. U.S. Geological Survey, Professional Paper 1439, pp.75-90.

———, et al., 1992. Middle Jurassic strata link Wallowa, Olds Ferry, and Izee terranes in the accreted Blue Mountains island arc, northeastern Oregon. Geology, v.20, pp.729-732.

White, R., and McKenzie, D., 1989. Magmatism at rift zones; the generation of volcanic continental margins and flood basalts. Journal Geophysical Research, v.94, pp.7685-7729.

Wicks, C., 2005. A possible link between volcanic eruptions in the Cascade graben and large Cascadia earthquakes. American Geophysical Union, Fall Meeting., 2005, Abstract V23C-02.

Wiley, T.J., 1991. Mining and exploration in Oregon during 1990. Oregon Geology, v.53, no.3, pp.63-70.

———, et al., 1993. Klamath Falls earthquakes, September 20, 1993. Oregon Geology, v.55, no.6, pp.127-134.

Wilkening, R. M., and Cummings, M.L., 1987. Mercury and uranium mineralization in the Clarno and John Day Formations, Bear Creek Butte area, Crook County, Oregon. Oregon Geology, v.49, no.9, pp.103-110.

Wilkins, D.E., and Clement, W.P., 2007. Paleolake shoreline sequencing using ground penetrating radar: Lake Alvord, Oregon, and Nevada. *In*: Baker, G.S., and Jol, H.M., eds., Stratigraphic analyses using GPR. Geological Society America, Special Paper 432, pp.103-110.

Wilkinson, W.D., and Oles, K. F., 1968. Stratigraphy and paleo-environments of Cretaceous rocks, Mitchell Quadrangle, Oregon. American Association Petroleum Geologists, Bull., v.52, no.l, pp.129-161.

———, et al., 1959. Field guidebook, geologic trips along Oregon Highways. Oregon Department Geology and Mineral Industries, Bull.50, 148p.

Williams, D.L., and Von Herzen, R.P., 1983. On the terrestrial heat flow and physical limnology of Crater Lake, Oregon. Journal Geophysical Research, v.88, no.B2, pp.1094-1104.

———, et al., 1982. The Mt. Hood region; volcanic history, structure, and geothermal energy potential. Journal Geophysical Research, v.87, no.B4, pp.2767-2781.

Williams, H., 1933. Mount Thielsen; a dissected Cascade volcano. University of California,

Publications in Geological Sciences, Bull., v.23, no.6, pp.195-214.

———, 1944. Volcanoes of the Three Sisters region, Oregon Cascades. University of California, Publications in Geological Sciences, Bull., v.27, no.3, pp.37-84.

———, 1961. The floor of Crater Lake, Oregon. American Journal Science, v.259, pp.81-83.

———, 1976. The ancient volcanoes of Oregon. Oregon State System of Higher Education, Condon Lectures, 70p.

———, and Compton, R.R., 1953. Quicksilver deposits of Steens Mountain and Pueblo Mountains southeast Oregon. U.S. Geological Survey, Bull.995-B., pp.B1-B76.

———, and Goles, G.G., 1968. Volume of the Mazama ash-fall and the origin of Crater Lake caldera. *In:* Dole, H.M., ed., Andesite conference guidebook – international mantle project; Science report 16-S. Oregon Department Geology and Mineral Industries, Bull.62, pp.37-41.

Williams, I.A., 1916. The Columbia River Gorge; its geologic history. Mineral Resources of Oregon, v.2, no.3, 130p.

———, 1923. The Lava River tunnels. Natural History, v.23, no.2, pp.162-171.

Williams, J.H., 1912. The guardians of the Columbia. Tacoma, Williams Publ., 142p.

Wilmoth, R., 1998. Volcanic hazards of Oregon. *In*: Burns, S., ed., Environmental, groundwater, and engineering geology; applications from Oregon. Belmont, Calif., Star Publ., p.317-324.

Wilson, D., and Cox, A., 1980. Paleomagnetic evidence for tectonic rotation of Jurassic plutons in Blue Mountains, eastern Oregon. Journal Geophysical Research., v.85, no.B7, pp.3681-3689.

Wilson, D.C., 1997. Post-middle Miocene geologic history of the Tualatin basin, Oregon, with hydrogeologic implications. PhD, Portland State University, 321p.

———, 1998. Post-middle Miocene geologic evolution of the Tualatin basin, Oregon. Oregon Geology, v.60, no.5, pp.99-116.

Winterer, E.L., Hussong, D.M., and Decker, R.W., eds., 1989. The Eastern Pacific Ocean and Hawaii. Geological Society America, The Geology of North America, v.N, 563p.

Wise, W. S., 1968. Geology of the Mount Hood volcano. *In:* Dole, Hollis M., ed., Andesite conference guidebook-international mantle project; science report 16-2. Oregon Department Geology and Mineral Industries, Bull.62, 81-98.

———, 1969. Geology and petrology of the Mt. Hood area: a study of High Cascade volcanism. Geological Society America, Bull.80, pp.969-1006.

Witter, R.C., 1999. Late Holocene paleoseismicity, tsunamis, and relative sea-level changes along the south-central Cascadia subduction zone, southern Oregon, USA. PhD, University of Oregon, 178p.

———, Allan, J.C., and Priest, G.R., 2007. Evaluation of coastal erosion hazard zones along dune and bluff-backed shorelines in southern Lincoln County. *In*: Burns, W.J., and Wang, Y., comp., Landslide symposium proceedings and field trip guide. Oregon Department Geology and Mineral Industries, Open-File Report 0-07-06, p.110.

———, Kelsey, H.M., and Hemphill-Haley, E., 2003. Great Cascadia earthquakes and tsunamis of the past 6700 years, Coquille River estuary, southern coastal Oregon. Geological Society America, Bull., v.15, no.10, pp.1289-1306.

Wolfe, Jack A., 1969. Neogene, floristic and vegetational history of the Pacific Northwest. Madrono, v.20, pp.83-110.

Wong, I.G., and Bott, J.D.J., 1995. A look back at Oregon's earthquake history, 1841-1994. Oregon Geology, v.57, no.6, pp.125-139.

———, Silva, W.J., and Madin, I.P., 1993. Strong ground shaking in the Portland, Oregon, metropolitan area. Oregon Geology, v.55, no.6, pp.137-143.

———, et al., 2000. Earthquake scenario and probabilistic ground shaking maps for the Portland, Oregon, metropolitan area. Oregon Department of Geology and Mineral Industries, Interpretive Map Series, IMS-16.

Wood, C.A., and Kienle, J., eds., 1990. Volcanoes of North America. New York, Cambridge University Press, 354p.

Woodward, D.G., Gannett, M.W., and Vaccaro, J.J., 1998. Hydrogeologic framework of the Willamette lowland aquifer system, Oregon and Washington. U.S. Geological Survey, Professional Paper 1424-B, 82p.

Woodward, J., White, J., and Cummings, R., 1990. Paleo-seismicity and the archaeological record: areas of investigation on the northern Oregon coast. Oregon Geology, v.52, no.3, pp.57-65.

Wright, H.E., and Frey, D.G., eds., 1965. The Quaternary history of the United States. Princeton University Press, Princeton, N.J., 723p.

Wright, J.E., and Shervais, J.W., 2008. Ophiolites, arcs, and batholiths. Geological Society America, Special Paper 438, 572p.

———, and Wyld, S.J., 1994. The Rattlesnake Creek terrane; Klamath Mountains, California. Geological Society America, Bull., v,106, pp.1033-1056.

———, and Wyld, S.J., 2007. Alternative tectonic model for late Jurassic through early Cretaceous evolution of the Great Valley Group, California. *In*: Cloos, M., et al., eds., Convergent margin terranes and associated regions: a tribute to W.G. Ernst. Geological Society America, Special Paper 419, pp.81-95.

Wright, T.L., Grolier, M.J., and Swanson, D.A.,1973. Chemical variation related to the stratigraphy of the Columbia River Basalt. Geological Society America, Bull., v.84, pp.371-386.

Wyld, S.J., and Wright, J.E., 1988. The Devils Elbow ophiolite remnant and overlying Galice Formation. Geological Society America, Bull.100, pp.29-44.

———, 2001. New evidence for Cretaceous slip-strike faulting in the United States Cordillera and implications for terrane-displacement, deformation patterns, and plutonism. American Journal Science, v.301, pp.150-181.

Wynn, J.C., and Hasbrouck, W.P., 1984. Geophysical studies of chromite deposits in the Josephine peridotite of northwestern California and southwestern Oregon. U.S. Geological Survey, Bull.1546-D, pp.D63-D86.

Yamaguchi, D.K., et al., 1997. Tree-ring dating the 1700 Cascadia earthquake. Nature, v.389, pp.922-923.

Yeats, R. S., 1989. Current assessment of earthquake hazard in Oregon. Oregon Geology, v.51, no.4, pp.90-92.

———, 1998. Living with earthquakes in the Pacific Northwest. Corvallis, Oregon State University Press, 309p.

———, Sieh, R.S., and Allen, C.R., 1997. The geology of earthquakes. New York, Oxford University Press, 568p.

———, et al., 1992. Tectonics of the Willamette Valley, Oregon. *In*: Rogers, A.M., et al., eds., Assessing earthquake hazards and reducing risk in the Pacific Northwest – Volume 1. U.S. Geological Survey, Professional Paper 1560, pp.183-222.

———, et al., 1998. Stonewall anticline; an active fold on the Oregon continental shelf. Geological Society America, Bull., v.110, no.5, pp.572-587.

Yelin, T.S., and Patton, H.J., 1991. Seismotectonics of the Portland, Oregon, region. Seismological Society America, Bull., v.81, no.1, pp.109-130.

Yuan, H., and Dueker, K., 2005. Teleseismic *P*-wave tomogram of the Yellowstone plume. Geophysical Research Letters, v.12, no.L07304.

Yule, J.D., Saleeby, J.B., and Barnes, C.G., 2006. A rift-edge facies of the late Jurassic Rogue-Chetco arc and Josephine ophiolite, Klamath Mountains, Oregon. *In*: Snoke, A.W., and Barnes, C.G., eds., Geological studies in the Klamath Mountains province, California and Oregon. Geological Society America, Special Paper 410, pp.53-76.

———, et al., 2009. Late Triassic to late Jurassic petrotectonic history of the Oregon Klamath Mountains. *In*: O'Connor, J.E., Dorsey, R.J., and Madin, I.P., eds., Volcanoes to vineyards; field trips through the . . . Pacific Northwest. Geological Society America, Field Guide 15, pp.165-185.

Zierenberg, R.A., et al., 1988. Mineralization, alteration, and hydrothermal metamorphism of the ophiolite-hosted Turner-Albright sulfide deposit, southwestern Oregon. Journal Geophysical Research, v.93, no.B5, pp.4657-4674.

Zoback, M.D., and Zoback, M.L., 1991. Tectonic stress field of North America with relative plate motions. *In*: Slemmons, D.B., et al., Neotectonics of North America. Geological Society America, Decade of North America, Map volume 1, pp.339-366.

Zoback, M.L., Anderson, R.E., and Thompson, G.A., 1981. Cainozoic evolution of the state of stress and style of tectonism of the Basin and Range province of the western United States. Philosophical Transactions, Royal Society London, no.300, pp.407-434.

———, et al., 1994. The northern Nevada rift; regional tectono-magmatic relations and middle Miocene stress direction. Geological Society America, Bull., v.106, pp.371-382.

Zwart, M.J., 1990. Groundwater conditions in the Stage Gulch area, Umatilla County, Oregon. Oregon Water Resources Department, Groundwater Report 35. Salem, 44p.

Index

Illustrations are indicated by *italics*.

C

D

E

H

M

N

O

P

T

U